Non-Associative Algebra and Its Applications

Mathematics and Its Applications

Volume 303

Non-Associative Algebra and Its Applications

edited by

Santos González
Department of Mathematics,
University of Oviedo,
Oviedo, Spain

SPRINGER-SCIENCE+BUSINESS MEDIA, B.V.

A C.I.P. Catalogue record for this book is available from the Library of Congress

ISBN 978-0-7923-3117-9 ISBN 978-94-011-0990-1 (eBook)
DOI 10.1007/978-94-011-0990-1

Printed on acid-free paper

TABLE OF CONTENTS

Preface

The Third International Conference on Non-associative Algebra and its Applications was held in Oviedo (Spain) from July 12th to July 17th, 1993. The two previous Conferences were held in Novosibirsk and Tashkent respectively.

The Organising Committee of the Conference was composed of Santos González Jiménez from Oviedo University and Alberto Elduque and Consuelo Martínez from Zaragoza University. The Scientific Committee was made up of the following members:

G. Benkart, University of Wisconsin, USA
C. Burgueño, University of La Frontera, CHILE
A. Galindo, University of Madrid, SPAIN
S. González, University of Oviedo, SPAIN
P. Holgate, University of London, ENGLAND
N. Jacobson, University of Yale, USA
W. Kaup, University of Tübingen , GERMANY
E. Kleinfeld, Univesity of Iowa, USA
A. J. Kostrikin, University of Moscow, RUSSIA
K. McCrimmon, University of Virginia, USA
A. Micali, University of Montpellier, FRANCE
R. Moody, University of Alberta, CANADA
H.C. Myung, University of Northern Iowa, USA
S. Okubo, University of Rochester, USA
M. Osborn, University of Wisconsin, USA
A. Pérez de Vargas, University of Madrid, SPAIN
H. Petersson, Fern-University, GERMANY
M. Racine, University of Otawa, CANADA
A. Rodríguez, University of Granada, SPAIN
I. Shestakov, University of Novosibirsk, RUSSIA
A. Slinko, University of Moscow, RUSSIA
E. Taft, University of Rutgers, USA
E. Zelmanov, University of Wisconsin, USA

Unfortunately, one of them, Professor Philip Holgate, died some months before the Conference took place.

The central topic of the Conference was Non-associative Algebra, a very active area at present, concerning its intrinsical interest as well as its applications in and outside the mathematical field, for instance in Physics or Biology.

During the Conference 17 plenary talks were given, by members of the Scientific Committee, and five parallel sections of communications were presented every day.

The lectures ranged from presentations of current research and their results to talks concerning surveys and expositions. This variety is reflected in the present volume. The atmosphere of the Conference was very stimulating. All researchers were able to comment on problems in Non-associative Algebra, exchange ideas and results and plan future research topics.

Two hundred and fifty scientists from universities of very different countries, all over the world, took part in the Conference which created a great deal of attention, not only in Asturias, but throughout the whole of Spain, since radio, TV and press paid special attention to this event.

The Editor wants to thank Professors Alberto Elduque and Consuelo Martínez for their dedication and efforts for the success of the Conference. He also would like to thank the members of the Scientific Committee for the confidence shown in letting us organize this important scientific meeting which was warmly received by Oviedo University, a 400 year-old institution which has never been the seat of such an important scientific event.

We specially want to mention Professor Natan Jacobson who, due to health problems, couldn't participate in the Conference. Until the very last moment he still hoped to be able to make the journey, but medical advice wouldn't let us enjoy his presence. However, he and his wife were present in the minds of all participants and his memory is a constant incentive for the future development of Non-associative Algebra.

The Editor wants also to express his thanks to the following Institutions for their financial support:

The Local Government of the Principado de Asturias,

The University of Oviedo,

The Foundation for Scientific and Technical Research (FICYT),

The General Direction of Scientific and Technical Research (DGICYT),

The Directorate General for Science, Research and Development, Commission of the European Communities,

The International Science Foundation and

The City Halls of Oviedo and Gijón.

The Editor is specially indebted to the President and the Research Vice-President of Oviedo University for their continuous support, help and encouragement for the celebration of this event.

Finally, the Editor would like to express his gratitude to the lecturers for their good and stimulating talks as well as for the careful preparation of their contributions to these Proceedings.

Santos González

Oviedo, April 1994

MALCEV SUPERALGEBRAS

HELENA ALBUQUERQUE
Departamento de Matemática, Universidade de Coimbra
3000 Coimbra, Portugal

Abstract. Most of the results in [1] concerning Malcev superalgebras are briefly summarized.

An anticomutative algebra is said to be a Malcev Algebra if it satisfies the identity

$$(xz)(yt) = ((xy)z)t + ((yz)t)x + ((zt)x)y + ((tx)y)z. \tag{1}$$

Let now $M = M_0 \oplus M_1$ be a superalgebra ($\mathbb{Z}_2$ graded algebra), and $G = G_0 \oplus G_1$ the Grassmann algebra over a countable set of generators, $G =< e_i, i = 1, \cdots, n, \cdots, e_ie_j = -e_je_i, \forall i, j >$, with the standard $\mathbb{Z}_2$ gradation in which G_0 (respectively G_1) is spanned by the words of even (respectively odd) length in the generators.

The Grassman envelope of the superalgebra M is the algebra

$$G(M) = M_0 \otimes G_0 \oplus M_1 \otimes G_1 \tag{2}$$

with the multiplication determined by

$$\begin{aligned}(x_0 \otimes e_0 + x_1 \otimes e_1)(y_0 \otimes f_0 + y_1 \otimes f_1) = \\ (x_0y_0 \otimes e_0f_0 + x_1y_1 \otimes e_1f_1) + (x_0y_1 \otimes e_0f_1 + x_1y_0 \otimes e_1f_0).\end{aligned} \tag{3}$$

If $\mathcal{P}$ is an homogeneous variety of algebras, M is said to be a $\mathcal{P}$-superalgebra if the algebra $G(M)$ belongs to the variety $\mathcal{P}$. In [13], it is explained how one obtains the graded identities that define a $\mathcal{P}$ superalgebra from the known identities that satisfies its Grassman envelope algebra.

Therefore, a superalgebra $M = M_0 + M_1$ is said to be a Malcev superalgebra if and only if $G(M)$ is a Malcev algebra and this is clearly equivalent to the following identities being satisfied :

$$\begin{aligned}&1) \quad xy = -(-1)^{\bar{x}\bar{y}}yx,\\ &2) \quad (-1)^{\bar{y}\bar{z}}(xz)(yt) = ((xy)z)t + (-1)^{\bar{x}(\bar{y}+\bar{z}+\bar{t})}((yz)t)x + \\ &\qquad (-1)^{(\bar{x}+\bar{y})(\bar{z}+\bar{t})}((zt)x)y + (-1)^{\bar{t}(\bar{x}+\bar{y}+\bar{z})}((tx)y)z,\end{aligned} \tag{4}$$

for any homogeneous elements $x, y, z, t \in M_0 \cup M_1$ where $\bar{a}$ stands for $i \pmod 2$ if $a \in M_i$.

A Malcev superalgebra is said to be trivial if $M_1^2 = 0$.

If $M = M_0 + M_1$ is a Malcev superalgebra, then M_0 is a Malcev algebra and M_1 is an M_0-module. Therefore redefining the multiplication in M so as to make 0 the

S. González (ed.), Non-Associative Algebra and Its Applications, 1–7.

product of any two odd elements, we get a Malcev algebra that we will be denoted by M'. Hence the multiplication in M' is given by

$$(x_0 + x_1)(y_0 + y_1) = x_0 y_0 + x_1 y_0 + x_0 y_1 \tag{5}$$

for $x_i, y_i, \in M_i$ $i = 1, 2$.

For homogeneous elements x, y, z in M the SuperJacobian is defined as

$$\tilde{J}(x, y, z) = (xy)z - x(yz) - (-1)^{\bar{y}\bar{z}}(xz)y \tag{6}$$

As Malcev algebras, we can define in M the Jacobian ideal and the nucleus of M.

The subspace of M spanned by the Jacobians is a graded ideal and is denoted by $\tilde{J}(M, M, M)$. The nucleus is the graded subspace de M:

$$\tilde{N}(M) = \{x \in M : \tilde{J}(x, M, M) = 0\} \tag{7}$$

These two ideals are of great importance in the study of the theory of Malcev superalgebras.

In what follows, all the algebras and superalgebras considered will be defined over fields of characteristic not two.

1. Malcev superalgebras of dimension ≤ 4

Despite the quite sucessful development of the theory of Malcev superalgebras over the last few years, few examples are yet known with the exception of Lie superalgebras and Malcev algebras. With the main purpose to get some examples that take us to a better understanding of the theory, we classify the Malcev non-Lie superalgebras with $M_1 \neq 0$, of dimension ≤ 4.

We recall that any Malcev algebra of dimension ≤ 3 is a Lie algebra and that there is, up to isomorphism, a unique non-Lie Malcev algebra of dimension 4. The Lie superalgebras of dimension ≤ 4 are classified in [4, 11].

We prove that if dimension is ≤ 4 then M is trivial or M' is a Lie superalgebra. Besides M is a solvable algebra (nilpotent or not), in which the ideals $\tilde{J}(M, M, M)$ and $\tilde{N}(M)$, are abelian.

If $\dim M \leq 2$ then M is a Lie superalgebra but contrarily to what happens with Malcev algebras there are 3 types of non-Lie Malcev indecomposable superalgebras with $M_1 \neq 0$ of dimension 3, and 19 types of non-Lie indecomposable Malcev superalgebras of dimension 4.

Our results are listed in Table I. We will denote the elements of M_0 by the first letters of the Latin alphabet and the elements of M_1 by the last letters. The elements of the ground field K will be taken to be Greek letters. We say that M is of type (m, n) if $\dim M_0$=m and $\dim M_1 = n$. All zero products are ommited.

TABLE I

Type	K	M	Multiplication table	Comments	$\tilde{J}(M,M,M)$	$\tilde{N}(M)$	Nilp. index	Solv. index
$(1,2)$	$\neq 2,3$	$M(1,2)$	$au=v;\ u^2=a$	—	$K(3v)$	$\langle a,v\rangle$	4	2
	$\neq 2$	$M(1,2,\mu)$	$au=v;\ uv=\mu a\ (\mu\neq 0)$	$M(1,2,\mu)\cong M(1,2,\beta)$ iff $\mu\beta^{-1}=\gamma^2$, $\gamma\in K\setminus\{0\}$	$K(2\mu a)\oplus K(2\mu v)$	0	—	2
$(2,1)$	$\neq 2,3$	$M(2,1)$	$ab=b;\ au=-u;\ u^2=b$	—	$K(-3b)$	$\langle b\rangle$	—	3
$(3,1)$	$\neq 2$	$M^1(3,1)$	$bc=a;\ u^2=c$	—	$K(a)$	$\langle a,c\rangle$	4	2
	$\neq 2,3$	$M^2(3,1)$	$ac=a;\ bc=a+b;\ cu=u;\ u^2=a$	—	$K(3a)$	$\langle a,b\rangle$	—	3
	$\neq 2,3$	$M^3(3,1)$	$ac=a;\ bc=-2b;\ cu=u;\ u^2=a+b$	—	$K(3a)$	$\langle a,b\rangle$	—	3
	$\neq 2,3$	$M(3,1,\mu)$	$ac=a;\ bc=\mu b;\ cu=u;\ u^2=a\ (\mu\neq 0)$	$M(3,1,\mu)\cong M(3,1,\beta)$ iff $\mu=\beta$	$K(3a)$	$\langle a,b\rangle$	—	3
	3	$M^4(3,1)$	$ac=a+b;\ bc=b;\ cu=u;\ u^2=a$	—	$K(b)$	$\langle a,b\rangle$	—	3
$(2,2)$	$\neq 2,3$	$M^0(2,2)$	$ab=b;\ au=u;\ av=-v;\ ub=v$	Trivial superalgebra	$K(-3v)$	$\langle v\rangle$	—	3
	$\neq 2$	$M^1(2,2)$	$bv=u;\ uv=a$	—	$K(2a)$	$\langle a,u\rangle$	4	2
	$\neq 2$	$M(\mu,2,2)$	$bv=u;\ uv=\mu b;\ v^2=a\ (\mu\neq 0)$	$M(\mu,2,2)\cong M(\beta,2,2)$ iff $\mu\beta^{-1}=\gamma^2$, $\gamma\in K\setminus\{0\}$	$K(2\mu b)\oplus K(2\mu u)$	$\langle a\rangle$	—	2
	$\neq 2$	$M(2,2,1)$	$bv=u;\ uv=a;\ v^2=b$	—	$K(2b)\oplus K(3u)$	$\langle a,u\rangle$	5	2
	$\neq 2,3$	$M^2(2,2)$	$ab=b;\ av=-v;\ bv=u;\ v^2=b$	—	$K(-3b)\oplus K(3u)$	$\langle b,u\rangle$	—	3
	$\neq 2,3$	$M(2,2,\mu)$	$ab=b;\ av=-v;\ au=-u;\ u^2=b;$ $v^2=\mu b\ (\mu\neq 0)$	$M(2,2,\mu)\cong M(2,2,\beta)$ iff $\exists_{\epsilon\in K\setminus\{0\}}$: $\begin{pmatrix}1&0\\0&\mu\end{pmatrix}$ and $\epsilon\begin{pmatrix}1&0\\0&\beta\end{pmatrix}$ are congruent	$K(-3b)\oplus K(-3\mu b)$	$\langle b\rangle$	—	3
	$\neq 2,3$	$M^1(2,2,\gamma)$	$ab=b;\ av=-v;\ au=\gamma u;\ v^2=b$ $(\gamma\neq 2,\frac{1}{2})$	—	$K(-3b)$	$\langle b,u\rangle$	—	3
	$\neq 2,3$	$M(2,2,2,1)$	$ab=b;\ av=-v;\ au=2u;\ uv=b;$ $v^2=b$	—	$K(3b)$	$\langle b,u\rangle$	—	3
	$\neq 2,3$	$M(2,2,2,0)$	$ab=b;\ av=-v;\ au=2u;\ v^2=b$	—	$K(-3b)$	$\langle b,u\rangle$	—	3
	$\neq 2,3$	$M(2,2,\frac{1}{2},\delta)$	$ab=b;\ av=-v;\ au=u/2;$ $v^2=b;\ u^2=\delta b\ (\delta\neq 0)$	$M(2,2,\frac{1}{2},\mu)\cong M(2,2,\frac{1}{2},\beta)$ iff $\exists_{\gamma\in K\setminus\{0\}}:\ \mu\beta^{-1}=\gamma^2$	$K(-3b)$	$\langle b,u\rangle$	—	3
	$\neq 2,3$	$M(2,2,\frac{1}{2},0)$	$ab=b;\ av=-v;\ au=u/2;\ v^2=b$	—	$K(-3b)$	$\langle b,u\rangle$	—	3
	$\neq 2,3$	$M^3(2,2)$	$ab=b;\ av=u-v;\ au=-u;\ v^2=b$	—	$K(-3b)$	$\langle b,u\rangle$	—	3
	3	$M^4(2,2)$	$ab=b;\ av=u-v;\ au=-u;\ uv=b$	—	$K(2b)$	$\langle b,u\rangle$	—	3
$(1,3)$	$\neq 2$	$M(1,3)$	$au=v;\ w^2=a$	—	$K(v)$	$\langle a,v\rangle$	4	2
	$\neq 2$	$M(1,3,\mu)$	$au=v;\ w^2=\mu a;\ u^2=a\ (\mu\neq 0)$	$M(1,3,\mu)\cong M(1,3,\gamma)$ iff $\exists_{\alpha\in K\setminus\{0\}}:\mu\gamma^{-1}=\alpha^2$	$K(3v)\oplus K(\mu v)$	$\langle a,v\rangle$	4	2
	$\neq 2$	$M(1,3,0)$	$au=v;\ uw=a$	—	$K(2v)$	$\langle a,v\rangle$	4	2

2. Engel's theorem for Malcev superalgebras

A representation of a Malcev superalgebra M is an even linear map $\rho : M \to End(V)$ with $V = V_0 \oplus V_1$ and $End(V)_i = \{\phi \in End(V) : \phi(V_j) \subset V_{j+i}(\text{modulo } 2), j = 0, 1\}$, such that the superalgebra $A = V \oplus M$ with $A_i = V_i \oplus M_i, i = 0, 1$ and multiplication determined by $V^2 = 0, vx = -(-1)^{\bar{v}\bar{x}}xv = \rho_x(v)$ for elements $v \in V$ and $x \in M$, and the multiplication in M, is a Malcev superalgebra. This is equivalent to the identity

$$\rho_{(xy)z} = \rho_x\rho_y\rho_z - (-1)^{(\bar{x}+\bar{y})\bar{z}}\rho_z\rho_x\rho_y - (-1)^{\bar{x}(\bar{y}+\bar{z})}(\rho_{yz}\rho_x - \rho_y\rho_{zx}) \tag{8}$$

for homogeneous elements $x, y, z \in M$.

The representation ρ is called faithful in case $\ker \rho = 0$ and is called almost faithful in case $\ker \rho$ does not contain any non zero ideal of M.

Classical Engel's Theorem asserts that if L is a Lie algebra of linear transformations on a finite dimensional vector space V over a field K consisting of nilpotent transformations, then the associative subalgebra of $End(V)$ generated by L is nilpotent. This is a key tool in the study of finite dimensional Lie algebras [9].

For Malcev algebras, a natural generalization of this theorem is stated without proof in [10]:

"If ρ is an almost faithful finite dimensional representation of a finite dimensional Malcev algebra M by nilpotent operators, then the associative subalgebra generated by $\rho(M)$ and M itself are nilpotent"

Stitzinger in [14] has proved an extension of the last result. In his version he doesn't assume that ρ_x is nilpotent for all $x \in M$, but he imposes a stronger condition supposing that ρ is faithful.

Theorem 1. *Let $\rho : M \to End(V)$ be a faihtful representation of the Malcev algebra M. Consider U a generating subset of M, closed for the product and such that ρ_x is nilpotent for each $x \in U$. Then the associative subalgebra of $End(V)$ generated by $\rho(M)$ and the Malcev algebra M are nilpotent.*

Stitzinger in his proof considers the representation ρ faithful to prove that if ρ_x is nilpotent for all $x \in M$ so is R_x. However, one can prove that this property is still verified if ρ is almost faithful.

For a Malcev superalgebra M, we have:

Theorem 2. *Let ρ be a representation of the Malcev superalgebra M and assume that ρ_x is nilpotent for any homogeneous $x \in M_0 \cup M_1$. Then the associative subalgebra of $End(V)$ generated by $\rho(M)$ is nilpotent. Moreover, if ρ is almost faithful, then M itself is also nilpotent.*

As an extension of this result and of the known versions of Engel's theorem for Lie superalgebras and Malcev algebras we obtain the following version of Engel's theorem for Malcev superalgebras.

Theorem 3. *Let ρ be a representation of the Malcev superalgebra $M = M_0 \oplus M_1$ and let U be a subset of homogeneous elements of M closed under multiplication, which spans M, and such that ρ_x is nilpotent for all $x \in U$. Then the associative subalgebra of $End(V)$ generated by $\rho(M)$ is nilpotent. Moreover, if ρ is almost faithful, then M itself is nilpotent.*

3. Malcev superalgebras with trivial nucleus

This seccion is devoted to Malcev superalgebras with trivial nucleus. We will extend the results of [6] and [7]. We know that M is a Lie superalgebra if and only if $\tilde{N}(M) = M$. So, in a sense, the Malcev superalgebras with trivial nucleus are the most distant from being Lie superalgebras.

In Table I there is an interesting example of a solvable Malcev superalgebra $M(1,2;\mu)$ with trivial nucleus.

Indeed in [7] it is proved that any finite dimensional solvable Malcev algebra over a field of characteristic $\neq 2$ or any finite dimensional Malcev algebra over a field of characteristic 3 has non trivial nucleus.

However, the Malcev superalgebra $M(1,2,\mu) =< a, u, v >;\ au = v; uv = \mu a$ is exactly an example that shows that none of the preceding properties are verified in Malcev superalgebras.

Moreover, with $M = M(1,2,\mu)$, $G(M)$ gives an example of infinite dimensional solvable Malcev algebra with trivial nucleus.

The solvable Malcev superalgebras with trivial nucleus are determined by the following assertion.

Theorem 4. *Let M be a finite dimensional Malcev superalgebra with trivial nucleus. Then M is solvable if and only if M' is nilpotent.*

Let M be a finite-dimensional Malcev superalgebra $M = M_0 \oplus M_1$, over an algebraically closed field K and let H be a nilpotent subalgebra of M_0. M decomposes as

$$M = \oplus_{\alpha\in\Phi} M^\alpha \tag{9}$$

with

$$M^\alpha = M_0^\alpha \oplus M_1^\alpha \tag{10}$$

and

$$M_i^\alpha = \{x \in M_i : \forall h \in H, x(R_h - \alpha(h)I)^n = 0, \text{ for some } n\} \tag{11}$$

where Φ is the set of all weights of $\{R_h\}$ $(h \in H)$.

Let Φ^+ be a set of representatives of the sets $\{\alpha, -\alpha\}$, for $\alpha \in \Phi \setminus \{0\}$. Let us consider now in M the subspaces $s(\alpha) = M^\alpha \oplus M^{-\alpha} \oplus M^\alpha M^{-\alpha}$, with $\alpha \in \Phi^+$. If $\tilde{N}(M) = 0$, $s(\alpha)$ is an ideal of M.

We define recursively $M^{(0)} = M$ and $M^{(n+1)} = M^{(n)}M^{(n)}$. Let $M^{(\infty)} = \cap_{n\geq 0} M^{(n)}$ and consider the graded ideal

$$P = \{x \in M : xM^{(\infty)} = 0\} \tag{12}$$

Theorem 5. *Assume $\tilde{N}(M) = 0$. Then, the graded ideal P is solvable, $(M^0)^2 \subset P$ and $\tilde{N}(M/P) = 0$. Moreover, P is the largest ideal of M contained in M^0.*

These properties of the ideal P and the fact that the quotient superalgebra M/P has trivial nucleus are fundamentals for the demonstration of our results.

Theorem 6. *Assume $\tilde{N}(M) = 0$. Then,*

1) *The sum $P + \bigoplus_{\alpha\in\Phi^+} s(\alpha)$ is direct;*

2) *For each* $\alpha \in \Phi^+$ *the ideal* $s(\alpha)$ *satisfies* $(s(\alpha))^2 = s(\alpha)$ *and has a trivial nucleus.*

3) *The quotient algebra* $s(\alpha)/\operatorname{rad} s(\alpha)$ *is a simple Malcev algebra, isomorphic to* $sl(2, K)$ *in case of being a Lie algebra.*

The solvable radical of a Malcev superalgebra with trivial nucleus can be determined by the following theorem

Theorem 7. *Assume* $\tilde{N}(M) = 0$ *and let* $B = \bigcap_{\alpha\in\Phi\setminus\{0\}} \ker\alpha \oplus M_1^0$. *Then, the solvable radical of* M, $\operatorname{rad} M$, *is* $B + \oplus_{\alpha\in\Phi^+} \operatorname{rad} s(\alpha)$. *Moreover,* $M/\operatorname{rad} M$ *is isomorphic to* $\oplus_{\alpha\in\Phi^+} s(\alpha)/\operatorname{rad} s(\alpha)$.

As a consequence of all we said we can enunciate the two main results of our work:

Theorem 8. *Let* M *be a finite-dimensional Malcev superalgebra with trivial nucleus over an an algebraically closed field* K *of characteristic* $\neq 2, 3$. *Then* M_1 *is contained in the solvable radical of* M.

Shestakov in [13] has proved that any prime Malcev superalgebra over a field of characteristic $\neq 2, 3$, with nonzero odd part, is a Lie superalgebra. That is any prime Malcev superalgebra over such a field is either a Lie superalgebra or a prime non-Lie algebra. Actually, the assumption of the characteristic being diferent from 3 can be removed [5]. Shestakov's result is fundamental to prove the preceding theorem, and this one is an extension of it for finite dimensional algebras.

Theorem 9. *Let* M *be a finite-dimensional Malcev superalgebra with trivial nucleus over an algebraically closed field* K *of characteristic* $\neq 2, 3$. *Then there is a subalgebra* U *of* M_0 *such that* $M = \operatorname{rad} M \oplus U$ *and* U *is a direct sum of simple ideals which are either isomorphic to* $sl(2, K)$ *or to the unique simple non-Lie Malcev algebra over* K.

A.Elduque has generalized these results to Malcev superalgebras defined over an arbitrary field of characteristic $\neq 2, 3$ concluding that

Theorem 10. *Let* M *be a finite-dimensional Malcev superalgebra with trivial nucleus over a field of characteristic* $\neq 2, 3$. *Then* M_1 *is contained in the solvable radical of* M *and* $M/\operatorname{rad} M$ *is a semisimple Malcev algebra, which is a direct sum of simple ideals of two types: three-dimensional over their centroids or non-Lie Malcev ideals (seven dimensional over their centroids). Moreover, if* $M/\operatorname{rad} M$ *is separable, then there is a subalgebra* U *of* M_0 *such that* $M = \operatorname{rad} M \oplus U$.

A.Elduque and I.Shestakov [8] have extended recently the first part of the result above by proving that the odd part of any Malcev superalgebra with trivial nucleus over a ring ϕ, with $1/6 \in \phi$, is contained in the Baer radical.

References

1. H. Albuquerque, *Contribuções para a teoria das superálgebras de Malcev*, (Portuguese). Doctoral Thesis, Coimbra 1993.
2. H. Albuquerque and A. Elduque, Engel's Theorem for Malcev Superalgebras, Commun. Algebra, to appear.

3. H. Albuquerque and A. Elduque, Malcev Superalgebras with trivial nucleus, Commun. Algebra **21** (1993), 3147–3164.
4. N. Backhouse,A classification of four dimensional Lie superalgebras, J. Math. Phys. **19** (11) (1978), 2400–2402.
5. A. Elduque, A note on semiprime Malcev superalgebras, Proc. Royal Soc. Edinburgh **123A** (1993), 887–891.
6. A. Elduque, On semisimple Malcev Algebras, Proc .Amer. Math. Soc. **107** (1989), 73–82.
7. A. Elduque and A.A. El-Malek, On the J-nucleus of a Malcev algebra, Algebras Groups Geom. **3** (1986), 493–503.
8. A. Elduque and I.P.Shestakov, On Malcev superalgebras with trivial nucleus, to appear in Nova J. Algebra Geom.
9. N. Jacobson, *Lie Algebras*, Interscience, New York 1962.
10. E.N. Kuzmin, Malcev algebras and their representations,Algebra and Logic **7** (1968), 233–244.
11. J. Patera, R.T. Sharp and P. Winternitz, Invariants of real dimensions Lie Algebras, J. Math. Phys. **17** (6) (1976), 986–994.
12. A.A. Sagle, Malcev Algebras, Trans. Amer. Math. Soc. **101** (1961), 426–458.
13. I.P. Shestakov, Prime Malcev Superalgebras, Math. USSR. Sb. **74** (1993), 101-110.
14. E.T. Stitzinger, On nilpotent and solvable Malcev algebras, Proc. Amer. Math. Soc. **92** (1984), 157–163.

THE FIRST TITS CONSTRUCTION OF ALBERT ALGEBRAS OVER LOCALLY RINGED SPACES

GÜNTER ACHHAMMER
Fachbereich Mathematik, FernUniversität - Gesamthochschule -
Lützowstraße 125
D-58084 Hagen, Germany

Abstract. The first Tits construction of Albert algebras over fields is rephrased in the theory of locally ringed spaces.

Key words: Primary 17C40, Secondary 17C20, 14F05 (1991 AMS Mathematics Subject Classification)

1. Introduction

Tits gave two constructions of simple exceptional linear Jordan algebras of dimension 27. McCrimmon [M1, M2] extended these to quadratic Jordan algebras over fields of arbitrary characteristic and showed that all finite-dimensional exceptional central simple quadratic Jordan algebras (Albert algebras) are given by one of these two constructions. We wish to extend Tits' first construction to Jordan algebras over locally ringed spaces. First we recall this construction over fields.

2. The first Tits construction over fields

Let k be an arbitrary field. J is called an Albert algebra over k, iff it satisfies one of the following equivalent conditions:

a) J is a finite-dimensional exceptional central simple quadratic Jordan algebra over k.

b) J is a form of $H_3(C)$ with C the split Octonion algebra over k, i.e. there is a field extension K of k with $J_K \cong H_3(C_K)$.

By $J(N, \#, 1)$ we mean the quadratic Jordan algebra that comes out from the cubic form $N : V \longrightarrow V$ ("norm") with the quadratic map $^{\#} : V \longrightarrow V$, $x \longmapsto x^{\#}$ ("adjoint") and the base point $1 \in V$, V the underlying finite-dimensional vector space over k [M1]. With T we denote the associated trace form $T : V \times V \longrightarrow k$. McCrimmon has shown the following theorem.

Theorem A. (Tits' first construction, [M1, M2]):
a) Let A be a unital associative algebra over k such that $A^+ = J(N, \#, 1)$ and $N(xy) = N(x)N(y)$ holds for all scalar extensions of k. Let $\mu \in k^\times$ and

$$\widetilde{A} := A_0 \oplus A_1 \oplus A_2 \text{ with } A_i \cong A,$$

S. González (ed.), Non-Associative Algebra and Its Applications, 8–11.

$\widetilde{N} : \widetilde{A} \longrightarrow k$ defined by

$\widetilde{N}(a_0, a_1, a_2) := N(a_0) + \mu N(a_1) + \mu^{-1} N(a_2) - T(a_0 a_1 a_2)$

$\widetilde{\#} : \widetilde{A} \longrightarrow \widetilde{A}$ defined by

$(a_0, a_1, a_2)^{\widetilde{\#}} := (a_0^{\#} - a_1 a_2, \mu^{-1} a_2^{\#} - a_0 a_1, \mu a_1^{\#} - a_2 a_0)$

$\widetilde{1} := (1, 0, 0).$

Then $J(A, \mu) := J(\widetilde{N}, \widetilde{\#}, \widetilde{1})$ defines a quadratic Jordan algebra over k.
b) Let J be an Albert algebra over k containing a subalgebra A^+ with A an Azumaya algebra over k of degree 3. Then there exists a $\mu \in k^\times$ with $J(A, \mu) \cong J$.

Remark. For an Azumaya algebra over k of degree 3 we have $A^+ = J(n, \#, 1)$ with n and $\#$ the usual norm and adjoint.

We now carry over this theorem to Jordan algebras over locally ringed spaces.

3. The first Tits construction over locally ringed spaces

Let X be a locally ringed space with structure sheaf $\mathcal{O}_X$. For $P \in X$ the local ring of $\mathcal{O}_X$ a P is denoted by $\mathcal{O}_P$, the maximal ideal of $\mathcal{O}_P$ by m_P and the residue class field by $\kappa(P) = \mathcal{O}_P / m_P$.
Algebras over X are assumed to be locally free of finite rank as $\mathcal{O}_X$-modules.
First of all we define Azumaya and Albert algebras over X.
An associative unital algebra over $\mathcal{O}_X$ is called an Azumaya algebra, iff $\mathcal{A}_P \otimes_{\mathcal{O}_P} \kappa(P)$ is an Azumaya algebra over $\kappa(P)$ for every $P \in X$.
A Jordan algebra $\mathcal{J}$ over X is called an Albert algebra, iff $\mathcal{J}_P \otimes \kappa(P)$ is an Albert algebra over $\kappa(P)$ for every $P \in X$.
It is easy to see that the general construction of Jordan algebras by cubic norm structures (cf. [M1]) works also in the context of locally ringed spaces. So by $\mathcal{J}(N, \#, 1)$ we mean the Jordan algebra over $\mathcal{O}_X$ that comes out from the cubic form $N : \mathcal{J} \longrightarrow \mathcal{O}_X$ with the adjoint $^{\#} : \mathcal{J} \longrightarrow \mathcal{J}$ and the base point $1 \in \Gamma(X, \mathcal{J}), \mathcal{J}$ the underlying $\mathcal{O}_X$-module. $T : \mathcal{J} \times \mathcal{J} \longrightarrow \mathcal{O}_X$ denotes the associated trace form.

Remark. In this setting N and $\#$ are viewed as polynomial maps. Therefore every polynomial identity f in this context holds strictly, i. e. $f(U)$ holds for every scalar extension of $\Gamma(U, \mathcal{O}_X)$ for every open subset $U \subset X$.

Now let $\mathcal{A}$ be an associative unital algebra over $\mathcal{O}_X$ such that $\mathcal{A}^+ = \mathcal{J}(N_{\mathcal{A}}, \#_{\mathcal{A}}, 1)$ and $N_{\mathcal{A}}(xy) = N_{\mathcal{A}}(x) N_{\mathcal{A}}(y)$ holds strictly for all sections x, y in $\mathcal{A}$. Denoting by $\mathcal{A}^\times$ the sheaf of units of $\mathcal{A}$ we know from the general construction that $N_{\mathcal{A}}(\mathcal{A}^\times) \subset \mathcal{O}_X^\times$. Restricting $N_{\mathcal{A}}$ we therefore get a morphism of groups sheaves

$$N_{\mathcal{A}} : \mathcal{A}^X \longrightarrow \mathcal{O}_X^\times.$$

This induces a morphism (again denoted by $N_{\mathcal{A}}$).

$$N_{\mathcal{A}} : \check{H}^1(X, \mathcal{A}^\times) \longrightarrow \check{H}^1(X, \mathcal{O}_X^\times).$$

Let $\mathcal{M}$ be a locally free right $\mathcal{A}$-Modul of rank one. Identifying $\check{H}^1(X, \mathcal{A}^\times)$ with the set of isomorphism classes of locally free right $\mathcal{A}$-modules of rank one and $\check{H}^1(X, \mathcal{O}_X^\times)$ with the isomorphism classes
of locally free $\mathcal{O}_X$-moduls of rank one $N_{\mathcal{A}}(\mathcal{M})$ is an invertible sheaf of $\mathcal{O}_X$, called the norm of $\mathcal{M}$. $\mathcal{M}$ is said to be of norm one, iff $N_{\mathcal{A}}(\mathcal{M}) \cong \mathcal{O}_X$.
Next we need the definition of multiplicative quadratic and cubic maps.
Let $\mathcal{F}$ be an $\mathcal{O}_X$-module and $\mathcal{G}$ be a left $\mathcal{A}$-module. A cubic map $N : \mathcal{M} \longrightarrow \mathcal{F}$ is said to be multiplicative iff

$$N(ma) = N_{\mathcal{A}}(a)N(m)$$

holds strictly for all sections m in $\mathcal{M}$, a in $\mathcal{A}$, and a quadratic map $Q : \mathcal{M} \longrightarrow \mathcal{G}$ is said to be multiplicative, iff

$$Q(ma) = a^{\#_{\mathcal{A}}}Q(m)$$

holds strictly for all sections m in $\mathcal{M}$ and a in $\mathcal{A}$. We use corresponding definitions in case $\mathcal{M}$ is a left $\mathcal{O}_X$-module and $\mathcal{G}$ is a right $\mathcal{A}$-module.
A multiplicative cubic form $N : \mathcal{M} \longrightarrow N_{\mathcal{A}}(\mathcal{M})$ is called a norm on $\mathcal{M}$.
Argueing in the same way as Petersson in [P, 2.4] one gets that for every locally right $\mathcal{A}$-module $\mathcal{M}$ of rank one there exists always a norm on $\mathcal{M}$, and two norms on $\mathcal{M}$ are equivalent up to an invertible factor in $\Gamma(X, \mathcal{O}_X)$.
The next lemma shows that part a) of Theorem A carries over to arbitrary (commutative, associative, unital) rings and part b) holds for local rings.

Lemma.

a) Part a) of Theorem A holds under the same conditions for algebras over commutative associative unital rings.

b) Let R be a (commutative associative unital) local ring and $J = J(N, \#, 1)$ be an Albert algebra which contains a subalgebra A^+ with A an Azumaya algebra of constant degree 3. Then there exists a $\mu \in R^\times$ such that

$$J(A, \mu) \cong J.$$

The proof of Theorem A carries over to this lemma under slight modifications.

Remark. If $\mathcal{M}$ is a locally free right $\mathcal{A}$-module of rank one, $\check{\mathcal{M}} = \mathcal{H}om(\mathcal{M}, \mathcal{A})$ canonically is a locally free left $\mathcal{O}_X$-module of rank one.

The lemma above plays a substantial part in the proof of the following theorem which corresponds to Theorem A in the setting of locally ringed spaces.

Theorem B.

a) Let $\mathcal{A}$ be an associative, unital $\mathcal{O}_X$-algebra such that $\mathcal{A}^+ = \mathcal{J}(N_\mathcal{A}, \#_\mathcal{A}, 1)$ and $N_\mathcal{A}(xy) = N_\mathcal{A}(x)N_\mathcal{A}(y)$ holds strictly for all sections x, y in $\mathcal{A}$. Let $\mathcal{M}$ be a locally free right $\mathcal{A}$-module

of rank one and norm one and $N : \mathcal{M} \longrightarrow \mathcal{O}_X$ a norm on $\mathcal{M}$. Then there exist a unique norm $\check{N}$ on $\check{\mathcal{M}} = \mathcal{H}om(\mathcal{M}, \mathcal{A})$, i.e. a multiplicative cubic form

$$\check{N} : \check{\mathcal{M}} \longrightarrow \mathcal{O}_X,$$

and unique multiplicative quadratic maps

$$\begin{aligned} \# &: \mathcal{M} \longrightarrow \check{\mathcal{M}} \\ \check{\#} &: \check{\mathcal{M}} \longrightarrow \mathcal{M} \end{aligned}$$

such that the identities

$$\begin{aligned} (m^\#)^{\check{\#}} &= N(m)m \quad , \quad (\check{m}^{\check{\#}})^\# = \check{N}(\check{m})\check{m} \\ m^\#(m) &= N(m) \quad\;\; , \quad \check{m}(\check{m}^{\check{\#}}) = \check{N}(\check{m}) \end{aligned}$$

hold strictly for all sections m in M, $\check{m}$ in $\check{\mathcal{M}}$.
Also, if

$$\widetilde{N} : \mathcal{A} \oplus \check{\mathcal{M}} \oplus \mathcal{M} \longrightarrow \mathcal{O}_X \qquad \text{is defined by}$$

$$\widetilde{N}(a, \check{m}, m) = N_\mathcal{A}(a) + \check{N}(\check{m}) + N(m) - T_\mathcal{A}(a, \check{m}(m))$$

$$\widetilde{\#} : \mathcal{A} \oplus \check{\mathcal{M}} \oplus \mathcal{M} \longrightarrow \mathcal{A} \oplus \check{\mathcal{M}} \oplus \mathcal{M} \qquad \text{is defined by}$$

$$(a, \check{m}, m)^{\widetilde{\#}} = (a^{\#_\mathcal{A}} - \check{m}(m), m^\# - a\check{m}, \check{m}^{\check{\#}} - ma)$$

and $\widetilde{1} \in \Gamma(X, \mathcal{A} \oplus \check{\mathcal{M}} \oplus \mathcal{M})$ is defined by $\widetilde{1} := (1, 0, 0)$, then $\mathcal{J}(\mathcal{A}, \mathcal{M}, N) := \mathcal{J}(\widetilde{N}, \widetilde{\#}, \widetilde{1})$ is a Jordan algebra over $\mathcal{O}_X$.

b) Let $\mathcal{J} = \mathcal{J}(N, \#, 1)$ be an Albert algebra over $\mathcal{O}_X$ containing a subalgebra $\mathcal{A}^+$ with $\mathcal{A}$ an Azumaya algebra over $\mathcal{O}_X$ of constant degree 3. Then there exist a locally right $\mathcal{A}$-module of rank one and norm one and a norm on $\mathcal{M}$ such that the embedding $\mathcal{A} \hookrightarrow \mathcal{J}$ extends to an isomorphism

$$\mathcal{J}(\mathcal{A}, \mathcal{M}, \mathcal{N}) \xrightarrow{\sim} \mathcal{J}$$

References

[M1] K. McCrimmon. *The Freudenthal Springer-Tits constructions of exceptional Jordan algebras.* Trans. Amer. Math. Soc. 139 (1969), 495 - 510.

[M2] – *The Freudenthal-Springer-Tits constructions revisited.* Trans. Amer. Math. Soc. **148** (1970), 293 - 314.

[P] H. P. Petersson. *Composition algebras over algebraic curves of genus zero.* Trans. Amer. Math. Soc. 337 (1993) 1, 473 - 493.

THE IDENTITY $(X^2)^2 = W(X)X^3$ IN BARIC ALGEBRAS

RAÚL ANDRADE and ALICIA LABRA*
Departamento de Matemáticas. Facultad de Ciencias. Universidad de Chile. Casilla 653. Santiago-Chile.

and

ABDÓN CATALÁN
Departamento de Matemáticas. Universidad de la Frontera. Casilla 54-D. Temuco-Chile.

Abstract In this paper we study the identity $(x^2)^2 = w(x)x^3$ and we prove that this identity characterizes power-associative Bernstein algebras of order 2, $A = Ke \oplus U \oplus V_2$ with $v^3 = 0$ for every $v \in V_2$. Moreover, we study the generalized Etherington's ideals of A.

1. Introduction

In what follows, K is an infinite field of characteristic not 2 and A is a commutative non necessarily associative algebra over K.

We recall that A is a power-associative algebra if every subalgebra generated by only one element is associative. If $w : A \to K$ is a nonzero algebra homomorphism, then the ordered pair (A, w) is called a baric algebra and w its weight homomorphism. If the baric algebra (A, w) satisfies the identity $x^{[n+2]} = (w(x)x)^{[n+1]}$, it is called a Bernstein algebra of order n where n is the minimun integer for which the identity holds and $x^{[1]} = x, .., x^{[k+1]} = x^{[k]}x^{[k]}$, $k \geq 1$ are the plenary powers of x. For references, see [7]. If the baric algebra (A, w) satisfies the equation $x^r + \gamma_1 w(x)x^{r-1} + ... + \gamma_{r-1}w(x)^{r-1}x = 0$ (train equation), it is called a train algebra of rank r, where r is the minimun integer for which the above identity holds, $\gamma_1, ..., \gamma_{r-1}$ are fixed elements in K and $x^1 = x$, ..., $x^{k+1} = x^k x$ are the principal powers of x. The baric algebra (A, w) is a special train algebra if $Ker(w)^k$ is an ideal of A for every $k \in \mathbb{N}$ and $Ker(w)$ is nilpotent. Moreover, every special train algebra is a train algebra, (for details see [8]).

It is well known that for Bernstein algebras A of order 2, the homomorphism w is uniquely determined and the set of idempotents is given by $I_p(A) = \{(x^2)^2/x \in A, w(x) = 1\}$. Moreover, A splits as the direct sum relative to one idempotent $e \neq 0$, $A = Ke \oplus Ker(w) = Ke \oplus U \oplus V_2$, where $U = \{x \in Ker(w)/ex = \frac{1}{2}x\}$ and $V_2 = \{x \in Ker(w)/e(ex) = 0\}$ and $U^2 \subseteq V_2$.

---

* Partially supported by Fondecyt 1930843 and D.T.I. Universidad de Chile, E-3276/9322.
A.M.S. (1991) Subject classification: 17 A 30, 17 D 92.

S. González (ed.), Non-Associative Algebra and Its Applications, 12–16.

2. The identity $(x^2)^2 = w(x)x^3$

The idea of studying the identity

$$(x^2)^2 = w(x)x^3 \tag{1}$$

in baric algebras arise of the study of power-associative Bernstein algebras of order 2, i.e. power-associative, baric algebras satisfying $((x^2)^2)^2 = w(x^4)(x^2)^2$.

In [6], it is proved the following result that characterize power-associative Bernstein algebras of order 2.

Theorem 2.1 *For a baric algebra (A, w), of arbitrary dimension, the following conditions are equivalent:*

1. *(A, w) is a power-associative Bernstein algebra of order 2.*
2. *There exist a decomposition of A relative to an idempotent $e \neq 0$, $A = Ke \oplus U \oplus V_2$ where $Ker(w) = U \oplus V_2$, $U = \{x \in Ker(w)/ex = \frac{1}{2}x\}$, $V_2 = \{x \in Ker(w)/ex = 0\}$, $U^2 \subseteq V_2$, $V_2^2 \subseteq V_2$, $UV_2 \subseteq U$ and every element $u \in U$ and $v \in V_2$ satisfy the relations: $u^3 = 0$, $v^4 = (v^2)^2 = 0$, $uv^2 = 2v(uv)$, $u^2v = 2u(uv)$.*

Moreover, for every $u, u' \in U$, $v, v' \in V$ we have

$$\begin{aligned}
(u^2)^2 &= 0, && (2)\\
u(vv') &= (uv)v' + (uv')v, && (3)\\
v(uu') &= (vu)u' + (vu')u, && (4)\\
4(uv)^2 + u^2v^2 &= 2v(vu^2). && (5)
\end{aligned}$$

On the other hand, in [1], it is proved that algebras satisfying the identity (1) always have idempotent elements and every linear form $w : A \to K$ is also a multiplicative map. Moreover, A admits a Peirce decomposition $A = Ke \oplus N_{\frac{1}{2}} \oplus N_0$, where $N_i = \{x \in Ker(w)/ex = ix\}$, $i = 0, \frac{1}{2}$ and $N_{\frac{1}{2}}^2 \subseteq N_0$, $N_{\frac{1}{2}}N_0 \subseteq N_{\frac{1}{2}}$, $N_0^2 \subseteq N_0$ and for every $u \in N_{\frac{1}{2}}$, $v \in N_0$: $u^3 = 0$, $v^3 = 0$, $uv^2 = 2v(uv)$, $uv^2 = 2u(uv)$.

Thus, we have

Proposition 2.2 *Every baric algebra satisfying the identity (1) is a power-associative Bernstein algebra of order 2.*

Remark The converse of Proposition 2.2 is not true. For example, $A = Ke \oplus V_2$, $V_2 = < v_1, v_2, v_3 >$ and multiplication table $e^2 = e$, $v_1^2 = v_2$, $v_1v_2 = v_3$, all other products being zero is a Bernstein algebra of order 2. Moreover, it is Jordan, so, it is a power-associative algebra and if $x = e + v_1 + v_2 + v_3$ then $(x^2)^2 = e$ and $w(x)x^3 = e + v_3$.

Theorem 2.3 *For a baric algebra (A, w), of arbitrary dimension, the following conditions are equivalent:*

1. *$A = Ke \oplus U \oplus V_2$ is a power-associative Bernstein algebra of order 2 with $v^3 = 0$ for every $v \in V_2$.*
2. *The identity $(x^2)^2 = w(x)x^3$ holds in A.*

Proof: We only need to prove that 1. implies 2. Let $A = Ke \oplus U \oplus V_2$ be a power-associative Bernstein algebra of order 2 with $v^3 = 0$ for every $v \in V_2$. Let $x = \lambda e + u + v$ be un element of A. Then, by using relations (2), (3) and (4) we have $(x^2)^2 - w(x)x^3 = 2u^2v^2 + 4(uv)^2$, because $u^2(uv) = 0$ and $v^2(uv) = 0$. Moreover, the relation (5) implies that $2u^2v^2 + 4(uv)^2 = 2v(vu^2) + u^2v^2$. By using $U^2 \subseteq V_2$ and Jacobi's identity in V_2 we have, $2v(vu^2) = -v^2u^2$. So, $2u^2v^2 + 4(uv)^2 = 0$ and $(x^2)^2 - w(x)x^3 = 0$.

Moreover, in [2], it is proved that every baric algebra satisfying (1) is a special train algebra, hence a Gonshor genetic algebra; that is, a conmutative, non necessarely associative algebra which admits a basis $\{e_1, ..., e_n\}$ such that if $e_ie_j = \sum_{k=1}^{n} \gamma_{ijk}e_k (i, j = 1, ..., n)$ then, the multiplication constants γ_{ijk} verify the following conditions: $\gamma_{111} = 1$, $\gamma_{1jk} = 0$ for $k < j$ and $\gamma_{ijk} = 0$ for $k \leq max(i, j)$, $i \geq 2$. The basis $\{e_1, ..., e_n\}$ is called a canonical basis, (for details see [4], [5] and [8]).

Remark We may note that there exist Bernstein algebras of order 2 which are Gonschor genetic algebras and are not special train algebras. For example, $A =< c_1, ..., c_6 >$ and multiplication table $c_1^2 = c_1$, $c_1c_2 = \frac{1}{2}c_1$, $c_1c_3 = \frac{1}{4}c_4$, $c_2^2 = \frac{1}{4}c_3$, $c_2c_3 = \frac{1}{8}c_5$, $c_3^2 = \frac{1}{16}c_6$, all other products being zero is an algebra and $\{c_1, ..., c_6\}$ is a canonical basis of A, (see [8]). Then A is a genetic algebra. Moreover, A is a baric algebra with weight homomorphism $w : A \to \mathbb{R}$ defined by $w(c_1) = 0$, $w(c_i) = 0$ for $i \in \{2, ..., 6\}$ and $((x^2)^2)^2 = w(x^4)(x^2)^2$ for every $x \in A$. Then, A is a Bernstein algebra of order 2. If $N = Ker(w)$ then $N^2 =< c_3, c_5, c_6 >$, $N^3 =< c_5, c_6 >$, $N^4 = 0$ but $c_1c_3 = \frac{1}{4}c_4 \notin N^2$. Thus N^2 is not an ideal of A and A is not a special train algebra.

3. E-ideals

Let (A, w) be a baric algebra and let $\gamma_1, \cdots, \gamma_{n-1}$ be arbitrary elements in the field K such that $1 + \gamma_1 + \cdots + \gamma_{n-1} = 0$.

The formal expression

$$p(x) = x^n + \gamma_1 w(x)x^{n-1} + \cdots + \gamma_{n-1}w(x)^{n-1}x \tag{6}$$

is called a train polynomial with coefficients $\gamma_1, \cdots, \gamma_{n-1}$ and degree n.

Definition 3.1 *The generalized Etherington's ideal, in short E-ideal of a baric algebra (A, w), associated to the train polynomial $p(x)$, is the ideal of A generated by all $p(a) = a^n + \gamma_1 w(a)a^{n-1} + \cdots + \gamma_{n-1}w(a)^{n-1}a$, $a \in A$.*

Let us denote this ideal as $E_A(1, \gamma_1, \cdots, \gamma_{n-1})$ or $E_A(p)$. The ideal $E_A(1, -1)$ is called the Etherington's ideal. The concept of E-ideal was introduced in [3]. In this section, we shall describe the equivalence classes of baric algebras satisfying the identity (1).

Proposition 3.2 $E_A(1,-1) = N_{\frac{1}{2}}N_0 \oplus N_0$.

Proof: It is easy to prove that $N_{\frac{1}{2}}N_0 \oplus N_0$ is an ideal of A, Now, let $x = \lambda e + u + v$ be an element of A, with $\lambda = w(x)$. Then $x^2 - w(x)x = 2uv + (u^2 + v^2 - \lambda v)$. Thus, $E_A(1,-1) \subseteq N_{\frac{1}{2}}N_0 \oplus N_0$. Conversely, if $v \in N_0$ then $(e+v)^2 - (e+v) = -2v + v^2$. So, $v = \frac{1}{2}[(e+v)^2 - (e+v) + v^2]$ belongs to the Etherington's ideal, which implies that N_0 and hence $N_{\frac{1}{2}}N_0 \oplus N_0$ is contained in that ideal. Thus, $E_A(1,-1) = N_{\frac{1}{2}}N_0 \oplus N_0$.

Proposition 3.3 $E_A(1,-1,0) = N_{\frac{1}{2}}N_0^2 \oplus N_0^2$.

Proof: First we prove that $N_{\frac{1}{2}}N_0^2 \oplus N_0^2$ is an ideal of A.

Let $u_1, u_2 \in N_{\frac{1}{2}}$ and $v_1, v_2 \in N_0$. Then, since $N_{\frac{1}{2}}N_0 \subseteq N_{\frac{1}{2}}$ and by using relations (3), (4) we have $u_1[u_2(v_1v_2)] = u_1[(u_2v_1)v_2] + u_1[(u_2v_2)v_1] = [u_1(u_2v_1)]v_2 - (u_2v_1)(u_1v_2) + [u_1(u_2v_2)]v_1 - (u_1v_1)(u_2v_2)$. But $(u_2v_1)(u_1v_2) + (u_1v_1)(u_2v_2) = -(u_1u_2)(v_1v_2) \in N_0^2$.

Moreover, $[u_1(u_2v_1)]v_2$, $[u_1(u_2v_2)]v_1 \in N_0^2$. Therefore, $N_{\frac{1}{2}}(N_{\frac{1}{2}}N_0^2) \subseteq N_0^2$ and $N_{\frac{1}{2}}(N_{\frac{1}{2}}N_0^2 + N_0^2) \subseteq N_{\frac{1}{2}}N_0^2 \oplus N_0^2$.

Since, for every $u \in N_{\frac{1}{2}}$ and $v_1, v_2, v_3 \in N_0$, $[u(v_1v_2)]v_3 = u[(v_1v_2)v_3] - (uv_3)(v_1v_2)$, we have that, $(N_{\frac{1}{2}}N_0^2)N_0 \subseteq N_{\frac{1}{2}}N_0^2$ and $(N_{\frac{1}{2}}N_0^2 \oplus N_0^2)N_0 \subseteq N_{\frac{1}{2}}^2 N_0^2 \oplus N_0^2$. Therefore, $N_{\frac{1}{2}}N_0^2 \oplus N_0^2$ is an ideal of A.

Moreover, $x^3 - w(x)x^2 = 2uv^2 + 2u^2v - v^2$. Thus, $E(1,-1,0) \subseteq N_{\frac{1}{2}}N_0^2 \oplus N_0^2$.

Next we prove that $N_0^2 \subseteq E_A(1,-1,0)$. Let $v \in N_0$. Then $(e+v)^3 - (e+v)^2 = -v^2 \in E_A(1,-1,0)$. Since, $v_1v_2 = \frac{1}{2}[(v_1+v_2)^2 - v_1^2 - v_2^2]$ for every $v_1, v_2 \in N_0$ we have $N_0^2 \subseteq E_A(1,-1,0)$. Thus, $N_{\frac{1}{2}}N_0^2 \oplus N_0^2 \subseteq E_A(1,-1,0)$ and Proposition 3.3 follows.

Theorem 3.4 *Suppose that A satisfies (1). Then the E-ideals of A are, $0 = E_A(1,-1,0,0)$, $E_A(1,-1,0)$ and $E_A(1,-1)$.*

Proof: Let $x = w(x)e + u + v$ be an element of A. Then, $x^2 = w(x^2)e + w(x)u + 2uv + v^2$, $x^3 = w(x^3)e + w(x^2)u + w(x)u^2 + 2w(x)uv + 2uv^2 + 2u^2v$ and $x^n = w(x)^{n-3}x^3$ for $n \geq 4$.

Now, let us consider the train polynomial

$$p(x) = x^n + \gamma_1 w(x)x^{n-1} + \cdots + \gamma_{n-1}w(x)^{n-1}x$$

and replacing x and his powers and by using $\gamma_1 + \cdots + \gamma_{n-1} = 0$ we have

$$(*) \quad p(x) = \gamma_{n-1}[\lambda^{n-1}v - \lambda^{n-2}(u^2 + (uv) - 2\lambda^{n-3}(uv^2 + u^2v)] + \gamma_{n-2}[\lambda^{n-2}v^2 - 2\lambda^{n-3}(uv^2 + u^2v)]$$

If we take $\gamma_{n-1} \neq 0$ in (*), we have $E_A(1,-1) = N_{\frac{1}{2}}N_0 \oplus N_0$. By setting, $\gamma_{n-1} = 0$ and $\gamma_{n-2} \neq 0$ in (*), we obtain $E_A(1,-1,0) = N_{\frac{1}{2}}N_0^2 \oplus N_0^2$.

Finally , by setting, $\gamma_{n-1} = 0 = \gamma_{n-2}$ in (*), we have $E_A(1,-1,0,0) = 0$.

References

1. M.T. Alcalde, C. Burgueño, C. Mallol, Les Pol(n,m)-algèbres, *Linear algebra and its Applications.* 9 (1993) 215-234
2. R. Andrade, A. Labra, On a class of Baric algebras , submitted (1993)
3. R. Costa, A. Catalan, E-ideals in baric algebras, submitted (1993)
4. H. Gonshor, Constributions to Genetics algebras, *Proc, Edinburg Math. Soc.* 17 (1971) 289-298
5. H. Gonshor, Constribution to Genetics algebras II, *Proc. Edinburg Math. Soc.* 18 (1973) 273-279
6. A. Labra, C. Mallol, A. Suazo, A characterization of power- associative Bernstein algebra of order 2, To appear in Nova Journal of Algebra and Geometry
7. C. Mallol, A. Micali, M. Ouattara, Sur les algèbres de Bernstein IV, *Linear algebra and its Aplc.* 158 (1991) 1-26
8. A. Wörz-Busekros, *Algebras in Genetics.* Lecture Notes in Biomathematics 36, Springer, New York, 1980.

CAYLEY–KLEIN LIE ALGEBRAS AND THEIR QUANTUM UNIVERSAL ENVELOPING ALGEBRAS

A. BALLESTEROS, F.J. HERRANZ, M.A. DEL OLMO and M. SANTANDER
Departamento de Física Teórica, Universidad de Valladolid, E-47011, Valladolid, Spain.

Abstract. The N-dimensional Cayley-Klein scheme allows a simultaneous description of 3^N symmetric orthogonal homogeneous spaces by means of a set of Lie algebras depending on N real parameters. We present here a quantum deformation of the Lie algebras generating the groups of motion of the two and three dimensional Cayley-Klein geometries. This Hopf algebra structure is presented in a compact form by using a formalism developed for the case of (quasi)free Lie algebras. Their quasitriangularity (i.e., the usual way to study the associativity of their dual objects, the quantum groups) is also discussed.

1. Introduction

We study here a certain type of Lie algebra deformations (so called "quantum" ones), that have recently appeared in the context of the Quantum Inverse Scattering Method. They are defined as deformations of the corresponding universal enveloping algebra Ug and their dual objects (in certain restricted sense) generate the "quantum groups" –deformations of the algebra of functions on the group, in the spirit of non-commutative geometry [1, 2]–. Their underlying algebraic structure (mainly Hopf algebra properties [3]) is rather rich and was soon described for the classical simple Lie algebras [4, 5, 6].

However, many physically interesting groups are not simple groups: for instance, the groups of inertial transformations of space–time such as Galilei or Poincare ones. Some quantum deformations have been built for their associated Lie algebras
[7, 8] which also arises as symmetries of certain physical problems [9]. We present here an attempt towards their characterization based on a Cayley-Klein (CK) geometrical scheme that includes all these groups as well as their transformed by Inönü–Wigner contractions [10]. We also discuss the problems arising in the definition of the quantum groups as dual objects of these quantum algebras, mainly in connection with the way in which the associativity of the deformed algebra of functions on the group is guaranteed (the R–matrix problem).

2. The Cayley-Klein Lie Algebras

From a physical point of view, some interesting homogeneous symmetric spaces can be simultaneously described in the framework of CK geometries. To do this, we consider an N-dimensional (N-d) symmetric orthogonal geometry as a group G of dimension $N(N+1)/2$ and a set of N commuting involutions $S^{(i)}$ in g, the Lie

S. González (ed.), Non-Associative Algebra and Its Applications, 17–23.

algebra of G. If we denote by $h^{(i)}$ the Lie subalgebras of elements invariant under $S^{(i)}$, their corresponding groups $H^{(i)}$ have to be taken as the isotropy groups of a point, a line, ..., an $(N-1)$-flat, so the homogeneous spaces $\mathcal{X}^{(i)} \equiv G/H^{(i)}$ turn out to be the spaces of points, lines, ... of the geometry. The involutions must satisfy certain requirements (specially on the dimensions of the subgroups $H^{(i)}$) and the group G is also required to act effectively on all the

$\mathcal{X}^{(i)}$. Without entering into details, the main result is that the CK Lie algebra depends on N real parameters $(\kappa_1, \ldots, \kappa_N)$ that can take any real value (however, these parameters can be scaled to -1, 0 or $+1$), and we have [11].

Let G be the CK Lie group corresponding to an N-d CK geometry.

a) The Lie algebra of G has dimension $N(N+1)/2$ with generators J_{ij} $(i < j;\ i,j = 0,1,\ldots,N)$ and is characterized by N real parameters $(\kappa_1,\ldots,\kappa_N)$.

b) The isotopy subgroups of i-flats $H^{(i)}$ $(i = 0,1,\ldots,N-1)$, correspond to the subalgebras $h^{(i)} = \langle J_{ab}, J_{cd}\rangle$ $(a < b,\ a,b = 0,1,\ldots,i;\ c < d;\ c,d = i+1,\ldots,N)$.

c) The Lie brackets of the basis elements J_{ij} can be written in terms of the parameters κ_{ij} $(i < j;\ i,j = 0,1,\ldots,N)$ defined by

$\kappa_{ij} = \kappa_{i+1}\kappa_{i+2}\ldots\kappa_j$, and are

$$[J_{ij}, J_{lm}] = \delta_{im}J_{lj} - \delta_{jl}J_{im} + \delta_{jm}\kappa_{lm}J_{il} + \delta_{il}\kappa_{ij}J_{jm}, \quad (i \le l,\ j \le m). \tag{1}$$

d) The CK algebras $g_{(\kappa_1,\ldots,\kappa_N)}$ can be realized in terms of real matrices:

$$\mathcal{D}(J_{ij}) = -\kappa_{ij}e_{ij} + e_{ji}, \tag{2}$$

where e_{ij} are the standard $(N+1)\times(N+1)$ matrices with elements $(e_{ij})_{kl} = \delta_{ik}\delta_{jl}$ and commutation relations

$[e_{ij}, e_{lm}] = \delta_{jl}e_{im} - \delta_{im}e_{lj}$.

A representation of the oneparameter subgroups associated to the generators

J_{ij} is obtained by exponentiation of the matrices $\mathcal{D}(J_{ij})$. The matrix entries of these subgroup elements are easily written by using the "generalized"

trigonometric functions $\text{sine}_\kappa(x) \equiv \mathrm{S}_\kappa(x)$ and $\text{cosine}_\kappa(x) \equiv \mathrm{C}_\kappa(x)$, defined in terms of power series as follows:

$$\mathrm{S}_\kappa(x) = \sum_{l=0}^{\infty}(-\kappa)^l\frac{x^{2l+1}}{(2l+1)!}, \quad \mathrm{C}_\kappa(x) = \sum_{l=0}^{\infty}(-\kappa)^l\frac{x^{2l}}{(2l)!}. \tag{3}$$

These functions constitute an intrinsic tool throughout any explicit computation within the CK scheme (either classical or quantum), and can be considered as deformations of the "galilean" or "parabolic" functions 1 and x, to which they tend in the limit $\kappa \to 0$:

$$\mathrm{C}_\kappa(x) = \begin{cases} \cos\sqrt{\kappa}x & \text{if } \kappa > 0 \\ 1 & \text{if } \kappa = 0 \\ \cosh\sqrt{-\kappa}x & \text{if } \kappa < 0 \end{cases}, \quad \mathrm{S}_\kappa(x) = \begin{cases} \frac{1}{\sqrt{\kappa}}\sin\sqrt{\kappa}x & \text{if } \kappa > 0 \\ x & \text{if } \kappa = 0 \\ \frac{1}{\sqrt{-\kappa}}\sinh\sqrt{-\kappa}x & \text{if } \kappa < 0 \end{cases} \tag{4}$$

3. Quantum Universal Enveloping Algebras

Following Drinfel'd [4], a "quantized universal enveloping algebra"

(QUE algebra) of a Lie algebra g is a Hopf algebra $\mathcal{A}$ over the formal power series $\mathbf{C}[[z]]$ on a deformation indeterminate z such that $\mathcal{A}$ is topologically free $\mathbf{C}[[z]]$–module and $\mathcal{A}/z\mathcal{A}$ is isomorphic (as Hopf algebra) to Ug.

Recall that a $\mathbf{C}$–algebra is a Hopf algebra if there exist two homomorphisms called coproduct ($\Delta : \mathcal{A} \longrightarrow \mathcal{A} \otimes \mathcal{A}$) and counit ($\epsilon : \mathcal{A} \longrightarrow \mathbf{C}$), as well as an antihomomorphism (the antipode $\gamma : \mathcal{A} \longrightarrow \mathcal{A}$) such that, $\forall a \in \mathcal{A}$:

$$
\begin{aligned}
(id \otimes \Delta)\Delta(a) &= (\Delta \otimes id)\Delta(a), & (5)\\
(id \otimes \epsilon)\Delta(a) &= (\epsilon \otimes id)\Delta(a) = a, & (6)\\
m((id \otimes \gamma)\Delta(a)) &= m((\gamma \otimes id)\Delta(a)) = \epsilon(a)1, & (7)
\end{aligned}
$$

where m is the usual multiplication $m(a \otimes b) = ab$. Counit and antipode are derived from the coproduct in a unique way.

Roughly speaking, we can think of the elements in $\mathcal{A}$ as formal power series in z with coefficients in Ug. Provided we are not going to take into account topological properties, the topological freeness is translated into algebraic terms by imposing that the QUE algebra must be free, or at least torsion–free as a

$\mathbf{C}[[z]]$–module [12]. Both the Hopf homomorphisms and the torsion condition restrict the formal power series suitable for quantization [13].

The "classical limit" property $\mathcal{A}/z\mathcal{A} \simeq Ug$ is also a very

strong constrain on the quantization. Our aim is to obtain quantum CK algebras, so the $z \rightarrow 0$ limit must lead to the Lie brackets in (1). The standard quantization of the classical Cartan series of simple Lie algebras uses the A_1 quantum structure as building block [4, 5]. For the quantum CK algebras already obtained [14, 15], the generalization seems not to be immediate. However, CK geometries of a given dimension do contain lower dimension subgeometries. Any compact way of writing the Hopf homomorphisms embodying somehow this embedding would help to obtain the general structure. In this sense, the following proposition [16] ensures the consistency of the deformation for a "quasi"–free algebra (the only condition is the commutativity of the primitive generators) and will be used to write the quantum CK algebras in the next section.

Proposition 3.1 *Let $\{1, H_1, \ldots, H_n, X_1, \ldots, X_m\}$ be the generators of a "free" (up to the conditions $[H_i, H_j] = 0\ \forall i, j$) associative algebra E over $\mathbf{C}$. Let $\alpha_i, \beta_j\ i, j = 1, \ldots, n$ be a set of $(m \times m)$ matrices with entries in $\mathbf{C}[[z]]$ such that $[\alpha_i, \beta_j] = [\alpha_i, \alpha_j] = [\beta_i, \beta_j] = 0\ \forall i, j$. Let $\vec{X}$ be a "vector" with components X_l, $l = 1, \ldots, m$. The coproduct*

$$
\begin{aligned}
\Delta 1 &= 1 \otimes 1, \qquad \Delta H_i = 1 \otimes H_i + H_i \otimes 1, & (8)\\
\Delta \vec{X} &= \exp(\sum_{i=1}^{n} \alpha_i H_i)\dot{\otimes}\vec{X} + \sigma\left(\exp(\sum_{i=1}^{n} \beta_i H_i)\dot{\otimes}\vec{X}\right),
\end{aligned}
$$

turns the completion B of formal power series on z with coefficients in E into a Hopf algebra.

Here σ is the permutation map $\sigma(a\otimes b)=b\otimes a$ and, if $P\equiv(p_{kl})$ is a $(m\times m)$ matrix with entries in B, the k-th component of $(P\dot{\otimes}\vec{X})_k$ is $(P\dot{\otimes}\vec{X})_k:=\sum_{l=1}^{m}p_{kl}\otimes X_l$.

4. QUE Cayley-Klein Algebras

We restrict in this section to the study of the algebras generating the two and three dimensional CK systems. The former is a family of 3-d Lie algebra depending on two parameters which contains as particular cases $so(3), so(2,1)$, the 2-d euclidean $e(2)$ and the $(1+1)$ Galilei and Poincaré algebras. A simultaneous quantization of all of these systems is:

Theorem 4.1 *Let $g_{(\kappa_1,\kappa_2)}$ be the Lie algebra generating the 2-d CK systems and whose infinitesimal generators are $\{J_{12},P_1,P_2\}$. The coproduct*

$$\begin{aligned}\Delta P_2 &= 1\otimes P_2+P_2\otimes 1,\\ \Delta\begin{pmatrix}P_1\\ J_{12}\end{pmatrix} &= \exp\left\{\begin{pmatrix}-\frac{z}{2}P_2 & 0\\ 0 & -\frac{z}{2}P_2\end{pmatrix}\right\}\dot{\otimes}\begin{pmatrix}P_1\\ J_{12}\end{pmatrix}\\ &\qquad +\sigma\left(\exp\left\{\begin{pmatrix}\frac{z}{2}P_2 & 0\\ 0 & \frac{z}{2}P_2\end{pmatrix}\right\}\dot{\otimes}\begin{pmatrix}P_1\\ J_{12}\end{pmatrix}\right),\end{aligned} \tag{9}$$

and the commutation relations

$$[J_{12},P_1]=\mathrm{S}_{-z^2}(P_2),\quad [J_{12},P_2]=-\kappa_2P_1,\quad [P_1,P_2]=\kappa_1J_{12}. \tag{10}$$

define the QUE algebra $U_zg_{(\kappa_1,\kappa_2)}$.

Corollary 4.1 *Counit and antipode are deduced from (9); if $X\in\{P_1,P_2,J_{12}\}$, read*

$$\epsilon(X)=0,\qquad \gamma(X)=-e^{\frac{z}{2}P_2}\,X\,e^{-\frac{z}{2}P_2}. \tag{11}$$

Corollary 4.2 *The center of $U_zg_{(\kappa_1,\kappa_2)}$ is generated by*

$$C_z=4\,\mathrm{C}_{\kappa_1\kappa_2}\left(\frac{z}{2}\right)\left[\mathrm{S}_{-z^2}\left(\frac{1}{2}P_2\right)\right]^2+\frac{2}{z}\,\mathrm{S}_{\kappa_1\kappa_2}\left(\frac{z}{2}\right)\left\{\kappa_2P_1^2+\kappa_1J_{12}^2\right\}. \tag{12}$$

Proposition 4.1 *The fundamental representation D_q of $U_zg_{(\kappa_1,\kappa_2)}$ in terms of the "classical" one D is defined as follows*

$$D_q(P_2)=D(P_2),\quad D_q(P_1)=\sqrt{\frac{\mathrm{S}_{\kappa_1\kappa_2}(z)}{z}}D(P_1),\quad D_q(J_{12})=\sqrt{\frac{\mathrm{S}_{\kappa_1\kappa_2}(z)}{z}}D(J_{12}). \tag{13}$$

Note that the "generalized trigonometric functions" appear as natural deformation functions in this context. Moreover, they are consistent formal power series in the sense that, for instance, $\Delta(\mathrm{S}_\kappa(P_2))=\mathrm{S}_\kappa(P_2)\otimes\mathrm{C}_\kappa(P_2)+\mathrm{C}_\kappa(P_2)\otimes\mathrm{S}_\kappa(P_2)$. The classical limit $z\to 0$ is always well defined and straightforwardly leads to the algebra defined in (1). It is also worth remarking that different algebras are obtained by specialization of the κ_i parameters. This deformation preserves always non–Lie character in (10), since the deformed bracket cannot vanish whatever the κ_i are.

For the case of 3-d geometries, the CK Lie algebra is now 6-d and depends on three measure coefficients $(\kappa_1, \kappa_2, \kappa_3)$. The number of different geometries included is now 3^3. The algebras

$so(4), so(3,1), so(2,2)$, the 3-d Euclidean algebra $e(3)$, the $(2+1)$–d versions of the Newton-Hooke, Galilei and Poincaré algebras can be found among the set $g_{(\kappa_1,\kappa_2,\kappa_3)}$.

Theorem 4.2 *Let $g_{(\kappa_1,\kappa_2,\kappa_3)}$ be the Lie algebra generating the 3-d CK systems, with infinitesimal generators J_{ij}, $(i<j;\ i,j=0,1,2,3)$. The coproduct*

$$\begin{aligned}\Delta(J_{03}) &= 1\otimes J_{03} + J_{03}\otimes 1, \qquad \Delta(J_{12}) = 1\otimes J_{12} + J_{12}\otimes 1,\\ \Delta\begin{pmatrix} J_{01}\\ J_{02}\\ J_{13}\\ J_{23}\end{pmatrix} &= \exp\{\alpha_1 J_{03}+\alpha_2 J_{12}\}\dot{\otimes}\begin{pmatrix} J_{01}\\ J_{02}\\ J_{13}\\ J_{23}\end{pmatrix} + \sigma\left(\exp\{\beta_1 J_{03}+\beta_2 J_{12}\}\dot{\otimes}\begin{pmatrix} J_{01}\\ J_{02}\\ J_{13}\\ J_{23}\end{pmatrix}\right)\end{aligned} \tag{14}$$

$$\alpha_1 = -\beta_1 = \begin{pmatrix} -\frac{z}{2} & 0 & 0 & 0\\ 0 & -\frac{z}{2} & 0 & 0\\ 0 & 0 & -\frac{z}{2} & 0\\ 0 & 0 & 0 & -\frac{z}{2}\end{pmatrix}, \quad \alpha_2 = -\beta_2 = \begin{pmatrix} 0 & 0 & 0 & -\frac{z}{2}\kappa_1\\ 0 & 0 & \frac{z}{2}\kappa_1 & 0\\ 0 & \frac{z}{2}\kappa_3 & 0 & 0\\ -\frac{z}{2}\kappa_3 & 0 & 0 & 0\end{pmatrix},$$

and the following non–vanishing commutation relations define $U_z g_{(\kappa_1,\kappa_2,\kappa_3)}$

$$\begin{aligned}[J_{12},J_{01}] &= J_{02},\quad [J_{12},J_{02}] = -\kappa_2 J_{01},\quad [J_{01},J_{02}] = \kappa_1\,\mathrm{S}_{-z^2\kappa_1\kappa_3}(J_{12})\,\mathrm{C}_{-z^2}(J_{03}),\\ [J_{13},J_{01}] &= \mathrm{S}_{-z^2}(J_{03})\,\mathrm{C}_{-z^2\kappa_1\kappa_3}(J_{12}),\quad [J_{13},J_{03}] = -\kappa_2\kappa_3 J_{01},\quad [J_{01},J_{03}] = \kappa_1 J_{13},\\ [J_{23},J_{02}] &= \mathrm{S}_{-z^2}(J_{03})\,\mathrm{C}_{-z^2\kappa_1\kappa_3}(J_{12}),\quad [J_{23},J_{03}] = -\kappa_3 J_{02},\quad [J_{02},J_{03}] = \kappa_1\kappa_2 J_{23},\\ [J_{23},J_{12}] &= J_{13},\quad [J_{23},J_{13}] = -\kappa_3\,\mathrm{S}_{-z^2\kappa_1\kappa_3}(J_{12})\,\mathrm{C}_{-z^2}(J_{03}),\quad [J_{12},J_{13}] = \kappa_2 J_{23}.\end{aligned} \tag{15}$$

Corollary 4.3 *Counit and antipode are*

$$\epsilon(J_{ij}) = 0, \qquad \gamma(J_{ij}) = -e^{zJ_{03}}\, J_{ij}\, e^{-zJ_{03}}. \tag{16}$$

Corollary 4.4 *The following deformed second order elements belong to the center of $U_z g_{(\kappa_1,\kappa_2,\kappa_3)}$*

$$\begin{aligned}\mathcal{C}_1^q &= 4\,\mathrm{C}_{\kappa_{03}}(z)\big[\mathrm{S}^2_{-z^2}(J_{03}/2)\,\mathrm{C}^2_{-z^2\kappa_1\kappa_3}(J_{12}/2) + \kappa_1\kappa_3\,\mathrm{S}^2_{-z^2\kappa_1\kappa_3}(J_{12}/2)\,\mathrm{C}^2_{-z^2}(J_{03}/2)\big]\\ &\quad + \frac{1}{z}\,\mathrm{S}_{\kappa_{03}}(z)\left[\kappa_2\kappa_3 J_{01}^2 + \kappa_3 J_{02}^2 + \kappa_1 J_{13}^2 + \kappa_1\kappa_2 J_{23}^2\right],\\ \mathcal{C}_2^q &= \mathrm{C}_{\kappa_{03}}(z)\,\mathrm{S}_{-z^2}(J_{03})\,\mathrm{S}_{-z^2\kappa_1\kappa_3}(J_{12}) + \frac{1}{z}\,\mathrm{S}_{\kappa_{03}}(z)[\kappa_2 J_{01}J_{23} - J_{02}J_{13}].\end{aligned} \tag{17}$$

Proposition 4.2 *The fundamental representation D_q of $U_z g_{(\kappa_1,\kappa_2,\kappa_3)}$ is*

$$\begin{aligned}D_q(J_{ij}) &= \sqrt{\frac{\mathrm{S}_{\kappa_{03}}(z)}{z}}\, D(J_{ij}), &&\text{if } ij = 01, 02, 13, 23,\\ D_q(J_{ij}) &= D(J_{ij}), &&\text{if } ij = 03, 12.\end{aligned} \tag{18}$$

5. Quasitriangular Hopf Algebras

A quasitriangular Hopf algebra [4] is a pair $(\mathcal{A}, \mathcal{R})$ where $\mathcal{A}$ is a Hopf algebra and $\mathcal{R} \in \mathcal{A} \otimes \mathcal{A}$ is invertible and obeys

$$\begin{aligned} \sigma \circ \Delta h &= \mathcal{R}(\Delta h)\mathcal{R}^{-1}, \qquad \forall\, h \in \mathcal{A} \\ (\Delta \otimes id)\mathcal{R} &= \mathcal{R}_{13}\mathcal{R}_{23}, \qquad (id \otimes \Delta)\mathcal{R} = \mathcal{R}_{13}\mathcal{R}_{12}, \end{aligned} \tag{19}$$

where, if $\mathcal{R} = \sum_i a_i \otimes b_i$, we denote $\mathcal{R}_{12} \equiv \sum_i a_i \otimes b_i \otimes 1$, $\mathcal{R}_{13} \equiv \sum_i a_i \otimes 1 \otimes b_i$ and $\mathcal{R}_{23} \equiv \sum_i 1 \otimes a_i \otimes b_i$. If $\mathcal{A}$ is a quasitriangular Hopf algebra, then $\mathcal{R}$ is called a "universal" $\mathcal{R}$–matrix and satisfies the *Quantum Yang–Baxter Equation*:

$$\mathcal{R}_{12}\mathcal{R}_{13}\mathcal{R}_{23} = \mathcal{R}_{23}\mathcal{R}_{13}\mathcal{R}_{12}. \tag{20}$$

Given a matrix representation $\rho : \mathcal{A} \to Mat(n, \mathbf{C})$, the matrix elements t_{ij} of its dual $\mathcal{A}^*$ (the "quantum group") satisfy the commutation relations

$$RT_1\,T_2 = T_2\,T_1\,R, \tag{21}$$

where $R = (\rho \otimes \rho)(\mathcal{R})$ and $T = (t_{ij})$, $T_1 = T \otimes 1_n$ and $T_2 = 1_n \otimes T$. These matrix solutions R are relevant for physical applications. If R satisfies (20), this equation ensures that the non–commutative algebra generated by the t_{ij} is associative (third order commutation relations are derived from second order ones). However, for non–semisimple quantum algebras (in our scheme, if some $\kappa_i = 0$), neither the universal $\mathcal{R}$ matrices nor even some of their particular realizations are easy to find [7, 17]. In fact, they could not exist.

Acknowledgements

This work has been partially supported by DGICYT (Spain) under project PB92–0255.

References

1. Manin Y.I.: 1989 *Quantum groups and non–commutative geometry* (Montreal: CRM)
2. Woronowicz S.L.: 1987 *Comm. Math. Phys.* **111** 613
3. Abe E.: 1980 *Hopf Algebras* (*Cambridge Tracts in Mathematics 74*) (Cambridge: Cambridge University Press)
4. Drinfeld V.G.: 1986 Quantum Groups *Proc. Int. Congr. of Mathematics MRSI Berkeley* p. 798
5. Jimbo M.: 1985 *Lett. Math. Phys.* **10** 63; 1986 *Lett. Math. Phys.* **11** 247
6. Faddeev L.D., Reshetikhin N. Yu. and Takhtadzhyan L.A.: 1990 *Leningrad Math. J.* **1** 193
7. Celeghini E., Giachetti R., Sorace E. and Tarlini M.: 1992 *Contractions of quantum groups* (*Lecture Notes in Mathematics 1510*) (*Berlin: Springer*) p. 221
8. Lukierski J., Ruegg H. and Nowicky A.: 1992 *Phys. Lett. B.* **293** 344
9. Bonechi F., Celeghini E., Giachetti R., Sorace E. and Tarlini M.: 1992 *Phys. Rev. Lett.* **68** 3178
10. Inönü E. and Wigner E P.: 1953 *Proc. Natl. Acad. Sci. U. S.* **39** 510
11. Santander M., Herranz F.J. and del Olmo M.A.: 1993 *Proc. XIX ICGTMP CIEMAT/RSEF (Madrid)* Vol. I, p. 455.
12. Majid S.: 1990 *Int. J. Mod. Phys. A* **5** 1

13. Truini P. and Varadarajan V.S.: 1993 *Rev. Math. Phys.* **5** 363
14. Ballesteros A., Herranz F.J., del Olmo M.A. and Santander M.: 1993 *J. Phys. A: Math. Gen.* **26** 5801
15. Ballesteros A., Herranz F.J., del Olmo M.A. and Santander M.: 1994
Quantum (2+1) kinematical algebras: a global approach *J. Phys. A: Math. Gen.* **27** ...
16. Lyakhovsky V. and Mudrov A.: 1992 *J. Phys. A: Math. Gen.* **25** 1139
17. Ballesteros A., Celeghini E., Giachetti R., Sorace E. and Tarlini M.: 1993 *J. Phys. A: Math. Gen.* **26** 7495

ON CONSTRUCTIONS OF NONSOLVABLE LIE ALGEBRAS WHOSE IDEALS ARE IN CHAIN

M. PILAR BENITO
Departamento de Matemáticas y Computación, Universidad de La Rioja, Edificio de Magisterio y Matemáticas, Luis de Ulloa s/n, 26004 Logroño, Spain

Abstract. The principal task in this paper is giving some explicit constructions of nonsolvable Lie algebras —over fields of characteristic zero— in which the ideals are a n-element chain. Two different procedures are used in order to get the constructions: the first one depends on the radical of the Lie algebra and the second on the semisimple Levi factor.

1. Introduction and preliminaries

In the present work we will consider finite dimensional Lie algebras over a field F of characteristic zero. Given such an algebra L, $\mathrm{Rad}(L)$ (resp. $\mathrm{Nil}(L)$) denotes the largest solvable (resp. nilpotent) ideal of L. We denote the terms of the lower central series (l.c.s.) of L by $L = L^1$ and $L^i = [L, L^{i-1}]$ for $i > 1$. The center of L is denoted by $Z(L)$. We define the upper central series (u.c.s.) of L by letting $Z_0(L) = 0$ and $Z_i(L)$ be the ideal of L such that $Z(L/Z_{i-1}(L)) = Z_i(L)/Z_{i-1}(L)$ for $i \geq 1$. We shall say that L has nilpotency index n if $L^n = 0$ but $L^{n-1} \neq 0$. Algebra direct sums are denoted by $\oplus$ whereas direct sums of vector space structures are denoted by $\dot{+}$.

The problem of determining the Lie algebras in which the ideals are totally ordered by set inclusion was first posed in [1], where a complete classification was given of the supersolvable algebras in this class; in particular, the solvable Lie algebras whose ideals are in chain are completely classify when the based field is algebraically closed. In the case of nonsolvable algebras of this type, it was also obtained the following basic structure result:

Theorem 1.1. *[1, Theorem 2.2] Let L be a nonsolvable Lie algebra over a field of characteristic zero. Then, the ideals of L are in chain if and only if L is a simple algebra or a semidirect sum of a nonzero nilpotent ideal N and a simple algebra S such that N/N^2 is a faithful S-module and $Z_i(N)/Z_{i-1}(N)$ are irreducible S-modules via the adjoint representation.*
Moreover, in that case, the terms of the l.c.s. of N coincide with the terms of the u.c.s. and, if n is the nilpotency index of N, the ideals of L are the following $(n+1)$-element chain:

$$0 < N^{n-1} < \ldots < N^i < \ldots < N < L.$$

The easier constructions —apart from simple algebras— of nonsolvable Lie algebras in which the ideals are in chain arise from Theorem 1.1 in a very simple way:

S. González (ed.), Non-Associative Algebra and Its Applications, 24–30.

Corollary 1.2. *(Characteristic 0) The nonsolvable Lie algebras whose ideals are a 3-element chain are a semi-direct sum of an abelian ideal N of dimension at least 2 and a simple subalgebra S such that N is an irreducible S-module.*

The principal task in this paper is giving some explicit constructions of nonsolvable Lie algebras in which the ideals are a n-element chain with $n = 4$. In view of Theorem 1.1, we can use two methods in order to get the construction of an algebra L of this type:

I) *Suppose the radical of L is known, identify the simple Lie algebras that can occur as its semisimple Levi factor.*

II) *Suppose the semisimple Levi factor of L is known, identify the nilpotent algebras that can occur as its radical.*

In section 2, we shall give constructions following the procedure I) when the radical is supposed to be an algebra in which the derived algebra is central of dimensionality one. In section 3, the alternative procedure II) is carried out when the semisimple Levi factor is supposed to be the split 3-dimensional simple algebra. The final section provides examples showing the existence of nonsolvable Lie algebras whose ideals are a chain of arbitrary length.

2. Making constructions from the radical

The construction we shall give in this section depends on Lie algebras in which the derived algebra is one-dimensional and equal to the center. Before giving the construction, we shall need some preliminary results about this type of algebras.

Lemma 2.1. *Let L be a Lie algebra such that $\dim L^2 = 1$ and $L^2 = Z(L)$ and z be a nonzero element of L^2. Then:*
1) For every V complementary subspace of L^2 in L, the bilinear form q_V defined in V by $q_V(x, y) = \lambda_{[x,y]}$ where $[x, y] = \lambda_{[x,y]}z$ is skew-symmetric and nondegenerate. In particular, the dimension of V is even and there exists $v_1, \ldots, v_n, w_1, \ldots, w_n$ basis of V for which $[v_i, w_i] = z$ and all other products being zero.
2) L has a basis $a_1, \ldots, a_{2n+1}$ with nonzero products $[a_i, a_{n+i}] = a_{2n+1}$, $1 \leq i \leq n$.
3) Suppose $\mathcal{B} = \{a_i\}_{1\leq i\leq 2n+1}$ is a basis of L as in 2) and $\partial: L \to L$ is a linear transformation such that the matrix of ∂ respect of $\mathcal{B}$ is of the form

$$\begin{pmatrix} & & & 0 \\ & A & & \vdots \\ & & & 0 \\ \alpha_1 & \cdots & \alpha_{2n} & \alpha \end{pmatrix}$$

where A is an $2n \times 2n$ matrix. Then, $\partial \in \mathrm{Der}(L)$ if and only if $A = \begin{pmatrix} \mathsf{a}_1 & \mathsf{a}_2 \\ \mathsf{a}_3 & \mathsf{a}_4 \end{pmatrix}$ where each a_i is an $n \times n$ matrix; $\mathsf{a}_2^T = \mathsf{a}_2$, $\mathsf{a}_3^T = \mathsf{a}_3$ and $\mathsf{a}_1^T + \mathsf{a}_4 = \alpha I_n$.

Proof. The assertion in 1) is straightforward (see [3]) and the part 2) follows from 1). To prove 3), consider the vector space $V = Fa_1 + \ldots + Fa_{2n}$ and let q_V be the skew bilinear form defined in 1) when $z = a_{2n+1}$. We have that $\partial([a_i, a_j]) =$

$\alpha q_V(a_i,a_j)a_{2n+1}$ and the matrix $B = (q_V(a_i,a_j))$ is $B = \begin{pmatrix} 0 & I_n \\ -I_n & 0 \end{pmatrix}$. Clearly, $\partial \in \mathrm{Der}(L)$ iff $\partial([a_i,a_j]) = [\partial(a_i),a_j] + [a_i,\partial(a_j)]$ for $i,j = 1,\dots,2n$. If we denote $A = (\tau_{ij})$, these conditions are that

$$\alpha q_V(a_i,a_j) = \sum_{1\le k\le 2n} q_V(a_i,a_k)\tau_{kj} + \sum_{1\le k\le 2n} \tau_{ki}q_V(a_k,a_j)$$

or in matrix form

$$\alpha B = BA + A^T B. \tag{2.1.1}$$

If we partion A in the same way as B, $A = \begin{pmatrix} \mathfrak{a}_1 & \mathfrak{a}_2 \\ \mathfrak{a}_3 & \mathfrak{a}_4 \end{pmatrix}$ where each $\mathfrak{a}_i$ is an $n\times n$ matrix, a simple computation shows that (2.1.1) holds if and only if $\mathfrak{a}_2^T = \mathfrak{a}_2$, $\mathfrak{a}_3^T = \mathfrak{a}_3$ and $\mathfrak{a}_1^T + \mathfrak{a}_4 = \alpha I_n$. ∎

In the sequel, for each $n \ge 1$ we shall denote by $L(n)$ the Lie algebra with basis $a_1,\dots,a_{2n+1}$ and nonzero products as it is described in 2) of Lemma 2.1. Let us consider the symplectic Lie algebra $\mathrm{sp}(2n,F)$, which by definition consists of all matrices $A = \begin{pmatrix} \mathfrak{a}_1 & \mathfrak{a}_2 \\ \mathfrak{a}_3 & \mathfrak{a}_4 \end{pmatrix}$ where each $\mathfrak{a}_i$ is an $n\times n$ matrix, $\mathfrak{a}_2^T = \mathfrak{a}_2$, $\mathfrak{a}_3^T = \mathfrak{a}_3$ and $\mathfrak{a}_1^T + \mathfrak{a}_4 = 0$ (see [4, p. 3]). This algebra is simple split of Type C_n and $(2n^2+n)$-dimensional —for $n = 1$ coincides with the split simple 3-dimensional, $\mathrm{sl}(2,F)$. By means of the homomorphism $A \to A^\rho$ of $\mathrm{sp}(2n,F)$ into $\mathrm{gl}(L(n))$, where A^ρ denotes the linear transformation in $L(n)$ with matrix relative to the basis $a_1,\dots,a_{2n+1}$:

$$\begin{pmatrix} & & 0 \\ & A & \vdots \\ & & 0 \\ 0 & \cdots & 0\ 0 \end{pmatrix},$$

the algebra $L(n)$ can be considered as an $\mathrm{sp}(2n,F)$-module. Now let denote by $\mathcal{L}(n)$ the direct sum of the two vector spaces $L(n)$ and $\mathrm{sp}(2n,F)$. We introduce in $\mathcal{L}(n)$ a multiplication $[u,v]$ by means of the formula

$$[x+A,y+B] = [x,y] + A^\rho(y) - B^\rho(x) + [A,B].$$

We have the following result:

Theorem 2.2. *Let $L(n)$, $\mathrm{sp}(2n,F)$ and $\mathcal{L}(n) = L(n) \dot{+} \mathrm{sp}(2n,F)$ as it is described in the preceding paragraph. Then:*
1) $\mathcal{L}(n)$ is a Lie algebra whose ideals are the following 4-element chain:

$$0 < L(n)^2 < L(n) < \mathcal{L}(n).$$

2) Suppose M is a nonsolvable Lie algebra such that $\mathrm{Rad}(M) \cong L(n)$. If the ideals of M are in chain, there exists a monomorphism embedding M in $\mathcal{L}(n)$. In particular, the semisimple Levi factors of M are isomorphic to a simple subalgebra of $\mathrm{sp}(2n,F)$.

Proof. 1): From Lemma 2.1, it is immediate that $A^\rho \in \mathrm{Der}(L(n))$ for each $A \in \mathrm{sp}(2n,F)$ and therefore $L(n)$ is a Lie algebra. Now denote by V the subspace spanned by a $a_1, \ldots, a_{2n}$. If $n = 1$, it is clear that V is an $\mathrm{ad}_{L(n)}\,\mathrm{sp}(2n, F)$-irreducible module. Suppose then $n \geq 2$ and consider q_V the bilinear form in V described in 1) of Lemma 2.1 for $z = a_{2n+1}$. The Lie algebra of linear transformations f in V which are skew relative to q_V, that is $q_V(f(x), y) = -q_V(x, f(y))$, coincides with the set $\{A^\rho|_V : A \in \mathrm{sp}(2n, F)\}$. Then from [2; Lemma 2, p. 137] we get that V is $\mathrm{ad}_{L(n)}\,\mathrm{sp}(2n, F)$-irreducible and therefore $L(n) = V \dot{+} Fa_{2n+1}$ is a decomposition of $L(n)$ into $\mathrm{sp}(2n, F)$-irreducible modules via the adjoint representation. Now the result follows from Theorem 1.1.

2): Since the ideals of M are in chain, Theorem 1.1 implies that $M = \mathrm{Rad}(M) \dot{+} S$ where S is simple and $\mathrm{Rad}(M)/\mathrm{Rad}(M)^2$, $\mathrm{Rad}(M)^2$ are irreducible $\mathrm{ad}_M\, S$-modules. Write $\mathrm{Rad}(M)^2 = Fz$ and note that $[\mathrm{Rad}(M)^2, S] = 0$. Moreover, there exists V S-irreducible module such that $\mathrm{Rad}(M) = V \dot{+} Fz$. ¿From 1) of Lemma 2.1, we can pick $v_1, \ldots, v_n, w_1, \ldots, w_n$ basis for V such that $[v_i, w_i] = z$ and all other products being zero. For each $s \in S$, the linear transformation $\mathrm{ad}_M\, s\colon x \mapsto [s, x]$ belongs to $\mathrm{Der}(\mathrm{Rad}(M))$; thus applying 3) of Lemma 2.1 we have that the matrix of $\mathrm{ad}_M\, s$ respect to the basis $\{v_i, w_j, z\}$ is of the form:

$$\begin{pmatrix} & & 0 \\ & C_s & \vdots \\ & & 0 \\ 0 & \cdots\ 0 & 0 \end{pmatrix}$$

where $C_s = \begin{pmatrix} \mathfrak{c}_1 & \mathfrak{c}_2 \\ \mathfrak{c}_3 & \mathfrak{c}_4 \end{pmatrix}$ and each $\mathfrak{c}_i$ is an $n \times n$ matrix with $\mathfrak{c}_2^T = \mathfrak{c}_2$, $\mathfrak{c}_3^T = \mathfrak{c}_3$ and $\mathfrak{c}_1^T + \mathfrak{c}_4 = 0$. Then, the mapping $\Phi\colon M \to \mathcal{L}(n)$ defined by:

$$\Phi(t_1v_1 + \ldots + t_nv_n + t_{n+1}w_1 + \ldots + t_{2n}w_n + t_{2n+1}z + s) = t_1a_1 + \ldots + t_{2n+1}a_{2n+1} + C_s$$

is a monomorphism, which proves the result. ∎

Corollary 2.3. *Up to isomorphism, the unique nonsolvable Lie algebra whose radical is isomorphic to* $L(1)$ *and such that the their ideals are in chain is* $L = Fa_1 + Fa_2 + Fa_3 + Fx + Fy + Fh$ *with nonzero products* $[a_1, a_2] = a_3$; $[x, a_2] = a_1$; $[y, a_1] = a_2$; $[h, a_1] = a_1$; $[h, a_2] = -a_2$; $[x, y] = h$; $[h, x] = 2x$ *and* $[h, y] = -2y$.

3. Making constructions from the Levi factor

Let $\mathrm{sl}(2, F) = Fx + Fy + Fh$ be the split 3-dimensional simple Lie algebra with nonzero products $[x, y] = h$, $[h, x] = 2x$ and $[h, y] = -2y$. We shall denote by $V(n)$ the $\mathrm{sl}(2, F)$-irreducible module with dimension $n + 1$ and basis $v_0, \ldots, v_n$ such that the action of $\mathrm{sl}(2, F)$ is given explicitly by the formulas (see [4, p. 32]):

$$\begin{aligned} h \cdot v_i &= (n - 2i)v_i \quad \text{for} \quad 0 \leq i \leq n, \\ x \cdot v_0 &= 0 \quad \text{and} \quad x \cdot v_i = (n - (i - 1))v_{i-1} \quad \text{for} \quad 0 < i \leq n, \\ y \cdot v_n &= 0 \quad \text{and} \quad y \cdot v_i = (i + 1)v_{i+1} \quad \text{for} \quad 0 \leq i < n. \end{aligned} \tag{3.0.1}$$

Consider the $\mathrm{sl}(2, F)$-modules $V(n)$ and $V(m)$ with basis $\{v_j\}_{0 \leq j \leq n}$ and $\{w_j\}_{0 \leq j \leq m}$ respectively as it is described by the above formulas (3.0.1). We are interested in determining the products $\odot$ from $V(n) \times V(n)$ into $V(m)$ which satisfy ($\alpha, \beta \in F$; $u, v, w \in V(n)$ and $a \in \mathrm{sl}(2, F)$):

$$\begin{aligned}
&\text{i)}\ (\alpha u + \beta v) \odot w = \alpha(u \odot w) + \beta(v \odot w),\\
&\text{ii)}\ v \odot (\alpha u + \beta w) = \alpha(v \odot u) + \beta(v \odot w),\\
&\text{iii)}\ a \cdot (v \odot u) = (a \cdot v) \odot u + v \odot (a \cdot u).
\end{aligned}$$

Lemma 3.1. *Let $V(n)$, $V(m)$, $\mathrm{sl}(2, F)$ and $\odot$ as it is described in the preceding paragraph. Then:*
1) If $m \notin \{2j : 0 \leq j \leq n\}$, $\odot$ is the trivial product.
2) If $m \in \{2j : 0 \leq j \leq n\}$, there exists $\alpha \in F$ such that $v_0 \odot v_n = \alpha w_{m/2}$. Moreover, for each $\alpha \in F$ and m there exists a unique product $\odot_\alpha$ satisfying i), ii) and iii) which is completely determined by $v_0 \odot_\alpha v_n = \alpha w_{m/2}$.
3) If $\odot$ is nontrivial, $v \odot u = -v \odot u$ if and only if m is of the form $2n - 2(1 + 2j)$ where $0 \leq 1 + 2j \leq n$; otherwise, $v \odot u = u \odot v$.

Proof. First of all we observe that the problem of determining all the products $\odot$ is equivalent to determine the elements of $\mathrm{Hom}_{\mathrm{sl}(2,F)}(V(n) \otimes_F V(n), V(m))$. From the Clebsch-Gordan formula (see [4, p. 126]), $V(n) \otimes_F V(n)$ is isomorphic, as $\mathrm{sl}(2, F)$-module, to $\oplus_{0 \leq j \leq n} V(2j)$ and consequently:

$$\mathrm{Hom}_{\mathrm{sl}(2,F)}(V(n) \otimes_F V(n), V(m)) \cong \oplus_{0 \leq j \leq n} \mathrm{Hom}_{\mathrm{sl}(2,F)}(V(2j), V(m)).$$

On the other hand, it is clear that $\mathrm{Hom}_{\mathrm{sl}(2,F)}(V(2j), V(m)) = 0$ if $m \neq 2j$ and, from Schur's Lemma, $\mathrm{Hom}_{\mathrm{sl}(2,F)}(V(2j), V(2j))$ is isomorphic to F. Then, if $m \in \{2j : 0 \leq j \leq n\}$, $\dim_F \mathrm{Hom}_{\mathrm{sl}(2,F)}(V(n) \otimes_F V(n), V(m)) = 1$ and, otherwise, $0 = \mathrm{Hom}_{\mathrm{sl}(2,F)}(V(n) \otimes_F V(n), V(m))$. This proves the assertion in 1).

Now, assume that $m \in \{2j : 0 \leq j \leq n\}$. We shall describe the elements $\phi \in \mathrm{Hom}_{\mathrm{sl}(2,F)}(V(n) \otimes_F V(n), V(m))$ and, consequently, all the products $\odot$. Since $v_0 \otimes v_n$ generates the $\mathrm{sl}(2, F)$-module $V(n) \otimes_F V(n)$, ϕ is completely determined by $\phi(v_0 \otimes v_n)$. Note that $h \cdot (v_0 \otimes v_n) = (h \cdot v_0) \otimes v_n + v_0 \otimes (h \cdot v_n) = 0$. Therefore, $h \cdot \phi(v_0 \otimes v_n) = 0$ which implies $\phi(v_0 \otimes v_n) = \alpha_\phi w_{m/2}$ for some $\alpha_\phi \in F$. Then, we have that the application $\phi \mapsto \alpha_\phi$ from $\mathrm{Hom}_{\mathrm{sl}(2,F)}(V(n) \otimes_F V(n), V(m))$ in F is F-lineal and bijective. This means that for every $\alpha \in F$ there exists a unique product $\odot_\alpha : V(n) \times V(n) \to V(m)$ satisfying i), ii) and iii) which is completely determined by $v_0 \odot_\alpha v_n = \alpha w_{m/2}$ (from the above facts, the rest of the products $v_i \odot_\alpha v_k$ are obtained by expressing $v_i \odot_\alpha v_k$ in terms of the action of $\mathrm{sl}(2, F)$ on $v_0 \odot_\alpha v_n$). This completes the proof of 2). Finally, we observe that $V(n) \otimes_F V(n) = S(V(n) \otimes_F V(n)) \oplus A(V(n) \otimes_F V(n))$ where $S = S(V(n) \otimes_F V(n))$ ($A = A(V(n) \otimes_F V(n))$) is the set of all symmetric (skewsymmetric) tensors, thus:

$$\mathrm{Hom}_{\mathrm{sl}(2,F)}(V(n) \otimes_F V(n), V(m)) \cong \mathrm{Hom}_{\mathrm{sl}(2,F)}(S, V(m)) \oplus \mathrm{Hom}_{\mathrm{sl}(2,F)}(A, V(m))$$

and, therefore, the products $\odot$ are either symmetric or skewsymmetric because the dimensionality of $\mathrm{Hom}_{\mathrm{sl}(2,F)}(V(n) \otimes_F V(n), V(m))$ is 1. From the Clebsch-Gordan formula we deduce that:

$$S = \oplus_{0 \leq j,\, 0 \leq 2n-4j} V(2n - 4j) \quad \text{and} \quad A = \oplus_{0 \leq j,\, 0 \leq 2n-2-4j} V(2n - 2 - 4j)$$

and therefore 3) holds. ■

Now, we have all the necessary information in order to get the construction. For every $n \geq 1$, let k be an arbitrary odd number less or equal than n and denote m by $2(n-k)$. Consider the $\mathrm{sl}(2,F)$-modules $V(n)$ and $V(m)$ with basis $\{v_j\}_{0\leq j\leq n}$ and $\{w_j\}_{0\leq j\leq m}$ respectively as in (3.0.1). Write as $\mathcal{L}(n,k)$ the direct sum of the vector spaces $V(n)$, $V(m)$ and $\mathrm{sl}(2,F)$. We introduce a product in $\mathcal{L}(n,k)$ giving the product table for the basis $\{v_i, w_j, x, y, h\}$ as follows:

TABLE (3.0.2)

$$[x,y]=-[y,x]=h;\ [h,x]=-[x,h]=2x;\ [h,y]=-[y,h]=-2y;$$
$$[u,u]=0 \quad \text{for} \quad u=x,y,h;$$
$$[h,w_i]=-[w_i,h]=h\cdot w_i \text{ for } 0\leq i\leq m; \quad [h,v_i]=-[v_i,h]=h\cdot v_i \text{ for } 0\leq i\leq n;$$
$$[x,w_i]=-[w_i,x]=x\cdot w_i \text{ for } 0\leq i\leq m; \quad [x,v_i]=-[v_i,x]=x\cdot v_i \text{ for } 0\leq i\leq n;$$
$$[y,w_i]=-[w_i,y]=y\cdot w_i \text{ for } 0\leq i\leq m; \quad [y,v_i]=-[v_i,y]=y\cdot v_i \text{ for } 0\leq i\leq n;$$
$$[w_i,w_j]=[w_i,v_j]=[v_i,w_j]=0; \quad [v_0,v_i]=0,\ 0\leq i<k, \text{ and } [v_0,v_k]=w_0;$$
$$[v_0,v_i]=\tfrac{(n-k)(n-k-1)\cdots(n-i+1)}{m(m-1)\cdots(m-i+k+1)}w_{i-k} \quad \text{for} \quad k<i\leq n$$

and the products $[v_i,v_j]$ for $1\leq i\leq n$, $0\leq j\leq n$ are obtained recursively by means of the formula:

$$[v_i,v_j]=\tfrac{1}{i}[y,[v_{i-1},v_j]]-\tfrac{j+1}{i}[v_{i-1},v_{j+1}] \qquad (F-3.0.2)$$

with $v_{n+1}=0$. Then, we have the following result:

Theorem 3.2. *Let $V(n)$, $V(m)$, $\mathrm{sl}(2,F)$ and $\mathcal{L}(n,k)=V(m)\dotplus V(n)\dotplus\mathrm{sl}(2,F)$ be as it is described in the preceding paragraph. Then:*
1) $\mathcal{L}(n,k)$ is a Lie algebra whose ideals are the following 4-element chain:

$$0<V(m)<V(m)\dotplus V(n)<\mathcal{L}(n,k).$$

2) Suppose M is a Lie algebra such that $M/\mathrm{Rad}(M)\cong\mathrm{sl}(2,F)$ and denote by $n=\dim\mathrm{Rad}(M)/\mathrm{Rad}(M)^2$. If the ideals of M are a 4-element chain, $\dim\mathrm{Rad}(M)^2=2(n-k)-1$ where k is an odd number, $0\leq k\leq n-1$, and M is isomorphic to $\mathcal{L}(n-1,k)$.

Proof. 1): From 2) of Lemma 3.1, the product $[,]|_{V(n)\times V(n)}$ as defined in TABLE (3.0.2) is the unique product from $V(n)\times V(n)$ into $V(m)$ satisfying i), ii) and iii) that there exists for $\alpha=\frac{(n-k)(n-k-1)\cdots(1)}{m(m-1)\cdots(m-n+k+1)}$. As $m=2(n-k)$ with k an odd number less or equal than n, 3) of Lemma 3.1 implies that $[v_i,v_j]=-[v_j,v_i]$. Then, it is immediate that $V(m)\dotplus V(n)$ is a Lie algebra. On the other hand, the properties of $[,]|_{V(n)\times V(n)}$ show that for $a\in\mathrm{sl}(2,F)$, $u\in V(m)\dotplus V(n)$, the mappings $u\mapsto[a,u]$ are derivations in $V(m)\dotplus V(n)$. Consequently, $\mathcal{L}(n,k)$ is a Lie algebra and the rest is an immediate consequence of Theorem 1.1.

2): From Theorem 1.1, $M=\mathrm{Nil}(M)\dotplus S$ with $\mathrm{Nil}(M)$ of nilpotency index 3 and $\mathrm{Nil}(M)/\mathrm{Nil}(M)^2$, $\mathrm{Nil}(L)^2$ irreducible ad_M S-modules and the action of S on $\mathrm{Nil}(M)/\mathrm{Nil}(M)^2$ is faithful. Since S is isomorphic to $\mathrm{sl}(2,F)$, we can pick x,y,h basis of S such that $[x,y]=h$, $[h,x]=2x$ and $[h,y]=-2y$. Moreover, $\mathrm{Nil}(M)=$

$V(n) \oplus V(m)$ where $\dim \mathrm{Nil}(L)/\mathrm{Nil}(L)^2 = n+1$, $\dim \mathrm{Nil}(M)^2 = m+1$; $V(n)$ and $V(m)$ are S-irreducible modules with $n \geq 1$ and $Z(\mathrm{Nil}(M)) = \mathrm{Nil}(L)^2 = [V(n), V(n)] = V(m)$. Then, there exist $\{v_i\}_{0\leq i\leq n}$ and $\{w_i\}_{0\leq i\leq m}$ basis for $V(n)$ and $V(m)$ respectively satisfying:

$[h, v_i] = (n-2i)v_i$ for $0 \leq i \leq n$;
$[x, v_i] = (n-(i-1))v_{i-1}$ for $0 < i \leq n$;
$[y, v_i] = (i+1)v_{i+1}$ for $0 \leq i < n$;
$[x, v_0] = 0$; $[y, v_n] = 0$;

$[h, w_i] = (m-2i)w_i$ for $0 \leq i \leq n$;
$[x, w_i] = (m-(i-1))w_{i-1}$, $0 < i \leq m$;
$[y, w_i] = (i+1)w_{i+1}$ for $0 \leq i < m$;
$[x, w_0] = 0$ and $[y, w_m] = 0$.

Since $[\,,]|_{V(n)\times V(n)}$ is a nontrivial skewsymmetric product from $V(n) \times V(n)$ into $V(m)$ satisfying i), ii) and iii), from Lemma 3.1, $m = 2n-2k$ for some $0 \leq k \leq n$ with k an odd number and there exists $0 \neq \alpha \in F$ such that $[v_0, v_n] = \alpha w_{m/2}$. If we denote by $\delta = \frac{(n-k)(n-k-1)\cdots(1)}{m(m-1)\cdots(m-n+k+1)}$, taking $u_i = \alpha/\delta\, w_i$, the set $\{x, y, h, v_i, u_j\}_{0\leq i\leq n, 0\leq j\leq m}$ is a basis for M for which the multiplication table is as it is described in TABLE (3.0.2). Therefore, M is isomorphic to $\mathcal{L}(n,k)$. ■

Corollary 3.3. *For every $n \geq 1$, the unique nonsolvable Lie algebra, up to isomorphisms, whose radical and the semisimple Levi factors are isomorphic to $L(n)$ and $\mathrm{sl}(2,F)$ respectively, and such that their ideals are in chain is $L = Fv_0 + \ldots + Fv_{2n-1} + Fw_0 + Fx + Fy + Fh$ with nonzero products $[x,y] = h$; $[h,x] = 2x$; $[h,y] = -2y$; $[x, v_i] = (2n-i)v_{i-1}$ for $0 < i \leq 2n-1$; $[y, v_i] = (i+1)v_{i+1}$ for $0 \leq i < 2n-1$; $[h, v_i] = (2n-1-2i)v_i$ for $0 \leq i \leq 2n-1$ and $[v_i, v_{2n-i-1}] = (-1)^i\binom{2n-1}{i}w_0$ for $0 \leq i \leq n-1$.*

Proof. Given such an algebra M, 2) of Theorem 3.2 implies that M is isomorphic to $\mathcal{L}(2n-1, 2n-1)$ and the result follows from TABLE (3.0.2). ■

4. Final examples

It is easy to obtain explicitly the multiplication table of the Lie algebras $\mathcal{L}(n,k)$ for $n = 1,2,3$. The following example guarantee the existence of nonsolvable Lie algebras whose ideals are an n-element chain when $n \geq 4$:

$L = F\langle a_0, a_1, z_{i,j}, x, y, h : 0 \leq i \leq m-2, 0 \leq j \leq i\rangle$ where $m \geq 2$ and with nonzero products: $[x,y] = h$; $[h,x] = 2x$; $[h,y] = -2y$; $[a_0, a_1] = z_{00}$; $[a_0, z_{i,j}] = \frac{(i+1-j)}{(i+1)} z_{i+1,j}$ for $0 \leq i \leq m-3$ and $0 \leq j \leq i$; $[a_1, z_{i,j}] = \frac{(j+1)}{(i+1)} z_{i+1,j+1}$ for $0 \leq i \leq m-3$ and $0 \leq j \leq i$; $[x, a_1] = a_0$; $[y, a_0] = a_1$; $[h, a_0] = a_0$; $[h, a_1] = -a_1$; $[x, z_{i,j}] = (i-j+1)z_{i,j-1}$ for $0 \leq i \leq m-2$ and $0 < j \leq i$; $[y, z_{i,j}] = (j+1)z_{i,j+1}$ for $0 \leq i \leq m-2$ and $0 \leq j < i$; $[h, z_{i,j}] = (i-2j)z_{i,j}$ for $0 \leq i \leq m-2$ and $0 \leq j \leq i$.

References

[1] Benito, M.P.: Lie algebras in which the lattice formed by the ideals is a chain, *Comm. Alg.* 20 (1992), 93–108.
[2] Jacobson, N.: *Lie Algebras*, Wiley-Interscience, New York, 1962.
[3] Jacobson, N.: *Lectures in Abstract Algebras*, Vol. II, Springer-Verlag, 1975.
[4] Humphreys, J.E.: *Introduction to Lie Algebras and Representation Theory*, Springer-Verlag, 1972.

LIE ALGEBRAS GRADED BY ROOT SYSTEMS

GEORGIA BENKART* and EFIM ZELMANOV†
Department of Mathematics
University of Wisconsin, Madison
Madison, Wisconsin 53706-1388 U.S.A.

Let Δ be a finite irreducible reduced root system. Assume Γ is the integer lattice generated by Δ. In [BM] S. Berman and R. Moody initiated the study of Lie algebras graded by the root system Δ. Following them we say that a Lie algebra is graded by Δ or is Δ-*graded* if

(i) L has a Γ-gradation $L = \sum_{\gamma\in\Gamma} L_\gamma$ in which $L_\gamma \neq (0)$ if and only if $\gamma \in \Delta\cup\{0\}$;

(ii) the split simple Lie algebra $\mathcal{G}$ whose root system is Δ is a subalgebra of L, and relative to some split Cartan subalgebra $\mathcal{H}$ of $\mathcal{G}$ we have $L_\gamma \supseteq \mathcal{G}_\gamma$ for all $\gamma \in \Delta\cup\{0\}$;

(iii) for all $h \in \mathcal{H}$ the operator $ad(h)$ acts diagonally on L_γ with eigenvalue $\gamma(h)$; and

(iv) L is generated by its nonzero root spaces L_γ where $\gamma \in \Delta$.

The conditions for being a Δ-graded Lie algebra imply that L is a direct sum of finite dimensional irreducible $\mathcal{G}$-modules whose highest weights are roots, hence are either the highest long or highest short root or are zero. Thus, condition (iii) in the definition of a Δ-graded Lie algebra can be replaced by:

(iii)′ As a $\mathcal{G}$-module L is a direct sum of adjoint modules (modules isomorphic to $\mathcal{G}$), *little adjoint modules*, whose highest weight is the highest short root, or one-dimensional $\mathcal{G}$-modules; the latter being contained in L_0.

It is easy to see that a Δ-graded Lie algebra L is *perfect* ($L = [L, L]$), and in particular, $L_0 = \sum_{\gamma\in\Delta}[L_{-\gamma}, L_\gamma]$. Since conditions (i)-(iv) concern only the root spaces corresponding to nonzero weights, it is appropriate to classify Δ-graded Lie algebras up to *central isogeny*: Any perfect Lie algebra L has a universal central extension which is also perfect, called a *universal covering algebra* (u.c.a.) of L (see [G]). Since any two u.c.a.'s of L are isomorphic, we will refer to *the* universal covering algebra of L. Two perfect Lie algebras L_1 and L_2 are said to be *centrally isogenous* if they have the same u.c.a. (up to isomorphism).

Berman and Moody [BM] present the following examples of Δ-graded Lie algebras:

(1) Let A be an associative F-algebra with identity. For any positive integer $n > 1$ the algebra of $n \times n$ matrices with coefficients in A forms a Lie algebra $gl_n(A)$

* Supported in part by National Science Foundation Grant #DMS-9300523
† Supported in part by National Science Foundation Grant #DMS-9212608

S. González (ed.), Non-Associative Algebra and Its Applications, 31–38.

under the commutator product. The subalgebra $e_n(A)$ generated by the elements $ae_{i,j}$ for $a \in A$ and $i \neq j$ is an ideal of $gl_n(A)$ which is perfect. The algebra $e_n(A)$ is graded by the root system A_{n-1}. The u.c.a. $st_n(A)$ of $e_n(A)$ is the Lie algebra analogue of the Steinberg group $St_n(A)$, (see [AF]).

(2) Let A be a unital, commutative, associative F-algebra and let $\mathcal{G}$ be the split simple Lie algebra with root system Δ. Then $\mathcal{G} \otimes A$ is obviously a Δ-graded Lie algebra. It is perfect, and the u.c.a. of $\mathcal{G} \otimes A$ is a generalization of the affine Kac-Moody algebra determined by $\mathcal{G}$.

Berman and Moody classify the Lie algebras graded by simply-laced finite root systems of rank ≥ 2, i.e. root systems of types $A_n, n \geq 2$, $D_n, n \geq 4$, E_6, E_7, E_8.

Recognition Theorem for Types A,D,E. ([BM]) Let L be a Lie algebra graded by a simply-laced finite root system Δ of rank $n \geq 2$.

(i) If $\Delta = D_n$, $n \geq 4$, or if $\Delta = E_6, E_7, E_8$, then there exists a unital, commutative, associative F-algebra A such that L is centrally isogenous with $\mathcal{G} \otimes A$, where $\mathcal{G}$ is the split simple Lie algebra with root system Δ.

(ii) If $\Delta = A_n$, $n \geq 3$, then there exists a unital associative F-algebra A such that L is centrally isogenous with $e_{n+1}(A)$.

(iii) If $\Delta = A_2$, then L is centrally isogenous with $st_3(A)$, where A is a unital alternative F-algebra.

At the Oviedo Conference we announced a recognition theorem for each of the remaining irreducible reduced root systems i.e. the root systems A_1, B_n, C_n, F_4, and G_2. We describe these results below, but the details of the proofs will appear elsewhere.

The result for Lie algebras graded by root systems of type C_n is very similar to the theorem for type A. In fact, it follows immediately from N. Jacobson's Coordinatization Theorem for Jordan algebras (see [J1]), if properly translated into the language of Jordan algebras. Here the following Lie algebras arise:

Let $(J, \circ)$ be a unital Jordan algebra and L_a the left multiplication operator determined by $a \in J$. The commutators $D = [L_b, L_c]$ form derivations of J, the so-called *inner derivations*. Then $\mathcal{L}(J) = \{L_a + \sum [L_{b_i}, L_{c_i}] \mid a, b_i, c_i \in J\}$ is a Lie algebra under the multiplication $[L_a + D, L_b + E] = [L_a, L_b] + L_{Db} - L_{Ea} + [D, E]$, and $\mathcal{L}(J)$ has an automorphism of order two $\eta : L_a + D \longrightarrow \overline{L_a + D} = -L_a + D$. If $\overline{J}$ denotes a second copy of J, then $\mathcal{K}(J) = \overline{J} \oplus \mathcal{L}(J) \oplus J$ can be endowed with the structure of a Lie algebra,

$$[a + X + \overline{c}, b + Y + \overline{d}] = Xb - b \diamond c - Ya + [X, Y] - \overline{\overline{Y}c} + a \diamond d + \overline{\overline{X}d}$$

where $b \diamond c = L_{b \circ c} + [L_b, L_c]$. The elements $e = 1 \in J$, $f = 2(\overline{1}) \in \overline{J}$, and $h = [e, f] = 2L_1$ generate a subalgebra isomorphic to $A_1 = sl_2(F)$, and the algebra $\mathcal{K}(J)$ decomposes into eigenspaces relative to *adh*: $\mathcal{K}_{-2} = \overline{J}$, $\mathcal{K}_0 = \mathcal{L}(J)$, $\mathcal{K}_2 = J$, corresponding to the eigenvalues $-2, 0, 2$ respectively. Thus, $\mathcal{K}(J)$ is graded by the root system A_1. The algebra $\mathcal{K}(J)$, now commonly referred to as the *Tits-Kantor-Koecher construction*, was introduced by Tits in [T1] in a more general form, and independently discovered by Kantor [Ka] and Koecher [Ko].

Let A be a unital associative F-algebra with an involution $* : A \longrightarrow A$. Consider the algebra $M_n(A)$ of $n \times n$ matrices over A and the induced involution $\sigma : (a_{i,j}) \longrightarrow (a^*_{j,i})$ on $M_n(A)$. The space $H(M_n(A), \sigma)$ of all σ-symmetric matrices has the structure of a Jordan algebra relative to the product $x \circ y = (1/2)(xy + yx)$, (see [Sc] or [J1]). Consider also the algebra $M_{2n}(A)$ of $(2n) \times (2n)$ matrices over A. Then the mapping

$$\tau : \begin{pmatrix} a & b \\ c & d \end{pmatrix} \longrightarrow \begin{pmatrix} d^\sigma & -b^\sigma \\ c^\sigma & a^\sigma \end{pmatrix} \qquad a, b, c, d \in M_n(A),$$

is an involution on $M_{2n}(A)$. The Lie algebra $\widetilde{S}$ of skew-symmetric elements in $M_{2n}(A)$ with respect to τ consists of the matrices

$$\begin{pmatrix} a & b \\ c & d \end{pmatrix}$$

such that $d = -a^\sigma$ and $b, c \in H(M_n(A), \sigma)$. Let $\Delta = \{\pm\epsilon_i \pm \epsilon_j \mid 1 \le i, j \le n\}$ be the root system of type C_n, and let Γ be the integer lattice generated by Δ. The algebra $\widetilde{S}$ is graded by Γ:

$$\begin{aligned}
\widetilde{S}_{-\epsilon_i-\epsilon_j} &= \{ae_{i,j+n} + a^* e_{j,i+n} \mid a \in A\} \quad 1 \le i < j \le n, \\
\widetilde{S}_{\epsilon_i+\epsilon_j} &= \{ae_{i+n,j} + a^* e_{j+n,i} \mid a \in A\} \quad 1 \le i < j \le n, \\
\widetilde{S}_{-2\epsilon_i} &= \{ae_{i,i+n} \mid a \in A, a^* = a\} \quad 1 \le i \le n, \\
\widetilde{S}_{2\epsilon_i} &= \{ae_{i+n,i} \mid a \in A, a^* = a\} \quad 1 \le i \le n, \\
\widetilde{S}_{\epsilon_i-\epsilon_j} &= \{ae_{i,j} - a^* e_{j,i} \mid a \in A\} \quad 1 \le i \ne j \le n, \\
\widetilde{S}_0 &= \sum_i \{ae_{i,i} - a^* e_{i+n,i+n} \mid a \in A, a^* = a\}.
\end{aligned}$$

The Lie subalgebra $sp_{2n}(A, *)$ of $\widetilde{S}$ generated by the subspaces $\widetilde{S}_\gamma$, $\gamma \ne 0$ clearly is Δ-graded.

For $n = 3$ this construction is applicable to more general rings of coefficients. Let A be a unital alternative algebra with an involution $* : A \longrightarrow A$ such that every symmetric element $a \in A$, $a^* = a$, lies in the associative center $\{a \in A \mid (ax)y = a(xy) \text{ for any } x, y \in A\}$ of A. Consider the algebra of 3×3 matrices over A and the involution $\sigma : (a_{i,j}) \longrightarrow (a^*_{j,i})$ on $M_n(A)$. The space $H(M_3(A), \sigma)$ of σ-symmetric matrices is a Jordan algebra with respect to the multiplication $x \circ y = (1/2)(xy + yx)$. (See for example [J1].) The Tits-Koecher-Kantor construction $\mathcal{K}(H(M_3(A), \sigma))$ is a Lie algebra graded by the root system C_3. In particular, if A is the split octonion algebra with the standard involution $*$, then $\mathcal{K}(H(M_3(A), \sigma))$ is the split simple Lie algebra of type E_7. Thus, E_7 is C_3-graded. The universal covering algebra of $\mathcal{K}(H(M_3(A), \sigma))$ is an analogue of the Steinberg construction for groups. We will denote it as $st\, sp_6(A, *)$.

Recognition Theorem for Type C. Let L be a Δ-graded Lie algebra.

(i) If $\Delta = C_n$, $n \geq 4$, then there exists a unital associative algebra A with an involution $* : A \longrightarrow A$ such that L is centrally isogenous with the algebra $sp_{2n}(A, *)$ of symplectic $(2n) \times (2n)$ matrices over A.

(ii) If $\Delta = C_3$, then L is centrally isogenous with the symplectic Steinberg algebra $st\, sp_6(A, *)$, where A is an alternative involutive algebra whose symmetric elements, $\{a \in A \mid a^* = a\}$, lie in the associative center of A.

(iii) If $\Delta = C_2$, then L is centrally isogenous with the Tits-Kantor-Koecher construction of a unital Jordan algebra J which contains the Jordan algebra of symmetric 2×2 matrices, and the identity of J lies in this subalgebra.

(iv) If $\Delta = C_1 = A_1$, then L is centrally isogenous with the Tits-Kantor-Koecher construction of a unital Jordan algebra J.

The key to the Recognition Theorems for types B_n, F_4, and G_2 lies in the celebrated Tits construction of Lie algebras. Let $\mathcal{C}$ denote the split octonion algebra (Cayley algebra) over the field F, and let J denote the split exceptional 27-dimensional Jordan algebra over F. It is known (see for example [Sc] or [J2]) that the derivation algebras $Der_F\mathcal{C}$ and $Der_F J$ are isomorphic to the split exceptional simple Lie algebras G_2 and F_4 respectively. Let $\mathcal{C}_0$ and J_0 denote the subspaces of trace zero elements in $\mathcal{C}$ and J. J. Tits ([T1], [T2]) proved that the space

$$\mathcal{T}(\mathcal{C}/F, J/F) = Der_F\mathcal{C} \oplus (\mathcal{C}_0 \otimes J_0) \oplus Der_F J$$

with a suitable multiplication becomes the exceptional simple Lie algebra of type E_8 (see [Sc], [M], [J2], [FF] for detailed discussions). Taking other degree one or three Jordan algebras J such as (I) the field F, (II) the algebra $H(M_3(F))$ of 3×3 symmetric matrices over F, (III) the Jordan algebra $M_3(F)^+$ over F, or (IV) the algebra $H(M_3(\mathcal{Q}))$ of 3×3 Hermitian matrices over a quaternion algebra $\mathcal{Q}$, provides realizations of the exceptional simple Lie algebras of types G_2, F_4, E_6, E_7 respectively. Moreover, if the quadratic alternative algebra $\mathcal{C}$ is varied too, this construction produces the entire "magic table" of Freudenthal (see [F] or [J1]).

The algebras $\mathcal{T}(\mathcal{C}/F, J/F)$ (with $\mathcal{C}$ the split octonions) have a natural G_2-grading coming from the adjoint action of $Der_F\mathcal{C} = G_2$ relative to which it decomposes into one copy of the adjoint module, $\dim J_0$ copies of the little adjoint module which is $\mathcal{C}_0$ in this case, and $\dim Der_F J$ copies of the trivial module. Thus, F_4, E_6, E_7 and E_8 are all G_2-graded. When we specialize the generically cubic Jordan algebra J to be $J = F1 \oplus J(W)$ where $J(W)$ is the Jordan algebra of a nondegenerate symmetric bilinear form on an m-dimensional vector space W, then $\mathcal{T}(\mathcal{C}/F, J/F)$ is a simple Lie algebra of type B_{n+3} if $m = 2n$ and of type D_{n+4} if $m = 2n+1$ for $n \geq 0$. Thus, the Lie algebras of types B and D possess G_2-gradings too.

The formula for $\mathcal{T}(\mathcal{C}/F, J/F)$ also provides an F_4-grading of the algebra of type E_8 coming from the adjoint action of $Der_F J = F_4$. If instead of the octonion algebra we take another degree 1 or 2 alternative algebra $\mathcal{C}$ such as (i) F, (ii) $F \oplus F$, or (iii) the quaternion algebra, then the construction above will yield F_4-gradings of F_4, E_6, and E_7.

We prove that (up to central isogeny) all the Lie algebras graded by root systems of types F_4 and G_2 arise from this general procedure. We introduce what we term the *generalized Tits construction*. Let A and $\mathfrak{A}$ be unital, commutative, associative algebras over F. Assume $(X,*)$ is an algebra over A associated with an A-module X_0 having a symmetric or skew-symmetric A-bilinear form $(\ ,\ )$. Thus, $X = A1 \oplus X_0$ with product $(a1+x)*(a'1+x') = aa'1+(x,x')1+ax'+a'x+x*x'$, where $x*x' \in X_0$. The trace on X is defined as $t(a1+x) = a$, and X_0 is the set of elements of trace zero. Let $\mathcal{D}(X)$ be a Lie subalgebra of the A-derivations of X which map X_0 to X_0, and assume $(Y,\circ)$, and $\mathcal{D}(Y)$ are similarly chosen over $\mathfrak{A}$. Assume further that there is a transformation $X \otimes X \longrightarrow \mathcal{D}(X)$, $\quad x \otimes x' \longrightarrow D_{x,x'}$, which is skew-symmetric (symmetric) if the form on X is symmetric (skew-symmetric), and let $Y \otimes Y \longrightarrow \mathcal{D}(Y)$, $\ y \otimes y' \longrightarrow d_{y,y'}$, be an analogous mapping for Y such that the following relations hold:

$$\begin{array}{lll} (i) & [E, D_{x,x'}] & = D_{Ex,x'} + D_{x,Ex'} \\ (ii) & [e, d_{y,y'}] & = d_{ey,y'} + d_{y,ey'} \\ (iii) & D_{ax,x'} & = D_{x,ax'} \\ (iv) & d_{\alpha y,y'} & = d_{y,\alpha y'} \end{array}$$

for all $E \in \mathcal{D}(X)$, $e \in \mathcal{D}(Y)$, $a \in A$, and $\alpha \in \mathfrak{A}$. Let

$$\mathcal{T}(X/A, Y/\mathfrak{A}) \stackrel{\text{def}}{=} \left(\mathcal{D}(X) \otimes \mathfrak{A}\right) \oplus \left(X_0 \otimes Y_0\right) \oplus \left(A \otimes \mathcal{D}(Y)\right)$$

be the anticommutative algebra with multiplication defined by

$$\begin{aligned} [D \otimes \alpha, D' \otimes \alpha'] &= [D, D'] \otimes \alpha\alpha' \\ [a \otimes d, a' \otimes d'] &= aa' \otimes [d, d'] \\ [D \otimes \alpha, a \otimes d'] &= 0 \\ [D \otimes \alpha, x \otimes y] &= Dx \otimes \alpha y = -[x \otimes y, D \otimes \alpha] \\ [a \otimes d, x \otimes y] &= ax \otimes dy = -[x \otimes y, a \otimes d] \\ [x \otimes y, x' \otimes y'] &= D_{x,x'} \otimes (y, y') + (x * x') \otimes (y \circ y') + (x, x') \otimes d_{y,y'} \end{aligned}$$

Under suitable restrictions which come from imposing the Jacobi identity, $\mathcal{T}(X/A, Y/\mathfrak{A})$ is a Lie algebra. Evidently, the Tits construction $\mathcal{T}(\mathcal{C}/F, J/F)$ above is just the special case that X is the algebra $\mathcal{C}$ of split octonions over F, $\mathcal{D}(X) = Der_F\mathcal{C}$, Y is the split exceptional Jordan algebra J, and $\mathcal{D}(Y) = Der_F J$. We then prove the following theorems:

Recognition Theorem for Type G_2. Let L be a G_2-graded Lie algebra. Then there exists a generically cubic Jordan algebra J over a unital, commutative, associative algebra $\mathfrak{A}$ such that L is centrally isogenous with

$$\begin{aligned} \mathcal{T}(\mathcal{C}/F, J/\mathfrak{A}) &= (Der_F\mathcal{C} \otimes \mathfrak{A}) \oplus (\mathcal{C}_0 \otimes J_0) \oplus < J, J > \\ &= (G_2 \otimes \mathfrak{A}) \oplus (\mathcal{C}_0 \otimes J_0) \oplus < J, J >, \end{aligned}$$

where $\mathcal{C}$ is split octonion algebra over F, $\mathcal{C}_0$ and J_0 are the trace zero elements of $\mathcal{C}$ and J respectively, and $< J, J >= [L_J, L_J]$, the Lie subalgebra of inner derivations of J.

Recognition Theorem for Type F_4. Let L be an F_4-graded Lie algebra. Then there exists a generically quadratic alternative algebra $\mathcal{C}$ over a unital, commutative, associative algebra A such that L is centrally isogenous with

$$\begin{aligned} \mathcal{T}(\mathcal{C}/A, J/F) &= < \mathcal{C}, \mathcal{C} > \oplus (\mathcal{C}_0 \otimes J_0) \oplus (A \otimes Der_F J), \\ &= < \mathcal{C}, \mathcal{C} > \oplus (\mathcal{C}_0 \otimes J_0) \oplus (A \otimes F_4), \end{aligned}$$

where J is the split exceptional 27-dimensional Jordan algebra J, $\mathcal{C}_0$ and J_0 are the trace zero elements of $\mathcal{C}$ and J respectively, and $< \mathcal{C}, \mathcal{C} >$ is the Lie subalgebra of inner derivations of $\mathcal{C}$.

Let V denote a vector space over F having a nondegenerate symmetric bilinear form $(\, , \,)$. The sum $J(V) = F1 \oplus V$ becomes a Jordan algebra under the operation $u \circ v = (u, v)1$; $u, v \in V$. Similarly for any unital, commutative, associative F-algebra $\mathfrak{A}$, and any unital $\mathfrak{A}$-module W with a symmetric $\mathfrak{A}$-bilinear form $(\, , \,) : W \times W \longrightarrow \mathfrak{A}$, the operation $w \circ x = (w, x)1$ defines the structure of a Jordan algebra on $J(W) = \mathfrak{A}1 \oplus W$.

Recognition Theorem for Type B_n. Let L be a B_n-graded Lie algebra for $n \geq 3$. Then there exists a unital, commutative, associative F-algebra $\mathfrak{A}$ and a unital $\mathfrak{A}$-module W with a symmetric $\mathfrak{A}$-bilinear form $(\, , \,) : W \times W \longrightarrow \mathfrak{A}$ such that L is centrally isogenous with

$$\begin{aligned} \mathcal{T}(J(V)/F, J(W)/\mathfrak{A}) &= (< V, V > \otimes \mathfrak{A}) \oplus (V \otimes W) \oplus < W, W >, \\ &= (B_n \otimes \mathfrak{A}) \oplus (V \otimes W) \oplus < W, W >, \end{aligned}$$

where V is a $(2n+1)$-dimensional F-vector space with a nondegenerate symmetric bilinear form (the defining representation for B_n), $< V, V >$ is the set of skew-symmetric transformations on V relative to the form on V which is a simple Lie algebra of type B_n, and $< W, W >$ is the set of skew-symmetric transformations on W relative to the form on W.

The preprint [Z], which was presented in Oberwolfach in 1992, gave the proof of the Recognition Theorem for Type C and treated Lie algebras graded by root systems of types B_n, C_n, F_4, and G_2 in which the algebras decompose into copies of the adjoint and trivial modules (no little adjoint modules allowed). Except in the C_n case where we appeal directly to the Jacobson's Coordinatization Theorem, the methods we use to establish the results discussed above are similar to those employed in [Se]; they allow arbitrary root graded algebras without restrictions on the summands, and they afford easier proofs than those in [Z]. The C_n case could be addressed similarly, but we have elected the other route because it is more direct. E.

Neher's preprint [N] provides an alternate uniform treatment of the algebras graded by root systems of types A_n, B_n, C_n, D_n, E_6, E_7 using the theory of 3-graded root systems, which is essentially the theory of certain Jordan pairs.

As an application of these Recognition Theorems, we determine the intersection matrix algebras of Slodowy (see [Sl1], [Sl2]) for types B_n, C_n, F_4, and G_2. The intersection matrix algebras which correspond to the simply laced root systems of rank ≥ 2 have been identified by Berman and Moody [BM]. They prove that the intersection algebras of type A_n, $n \geq 3$, are isomorphic to $st_n(R)$, where R is the group algebra of a free group. If $n = 2$, then for the coefficient ring R we should take the free product $*_{i=1}^{m} F[x_i, x_i^{-1}]$ in the variety of alternative algebras. For types D_n, $n \geq 4$, E_6, E_7, E_8 the intersection algebras are isomorphic to the u.c.a. of $\mathcal{G} \otimes R$ where R is an algebra of Laurent polynomials in several variables (see [BK] for a discussion of these central extensions).

From the Recognition Theorems above it follows that the intersection algebras of type C_n, $n \geq 4$, are isomorphic to $stsp_{2n}(R, *)$ where R is a group algebra of a free group where the involution $*$ fixes the free generators. For $n = 3$, the ring of coefficients is the same free product as in the A_2 case with the involution $*$ fixing the x_i's. An intersection algebra of type G_2 is the u.c.a. of $\mathcal{T}(\mathcal{C}/F, J/\mathfrak{A})$, where $\mathcal{C}/F$ is the split octonion algebra and $J/\mathfrak{A}$ is a universal generically cubic Jordan algebra, which is quite a complicated object. Similarly, for type F_4 we get the u.c.a. of $\mathcal{T}(\mathcal{C}/A, J/F)$, where $\mathcal{C}/A$ is a universal generically quadratic alternative algebra and J/F is the split 27-dimensional exceptional Jordan algebra. For type B an intersection algebra is the u.c.a. of $\mathcal{T}(J(V)/F, J/\mathfrak{A})$, where $J/\mathfrak{A}$ is a universal generically quadratic alternative algebra.

Further applications of the results of this paper to the study of Lie bialgebras have been explored by Montaner and Zelmanov [MZ].

References

[AF] B.N. Allison and J.R. Faulkner, Nonassociative coefficient algebras for Steinberg unitary Lie algebras, *J. Algebra* **161** (1993), 1-19.

[BK] S. Berman and Y. Krylyuk, Universal central extensions of twisted and untwisted Lie algebras extended over commutative rings, *J. Algebra*, to appear.

[BM] S. Berman and R.V. Moody, Lie algebras graded by finite root systems and the intersection matrix algebras of Slodowy, *Invent. Math.* **108** (1992), 323-347.

[FF] J.R. Faulkner and J.C. Ferrar, Exceptional Lie algebras and related algebraic and geometric structures, *Bull. London Math. Soc.* 9 (1977), 1-35.

[F] H. Freudenthal, Beziehungen der E_7 und E_8 zur Oktavenebene I, *Nederl. Akad. Weten. Proc. Ser. A* **57** (1954), 218-230.

[G] H. Garland, The arithmetic theory of loop groups, *Publ. Math. Inst. Hautes Étud. Sci.* **52** (1980), 5-136.

[J1] N. Jacobson, *Structure and Representations of Jordan Algebras*, Amer. Math. Soc. Colloquium Publ. **39** Providence, R.I. (1968).

[J2] N. Jacobson, *Exceptional Lie Algebras*, M. Dekker Lect. Notes in Pure and Appl. Math. **1** New York (1971).

[Ka] I.L. Kantor, Classification of irreducible transitively differential groups, *Dokl. Akad. Nauk SSSR* **158** (1964), 1271-1274.

[Ko] M. Koecher, Imbedding of Jordan algebras into Lie algebras, *Amer. J. Math.* **89** (1967), 787-815.

[M] K. McCrimmon, The Freudenthal-Springer-Tits constructions of exceptional Jordan algebras, *Trans. Amer. Math. Soc.* **139** (1969), 495-510.

[MZ] F. Montaner and E.I. Zelmanov, Bialgebra structures on current Lie algebras, preprint (1993).

[N] E. Neher, Lie algebras graded by 3-graded root systems, preprint (1993).

[Sc] R.D. Schafer, *Introduction to Nonassociative Algebras*, Academic Press **22** New York (1966).

[Se] G.B. Seligman, *Rational Methods in Lie Algebras*, M. Dekker Lect. Notes in Pure and Appl. Math. **17** New York (1976).

[Sl1] P. Slodowy, Beyond Kac-Moody algebras and inside, *Canad. Math. Soc. Conf. Proc.* **5** (1986), 361-371.

[Sl2] P. Slodowy, *Singularitäten, Kac-Moody Lie Algebren, assoziierte Gruppen und Verallgemeinerungen*, Habilitationsschrift, Universität Bonn (1984).

[T1] J. Tits, Une classe d'algèbres de Lie en relation avec les algèbres de Jordan, *Nederl. Akad. Weten. Proc. Ser. A*, **65** (1962), 530-535.

[T2] J. Tits, Algèbres alternatives, algèbres de Jordan, et algèbres de Lie exceptionelle, *Nederl. Akad. Weten. Proc. Ser. A*, **69** (1966), 223-237.

[Z] E. Zelmanov, Lie algebras graded by finite root systems, preprint, (1992).

1991 Mathematics Subject Classifications: Primary 17B20, 17B70, 17B25

BERNSTEIN REPRESENTATIONS

J. BERNAD, A. ILTYAKOV and C. MARTINEZ*
Departamento de Matemáticas, Universidad de Oviedo
33007 Oviedo, Spain

Abstract. In this paper is defined the notion of Bernstein representation and it is proved that every irreducible module over a nuclear Bernstein algebra is one-dimensional.

The notion of universal representation of a Bernstein algebra is also introduced and some properties of this algebra are estudied by using properties of the given algebra.

1. Introduction

Let B be a non associative, commutative algebra over a field K of characteristic $\neq 2$. B is said to be a Bernstein algebra if there exists a nonzero homomorphism $\omega: B \to K$, and it holds the identity

$$(x^2)^2 = \omega(x)^2 x^2 \qquad \forall x \in B$$

B has a pierce descomposition $B = Ke \oplus U_e \oplus Z_e$, where $U_e = \{x \in \ker\omega \mid ex = \frac{1}{2}x\}$, $Z_e = \{x \in \ker\omega \mid ex = o\}$

The following identities are valid in Bernstein algebras for all $u, u_1, u_2 \in U_e$ $z, z_1, z_2 \in Z_e$

$$(u_1u_2)u_3 + (u_2u_3)u_1 + (u_1u_3)u_2 = 0; \;\; u_1(u_2z) = 0; \;\; (z_1z_2)u = 0$$

Besides, $U_e^2 \subseteq Z_e \;\; U_eZ_e \subseteq U_e \;\; Z_e^2 \subseteq U_e$.

In this paper we deal with representations of algebras. Let A be an algebra which belongs to a class of algebras $\mathcal{C}$ over a field K and M be a K- module. A linear mapping $\mu: A \to End(M)$ is said to be a representation of A in the class $\mathcal{C}$, if the explicit algebra $A \dot{+} M$ with multiplication

$$(a + m)(a' + m') = aa' + m\mu(a') + m'\mu(a) \; \forall a, a' \in A \; \forall m, m' \in M \tag{1}$$

belongs to the class $\mathcal{C}$. We say that M is an A-module. We denote the action $m\mu(a)$ by $m \cdot a$.

An A-module M is irreducible if M has not non-zero proper submodules.

2. Bernstein representations

Let (B, ω) a Bernstein algebra over a field K of char$\neq 2$, M a K-module and $\mu: B \to End(M)$ a linear mapping.

* Partially supported by D.G.A. P. CB-6/91.

S. González (ed.), Non-Associative Algebra and Its Applications, 39–45.

We say that $\mu : B \to M$ is a Bernstein representation if the explit algebra

$$\widehat{B} = B\dot{+}M$$

with the homomorphism

$$\hat{\omega}: \to End(M)$$

defined by $\hat{\omega}(b+m) = \omega(b)$ is a Bernstein algebra.

The K-module M is called B-Bernstein module.

Lemma 2.1 *Let (B,ω) be a Bernstein algebra and $\mu: B \to End(M)$ a Bernstein representation. Then*

$$2\mu(b)\mu(b^2) = \mu(b)\omega(b^2) \quad \forall b \in B. \tag{2}$$

Proof. Let $b+m$ be an element from $B\dot{+}M$, then $(b+m)^2 = b^2 + 2m\mu(b)$ and $\left((b+m)^2\right)^2 = \left(b^2\right)^2 + 4m\mu(b)\mu(b^2)$. Since μ is a Bernstein representation,

$$\left(b^2\right)^2 + 4m\mu(b)\mu(b^2) = \left((b+m)^2\right)^2 = \hat{\omega}(b+m)^2(b+m)^2 = \left(b^2 + 2m\mu(b)\right)\omega(b)^2.$$

It follows $2m\mu(b)\mu(b^2) = m\mu(b)\omega(b^2) \quad \forall m \in M$. □

Proposition 2.1 *Let (B,ω) be a finite dimensional Bernstein algebra and $\mu: B \to End(M)$ a non trivial irreducible representation. Then $\mu(e) = \frac{1}{2}I$, $\mu(u) = 0 \ \forall u \in U$.*

Corollary 2.1 *Let (B,ω) be a nuclear, finite dimensional Bernstein algebra. Every irreducible module has dimension one over the ground field.*

3. Universal representation

Let C be an associative algebra and (B,ω) be a Bernstein algebra. A linear mapping $\mu: B \to C$ is a Bernstein representation if

$$2\mu(b)\mu(b^2) = \mu(b)\omega(b^2).$$

We call to a linear mapping $\epsilon: B \to U(B)$ universal representation, if for all representations $\mu: B \to C$, there exists a unique homomorphism of algebras

$$\psi: U(B) \to C$$

such that $\epsilon \circ \psi = \mu$.

The algebra $U(B)$ is unique up to isomorphism and it is called universal enveloping algebra.

We can construct $U(B)$ in the followig way: let us take a basis of B, $\{a_i\}$ and consider the free associative algebra $Ass[\epsilon(a_1), \ldots, \epsilon(a_n)]$. Let μ be a Bernstein representation $\mu: B \to C$, we define the homomorphism

$$I_\mu: Ass[\epsilon(a_1), \ldots, \epsilon(a_n)] \to C$$

$I_\mu\left(\epsilon(a_i)\right) = \mu(a_i)$.

The linear mapping $\epsilon: B \to Ass[\epsilon(a_1), \ldots, \epsilon(a_n)]/I$ where $I = \cap_\mu \ker I_\mu$ and μ representation, is a universal representation.

It is proved in [2] that for a finite dimensional Jordan algebra the universal enveloping algebra is finite dimensional, and for a Lie algebra it is noetherian.

Theorem 3.1 *Let (B,ω) be a Bernstein algebra and $\epsilon: B \to U(B)$ universal representation. Then,*

(1) $U(B)$ is finite dimensional if and only if $B = B^2$

(2) $U(B)$ is right Noetherian if and only if $\dim B/B^2 \leq 1$

(3) If $\dim B/B^2 \geq 2$, there exists a monomorphism $Ass[x_1,..,x_n,...] \to U(B)$.

Proof. If $B = B^2$ and $\mu: B \to C$ a representation, there exists a number $k \in \mathbb{N}$ such that for all μ representation

$$\mu(n_1)\cdots\mu(n_k) = 0 \; \forall n_i \in N = \ker\omega.$$

Linearizing the identity (2), we obtain

$$2\big(\mu(a)\mu(bc) + \mu(b)\mu(ac) + \mu(c)\mu(ab)\big) = \mu(a)\omega(bc) + \mu(b)\omega(ac) + \mu(c)\omega(ab) \quad (3).$$

It follows from (3)

$$\begin{gathered} 2\mu(e)^2 = \mu(e);\; \mu(u)\mu(e) = \tfrac{1}{2}\mu(u) - \mu(e)\mu(u);\; \mu(u_iu_j)\mu(e) = \mu(u_iu_j) \\ \mu(e)\mu(u_iu_j) = -\tfrac{1}{2}\big(\mu(u_i)\mu(u_j) + \mu(u_i)\mu(u_j) + \mu(u_j\mu(u_i)\big). \end{gathered} \quad (4)$$

Let $\{e, u_1, \ldots, u_r, \ldots, u_iu_j\}$ be a basis of B and $p(\epsilon(e), \ldots, \epsilon(u_iu - j))$ be a polynomial of $Ass[\epsilon(e), \ldots, \epsilon(u_iu_j)]$. It is enough to prove that there exists $n \in \mathbb{N}$ such that any polynomial with more than n elements from N belongs to $I = \cap \ker I_\mu$.

Let us suppose that $p \in Ass[\epsilon(e), \ldots, \epsilon(u_iu_j)]$ has more than k elements from N. By universal property of $U(B)$, $\psi(p(\epsilon(e), \ldots, \epsilon(u_iu_j)) = p(\psi\epsilon(e), \ldots, \psi\epsilon(u_iu_j)) = p(\mu(e), \ldots, \mu(u_iu_j))$.The element $\mu(e)$ disappears or it is at the end of the polynomial by identities (4). So, if we have more than k elements from N then $p \in \ker I_\mu$, since $\mu(n_1)\cdots\mu(n_k) = 0$.

Now suppose that $B \neq B^2$. It means that there exists $z \in Z \setminus U^2$. Let us define $\varphi: K[x] \to U(B)$ such that $\varphi(x) = \epsilon(z)$, where $K[x]$ is the polynomial algebra.

If $\ker\varphi \neq 0$, we can get $f(x) \in K[x]$ such that $f(\epsilon(z)) \in I$. By the universal property of $U(B)$, $f(\mu(z)) = 0$ for all μ representation.

Let us take a K-module M with basis $\{m_i\}_{i=1}^{\infty}$ and the linear mapping defined

$$\mu: \to End(M)$$

$\mu(e) = \frac{1}{2}I$ $\mu(u) = \mu(u^2) = 0$ $\mu(z) = \alpha$ $\forall u \in U$ where $\alpha \in End(M)$ $f(\alpha) \neq 0$.

μ is a Bernstein representation and $f(\mu(z)) \neq 0$. hence φ is monomorphism and $U(B)$ is infinite dimensional.

(2) Suppose $\dim B/B^2 \leq 1$, i.e, $\{e, u_1, \cdots u_r, \cdots u_iu_j, z\}$ $z \in Z \setminus U^2$ is a basis of B. Let $K[x]$ be the free associative, commutative algebra in one variable. We define an action over $U(B)$ as K- vector space,

$$bx = b \cdot \epsilon(z) \; \forall b \in U(B) \; \epsilon(z) \in U(B).$$

$U(B)$ is a module over $K[x]$. We can prove that $U(B)$ is a finitely generated module over $K[x]$ using the following identities

$$\begin{gathered} \epsilon(z)\epsilon(e) = \epsilon(z) \qquad \epsilon(z)\epsilon(u) = -\tfrac{1}{2}\epsilon(e)\epsilon(uz) \\ \epsilon(z)\epsilon(u_iu_j) = -\epsilon(u_i)\epsilon(u_jz) - \epsilon(u_j)\epsilon(u_iz) \end{gathered} \quad (5)$$

Besides, if A is associative, commutative, noetherian algebra and M is a finitely generated module over A, M is noetherian. Using this result, we obtain $U(B)$ is right noetherian.

Suppose $U(B)$ is right noetherian and $\dim B/B^2 \geq 2$. Let $\{e, u_1, \ldots, u_i u_j, z_1, z_2\}$ be a basis of B and M be a K-module with basis $\{m_i, n_j \mid i = 0, \ldots, \infty\ j = 1, \ldots \infty\}$. We define the following action

$$\begin{array}{lll} m_i \cdot e = \frac{1}{2} m_i & n_j \cdot e = \frac{1}{2} n_j & m_i \cdot (U + U^2) = 0 \\ n_j \cdot (U + U^2) = 0 & m_0 \cdot z_1 = n_1 & m_i \cdot z_1 = m_i\ \forall i \geq 1 \\ n_j \cdot z_1 = m_j\ \forall j & m_i \cdot z_2 = 0\ \forall i & n_j \cdot z_2 = n_{j+1}\ \forall j \end{array}$$

This action defines a Bernstein representation $\mu \colon B \to End(M)$.

Denote by I_i the right ideal $\epsilon(z_1)\epsilon(z_2)^i\epsilon(z_1)U(B)$ and $G_s = \sum_{i=1}^{s} I_i$. Consider the chain of ideals

$$G_1 \subseteq G_2 \subseteq \cdots \subseteq G_s \subseteq \cdots$$

Since $U(B)$ is right noetherian, $G_s = G_{s+1}$ and by the universal property of $U(B)$, $\mu(z_1)\mu(z_2)^{s+1}\mu(z_1) = \mu(z_1)\mu(z_2)\mu(z_1)f_1(\mu(e), \ldots, \mu(z_2)) + \cdots + \mu(z_1)\mu(z_2)^s$ $f_s(\mu(e), \ldots, \mu(z_2))$. Let us note that $m_i \cdot f(\epsilon(e), \ldots, \epsilon(z_2)) = \lambda m_i\ \forall i \geq 1$.

Hence $m_0\mu(z_1)\mu(z_2)^{s+1}\mu(z_1) = m_{s+2}$ and $m_0\mu(z_1)\mu(z_2)\mu(z_1)f_1(\mu(e), \ldots, \mu(z_2))$ $+ \cdots + \mu(z_1)\mu(z_2)^s f_s(\mu(e), \ldots, \mu(z_2)) = \sum_{i=1}^{s} \lambda_i m_i$. This is a contradiction with the linear independence of $\{m_i\}$.

(3) Let $\{e, u_1, \ldots, u_i u_j, z_1, z_2\}$ be a basis of B and $K[x, y]$ the free associative algebra. We define an homomorphism

$$\phi \colon K[x, y] \to U(B)$$

$\phi(x) = \epsilon(z_1)\ \phi(y) = \epsilon(z_2)$.

Let us check that ϕ is a monomorphism. Defining the same action as before, $f(x, y) \in \ker \phi$ and due to universal property, we obtain $f = 0$. □

Example: In general, $U(B)$ is not left noetherian when $dim B/B^2 \leq 1$.

Let B be a Bernstein algebra and $\{e, u_1, \ldots, u_r, u_i u_j, \ldots, z\}$ be a basis of B. Let us consider M a K-module with basis $\{m_i\}_{i=0}^{\infty}$. We define the action

$$\begin{array}{cc} m_i \cdot e = \frac{1}{2} m_i\ m_i \cdot U = 0\ \forall i \geq 1 & m_0 \cdot e = 0 \\ m_0 \cdot u_j = u_1\ m_i \cdot U^2 = 0\ m_i \cdot z = m_{i+1}\ faI \geq 1 & m_0 \cdot z = 0 \end{array}$$

M is a B module and this action defines a representation $\mu_1 \colon B \to End(M)$. Let $U(B)$ be the universal representation and the following chain of left ideals

$$U(B)\epsilon(u)\epsilon(z) \subseteq U(B)\epsilon(u)\epsilon(z) + U(B)\epsilon(u)\epsilon(z)^2 \subseteq \ldots \subseteq \sum_{k=1}^{i} U(B)\epsilon(u)\epsilon(z)^k \subseteq \ldots$$

Let us suppose $U(B)$ is left noetherian,

$$U(B)\epsilon(u)\epsilon(z) + \ldots + U(B)\epsilon(u)\epsilon(z)^s = U(B)\epsilon(u)\epsilon(z) + \ldots + U(B)\epsilon(u)\epsilon(z)^{s+1} \quad (6)$$

Let us note that $m_0\psi_1(f_i)\mu_1(u)\mu_1(z)^i = \alpha_i m_1$ by definition of the action. Hence, $m_0\mu_1(u)\mu_1(z)^{s+1} = m_{s+2}\ m_0\left[\psi_1(f_1)\mu_1(u)\mu_1(z) + \ldots + \psi_1(f_s)\mu_1(u)\mu_1(z)^s\right] = \alpha_1 m_2 + \ldots + \alpha_s m_{s+1}$.

But this is a contradiction because $\{m_2, \ldots, m_{s+2}\}$ are linear independent.

4. Irreducible representation in the non-finitely generated case

Yu.A.Medvedev and E.I.Zel'manov proved if J is a solvable Jordan algebra, then J^2 is nilpotent. E.I.Zel'manov and Skosyrkii proved if J is a special Jordan algebra without elements of additive order $\leq 2n$ and $x^n = 0 \quad \forall x \in J$, then J is solvable. See [3] and [4] for more details. Using these results, we can prove the following proposition.

Proposition 4.1 *Let (B,ω) be a Bernstein algebra over a field of characteristic $\neq 2, 3, 5$. The square of the barideal $N = \ker\omega$ is nilpotent.*

Proof. Let $B = Ke \oplus U \oplus Z$ be a Bernstein algebra. In [1] is proved $\overline{N} = N/Ann(U)$ is a Jordan algebra satisfying $x^3 = 0 \quad \forall x \in \overline{N}$. In view of the above remark, $(N^2)^k \in Ann(U)$ for some $k \in \mathbb{N}$. Besides, $N^2 = U_2 \dot{+} Z_2$ where $U_2 = UZ + Z^2$ and $Z_2 = U^2$, hence, by identities (1), $(N^2)^k N^2 N^2 = U^2 U^2 U^2 = 0$. □

Proposition 4.2 *Let (B,ω) be a nuclear Bernstein algebra characteristic $\neq 2, 3, 5$ and $\mu: B \to End(M)$ a irreducible representation. Then $\mu(N) = 0$ where $N = \ker\omega$.*

Proof. Let us consider $M \cdot N^2 = \{\sum m_i \cdot n_i | \ m_i \in M, \ n_i \in N^2\}$. This is a submodule of M. Thus $M \cdot N^2$ is a submodule and since M is irreducible, $M \cdot N^2 = 0$ or $M \cdot N^2 = M$.

The explit algebra $\widehat{B} = B \dot{+} M$ is a nuclear algebra ($\widehat{B}^2 = B^2 \dot{+} B \cdot M = B \dot{+} M$) and therefore $(\widehat{N}^2)^k = 0$.

But if $M \cdot N^2 = M$, $(M \cdot N^2) \cdot N^2 \cdots) \cdot N^2 = M$, $M \cdot N^2$, $N^2 \subseteq \widehat{N}^2$ what it is a contradiction with the nilpotency of $\widehat{N}^2$. Hence $M \cdot N^2 = 0$ and $\mu(U^2) = 0$.

Let us take an element $u_0 \in U$ and consider

$$L_{u_0} = \{m \in M | \ m \cdot u_0 = 0\}$$

L_{u_0} is a submodule of M.

It follows $L_{u_0} = 0$ or $L_{u_0} = M$. Besides, if $0 \neq m \in M$, $m \cdot u_0 \neq 0$ $(m \cdot u_0) \cdot u_0 = ((m_{\frac{1}{2}} + m_0) \cdot u_0) \cdot u_0 = -\frac{1}{2} m_{\frac{1}{2}} \cdot u_0^2 + (m_0 \cdot u_0) \cdot u_0 = 0$. Then, $L_{u_0} \neq 0$. Therefore, $L_u = M \ \forall u \in M$ and $\mu(U) = 0$. □

Corollary 4.1 *For every irreducible module M over a nuclear Bernstein algebra $\dim_K M = 1$.*

Proposition 4.3 *Let (B,ω) be a nuclear Bernstein algebra, char$\neq 2, 3, 5$. Then for all $u \in U$, $idl(R_u) \triangleleft M(B)$ is nilpotent.*

Proof. We will prove

$$R_u R_{a_1} \cdots R_{a_k} R_u R_{a_{k+1}} \cdots R_u \cdots = 0$$

where a_i is equal to e, $u_i \in U$ or $z_i \in Z$.

Let us suppose $a_i \in N = \ker\omega\ \forall i$. We can assume $a_i \in \overline{N} = N/ann(N)$, so

$$R_{x_1}R_{x_2} + R_{x_2}R_{x_1} + R_{x_1x_2} = 0\ \forall x_1x_2 \in \overline{N}. \tag{7}$$

By (7), $R_uR_{a_1}\cdots R_{a_k}R_u = -R_uR_{a_1}\cdots R_{a_k-1}R_uR_{a_k} - R_uR_{a_1}\cdots R_{a_k-1}R_{ua_k}$ Repeting (7) several times, we obtain

$$R_uR_{a_1}\cdots R_{a_k}R_u = R_uR_uR_{a_1}\cdots R_{a_k} + \sum R_uR_{c_1}\cdots R_{c_k-1}$$

where between $c_1,\ldots,c_{k-1}$ there is an element $ua_i \in N^2$.

Since $R_u^2 = -\frac{1}{2}R_{u^2}$,we have poved that $R_uR_{a_1}\cdots R_{a_k}R_u$ is a sum of products R_{b_i} with at least one element $b_i \in N^2$ in each sumand.

Let us see that if $T = R_u\Box R_u\Box\cdots R_u$ is a sum of products R_{b_i} and in each sumand there are k elements $ua_i \in N^2$, adding to T operators $\Box R_u$ we obtain $T' = T\Box R_u\cdots\Box R_u$ which is a sum of products R_b and in each sumand there are $k+1$ elements of N^2.

$T = R_U\Box R_u\cdots\Box = \sum R_uR_{c_1}\cdots R_{c_s} + \sum R_{u^2}R_{d_1}\cdots R_{d_p}$ where between $c_1\cdots,c_s$ there are k elements $ua \in N^2$ and between $d_1\cdots d_p$ there are $k-1$ elements $ua \in N^2$.

We define $T_1 \equiv T_2$ if $T_1 = T_2 + T'$ and in each sumand of T' there are $k+1$ elements of N^2.

Applying (7) several times

$$R_uR_{c_1}\cdots R_{c_s}\Box R_u \equiv \sum R_uR_{s_1}\cdots R_{s_n}$$

with k elements of N^2 and at least one of them $s_i = u(ua) \in N^2$.

Adding one more operator $\Box R_u$,

$$R_uR_{c_1}\cdots R_{c_s}\Box R_u\Box R_u \equiv \sum R_uR_{s_1}\cdots R_{s_n}\Box R_u \equiv \sum R_uR_{s'_1}\cdots R_{s'_m}$$

with k elements of N^2 and at least two of them have the form $u(ua) \in N^2$.

Adding k operators $\Box R_u$,

$$R_uR_{c_1}\cdots R_{c_s}\Box R_u\cdots\Box R_u \equiv \sum R_uR_{t_1}\cdots R_{t_j}$$

with k elements $u(ua)$ of N^2.

If we add another operator $\Box R_u$

$$R_uR_{c_1}\cdots R_{c_s}\Box R_u\cdots\Box R_u \equiv \sum R_uR_{t_1}\cdots R_{t_j}\Box R_u \equiv \sum R_{u^2}R_{t_1}\cdots R_{t_1}\cdots R_{t_j} \equiv 0$$

since $R_u, R_{u(ua)}$ anticommute.

In the same way, we can prove $R_{u^2}R_{d_1}\cdots R_{d_p}\Box R_u\cdots\Box R_u \equiv 0$.

Hence $T = R_u\Box\cdots\Box R_u = \sum R_{a'_0}\cdots R_{a'_s}$ where there are $k+1$ elements of N^2. Moreover, if T has k elements of N^2, after adding $k+1$ operators $\Box R_u$, T has $k+1$ elements of N^2.

On the other hand, $\overline{N}^2$ is nilpotent, $(\overline{N}^2)^r = 0$. Therefore T is zero when T contains r elements of N^2 and we obtain these r elements after $(r+2)(r+1)/2$ steps.

Now suppose that $T = R_u \cdots R_e R_a \cdots R_u \cdots$ and consider $\widehat{T} = R_u \cdots \widehat{R_e} R_a \cdots R_u$ where $\widehat{T}$ has the same right operators except R_e. It is easy to show that if $\widehat{T} = 0$, then $T = 0$. This means that we can remove operators R_e to prove the nilpotency of $idl(R_u)$. Then, by the first case, $idl(R_u)$ is nilpotent. □

Corollary 4.2 *Let (B, ω) be a nuclear Bernstein algebra, characteristic $\neq 2, 3, 5$, and $z \in U^2$. The ideal $idl(R_z) \triangleleft M(B)$ is nilpotent.*

Proof. Let us note that in $\overline{N} = ann(N)$it is true that $R_z = R_{u_1 u_2} = -R_{u_1} R_{u_2} - R_{u_2} R_{u_1}$. Hence by the above proposition , $idl(R_z)$ is nilpotent. □

Corollary 4.3 *If (B, ω) is a nuclear Bernstein algebra, characteristic $\neq 2, 3, 5$, then $idl(u) \triangleleft B$ is nilpotent.*

Theorem 4.1 *Let (B, ω) be a nuclear Bernstein algebra and $\epsilon: B \to U(B)$ universal representation of B. Then the Jacobson radical of B, $J(U(B))$, is a sum of nilpotent ideals.*

References

1. I.R.HENTZEL and L.A.PERESI, "Semiprime Bernstein algebras", Arch.math, 52 (1989) 534-543.
2. N.JACOBSON, "Structure and Representations of Jordan algebras", American Mathematical Society Colloquim Publications 1968.
3. Yu.A.MEDVEDEV and E.I.ZEL'MANOV, "Solvable Jordan algebras", Comm.Alg, 13(6), 1389-1414 (1985).
4. E.I.ZEL'MANOV and V.G.SKOSYRSKII, "Special Jordan nilalgebras of bounded index", Algebra i Logica, vol.22, n6, 626-635.
5. A.WÖRZ-BUSEKROS, "Algebras in Genetics", Lecture Notes in Biomathematics 36, Berlin-Heidelberg-New York 1990.

ON FREE DIFFERENTIALS ON ASSOCIATIVE ALGEBRAS

A. BOROWIEC *
Institute of Theoretical Physics
University of Wrocław, Poland, borowiec@plwruw11.bitnet

V. K. KHARCHENKO †
Institute of Mathematics
Novosibirsk, Russia, kharchen@math.nsk.su

and

Z. OZIEWICZ
Institute of Theoretical Physics
University of Wroclaw, Poland, oziewicz@plwruw11.bitnet

1. Introduction

A differential $d : R \to {}_RM_R$ is called free if the differential of any element v has a unique presentation of the form $dv = dx^i \cdot v_i$, where $x^1, \ldots, x^n$ are generators of the unitary algebra R and $dx^1, \ldots, dx^n$ their differentials. Any free differential defines a commutation formula $vdx^i = dx^k \cdot A(v)^i_k$, where $A : v \mapsto A(v)^i_k$ is an algebra homomorphism $A : R \to R_{n\times n}$. It is easy to see that for any homomorphism $R \to R_{n\times n}$ there exists not more than one free differential. We are going to consider the existence problem of such a differential. We will show that for a given commutation rule $vdx^i = dx^k \cdot A(v)^i_k$ a free algebra generated by the variables $x^1, \ldots, x^n$ has a related free differential. We will define an *optimal* algebra with respect to a fixed commutation rule. In the homogeneous case this algebra is characterized as the unique algebra which has no nonzero A-invariant subspaces with zero differentials. Finally, we will consider a number of examples of optimal algebras for different commutation rules. In particular, we will describe two variable commutation rules which define commutative optimal algebra.

This article is closely related to the known Wess and Zumino paper [1]. In our terms, they prove in particular that a system on n^2 quadratic forms vanish in the optimal algebra if the Yang-Baxter equation holds.

2. Free differential calculi

Recall that a differential is a linear mapping from an algebra R to a bimodule M satisfying the Leibniz rule:

$$d(uv) = d(u)v + ud(v)$$

* Supported by the State Research Committee KBN
† Supported by the Russian Fund of Fundamental Research No 93-011-16171

S. González (ed.), *Non-Associative Algebra and Its Applications*, 46–53.

Lemma 1.1 *A differential d has the uniqueness property iff* $\Omega_d(R) = Rd(R)R$ *is a free right R-module freely generated by* $dx^1, \ldots, dx^n$.
Due to the lemma the following definition is natural.
Definition 1.2 A differential is said to be *free* if it has the uniqueness property. That definition essentially depends on the generating space $V = \sum x^i F$, with F being a base field. Let us consider as example the case V is any supplement of the subspace $F \cdot 1$, i.e. $R = V \dot{+} F \cdot 1$. Of course, if R is not finite dimensional, then we have an infinite set of generators. Nevertheless, there exists a free differential with respect to that space of generators. It is exactly the universal derivation.

If d is a free differential, then linear maps $D_k : R \to R$ (partial derivatives) can be defined by the formula:

$$dv = dx^k \cdot D_k(v) \tag{1}$$

Those maps satisfy the relations

$$D_k(x^i) = \delta^i_k, \tag{2}$$

where δ^i_k is the Kronecker delta.
Lemma 1.3 *A linear map* $A_d : R \to R_{n \times n}$ *from the algebra R into the algebra of n by n matrices over R given by the formula*

$$A_d(v)^i_k = D_k(vx^i) - D_k(v)x^i \tag{3}$$

is an algebra homomorphism i.e.

$$A_d(uv)^i_k = A_d(u)^l_k A_d(v)^i_l \tag{4}$$

Proof: Let $v \in R$. The left multiplication $A(v) : \omega \mapsto v \cdot \omega$ is an endomorphism of the right module $\Omega_d(R)$. Ring of all endomorphisms of any free module of rank n is isomorphic to the ring of all n by n matrices. Therefore, we can find a homomorphism $A : R \to R_{n \times n}$ defined by the formulae

$$v\, dx^i = dx^k \cdot A(v)^i_k$$

By the Leibniz rule we have

$$v\, dx^i = d(vx^i) - d(v)x^i = dx^k[D_k(vx^i) - D_k(v)x^i]$$

therefore,

$$dx^k \cdot A_d(v)^i_k = dx^k \cdot A(v)^i_k$$

i.e. $A_d = A$ and A_d is also a homomorphism. □

Let us consider a linear map $D : R \to R^n$ of R to the space of columns of height n acted by the formula

$$D(v) = \begin{pmatrix} D_1(v) \\ \vdots \\ D_n(v) \end{pmatrix}, \quad i.e. \quad D = \begin{pmatrix} D_1 \\ \vdots \\ D_n \end{pmatrix}$$

Proposition 1.4 *The map D and homomorphism A_d are connected by the relation*

$$D(uv) = D(u)v + A_d(u)D(v) \tag{5}$$

Proof: We have

$$u\,dv = u\,dx^i \cdot D_i(v) = dx^k \cdot A_d(u)^i_k D_i(v)$$

and

$$dx^k \cdot D_k(uv) = D_k(u)v + A_d(u)^i_k D_i(v)$$

i.e. by the uniqueness condition

$$D_k(uv) = D_k(u)v + A_d(u)^i_k D_i(v)$$

□

The inverse statement is also valid

Proposition 1.5 *Let R be an unitary algebra generated by elements $x^1, \ldots, x^n$ and $A : R \to R_{n\times n}$ be an algebra homomorphism. If $D : R \to R^n$ is a linear map such that*

$$D_k(x^i) = \delta^i_k \tag{6}$$

$$D(uv) = D(u)v + A(u)D(v), \tag{7}$$

then the map $\Delta : v \mapsto dx^k \cdot D_k(v)$ is a free differential, where $\Omega_\Delta(R) = \sum dx^i \cdot R$ is a free right module with the left module structure defined by commutation rule $vdx^i = dx^k A(v)^i_k$, i.e. $A_\Delta = A$.

Proof: We have to prove $\Delta(x^i) = dx^i$ and the Leibniz formulae. First equality follows from (6) and definition of Δ. Finally,

$$\Delta(uv) = dx^k \cdot D(uv) = dx^k \cdot [D_k(u)v + A_d(u)^i_k D_i(v)] = \Delta(u)v + u\Delta(v).$$

□

A natural question concerning Proposition 1.5 arises here. If a homomorphism A is given, then formula (7) allows one to calculate partial derivatives of a product in terms of its factors. That fact and formula (6) show that for a given A there exists not more then one D satisfying formulas (6) and (7). It is not clear yet whether or not there exists at least one D of such a type. Thus, our first task is to describe these homomorphisms of A for which there exist free differentials with $A_d = A$.

Theorem1.6 *Let $R = F < x^1, \ldots, x^n >$ be a free unitary algebra generated by $x^1, \ldots, x^n$ and $A^1, \ldots, A^n$ be any set of $n \times n$ matrices over R. There exists the unique free differential d such that $A_d(x^i) = A^i$.*

Proof: The map $x^k \mapsto A^k$ can be uniquely extended to a homomorphism of algebras $A : R \to R_{n\times n}$. Let us define a map D on monomials in $x^1, \ldots, x^n$ by induction on its degree. Let $D_k(1) = 0$ and $D_k(x^i) = \delta^i_k$. We define

$$D_k(x^i v) = \delta^i_k v + A(x^i)^j_k D_j(v), \quad \textit{where } A(x^i) = A^i \tag{8}$$

We have to prove formulae (7) for arbitrary elements u, v. It can be done by induction on degree of a monomial u.

If it's degree is zero: $u = \alpha \in F$ then $D(\alpha v) = \alpha D(v) = D(\alpha)v + A(\alpha)v$ as $D(\alpha v) = \alpha D(1) = 0$ and $A(\alpha) = \alpha E$. If the degree of u is equal to one then (8) implies required result. Let $u = x^i u_1$. Then by the equality $A(u) = A(x^i u_1) = A(x^i)A(u_1)$ and by induction supposition we have

$$D_k(uv) = D_k(x^i u_1 v) = \delta^i_k u_1 v + A(x^i)^j_k D_j(u_1 v) =$$

$$= [\delta^i_k u_1 + A(x^i)^j_k D_j(u_1)]v + A(x^i)^l_k A(u_1)^j_l D_j(v) = D_k(u)v + A(u)^j_k D_j(v)$$

□

Let now R be a non-free unitary algebra defined by the set of generators $x^1, \ldots, x^n$ and the set of relations $f_m(x^1, \ldots, x^n) = 0$, $m \in \mathcal{M}$ i.e. $R = \hat{R}/\hat{I}$, where $\hat{R} = F < \hat{x}^1, \ldots, \hat{x}^n >$ is a free algebra with unite and $\hat{I}$ is its ideal generated by elements $f_m(\hat{x}^1, \ldots, \hat{x}^n)$, $m \in \mathcal{M}$.

Let us denote by π the natural projection $\hat{R} \to R$ such that $\pi(\hat{x}^i) = x^i$, $\pi(1) = 1$. Since $R_{n\times n} = R \otimes F_{n\times n}$, π defines an epimorphism $\hat{\pi} : \hat{R}_{n\times n} \to R_{n\times n}$ by the formula $\hat{\pi} = \pi \otimes id$, where $id : F_{n\times n} \to F_{n\times n}$ is the identity map.

If $A : R \to R_{n\times n}$ is any homomorphism of algebras, then we have the following diagram of algebra homomorphisms

$$\begin{array}{ccccc} \hat{R} & \xrightarrow{\hat{A}} & \hat{R} \otimes F_{n\times n} & = & \hat{R}_{n\times n} \\ \pi \downarrow & & \downarrow \pi \otimes id & = & \downarrow \hat{\pi} \\ R & \xrightarrow{A} & R \otimes F_{n\times n} & = & R_{n\times n} \end{array} \tag{9}$$

Let us choose for any generator x^i an arbitrary element $\hat{A}^i \in \hat{R}$ such that $\hat{\pi}(\hat{A}^i) = A^i$ (recall that $\hat{\pi}$ is epimorphism). Then the map $\hat{x}^i \mapsto \hat{A}^i$ can be extended to an algebra homomorphism $\hat{A} : \hat{R} \to \hat{R}_{n\times n}$ (recall that $\hat{x}^1, \ldots, \hat{x}^n$ are free variables). That homomorphism completes (9) to a commutative diagram. For any relation $f_m(\hat{x}^1, \ldots, \hat{x}^n)$ we have:

$$\hat{\pi}(\hat{A}(f_m(\hat{x}^1, \ldots, \hat{x}^n))) = A(\pi(f_m(\hat{x}^1, \ldots, \hat{x}^n))) = 0. \tag{10}$$

Furthermore,

$$ker\hat{\pi} = ker(\pi \otimes id) = ker\pi \otimes F_{n\times n} = I_{n\times n},$$

and finally,

$$\hat{A}(f_m(\hat{x}^1, \ldots, \hat{x}^n)) \in ker\hat{\pi} = \hat{I}_{n\times n}.$$

Theorem 1.6 claims that for the homomorphism $\hat{A} : \hat{R} \to \hat{R}_{n\times n}$ there exists a unique free differential $\hat{d}$ of the free algebra $\hat{R}$.

Definition 1.7 The differential $\hat{d}$ is called a *cover* differential with respect to the homomorphism $A : R \to R_{n\times n}$.

Thus we have proved:

Theorem 1.8 *For any homomorphism $A : R \to R_{n\times n}$ there there exists a cover differential $\hat{d}$ of the free unitary algebra $\hat{R}$.*

Proposition 1.9 *An unitary algebra R with generators $x^1, \ldots, x^n$ and the set of* *defining relations $\{f_m,\ m \in \mathcal{M}\}$ has a free differential with respect to a homomorphism $A : R \to R_{n\times n}$ if and only if*

$$\hat{D}_k(f_m(\hat{x}^1, \dots, \hat{x}^n)) \in \hat{I}$$

where $\hat{D}_k$ are partial derivatives of the cover differential $\hat{d}$, and $\hat{I}$ is the ideal generated by $\{f_m,\ m \in \mathcal{M}\}$.
Proof: Let the free differential exists. We claim that the diagram

$$\begin{array}{ccc} \hat{R} & \xrightarrow{\hat{D}} & \hat{R}^n \\ \pi \downarrow & & \downarrow \pi^n \\ R & \xrightarrow{D} & R^n \end{array}$$

is commutative. Indeed, the difference $\Delta = D \circ \pi - \pi^n \circ \hat{D}$ acts trivially on generators:

$$\Delta_k(\hat{x}^i) = D_k\pi(\hat{x}^i) - \pi\hat{D}_k(\hat{x}^i) = \delta^i_k - \pi(\delta^i_k) = 0$$

Commutativity of (9) implies $A(\pi(f)) = \hat{\pi}(A(f))$ and by (7) we have

$$\Delta(fh) = D(\pi f \cdot \pi h) - \pi^n\hat{D}(fh) =$$

$$= D(\pi f)\pi h + A(\pi f)D(\pi h) - \pi^n(\hat{D} \cdot h + \hat{A}f \cdot \hat{D}h) =$$
$$= \Delta(f)\pi h + A(\pi f)\Delta(h)$$

By evident induction, $\Delta = 0$.
Finally, for any relation f_m we have

$$\pi\hat{D}(f_m(\hat{x}^1, \dots, \hat{x}^n)) = D(\pi(f_m(\hat{x}^1, \dots, \hat{x}^m))) = 0$$

i.e. $\hat{D}_k(f_m) \in ker\pi = \hat{I}$.
Inversely, if $\hat{D}_k(f_m) \in ker\pi = \hat{I}$ then we have

$$\hat{D}(uf_m v) = \hat{D}(u)f_m v + \hat{A}(u)\hat{D}(f_m)v + \hat{A}(u)\hat{A}(f_m)\hat{D}(v)$$

$$\equiv \hat{A}(u)\hat{A}(f_m)\hat{D}(v) \qquad (mod\, \hat{I})$$

By (9) one has $\hat{\pi}\hat{A}(f_m) = A(\pi f_m) = 0$ and $\hat{A}(f_m) \in ker\,\hat{\pi} = \hat{I}_{n\times n}$. Therefore $\hat{D}_k : \hat{R} \to \hat{R}$ induce maps $D_k : \hat{R}/\hat{I} \to \hat{R} \xrightarrow{\pi} R$ in such a way that $D \circ \pi = \pi^n \circ \hat{D}$. Finally, for arbitrary $u = \pi f \in R$ and $v = \pi h \in R$ we have

$$D(uv) = D(\pi f \cdot \pi h) = \pi^n\hat{D}(f)h + \pi^n\hat{A}(f)\hat{D}(h) =$$

$$= D(\pi f)\pi h + A(\pi f)D(\pi h) = D(u)v + A(u)v$$

and by Proposition 1.5 the proposition is proved. □

Corollary 1.10 *Let an unitary algebra R be defined by generators $x^1, \dots, x^n$ and the set of homogeneous relations $\{f_m\}$ of the same degree. If $A : R \to R_{n\times n}$ acts linearly on generators $A(x^j)^i_k = \alpha^{ij}_{kl}x^l$, then for the pair (R, A) there exists a free differential iff for all m*

$$\hat{d}f_m = 0 \quad . \tag{11}$$

Definition 1.11 An ideal $I \neq \hat{R}$ of a free algebra $\hat{R} = F < \hat{x}^1, \ldots, \hat{x}^n >$ is said to be ***consistent*** with a homomorphism $A : \hat{R} \longrightarrow \hat{R}_{n \times n}$ if the factor algebra $\hat{R}/I$ has a free differential satisfying the commutation rules

$$x^j dx^i = dx^k \cdot A(x^j)^i_k \ . \tag{12}$$

If an ideal I is A-consistent, then Lemma 1.3 defines a homomorphism $A : r \mapsto A^i_k(r)$ from the factor algebra into the matrix algebra over it. Thanks to Proposition 1.8, it follows that I is A-invariant and A-stable in the sense of the following definition:

Definition 1.12 An ideal J of the algebra $\hat{R}$ is said to be *A-invariant* if $A^i_k(J) \subseteq J$, where $A : r \mapsto A^i_k(r)$ is a homomorphism. An ideal I is said to be *A-stable* if $D_k(I) \subseteq I$ for any of partial derivatives D_k defined by a differential d corresponding to A (see Theorem 1.6).

For any homomorphism A there exists the largest A-consistent ideal $I(A)$ contained in the ideal $\bar{R}$ of polynomials with zero constant terms ($\bar{R} \dot{+} F \cdot 1 = \hat{R}$) – the sum of all consistent ideals of such a type. It is is again A-consistent because a sum of invariant ideals is invariant and a sum of stable ideals is stable one.

Now, we are going to describe the ideal $I(A)$ in the homogeneous case. If a homomorphism A preserves a degree, then it must act linearly on generators $A^i_k(\hat{x}^j) = \alpha^{ij}_{kl} \hat{x}^l$. Therefore, the homomorphism A is defined by the 2-covariant 2-contravariant tensor $A = \alpha^{ij}_{kl}$.

Theorem 1.13 *For any 2-covariant 2-contravariant tensor $A = \alpha^{ij}_{kl}$ the ideal $I(A)$ can be constructed by induction as the homogeneous space $I(A) = I_1(A) + I_2(A) + I_3(A) + \cdots$ in the following way:*

1. $I_1(A) = 0$
2. *Assume that $I_{s-1}(A)$ has been defined and U_s be a space of all polynomials m of degree s such that $D_k(m) \in I_{s-1}(A)$ for all k. $1 \leq k \leq n$. Then $I_s(A)$ is the largest A-invariant subspace of U_s.*

The ideal $I(A)$ is a maximal A–consistent ideal in $\hat{R}$.

Proof: First of all, we should note that the ideal $I(A)$ has to be homogenous (graded). It is sufficient to prove that every A–consistent ideal in $\bar{R}$ is contained in the homogenous one. Since our free algebra $\hat{R} = F \cdot 1 \dot{+} \hat{R}_1 \dot{+} \hat{R}_2 \dot{+} \ldots = F \cdot 1 \dot{+} \bar{R}$ is graded, every element $u \in \bar{R}$ has unique decomposition $u = u_1 + u_2 + \ldots$ into homogenous components u_s. Let J be an arbitrary A–consistent ideal in $\bar{R}$. Define $J_s = \{u_s : u \in J\}$, $s \geq 1$. For $u \in J$ one has $A^i_k(u) = A^i_k(u_1) + A^i_k(u_2) + \ldots \in A^i_k(J) \subseteq J$ and $deg A^i_k(u_s) = deg u_s = s$. Therefore $A^i_k(J_s) \subseteq J_s$. Analogously, $D_k(u) = D_k(u_1) + D_k(u_2) + \ldots \in D_k(J) \subset J$ and $deg D_k(u_s) = s - 1$. So $D_k(J_s) \subset J_{s-1}$ and the sum $J_1 + J_2 + \ldots$ is an A–consistent subset. Similarly, $\hat{R}_t J_s \hat{R}_p \subseteq J_{t+s+p}$, hence $J_1 + J_2 + \ldots$ is an ideal in $\hat{R}$.

Next step is to prove that $I_1(A) + I_2(A) + \ldots$ is an ideal. It is sufficient to show that $I_{s-1}\hat{x}^i + \hat{x}^j I_{s-1} \subseteq I_s \ \forall i, j$. Let Vbe the space generated by the variables $\hat{x}^1, \ldots, \hat{x}^n$. Let us prove by induction that $I_{s-1}V + VI_{s-1} \subseteq I_s$. We have

$$D_k(I_{s-1}V + VI_{s-1}) \subseteq D_k(I_{s-1})V + A^j_k(I_{s-1})D_j(V) + D_k(V)I_{s-1} +$$

$$+A^j_k(V)D_j(I_{s-1}) \subseteq I_{s-2}\cdot V + I_{s-1} + I_{s-1} + V\cdot I_{s-2} \subset I_{s-1}$$

It follows that $I_{s-1}V + VI_{s-1} \subseteq U_s$. Finally, the space $I_{s-1}V + VI_{s-1}$ is A^j_k-invariant as so are I_{s-1} and V. Therefore $I_{s-1}V + VI_{s-1} \subset I_s$ and I is an ideal.
Let now $J = J_1 + J_2 + \ldots$ be an arbitrary (graded) $A-$ consistent ideal in $\bar{R}$. We are going to prove by induction that $J_s \subseteq I_s$. Let $u = \beta_k\hat{x}^k \in J_1$. Then $\beta_k x^k = 0$ in factor-ring $\hat{R}/J$. Therefore $\beta_k dx^k = 0$ and $\beta_k = 0$ by uniqueness condition. So $u = 0$ in the free algebra and $0 = J_1 = I_1$.
Let $J_{s-1} \subseteq I_{s-1}$. By the Proposition 1.9 one has $D_k(J_s) \subset J$. All elements from $D_k(J_s)$ have degree equal to $s-1$. Therefore $D_k(J_s) \subseteq J_{s-1} \subseteq I_{s-1}$ and by the definition of U_s we have $J_s \subseteq U_s$. Finally, J_s is A-invariant space and by the definition of the space I_s we obtained $J_s \subseteq I_s$. It means that $I_1 + I_2 + \ldots$ is a maximal A-consistent ideal in $\bar{R}$, so $I_1 + I_2 + \ldots = I(A)$. Now we are going to prove that $I(A)$ is a maximal A-consistent ideal in $\hat{R}$.
Let $J \supseteq I(A)$ be a consistent (non-homogeneous) ideal in $\hat{R}$. Let $f = f_m + f_{m-1} + \ldots + \alpha\cdot 1$be a polynomial of minimal possible degree from $J\setminus I(A)$ with homogeneous components f_s. The polynomials $D_k f$ have less degree and belong to J. It implies that $D_k f \in I(A)$. By the homogeneity of $I(A)$ one has that all the homogeneous components $D_k f_s$ belong to $I_s(A)$. Therefore $f_s \in U_s$. In particular $D_k(f_1) = 0$ and $f_1 = 0$.
Analogously $D_k(A^i_j(f_s)) \in I(A)$ for i, j, k, s. It means that invariant subspace generated by f_s is contained in U_s and therefore $f_s \in I(A) \subset J$, which implies that $\alpha\cdot 1 = f - f_m - f_{m-1} - \ldots - f_2 \in J$ and $\alpha = 0$. So $f = f_m + \ldots + f_2 \in I(A)$.
□

NOTE: It should be noted that the ideal $I(A)$ can be not the only maximal $A-$consistent ideal in $\hat{R}$.

Let us denote by $\hat{R}_A$ the factor algebra $\hat{R}/I(A)$. In some sense $\hat{R}_A$ is an *optimal* algebra which has a free differential with respect to the commutation rule A. Indeed, Theorem 1.13 shows in particular that if a homogeneous element is such that all elements of the invariant subspace generated by it have all partial derivatives equal to zero, then that element vanishes in the optimal algebra.

We also have proved that there exists maximal algebra which has a free differential with any given commutation rule (this is the free algebra, see Theorem 1.6). Of course, it is very interesting to consider a number of concrete commutation rules A and related algebras $\hat{R}_A$.

Example 1. Let us consider the diagonal commutation rule: $x^j dx^i = dx^i \cdot q^{ij}x^j$, with the symmetry condition $q^{ij}q^{ji} = 1, i \neq j$. If none of the coefficients q^{ij} is a root of a polynomial of the type $\lambda^{[m]} \doteq \lambda^{m-1} + \lambda^{m-2} + \cdots + 1$, then the optimal algebra $\hat{R}_A$ is equal to $F < x^1, \ldots, x^n > /\{q^{ij}x^i x^j = x^j x^i,\ i<j\}$.
If $(q^{ii})^{[m_i]} = 0,\ 1 \le i \le s$ with minimal m_i then
$\hat{R}_A = F < x^1, \ldots, x^n > /\{q^{ij}x^i x^j = x^j x^i,\ i<j,\ (x^i)^{m_i} = 0,\ 1 \le i \le s\}$.

Example 2. Let $A = 0$ i.e. $x^i dx^j = 0$. Then $\hat{d}$ is a homomorphism of right modules and the optimal algebra is free $\hat{R}_A = \hat{R}$.

Example 3. Let $x^i dx^j = -dx^i \cdot x^j$. Then the optimal algebra is the smallest possible algebra generated by the space V i.e. $\hat{R}_A = F < x^1, \ldots, x^n > /\{x^i x^j = 0\}$.

Example 4. Let $x^1 dx^1 = dx^1 \cdot (\alpha_2 x^2 + \cdots + \alpha_n x^n)$ and $x^i dx^j = -dx^i \cdot x^j$ if

$i \neq 1$ *or* $j \neq 1$. Then the optimal algebra is almost isomorphic to the ring of polynomials in one variable. More precisely, $\hat{R}_A = F < x^1, \ldots, x^n > /\{x^i x^j = 0,$ *unless* $i = j = 1\}$.

Example 5. If $n = 2$ and $x^1 dx^1 = dx^1 \cdot \mu x^2$, $x^1 dx^2 = -dx^1 \cdot x^2$, $x^2 dx^1 = -dx^2 x^1$, $x^2 dx^2 = dx^2 \cdot \lambda x^1$, then the optimal algebra is isomorphic to the direct sum of two copies of the polynomial algebra $\hat{R}_A = F < x^1, x^2 > /\{x^1 x^2 = x^2 x^1 = 0\}$.

Finally, we can formulate result which describes numbers of commutation rules in two variables for which the optimal algebra is commutative.

Theorem 1.14 *In the two variable case, the following five series have commutative optimal algebra:*

1. $x^1 dx^1 = dx^1 \cdot u + dx^2 \cdot s, \quad x^1 dx^2 = dx^1 \cdot w + dx^2 \cdot (\lambda s + x^1),$

$$x^2 dx^1 = dx^1 \cdot (w + x^2) + dx^2 \cdot (\lambda s),$$

$$x^2 dx^2 = dx^1 \cdot (\lambda w) + dx^2 \cdot (\lambda^2 s - \lambda u + w + \lambda x^1 + x^2);$$

2. $x^1 dx^1 = dx^1 \cdot (x^1 + \lambda w + s) + dx^2 \cdot w, \quad x^1 dx^2 = dx^1 \cdot \gamma w + dx^2 \cdot (x^1 + s),$

$$x^2 dx^1 = dx^1 \cdot (x^2 + \gamma w) + dx^2 \cdot s, \quad x^2 dx^2 = dx^1 \cdot \gamma s + dx^2 \cdot (x^2 + \gamma w - \lambda s);$$

3. $x^1 dx^1 = dx^1 \cdot (x^1 + \gamma w), \quad x^1 dx^2 = dx^1 \cdot w + dx^2 \cdot x^1,$

$$x^2 dx^1 = dx^1 \cdot (x^2 + w), \quad x^2 dx^2 = dx^1 \cdot s + dx^2 \cdot (x^2 + w - \gamma s);$$

4. $x^1 dx^1 = dx^1 \cdot u, \quad x^1 dx^2 = dx^2 \cdot x^1,$

$$x^2 dx^1 = dx^1 \cdot x^2, \quad x^2 dx^2 = dx^2 \cdot v;$$

5. $x^1 dx^1 = dx^1 \cdot u, \quad x^1 dx^2 = dx^2 \cdot u,$

$$x^2 dx^1 = dx^1 \cdot x^2 + dx^2 \cdot (u - x^1), \quad x^2 dx^2 = dx^1 \cdot w + dx^2 \cdot v.$$

where u, v, w, s *are arbitrary elements from the space* V *and* λ, γ *are parameters from the base field.*

Remark. The above series are not independent. For example, the standard Newton–Leibniz calculus ($x^i dx^j = dx^j \cdot x^i$) belongs, as a special case to each of them, by putting $s = w = 0, u = x^1, v = x^2$. More detailed discussion of the above examples and classification theorem for calculi with a commutative optimal algebra will be given elsewhere [2].

References

1. J. Wess and B. Zumino, *Covariant differential calculus on the quantum hyperplane*, Nuclear Physics B (*Proc. Supl. in Honour of R. Stora*) **18B** (1990), p. 303 ;
2. A. Borowiec, V. K. Kharchenko and Z. Oziewicz, *Differential with Uniqueness Property* – in preparation.

ON PRIMITIVE JORDAN BANACH ALGEBRAS

MIGUEL CABRERA GARCIA, ANTONIO MORENO GALINDO and ANGEL RODRIGUEZ PALACIOS
Departamento de Análisis Matemático, Facultad de Ciencias
Universidad de Granada, 18071-Granada, Spain

Abstract. We give a description of primitive Jordan Banach algebras J for which there exists an associative primitive algebra A such that J is a Jordan subalgebra of the two-sided Martindale ring of fractions $Q_S(A)$ of A containing A as an ideal. Precisely, we prove that there exists a Banach space X and a one-to-one homomorphism ϕ from $Q_S(A)$ into the Banach algebra $BL(X)$ of all bounded linear operator on X such that $\phi(A)$ acts irreducibly on X and the restriction of ϕ to J is continuous.

1. Introduction

Since the publication of Zel'manov prime theorem for Jordan algebras [17], several "normed" versions of it have appeared in the literature. Zel'manovian methods have been applied to normed simple algebras with a unit [3], prime JB- and JB^*-algebras [6] and nondegenerate ultraprime Jordan Banach algebras [4]. Following this line of work, we begin in this note with the consideration of the Zel'manovian treatment of primitive Jordan Banach algebras in the way suggested in [13]. The starting point in this direction would be the classification theorem of primitive Jordan algebras, provided independently by A. Anquela, F. Montaner and T. Corts [1] and V. G. Skosyrsky [15]. Acording to this theorem the primitive Jordan algebras over a field K are the following:

1. The simple exceptional 27-dimensional Jordan algebras over a field extension Γ of K.
2. The Jordan algebras of a nondegenerate symmetric bilinear form on a vector space X over a field extension Γ of K with $dim_\Gamma(X) \geq 2$.
3. Jordan subalgebras of $Q_s(A)$ containing A as an ideal, where A is a primitive associative algebra over K.
4. Jordan subalgebras of $Q_s(A)$ contained in $H(Q_s(A), *)$ and containing $H(A, *)$ as an ideal, where A is a primitive associative algebra over K with a linear algebra involution *.

Since the algebras in the case 1 and 2 in the statement have a unit, the field Γ which arises there can be imbedded into the centre of the algebra. Therefore, as a consequence of the Gelfand-Mazur theorem, all the complex primitive Jordan Banach algebras in these two first cases are central simple algebras (over $\mathbb{C}$), and so these algebras are the following:

S. González (ed.), Non-Associative Algebra and Its Applications, 54–59.

1. The simple exceptional 27-dimensional complex Jordan algebra $M_3^8(C)$ of all hermitian 3x3 matrices over the complex octonions.
2. The Jordan Banach algebras of a continuous nondegenerate symmetric bilinear form on a complex Banach space with $dim(X) \geq 2$.

Our main result, Theorem 1, gives a precise description of complex primitive Jordan Banach algebras J that are in case 3 in the above theorem, asserting that in such a case J can be seen "well" imbedded in the Banach algebra $BL(X)$ of all bounded linear operators on a suitable Banach space X. A partial result concerning the "purely" hermitian case is also included.

2. The main result

We will deal with (linear) Jordan algebras, i. e., algebras over a field of characteristic not two satisfying $a.b = b.a$ and the Jordan intentity $(a^2.b).a = a^2.(b.a)$.

We recall that every associative algebra A (whit product denoted by yuxtaposition) gives rise to a Jordan algebra A^+ under the new product defined by $a.b = \frac{1}{2}(ab + ba)$. Subalgebras of A^+ are called Jordan subalgebras of A.

Also we recall that if A is a prime associative algebra then the Martindale algebra of symmetric quotients of A, $Q_S(A)$, is the maximal algebra extension Q of A satisfying the following conditions:

1. for each q in Q there is a nonzero ideal I of A such that qI and Iq are contained in A, and
2. if q is in Q and I is a nonzero ideal of A satisfying $qI = 0$, then $q = 0$.

Theorem 1. *Let J be a complex Jordan Banach algebra and assume that there exists a primitive associative algebra A such that J is a Jordan subalgebra of $Q_S(A)$ containing A as an ideal. Then there exists a Banach space X and a one-to-one homomorphism ϕ from $Q_S(A)$ into the Banach algebra $BL(X)$ of all bounded linear operator on X such that $\phi(A)$ acts irreducibly on X and the restriction of ϕ to J is continuous.*

A first step for the proof of the theorem is a purely algebraic result from which it follows in particular the existence, for each prime associative algebra A, of a nonzero ideal I of A such that xI and Ix are contained in A whenever x lies in some Jordan subalgebra of $Q_S(A)$ containing A as an ideal.

Proposition 1. *Let B be an associative algebra and let A be a subalgebra of B. If J is a Jordan subalgebra of B containing A as an ideal, then $A^2J \subseteq A$ and $JA^2 \subseteq A$.*

Proof. We begin by observing that a^2x lies in A whenever a is in A and x is in J. This is a consequence of the fact that A is an ideal of J and the equalities $a^2x = a^2.x + \frac{1}{2}[a^2, x] = a^2.x + [a, a.x]$. By simple linearization, also $(a.b)x$ lies in A whenever a and b are in A and x is in J. Now, to prove $A^2J \subseteq A$, it is enouhg to see that $[a, b]x$ also lies in A. This follows from $[a, b]x = [a, b].x + \frac{1}{2}[[a, b], x] = [a, b].x + 2(b, x, a)^+$, where $(a, b, c)^+ := (a.b).c - a.(b.c)$. Analogously $JA^2 \subseteq A$.

Under the assumptions of the above proposition one can consider for each b in A^2 the mappings $\lambda_b : x \rightarrow bx$ and $\rho_b : x \rightarrow xb$ from J into A. Our next goal will be to show that, if J is semiprime and complete normed, then these mappings are continuous.

Proposition 2. *Let J be a semiprime Jordan Banach algebra, B an associative algebra, A be a subalgebra of B, and assume J can be seen as a Jordan subalgebra of B containing A as an ideal. Then for each b in A^2 the mappings λ_b and ρ_b from J into A defined by $\lambda_b(x) := bx$ and $\rho_b(x) := xb$ are continuous.*

Proof. In view of Proposition 1, for b in A^2, the mapping $D_b : J \rightarrow B$ defined by $D_b(x) := [b, x]$ actually is A-valued, hence it is a derivation of the Jordan algebra J. Since $D_b^2(x) = [b,[b,x]] = -4(x,b,b)^+$ for all x in J, D_b is a derivation of the semiprime complete normed algebra J such that D_b^2 is continuous. Since in a semiprime complete normed algebra, a derivation is continuous if its square is continuous [12; Corolario II.5] we have that D_b is continuous. Then it is enough to notice that for all x in J $\lambda_b(x) = b.x + \frac{1}{2}D_b(x)$ and $\rho_b(x) = b.x - \frac{1}{2}D_b(x)$.

Another tool necessary for the proof of Theorem 1 is the following proposition in which we collect in a Jordan context a result whose associative forerunnner is well-know [11; Theorem 2.2.6]. As usual, for a,b,x in a Jordan algebra J we write $U_{a,b}(x) := a.(x.b) + b.(x.a) - (a.b).x$ and $U_a := U_{a,a}$. If J is a Jordan subalgebra of an associative algebra, then $U_{a,b}(x) = \frac{1}{2}(axb+bxa)$ and $U_a(x) = axa$. Also we recall that an inner ideal of a Jordan algebra J is a subspace I such that $U_I(J') \subseteq I$ where J' is the unital hull of J. An inner ideal I of J is said to be e-modular for e in J if $U_{1-e}(J) \subseteq I$, $U_{1-e,I}(J') \subseteq I$ and $e - e^2 \in I$. In such a case the element e is called a modulus for I. We will say an inner ideal I is e-maximal if it is maximal among all proper e-modular inner ideals. We say I is maximal modular if it is e-maximal for some e.

Proposition 3. *Let X be a vector space, let J be a Jordan algebra of linear operators on X, and assume that J contains as an ideal some (associative) algebra A of linear operators on X acting irreducibly on X. Then for every nonzero element u in X the linear mapping $E_u : F \rightarrow F(u)$ from J into X is onto, and its kernel $ker_J(u) := \{F \in J : F(u) = 0\}$ is a maximal modular inner ideal of J with set of modulus $id_J(u) := \{F \in J : F(u) = u\}$. If moreover we assume that J is a Jordan Banach algebra, then $ker_J(u)$ is a closed subspace of J and therefore X becomes in a natural way a Banach space under the norm $|x| := Inf\{\|F\| : F \in J, F(u) = x\}$.*

Proof. Fix a nonzero element u in X. Since $A \subseteq J$ and A acts irreducibly on X the linear mapping $E_u : J \rightarrow X$ is onto and $id_J(u)$ is nomempty. It is routine to prove that $ker_J(u)$ is a modular inner ideal of J with set of modulus $id_J(u)$. To prove that $ker_J(u)$ is a maximal modular inner ideal of J, we fix H in $A \cap id_J(u)$, we consider an H-modular inner ideal P of J containing $ker_J(u)$ and we will prove that either $P = J$ or $P = ker_J(u)$. Since A is an ideal of J containing H, $P \cap A$ is a H-modular inner ideal of A^+ containing $ker_A(u) := \{F \in A : F(u) = 0\}$. But $ker_A(u)$ is a maximal H-modular left ideal of A, hence a maximal H-modular inner ideal of A^+ ([8; Example 3.3]). It follows either $P \cap A = A$ or $P \cap A = ker_A(u)$. If $P \cap A = A$, then P contains A and hence it also contains H, and so $P = J$ because it does not exclude some of its modula [8; Proposition 3.1]. If $P \cap A = ker_A(u)$, and if we assume $ker_J(u) \neq P$, then we can fix F in $P \setminus ker_J(u)$ and consider G in A such that $GF(u) = u$. Then $FGF(u) = F(GF(u)) = F(u) \neq 0$ and $FGF(u) = 0$ because $U_P(A) \subseteq P \cap A = ker_A(u)$, a contradiction. Finally, since maximal modular inner ideals of a Jordan Banach algebra are closed (see [5; Lemma 6.5] or [10; Proposition 6]), if J is a Jordan Banach algebra, then $ker_J(u)$ is a closed subspace of J, and so, via the canonical linear bijection from $J/ker_J(u)$ onto X induced by E_u, X becomes a Banach space under the norm given in the statement.

Now we are ready to prove our main result.

Proof of the theorem 1. Since A is a primitive algebra there exists a complex vector space X such that A can be seen as an algebra of linear operators on X acting irreducibly on X. By [16; Theorem 3.1] also $Q_s(A)$ can be seen as an algebra of linear operators on X including A, and so acting irreducibly on X. If we fix a nonzero element u in X, then by Proposition 3 the vector space X can and will be seen as a Banach space for the norm $|x| := Inf\{\|F\| : F \in J, F(u) = x\}$. Given G in A^2 and F in J, by Propositions 1 and 2, we have that GF lies in A and $\|GF\| \leq \|\lambda_G\| \, \|F\|$. Hence, for G in A^2, x in X, and F in J with $F(u) = x$, we have $|G(x)| = |GF(u)| \leq \|GF\| \leq \|\lambda_G\| \, \|F\|$, and therefore $|G(x)| \leq \|\lambda_G\| \, |x|$. So every element G in A^2 is a continuous linear operator on X. Now, given G in $Q_s(A)$ we will prove the continuity of G by showing that G has closed graph. By definition of $Q_s(A)$, we can choose a nonzero ideal $\mathcal{P}$ of A such that $\mathcal{P}G \subseteq A$, and note that from the irreducible actuation of A on X it follows that $\mathcal{P}^2$ is a nonzero ideal of A acting irreducibly on X. Since $\mathcal{P}^2$ and $\mathcal{P}^2 G$ are contained in A^2 and therefore in $BL(X)$, given a null sequence $\{x_n\}$ of X such that $\{G(x_n)\} \to y$, it follows that for all F in $\mathcal{P}^2$ the sequence $\{(FG)(x_n)\} = \{F(G(x_n))\}$ converges to 0 and to $F(y)$. Therefore $F(y) = 0$, and so $y = 0$.

Now only remains to prove that the embeding $(J, \|.\|) \hookrightarrow (BL(X), |.|)$ is continuous. Given F in $ker_J(u)$, x in X, and G in J with $G(u) = x$, we have

$$|F(x)| = |(FG + GF)(u)| \leq \|FG + GF\| = \|2\, F.G\| \leq 2\|F\| \, \|G\|,$$

so $|F(x)| \leq 2\|F\| \, |x|$, and so $|F| \leq 2\|F\|$. Finally we will use this last inequality together with the closed graph theorem to prove the continuity of the embedding $(J, \|.\|) \hookrightarrow (BL(X), |.|)$. Let $\{T_n\}$ be a null sequence in $(J, \|.\|)$ such that $\{T_n\}$ $|.|$-converges to some T in $BL(X)$. Then, for each G in A^2 such that $G(u) = u$, the sequence $\{T_n - T_n G\}$ $\|.\|$-converges to 0 (Proposition 2) and $|.|$-converges to $T - TG$. But, since $T_n - T_n G$ lies in $ker_J(u)$, we have $|T_n - T_n G| \leq 2\|T_n - T_n G\|$, and therefore $T - TG = 0$ for all G in A^2 with $G(u) = u$. Since A^2 acts irreducibly on X, the algebra $\mathcal{D}$ of all (possibly discontinuous) linear operators on X which commute with every element in A^2 is a division algebra [2; Proposition 24.6] and a normed algebra for a suitable norm [13; Lemma B.13], hence $\mathcal{D} \simeq \mathcal{C}$ by Gelfand-Mazur's theorem. Now, for each x in X $\mathcal{C}$-independent with u, the Jacobson density theorem assures the existence of elements G in A^2 such that $G(u) = u$ and $G(x) = 0$, and for such a G we have $T(x) = TG(x) = T(0) = 0$. It follows that T is zero on $X \setminus \mathbb{C}u$. Since obviously we may assume $dim(X) \geq 2$, we can choose an element x in X $\mathbb{C}$-independent with u. Then, writing $u = (u + x) - x$, and taking into account that $u+x$ and x both are $\mathbb{C}$-independent with u, we have also $T(u) = T(u+x) - T(x) = 0$. Therefore $T = 0$.

3. A partial result concerning the "purely" hermitian case

Recall that an associative algebra with involution $(A, *)$ is said to be a *-tight envelope of a Jordan algebra J included in $H(A, *)$ if A is generated by J and every nonzero *-ideal of A meets J.

Theorem 2. *Let J be a complex Jordan Banach algebra, and assume that there exists a complex primitive associative algebra A with an algebra involution $*$ such that $J = H(A, *)$ and $(A, *)$ is a *-tight envelope of J. Then there exists a Banach space X such that A can be seen as an algebra of bounded linear operators acting irreducibly on X and the embedding $J \hookrightarrow BL(X)$ is continuous.*

Proof. Since J is a semiprime Jordan Banach algebra and $(A, *)$ is a *-tight envelope of J, by [14; Theorem 2] (up to equivalent renorming of J if necesary) there exists an algebra norm $||.||$ on A which extends the norm of J, makes $*$ isometric, and, if $(\widehat{A}, *)$ denotes the $||.||$-completion of $(A, *)$, then $J = H(\widehat{A}, *)$ and every nonzero *-ideal of $\widehat{A}$ meets J. First we note that, if A is commutative, then A is a division algebra, so A equals $\mathbb{C}$ by the Gelfand-Mazur theorem, and so the theorem is true in this case. Assume that A is not commutative. Since A is generated by J and $J = H(\widehat{A}, *)$, A is a Lie ideal of $\widehat{A}$ [7; Example 2 in page 59]. On the other hand the product of two arbitrary nonzero $*$-ideals of $\widehat{A}$ is nonzero (hence $\widehat{A}$ is semiprime) because every nonzero *-ideal of $\widehat{A}$ meets $H(\widehat{A}, *)$ and $H(\widehat{A}, *)$ is a prime Jordan algebra. It follows from [7; Theorem 2.1.2] that A contains a nonzero ideal P of $\widehat{A}$. Replacing P by $P + P^*$ if necessary we can assume that P is a nonzero *-(hence essential) ideal of $\widehat{A}$. Then we can identify the Martindale algebra of symmetric quotients of P, with that of A and $\widehat{A}$ [9; Proposition 3.1]. Now, if we see A as an algebra of operators acting irreducibly on a suitable complex vector space X, then by [16] $\widehat{A}$ can also be seen as an algebra of operators acting irreducibly on X. Since $\widehat{A}$ is a Banach algebra, by [11; Theorem 2.2.6] X appears as a Banach space in such a way that all the elements in $\widehat{A}$ are bounded operators and the embedding of $\widehat{A}$ into $BL(X)$ is continuous.

Acknowledgements

The authors want to tank J. Martínez for several interesting suggestions concerning the topics in this note.

References

1. J. A. Anquela, F. Montaner and T. Cortés, On primitive Jordan algebras, J. Algebra (to appear).
2. F. F. Bonsall and J. Duncan, *Complete normed algebras,* Springer-Verlag, Berlin 1973.
3. M. Cabrera and A. Rodríguez, Zel'manov's theorem for normed simple Jordan algebras with a unit, Bull. London Math. Soc. **25** (1993), 59-63.
4. M. Cabrera and A. Rodriguez, Nondegenerately ultraprime Jordan Banach algebras: A Zel'manovian treatment, Proc. London Math. Soc. (to appear).
5. A. Fernández, Modular annihilator Jordan algebras, Commun. Algebra **13** (1985), 2597-2613.
6. A. Fernández, E. García and A. Rodríguez, A Zel'manov prime theorem for JB^*-algebras, J. London Math. Soc., **46** (1992), 319-335.
7. I. N. Herstein, *Rings with involution,* Chicago Lectures in Mathematics, The University of Chicago Press, Chicago 1976.
8. L. Hogben and McCrimmon, Maximal modular inner ideals and the Jacobson radical of a Jordan algebra, J. Algebra **68** (1981), 155-169.
9. D. S. Passman, Computing the symmetric ring of quotients, J. Algebra **105** (1987), 207-235.
10. J. Prez, L. Rico and A. Rodríguez, Full subalgebras of Jordan Banach algebras and algebra norms on JB^*-algebras, Proc. Amer. Math. Soc. (to appear).
11. C. E. Rickart, *General theory of Banach algebras.* Krieger, New York 1974.
12. A. Rodríguez, La continuidad del producto de Jordan implica la del ordinario en el caso completo semiprimo, In Contribuciones en Probabilidad, Estadística Matemática, Enseanza de la Matemática y Análisis 280-288, Secretariado de Publicaciones de la Universidad de Granada, 1979.
13. A. Rodríguez, Jordan structures in Analysis, In Proceedings of the 1992 Oberwolfach Conference on Jordan algebras (to appear).
14. A. Rodríguez, A. Slin'ko and E. Zel'manov, Extending the norm from Jordan Banach algebras of hermitian elements to their associative envelopes, Commun. Algebra (to appear).
15. V. G. Skosyrsky, Primitive Jordan algebras, Algebra & Logica, no. 2, 1992.
16. E. A. Whelan, The symmetric ring of quotients of a primitive ring is primitive, Commun. Algebra **18**, (1990), 615-633.
17. E. Zel'manov, On prime Jordan algebras II, Siberian Math. J. **24** (1983), 89-104.

ZEL'MANOV'S THEOREM FOR NONDEGENERATELY ULTRAPRIME JORDAN-BANACH ALGEBRAS

MIGUEL CABRERA GARCIA and ANGEL RODRIGUEZ PALACIOS
Departamento de Análisis Matemático, Facultad de Ciencias
Universidad de Granada, 18071-Granada, Spain

Abstract. This note is a review of the paper of the authors [8]. In this paper, nondegenerately ultraprime normed Jordan algebras are introduced and studied under the light of Zel'manov's prime theorem. With this aim, ultra-$*$-prime normed associative algebras are introduced and, for such an algebra A, a normed version of the symmetric Martindale ring of quotients of A (denoted by $Q_b(A)$) is provided. The main result asserts that, up to bicontinuous isomorphisms, the non-degenerately ultraprime Jordan-Banach algebras are the following: the only simple exceptional 27-dimensional Jordan complex algebra, the Jordan-Banach algebras of a regular continuous non-degenerate symmetric bilinear form on a complex Banach space, and the closed Jordan subalgebras of $Q_b(A)$ contained in $H(Q_b(A), *)$ and containing $H(A, *)$ as an ideal, where A is an ultra-$*$-prime associative complex Banach algebra such that $H(A, *)$ generates A as a Banach algebra.

The paper [8] tries to be a contribution to the line of work devoted to obtain normed versions of the Zel'manov prime theorem for Jordan algebras [15]. As fore-runners in this direction we know the zel'manovian treatment carried out for prime JB and JB*-algebras in [9], as well as the Zel'manov theorem for simple normed Jordan algebras with a unit given in [6]. The reader is referred to Section F in the survey [13] for a glance at the recent results, problems and future directions in this field. We also comment that the zel'manovian treatment of primitive Jordan-Banach algebras is being attacked (see [4]).

We begin by trying to give a presentation in a uniform way of those "ultra-concepts" involved in our development. We recall that given an ultrafilter $\mathcal{U}$ on a set I and a family $(A_i)_{i\in I}$ of normed algebras, the *(normed) ultraproduct of the family* $(A_i)_{i\in I}$ *with respect to the ultrafilter* $\mathcal{U}$ is defined as the quotient algebra $(A_i)_{\mathcal{U}} := \bigoplus_{i\in I}^{\infty} A_i/N_{\mathcal{U}}$, where $\bigoplus_{i\in I}^{\infty} A_i$ denotes the normed algebra l_∞-sum of the family (A_i) and $N_{\mathcal{U}}$ is the closed ideal of the l_∞-sum given by

$$N_{\mathcal{U}} := \{(a_i)_{i\in I} \in \bigoplus_{i\in I}^{\infty} A_i \ : \ lim_{\mathcal{U}}\|a_i\| = 0\} .$$

If by abuse of notation we denote by (a_i) the canonical projection in $(A_i)_{\mathcal{U}}$ of the element (a_i) in $\bigoplus_{i\in I}^{\infty} A_i$, then it is easy to verify that $\|(a_i)\| = lim_{\mathcal{U}}\|a_i\|$. In the case that for all i in I A_i equals a given normed algebra A, the ultraproduct $(A_i)_{\mathcal{U}}$ is called the *ultrapower* of A (with respect to $\mathcal{U}$) and is denoted by $A_{\mathcal{U}}$. Also we recall that an ultrafilter $\mathcal{U}$ on a set I is said to be *countably incomplete* if it is not closed under countable intersections of their elements.

S. González (ed.), Non-Associative Algebra and Its Applications, 60–65.

Given an algebra property P and fixed a class of normed algebras closed by ultrapower $\mathcal{C}$, we say that a normed algebra A in $\mathcal{C}$ is *ultra*-P if some ultrapower of A with respect to a countably incomplete ultrafilter has the property P. By considering the algebraic property P=PRIME ALGEBRA in the class of all normed algebras we obtain the concept of ultraprime normed (nonassociative) algebra, but if one consider the above algebraic property in the class of all normed associative algebras then one obtain the ultraprime normed associative algebras introduced by M. Mathieu [10]. If for each two elements a and b in an associative algebra, $M_{a,b}$ denotes the linear operator defined by $M_{a,b}(x) := axb$, then it is well-known that an associative algebra is prime if and only if $M_{a,b} = 0$ implies either $a = 0$ or $b = 0$. We emphasize that, by using this fact, M. Mathieu characterized the ultraprimeness of an associative algebra A by the existence of a positive number K such that $K\|a\|\,\|b\| \leq \|M_{a,b}\|$ for all a, b in the algebra.

When one consider the algebraic property P=$*$-PRIME ALGEBRA in the class of all normed associative algebras with an isometric involution the concept of *ultra-$*$-prime normed (associative) algebra* appear. By using the well-known fact that an associative algebra A with linear involution $*$ is $*$-prime if and only if, for a, b in A, the equalities $M_{a,b} = M_{a^*,b} = 0$ imply either $a = 0$ or $b = 0$, it is easy to prove the following characterization: *A normed associative algebra A with an isometric linear involution $*$ is ultra-$*$-prime if and only if there exists a positive number K such that*

$$K\|a\|\,\|b\| \leq Max\{\|M_{a,b}\|, \|M_{a^*,b}\|\}$$

for all a, b in A.

Following ideas previously developped by M. Mathieu for ultraprime normed algebras in [11], we give for ultra-$*$-prime normed associative algebras a construction of a normed version of the symmetric Martindale ring of quotients, which will appear as the appropiate universe for our zel'manovian treatment of nondegenerately ultraprime Jordan algebras of hermitian type.

Theorem 1. *Let $(A, \|.\|)$ be an ultra-$*$-prime associative algebra, K denote a positive number such that*

$$K\|a\|\,\|b\| \leq Max\{\|M_{a,b}\|, \|M_{a^*,b}\|\}$$

for all a, b in A, and $Q(A)$ denote the symmetric Martindale algebra of quotients of A. Let us consider the set $Q_b(A)$ of all elements q in $Q(A)$ for which there exists a nonzero $*$-ideal I of A such that $qI \subseteq A$ and the mapping $\lambda_q^I : x \longrightarrow qx$ from I into A is continuous, denote by $|q|$ the infimum of $\|\lambda_q^I\|$ for all the ideals I as above, and define

$$|||q||| := Inf\{\|x\| + \frac{1}{K}Max\{|q-x|, |(q-x)^*|\} \ : \ x \in A\}.$$

Then $Q_b(A)$ is a unital $*$-subalgebra of $Q(A)$ containing A, and $|||.|||$ is an algebra norm on $Q_b(A)$ extending the norm on A and such that $(Q_b(A), |||.|||)$ is ultra-$*$-prime.

The ultra-$*$-prime algebra $Q_b(A)$ given in the above theorem will be called the *normed Martindale algebra of symmetric quotients of the ultra-$*$-prime algebra A.*

By applying our general method of obtaining "ultraconcepts", from the algebraic property P=PRIME NONDEGENERATE JORDAN ALGEBRA in the class of all normed Jordan algebras we get *nondegenerately ultraprime normed Jordan algebras.* We recall the nice characterization of prime nondegenerate Jordan algebras in terms of U-operators obtained recently by K.I. Beidar, A.V. Mikhalev and A.M. Slin'ko [3]: *A Jordan algebra J is prime and nondegenerate if and only if $U_{a,b} = 0$ implies either $a = 0$ or $b = 0$,* where, as usual, the operator $U_{a,b}$ is defined by $U_{a,b}(x) := (a.x).b+(b.x).a-(a.b).x$. This ideal-free characterization of nondegenerate primeness for Jordan algebras allows us to give the following ultrapower-free characterization of nondegenerately ultraprime normed Jordan algebras: *A normed Jordan algebra J is nondegenerately ultraprime if and only if there exists a positive number K such that*

$$K||a||\,||b|| \leq ||U_{a,b}||$$

for all a, b in J.

At the moment, perhaps the reader thinks that, given an ultra-$*$-prime Banach algebra $(A,*)$ which is generated as a Banach algebra by its hermitian part $H(A,*)$, then closed Jordan subalgebras of $Q_b(A)$ contained in $H(Q_b(A),*)$ and containing $H(A,*)$ as an ideal are examples of nondegenerately ultraprime Jordan-Banach algebras. If you think so, then you are right, but the proof of this fact is not too easy. The reason is that, for an ultra-$*$-prime associative Banach algebra A, the normed Martindale algebra of symmetric quotients $Q_b(A)$ of A may be not complete. Then, to show that closed Jordan subalgebras of $Q_b(A)$ contained in $H(Q_b(A),*)$ and containing $H(A,*)$ as an ideal are complete, we have had to prove the following new result of a purely algebraic type.

Proposition 1. *If A is a $*$-prime associative algebra generated by $H(A,*)$, then there exists a nonzero $*$-ideal I of A such that, for every Jordan subalgebra J of $Q(A)$ contained in $H(Q(A),*)$ and containing $H(A,*)$ as an ideal, we have $JI + IJ \subseteq A$.*

More examples of not necessarily complete nondegenerately ultraprime normed Jordan algebras will be exhibited in what follows. Our next first construction involves the following auxiliary result.

Proposition 2. [1] *Let X, Y and Z be normed spaces, and $F : X \to Y$, $G : X \to Z$ be bounded linear operators. Then*

$$Sup\{\, ||F(x)||\,||G(x)|| \;:\; x \in X,\ ||x|| \leq 1\,\} \geq \frac{1}{4}||F||\,||G||\,.$$

Using these last two propositions we were able to prove the following theorem providing a general method to build nondegenerately ultraprime (complete) normed Jordan algebras from ultra-$*$-prime (complete) normed associative algebras.

Theorem 2. *Let A be an ultra-$*$-prime normed algebra. Then every Jordan subalgebra of $Q_b(A)$ contained in $H(Q_b(A),*)$ and containing $H(A,*)$ (not necessarily as an ideal) is a nondegenerately ultraprime normed Jordan algebra.*

Our second and last construction method of nondegenerately ultraprime normed Jordan algebras consists in determining those normed simple Jordan algebras of a bilinear form which are nondegenerately ultraprime. We recall that the complex simple Jordan algebra $J(X,<.,.>)$ build from a nondegenerate symmetric bilinear form $<.,.>$ on a complex vector space X of dimension ≥ 2 is defined as the vector space $C1 \oplus X$ with Jordan product given by

$$(\alpha\mathbf{1}+\mathrm{x}).(\beta\mathbf{1}+\mathrm{y}) := (\alpha\beta+<\mathrm{x},\mathrm{y}>)\mathbf{1}+(\alpha\mathrm{y}+\beta\mathrm{x}) .$$

It is easy to verify that if $\|.\|$ is an algebra norm on $J(X,<.,.>)$, then X is a closed subspace of J, and $<.,.>$ is continuous on $X\times X$ with $\|<.,.>\|\leq 1$. Pairs $(X,<.,.>)$ as above, where X is a normed space and $<.,.>$ is a continuous nondegenerate symmetric bilinear form on X, will be called *symmetric self-dual normed spaces.* Given such a space $(X,<.,.>)$, we have a canonical continuous imbedding $X \hookrightarrow X'$, where X' denotes the Banach dual space of X, consisting in identifying each element x in X with the bounded linear functional $<x,.>$. A symmetric self-dual normed space $(X,<.,.>)$ will be called *regular* if the above canonical imbedding is topological.

Theorem 3. *Let $J=J(X,<.,.>)$ be the simple quadratic Jordan algebra of the nondegenerate symmetric bilinear form $<.,.>$ on the complex vector space X of dimension ≥ 2, assume that $\|.\|$ is an algebra norm on J, and consider X as a normed subspace of J. Then the following assertions are equivalents:*
i) $(J,\|.\|)$ is nondegenerately ultraprime.
ii) The symmetric self-dual normed space $(X,<.,.>)$ is regular.

Our main result in [8] is the following classification theorem for nondegenerately ultraprime Jordan-Banach algebras.

Theorem 4. Classification theorem for nondegenerately ultraprime Jordan-Banach algebras. *Up to bicontinuous isomorphisms, the nondegenerately ultraprime Jordan-Banach complex algebras are the following:*
i) The simple exceptional 27-dimensional complex Jordan algebra $M_3^8(C)$ of all hermitian 3×3 matrices over the complex octonions.
ii) The Jordan-Banach algebras of the form $J(X,<.,.>)$, where $(X,<.,.>)$ is a regular symmetric self-dual complex Banach space with $dim(X)\geq 2$.
iii) The closed Jordan subalgebras of $Q_b(A)$ contained in $H(Q_b(A),*)$ and containing $H(A,*)$ as an ideal, where A is an ultra-$*$-prime complex Banach algebra such that $H(A,*)$ generates A as a Banach algebra, and $Q_b(A)$ denotes the normed Martindale algebra of symmetric quotients of A.

The proof in [8] of this theorem is quite long and difficult. As we already noted, the proof that all algebras listed above are nondegenerately ultraprime Jordan-Banach algebras is not too easy. However the actually dificult part of the theorem is to prove that all nondegenerately ultraprime Jordan-Banach algebras must arise in the above list. The proof of this part involves Zel'manov's theorem, a norm-extension theorem in a recent paper by A. Rodríguez, A.M. Slin'ko and E.I. Zel'manov [14], as well as the main results in two previous paper of the authors which read as follows:

- *Every (nonassociative) ultraprime normed algebra is centrally closed.* [5]
- *The centre $Z(J)$ of a nondegenerate Jordan algebra J can be characterized by*

$$Z(J) = \{x \in J \ : \ 2U_{x.y} = U_xU_y + U_yU_x \ for \ all \ y \ in \ J\} \ . [7]$$

Also for the proof of this theorem we have needed some new results, among which we emphasize the following one, which gives an "approximate" variant of the fact that, if A is a prime normed associative algebra with a unit $\mathbf{1}$ and an involution $*$, then for all $*$-skew element s such that $s^2 = \mathbf{1}$ we have $\|M_{\mathbf{1}+s,\mathbf{1}-s}\| \geq 4$.

Theorem 5. *Let A be a prime associative Banach complex algebra with a unit $\mathbf{1}$ and a (not necessarily continuous) involution $*$. Then for every $*$-skew element s in A satisfying $\|\mathbf{1} - s^2\| < 1$ we have*

$$\|M_{\mathbf{1}+s,\mathbf{1}-s}\| \geq \left(1 + \sqrt{1 - \|\mathbf{1} - s^2\|}\right)^2 .$$

To conclude this note we review two minor results in the paper [8]. The first one is the Jordan version of an associative result of P. Ara and M. Mathieu [2], and its proof uses the description theorem of prime nondegenerate Jordan-Banach algebras with nonzero socle and minimality of norm topology recently obtained by J. Prez, L. Rico, A. Rodríguez and A.R. Villena [12]. We recall that a normed algebra $(A, \|.\|)$ is said to have *minimality of norm topology* if every algebra norm $|.|$ on A satisfying $m|.| \leq \|.\|$ for some positive number m actually is equivalent to $\|.\|$, while A is said to have *minimum topology* if every algebra norm on A majorises the original norm.

Proposition 3. *For a prime nondegenerate Jordan-Banach complex algebra J with nonzero socle, the following assertions are equivalent:*
i) J is nondegenerately ultraprime.
ii) J has minimality of norm topology.
iii) J has minimun topology.

Our last result improves the one obtained by A. Fernández, E. García and A. Rodríguez [9] asserting that every prime JB*-algebra is (nondegenerately) ultraprime.

Proposition 4. *Every normed ultraproduct of prime JB*-algebras is prime.*

References

1. M.D. Acosta, An inequality for norm of operators. Preprint Universidad de Granada.
2. P. Ara and M. Mathieu, On ultraprime Banach algebras with nonzero socle, Proc. Royal Irish Academy **91A** (1991), 89-98.
3. B.I. Beidar, A.V. Mikhalev, and A.M. Slin'ko, Criteria for primeness of nondegenerate alternative and Jordan algebras, Tr. Mosk. Mat. O-va **50** (1987), 130-137. (Engl. Transl.: Trans. Moscow Math. Soc. (1988), 129-137.)
4. M. Cabrera, A. Moreno, and A. Rodríguez, On primitive Jordan Banach algebras. In Proceedings of the Third international conference on nonassociative algebras and its aplications, Oviedo 1993. Kluwer Academic Publishers (to appear).
5. M. Cabrera and A. Rodríguez, Nonassociative ultraprime normed algebras, Quart. J. Math. Oxford **43** (1992), 1-7.
6. M. Cabrera and A. Rodríguez, Zel'manov theorem for normed simple Jordan algebras with a unit, Bull. London Math. Soc. **25** (1993), 59-63.
7. M. Cabrera and A. Rodríguez, A characterization of the centre of a nondegenerate Jordan algebra, Commun. Algebra **21** (1993), 359-369.
8. M. Cabrera and A. Rodríguez, Nondegenerately ultraprime Jordan-Banach algebras: A Zel'manovian treatment, Proc. London Math. Soc. (to appear).
9. A. Fernández, E. García, and A. Rodríguez, A Zel'manov prime theorem for JB*-algebras, J. London Math. Soc. **46** (1992), 319-335.
10. M. Mathieu, Rings of quotients of ultraprime Banach algebras with applications to elementary operators, Proc. Centre Math. Anal. Austral. Nat. Univer. **21** (1989), 297-317.
11. M. Mathieu, The symmetric algebra of quotients of an ultraprime Banach algebra, J. Austral. Math. Soc. (Series A) **50** (1991), 75-87.
12. J. Pérez, L. Rico, A. Rodríguez, and A.R. Villena, Prime Jordan-Banach algebras with nonzero socle, Commun. Algebra **20** (1992), 17-53.
13. A. Rodríguez, Jordan structures in Analysis, In Proceedings of the 1992 Oberwolfach Conference on Jordan algebras (to appear).
14. A. Rodríguez, A. Slin'ko and E. Zel'manov, Extending the norm from Jordan-Banach algebras of hermitian elements to their associative envelopes, Commun. Algebra (to appear).
15. E.I. Zel'manov, On prime Jordan algebras II, Siberian Math. J. **24** (1983), 89-104.

JORDAN H*-TRIPLE SYSTEMS *

ALBERTO CASTELLÓN SERRANO,
JOSÉ ANTONIO CUENCA MIRA and CÁNDIDO MARTíN GONZÁLEZ
Departamento de Algebra, Geometría y Topología.
Universidad de Málaga, Aptdo. 59, 29080, Málaga. Spain.

Abstract. This work, jointly with [7], gives a complete classification of Jordan H^*-triple systems . There, the infinite–dimensional topologically simple special nonquadratic Jordan H^*-triple systems are fully described in terms of the odd part of a $\mathbb{Z}_2$–graded H^*–algebra. Here we complete the structure theory endowing to any simple finite–dimensional real Jordan triple system, of an H^*–structure, essentialy unique, and determining the ones of quadratic type.

1. Definitions and previous results

The basic references for definitions and notations will be : [7], [9], [6], [3], [2] and [5]. We recall also that an *involution* of a K-triple system $(V, < \;\; >)$ (with $K = \mathbf{R}$ or $\mathbf{C}$) is a map $* : V \to V$ $(x \mapsto x^*)$, which is conjugate-linear if $K = \mathbf{C}$ or K-linear if $K = \mathbf{R}$, and such that $(x^*)^* = x$, $< xyz >^* = < x^*y^*z^* >$ for any $x, y, z \in V$. Let $(V, < \;\; >)$ be a K-triple system ($K = \mathbf{R}$ or $\mathbf{C}$) with involution $*$, whose underlying vector space is a K-Hilbert space, V is said to be an *H^*-triple system* if

$$(< xyz > |t >= (x| < tz^*y^* >) = (y| < z^*tx^* >) = (z| < y^*x^*t >)$$

for arbitrary $x, y, z, t \in V$. The structure of Jordan H^*-triple systems is reduced to the topologically simple case as in [3]. Let V be an H^*-triple system . We define the *opposite* $-V$ (resp. the *twin* $V^\flat$) of V as the H^*-triple system with same Hilbert space and involution as V (resp. with same Hilbert space and triple product as V) and triple product (resp. involution) opposite of V. We define the *polarized* $\mathcal{P}(V)$ of V as the H^*-triple system with Hilbert space $V \perp V$ whose triple product and involution are given by

$$< (x, y)(z, t)(u, v) > := (< xtu >, < yzv >), \qquad (x, y)^* := (y^*, x^*)$$

for all $x, y, z, t, u, v \in V$. Let V be an H^*-triple system endowed with an *isotopy* s, that is an isometric involutive $*$-automorphism of V. We define the *s-isotope* $V^{(s)}$ of V, as the H^*-triple system with same Hilbert space as V but whose triple product and involution are given by $< xyz >^{(s)} := < x, sy, z >$, and $x^{(*,s)} := s(x^*)$. The isotopy and polarization are linked by the following

Theorem 1 ([2]). *Let V be an H^*-triple system , then*
1. $Ann(\mathcal{P}(V)) = \mathcal{P}(Ann(V))$.

* This work has been partially supported by the "Plan Andaluz de Investigación y Desarrollo Tecnológico" and the DGICYT with project no. PS89-0119.

S. González (ed.), Non-Associative Algebra and Its Applications, 66–72.

2. $\mathcal{P}(V) \neq 0$ *is topologically simple iff* V *is topologically simple and nonpolarized.*
3. If V *is topologically simple, then* V *is polarized iff* $V = W^{(s)}$ *for some non topologically simple* H^**-triple system* W *endowed with an isotopy* s*.*
4. $\mathcal{P}(V)$ *is isometrically* $*$*-isomorphic to* $\mathcal{P}(V^{(s)})$ *for each isotopy* s *of* V*.*
5. e) If V *is topologically simple and nonpolarized, then every* $V^{(s)}$ *of* V *is topologically simple and nonpolarized.*

The *metacentroid* $C(V)$ of a Φ–triple system V is the set of all linear maps $T : V \to V$ such that $T < xyz >=< Tx, y, z >=< x, y, Tz >$ for all $x, y, z \in V$. The *centroid* $Z(V)$ of a Φ-triple system V is the subset of $C(V)$ of the maps T such that $T < xyz >=< x, Ty, z >$ for all $x, y, z \in V$. In [4], it is proved that the centroid $Z(V)$ of a topologically simple H^*-triple system V over K, is $\mathbf{C}Id$ if $K = \mathbf{C}$ and $\mathbf{R}Id$ or $\mathbf{C}Id$ if $K = \mathbf{R}$. In the last case ($K = \mathbf{R}$, $Z(V) = \mathbf{C}Id$), the real H^*-triple system V is obtained from a complex structure of H^*-triple system by *reallyfication* of its Hilbert space. For any $T \in C(V)$, we denote by $T^\diamond$ the adjoint operator of T and by T^* the operator given by $T^* : x \mapsto (T(x^*))^*$. In [4] is proved that $[*, \diamond] = 0$ and the elements $T \in C(V)$ are continuous being $Z(V) = Sym(C(V), j)$ where $j = * \circ \diamond$. Moreover either $(C(V), j) \cong (\hat{K} \oplus \hat{K}, ex)$ or $C(V) = \hat{K}$ where $\hat{K} = \mathbf{R}$ or $\mathbf{C}$, $\hat{K} \supset K$. The first possibility holds only if V is polarized.

2. Examples of H^*-triple systems

1. Let V be an H^*-algebra endowed with an isometric involutive $*$-antiautomorphism s. The *isotope* $V^{(s)}$ of V is an H^*-triple system with same Hilbert space as J, triple product given by $(x, y, z) \mapsto < xyz >= (xs(y))z$, and involution $x^{(*,s)} := s(x^*)$. If V is an alternative (resp. Jordan) algebra then $V^{(s)}$ is an alternative (resp. Jordan) triple system. The map s is called an *isotopy* of V.
2. Let A be an associative H^*-triple system or a ternary H^*-algebra (see [8]). the *symmetrized* A^+ of A is defined as the Jordan H^*-triple system with same Hilbert space and involution as A and quadratic operator given by $P(x)y := \frac{1}{2} < xyx >$. The structure theory of associative H^*-triple system and ternary H^*-algebras has been fully achieved in [3] and [8]. We recall that if (D, s) is an associative H^*-algebra endowed with an isometric involutive $*$-antiautomorphism s, $\mathcal{A}$ and $\mathcal{B}$ nonempty sets, and $T_1 = (\lambda_{ij})$, $T_2 = (\mu_{ij})$, are $\mathcal{B} \times \mathcal{B}$ and $\mathcal{A} \times \mathcal{A}$ respectively diagonal matrices with entries in D such that the λ_{ii}'s, μ_{jj}'s are either 1 or -1. The set $\mathcal{M}_{\mathcal{A},\mathcal{B}}(D)$ of the $\mathcal{A} \times \mathcal{B}$ matrices (a_{ij}) with entries in D such that $\sum_{ij} \|a_{ij}\|^2 < \infty$ can be structured as an associative H^*-triple system with the triple product, inner product and involution given by

$$< (a_{ij})(b_{ij})(c_{ij}) >:= (a_{ij})(T_1(s(b_{ji}))T_2(c_{ij})) \tag{1}$$

$$((a_{ij})|(b_{ij})) := \sum (a_{ij}|b_{ij}) \tag{2}$$

$$(a_{ij})^* := T_2(s(a_{ij}^*))T_1. \tag{3}$$

These, will be denoted by $\mathcal{M}_{\mathcal{A},\mathcal{B}}(D, s, T_1, T_2)$. Obviously, the symmetrized of $\mathcal{M}_{\mathcal{A},\mathcal{B}}(D, s, T_1, T_2)$is a Jordan H^*-triple system.

3. The subsystem of the H^*-triple system of $\mathcal{M}_{\mathcal{A},\mathcal{A}}(C,s,T,T)^+$ of the alternate matrices a ($a = -a^t$), where C is a commutative associative H^*-algebra, is a Jordan H^*-triple system which will be denoted by $\mathcal{S}_{\mathcal{A}}(C,s,T)$.
4. Let (D,s) be an unital associative H^*-algebra with isometric involutive $*$-antiautomorphism s and j an isometric involutive $*$-automorphism commuting with s. Let $\mathcal{H}_{\mathcal{A}}(D,s)$ be the H^*-subalgebra of $\mathcal{M}_{\mathcal{A},\mathcal{A}}(D)^+$ (see [10]) of the *hermitian* $\mathcal{A}\times\mathcal{A}$-matrices $x = (x_{ij})$ (i.e. $s(x_{ij}) = x_{ji}$). The map $(x_{ij}) \mapsto (\mathrm{j}(x_{ij}))$ turns out to be an isotopy of $\mathcal{H}_{\mathcal{A}}(D,s)$. Let $T = (\lambda_{ij})$ be an $\mathcal{A}\times\mathcal{A}$ diagonal matrix with entries in D and $\lambda_{ii} \in \{-1,1\}$. The map $x \mapsto S(x) := Tj(x)T$ is also an isotopy of $\mathcal{H}_{\mathcal{A}}(D,s)$. The Jordan H^*-triple system S-isotope of $\mathcal{H}_{\mathcal{A}}(D,s)$ will be denoted by $\mathcal{H}_{\mathcal{A}}(D,s,j,T)$. The above H^*-triple system can be considered as the H^*-subsystem of $\mathcal{M}_{\mathcal{A},\mathcal{A}}(D,j\circ s,T,T)^+$ of the hermitian matrices.
5. Let X be a Hilbert Φ-space ($\Phi = \mathbf{R}$ or $\mathbb{C}$), $\sigma : \Phi \to \Phi$ given by $\sigma = Id_{\mathbf{R}}$ if $\Phi = \mathbf{R}$ and $\sigma = Id_{\mathbb{C}}$ or $^-$ if $\Phi = \mathbb{C}$, $* : X \to X$ an isometric involutive map, conjugate-linear if $(\Phi,\sigma) = (\mathbb{C}, Id)$ and Φ-linear in the other cases, and let $\eta : X \to X$ be a σ-semilinear operator with $\eta^2 = -\gamma Id_X$, $\eta^\diamond = -\gamma\eta$ and satisfying $\eta(x^*) = \gamma(\eta(x))^*$ for any $x \in X$ ($\gamma \in \{-1,1\}$). We can now define in X a structure of Jordan H^*-triple system over $Sym(\Phi,\sigma)$ with inner product $(x,y) \mapsto \frac{1}{2}((x|y) + \sigma(x|y))$, involution $x \mapsto x^*$ and quadratic operator
$$P(x)y := 2[x|y^*]x - [x|\eta(x^*)]\eta(y)$$
where $[x|y] := (x|y)$ if $\sigma = Id$ and $[x|y] := (x|y) - i(ix|y)$ otherwise. These Jordan H^*-triple system will be denoted by $J(X,\Phi,\sigma,\eta,\gamma,*)$.

All of the above Jordan H^*-triple systems are special in the well known sense that they can be embedded in the symmetrized of some associative triple system. The finite-dimensional real simple exceptional Jordan triple systems are described in [13]. We recall that the (real or complex) octonions D can be obtained by the Cayley-Dickson H^*-process (see [2]) from the quaternions D_0, taking $D = D_0 \perp v D_0$ with $v^2 = \gamma$ and $\gamma = \pm 1$. It is known that D can be endowed of a structure $D^{(t)}$ of alternative H^*-triple system which is given in [2] as certain t-isotopes of the H^*-algebra D. On the other hand, if D is a split real or complex octonions algebra and t is an involutive isometric $\mathbf{R}$–linear or conjugate-linear $*$–antiautomorphism which is not the Cayley antiautomorphism, the map $L_v : a + vb \mapsto b + va$ gives an isotopy of $D^{(t)}$ which provides another real alternative H^*-triple system .

6. Let D be some of the octonions real or complex H^*-algebras $\mathbb{O}_S$ (split real octonions), $\mathbb{O}_{\mathbf{R}}$ (division real octonions) and $\mathbb{O}_{\mathbb{C}}$ (complex octonions), and t an isometric involutive $*$-antiautomorphism of D (such a t is fully described, up to isotopy class, in Corollary 7 of [2]. Let T_1 and T_2 be 2×2 and 1×1 respectively diagonal matrices whose diagonal entries lie in $\{-1,1\}$. We denote by $\mathcal{M}_{1,2}(D,t,T_1,T_2)$ the Jordan exceptional H^*-triple system of the 1×2 matrices with entries in D with quadratic operator, inner product and involution given by (1), (2) and (3). If D is a split octonions algebra, t is a $\mathbf{R}$-linear or conjugate–linear involutive isometric $*$-antiautomorphism of D which is not the Cayley antiautomorphism, the maps $R : (x,y) \mapsto (R_v(x), R_v(y))$ and $S : (x,y) \mapsto (U_v(x), L_v(y))$ are isotopies of $\mathcal{M}_{1,2}(D,t,T_1,T_2)$ and therefore we have other exceptional real Jordan H^*-triple systems .

7. Isotopes of the exceptional H^*-algebras $\mathcal{H}_3(D,s)$ are also simple exceptional Jordan H^*-triple systems. Let D be an octonions H^*-algebra and s the Cayley antiautomorphism of D. Let j be an isometric involutive $*$-automorphism of D commuting with s and $T = (\lambda_{ij})$ a 3×3 diagonal matrix with $\lambda_{ii} \in \{-1, 1\}$. The map j is fully described in Lemma 6 of [2]. We denote by $\mathcal{H}_3(D,t,j,T)$ the Jordan H^*-triple system S-isotope of $\mathcal{H}_3(D,s)$ where S is given by $S(x_{ij}) := T(j(x_{ij})T$.

3. Jordan H^*-triple systems of quadratic type

In this section we study nonpolarized topologically simple Jordan H^*-triple systems of quadratic type. Annalogously to Section 2 (b) of [1], if J is a nonpolarized topologically simple Jordan H^*-triple system of quadratic type, there exist:

a) A nondegenerate bilinear form $q : J \times J \to C$ where C is the metacentroid of J which is endowed of an involutive automorphism σ such that $Sym(C,\sigma) = Z$ the centroid of J.

b) A σ-semilinear isomorphism $\eta : J \to \gamma J$ where $\eta^2 = \gamma Id$ and $\gamma^2 = 1$, being

$$P(x)y = q(x,\eta(y))x - q(x)\eta(y) \tag{4}$$

$$q(x,\eta(y)) = \sigma q(\eta(x),y) \tag{5}$$

for any $x, y \in J$.

Lemma 1 *Let V be a nonpolarized topologically simple H^*-triple system (not necessarily a Jordan triple system), (C,σ) the metacentroid of V and $Z = Sym(C,\sigma)$ the centroid of V. Then:*

a) (C,σ) is either $(\mathbf{C}, Id)$, $(\mathbf{C}, ^-)$ or $(\mathbf{R}, Id)$ and V is a C-Hilbert space with the inner product $(x,y) \mapsto [x|y]$ where $[x|y] = (x|y) - i(ix|y)$, if V is a real H^*-triple system with $C = \mathbf{C}$ and $[x|y] = (x|y)$ in the remaining cases.

b) The map $f : V \times V \to (C,\sigma)$ given by $f(x,y) = [x|y^*]$ is a nondegenerate sesquilinear map (relative to σ) and satisfies

$$f(< xyz >,t) = f(x,< tzy >) = \sigma f(y,< ztx >) = f(z,< yxt >)$$

for any $x,y,z,t \in V$.

c) If $g : V \times V \to (C,\sigma)$ is another nondegenerate sesquilinear map satisfying conditions similar to those of f given in b), then there exists a nonzero element $k \in Z$ such that $g = kf$.

The parts a) and b) follows from [4] and a straightforward verification. Now let g be as in c), and $a \in V$. The map $b \mapsto g(a,b)$ is σ-semilinear and therefore has an unique representation $b \mapsto [a'|b^*]$. This allow us to define the map $h : a \mapsto a'$ and we shall prove that h is in the centroid Z of V. Indeed, let $a,b,c,d \in V$. Then

$$[h < abc > |d^*] = g(< abc >,d) = g(a,< dcb >) = [h(a)| < dcb >^*] =$$

$$[h(a)| < d^*c^*b^* >] = [< h(a)bc > |d^*]$$

hence $h < abc >=< h(a)bc >$. Similar computations prove that $h \in Z$ and therefore (see [4]), $h = kId$ for some nonzero $k \in Z$. Thus $g(x, y) = [h(x)|y^*] = kf(x, y)$, and the lemma is proved.

In the conditions of the previous paragraph it can be shown that $g : J \times J \to C$ given by $g(x, y) = q(x, \eta(y))$ satisfies

$$g(< xyz >, t) = g(x, < tzy >) = \sigma g(y, < ztx >) = g(z, < yxt >)$$

and by the previous lemma there exists a nonzero $k \in Z$ such that $q(x, \eta(y)) = k[x|y^*]$. By (5) $\eta^\diamond = \eta^*$ where $\eta^\diamond$ is the adjoint operator of η and η^* is defined by $\eta^* : x \mapsto (\eta(x^*))^*$. On the other hand the operator P is described by

$$P(x)y = k[x|y^*]x - \frac{\gamma k}{2}[x|(\eta(x))^*]\eta(y) \tag{6}$$

Applying $*$ to (6) for any $x, y \in J$ with $[x^*|y] = 0$ we obtain

$$k^\tau[x^*|\eta(x)](\eta(y))^* = k[x^*|\eta^*(x)]\eta(y^*) \tag{7}$$

where τ is the complex conjugation if $(C, \sigma) = (\mathbb{C}, Id)$ and the identity in the remaining cases. If each $x \in J$ is an isotropic vector for g, then $P(x)y = k[x|y^*]x$ and by passing to the twin and to the opposite if it were necessary, we can take $e \in J$ a selfadjoint tripotent with $||e|| = 1$. Then any orthogonal element x to e is zero and $J = Ce$. Suppose now that there exists a nonisotropic vector $x \in J$. Then (7) implies that $\eta^* = \mu\eta$ and

$$\gamma\mu Id = \mu\eta^2 = \eta^*\eta = \eta^\diamond\eta.$$

Thus the positivity of the operator $\eta^\diamond\eta$ implies $\mu \in \mathbb{R}$. Hence k is also a real number. An easy calculation proves finally that $\gamma = \mu$ and we have shown the following

Theorem 2 *Let J be a nonpolarized topologically simple Jordan H^*-triple system of quadratic type. Then J is, up to twins, opposites and a positive factor of the inner product, isometrically $*$-isomorphic to a H^*-triple system $J(X, \mathbb{K}, \sigma, \eta, \gamma, *)$.*

4. Classification of Jordan H^*–triple systems

All through this section J will be a topologically simple Jordan H^*-triple system . As in Theorem E.28 of [14], it can be proved the following

Theorem 3 *Let V be a real triple system with zero annihilator such that its complexification can be structured as an H^*-triple system , then V can also be structured as an H^*-triple system .*

For completeness we disgress to prove this in a more simple way. The fact that the complexified $V_{\mathbb{C}}$ of V has zero annihilator jointly with Theorem 1.6 of [3] allow us to assume that $V_{\mathbb{C}}$ is topologically simple. Consider the mapping $\tau : x + iy \mapsto x - iy$ which is a conjugate–linear automorphism of $V_{\mathbb{C}}$. Aplying the main theorem of [6] and argueing as in the proof of theorem E.28 of [14], we obtain that either $\tau : V_{\mathbb{C}} \to V_{\mathbb{C}}$ or $-\tau : V_{\mathbb{C}}^{\flat} \to V_{\mathbb{C}}$ splits in a unique way $\sigma\tau = F_2 F_1$ ($\sigma \in \{1, -1\}$) where F_2 is an involutive conjugate–linear *-isomorphism and F_1 a positive automorphism of $V_{\mathbb{C}}$ with $(F_1^*)^{-1} = F_1$. We analyze only the case $\sigma = 1$ since the other case is similar. Let G be the only positive automorphism of $V_{\mathbb{C}}$ such that $G^2 = F_1$ (see Proposition 12 of [6]) and $(G^*)^{-1} = G$. It follows from the fact that $\tau^2 = 1$ that $G^{-1} = GF_2F_1F_2$. Let $H = GF_2G$. An easy computation shows that H is a conjugate–linear involutive $*$–automorphism of $V_{\mathbb{C}}$ with $HG = G\tau$. So V is isomorphic to the real H^*-triple system of the H–symmetric elements of $V_{\mathbb{C}}$, and the proof is completed.

Corollary 1 *Any simple finite-dimensional real Jordan (or alternative) triple system can be endowed, up to twins and positive factors of the inner product, with an unique H^*-triple system structure.*

The corollary follows from the above theorem, Corollary 14 of [6] and the well knowns structure theories [12], [11] and [2].

The above Corollary allows the obtention (in the finite–dimensional case) of structure H^*–theories from algebraic structure theories and viceversa. For instance, Theorem 8 of [2] gives a classification of alternative H^*–triple systems. So it allows us to give all the simple real finite–dimensional alternative triple systems. We note that in the last reference, a confussion of versions in the revision on the paper led us to an omission affecting the main result, which must be modified in the following sense: In page 3204 line 20, where it is read "$D = \mathbb{O}_S$ (the split real octonions)", must be read "$D = \mathbb{O}_S$ or $\mathbb{O}_{\mathbb{C}}$ (complex octonions)". And therefore in Theorem 8 page 3202, line 12, where it is read "corollary 7(a)iii) where $D = \mathbb{O}_S$", must be read "corollary 7(a)iii) (resp. 7(c)) where $D = \mathbb{O}_S$ (resp. $D = \mathbb{O}_{\mathbb{C}}$)".

The structure of topologically simple Jordan H^*–triple systems is now easily obtained. In fact, every topologically simple Jordan H^*–triple system J is trivially prime nondegenerate and, by Zel'manov's Theory ([15], [16] and [17]), is finite–dimensional over $\mathbb{R}$ or $\mathbb{C}$ if it is exceptional. Thus J is a simple finite–dimensional Jordan H^*–triple systems. The real simple finite–dimensional Jordan triple systems are classified by Neher in [13] and all of them can be endowed on an, essentially unique, H^*–structure as in the above examples. So we have the following

Theorem 4 *Every real nonpolarized exceptional topologically simple Jordan H^*-triple system is, up to twins and positive factors of the inner product, isometrically $*$-isomorphic to one of the examples 6 and 7 of the section 2.*

Acknowledgements

The authors wish to thank the Professor E. Neher for the talks began in Oviedo's Conference and posterior mail, as a consequence of which this work has been notably simplified.

References

1. Bouhya K. and Fernández, A.: , 'Jordan–∗–triples with minimal inner ideals and compact JB^*–triples', *to appear in Proc. London Math. Soc.*
2. Castellón A. and Cuenca J.A.: 1992, 'Alternative H^*-triple systems ', Comm. in Algebra **20**(11), 3191-3260.
3. Castellón A. and Cuenca J.A.: 1992, 'Associative H^*-triple systems ', Workshop on Nonassociative Algebraic Models (eds. González S. and Myung H.C.), New York.
4. Castellón A. and Cuenca J.A.: 1993, 'Centroid and metacentroid of an H^*-triple system ', Bulletin de la Societé Mathematique de Belgique vol. **45** fascicules 1-2 seria A, 85-91.
5. Castellón A. and Cuenca J.A.: 1990, 'Compatibility in Jordan H^*-triple systems ', Boll. UMI **7**(4-B), 433,477.
6. Castellón A. and Cuenca J.A.: 1992, 'Isomorphisms of H^*-triple systems ', Analli della Scuola Normale Superiore di Pisa, 507-514.
7. Castellón A., Cuenca J.A. and Martín C.: , 'Special Jordan H^*-triple systems ', III International Conference on Nonassociative Algebras and its Applications, Oviedo, 1993, preprint.
8. Castellón A., Cuenca J.A. and Martín C.: 1992, 'Ternary H^*–algebras', Boll UMI **7**(6-B), 217-228.
9. Cuenca J.A. and Martín C.: 1992, 'Jordan two-graded H^*–algebras', Workshop on Nonassociative Algebraic Models (eds. González S. and Myung H.C.), Nova Science Publishers, New York.
10. Cuenca J.A. and Sánchez A.: , 'Structure theory for real noncommutative Jordan H^*–algebras', to appear in Journal of Algebra.
11. Loos O.: 1972, 'Alternative tripelsysteme', Math. Ann. **198**, 205-238.
12. Loos O.:1971, 'Lectures on Jordan triples', *Lecture Notes*, University of British Columbia, Vancouver.
13. Neher E.: 1981, 'Klassifikation der einfachen Ausnahme Jordan-Tripelsysteme', J. Reine Angew. Math. 322.
14. Rodríguez A.: , 'Jordan Structures in Analysis', In Proceedings of the 1992 Oberwolfach Conference on Jordan Algebras (to appear).
15. Zel'manov E.I.: 1983, 'Primary Jordan triple systems I', Sib. Math. Zh. **24** No. 4, 23-27.
16. Zel'manov E.I.: 1984, 'Primary Jordan triple systems II', Sib. Math. Zh. **25** No. 5, 60-51.
17. Zel'manov E.I.: 1985, 'Primary Jordan triple systems III', Sib. Math. Zh. **26** N0. 1, 71-82.

PRIME ALTERNATIVE TRIPLE SYSTEMS *

ALBERTO CASTELLÓN SERRANO and
CÁNDIDO MARTÍN GONZÁLEZ
Departamento de Algebra, Geometría y Topología.
Universidad de Málaga, Aptdo. 59, 29080, Málaga. Spain.

Abstract. In [8] Loos obtains a structure theorem for finite dimensional simple alternative triple systems over an algebraically closed field of characteristic $\neq 2$. This theorem has been generalized in [10] (resp. [9]) for simple alternative triple systems (resp. alternative pairs) containing a maximal tripotent (resp. a maximal idempotent). Other classes of alternative triple systems without maximal tripotent over a Hilbert space (real or complex) have been characterized in [2] and [3] where the simplicity is replaced by the topological simplicity and other hypothesis involving an involution and the inner product. In this paper, we obtain a description of nondegenerate prime alternative triple systems containing a maximal tripotent in terms of its central closure. A socle theory for prime nondegenerate alternative triple systems with maximal tripotent is established in a forthcoming paper [4].

1. Preliminary definitions and results

Let A be a module over the commutative unital ring R. We say that A is a *triple system* (over R) if it is endowed with a trilinear composition $(x, y, z) \mapsto <xyz>$ of $A \times A \times A$ to A. We denote by L, M, R the bilinear mappings from $A \times A$ to $End_R(A)$ given by $L(x,y)z = M(x,z)y = R(z,y)x = <xyz>$. We recall that a triple system A is said to be *associative* if it satisfies the identity

$$<uv <xyz>> = <u <yxv> z> = <<uvx> yz>,$$

and A is *alternative* if the following identities hold

$$<uv <xyz>> + <xy <uvz>> = <<uvx> yz> + <x <vuy> z>,$$

$$<uv <xyx>> = <<uvx> yx>,$$

$$<xy <xyz> = <<xyx> yz> .$$

An associative triple system is alternative. An alternative non associative triple system is called *properly alternative.* A submodule I of a triple system A is said to be an *ideal* if $<IAA> + <AIA> + <AAI> \subset I$. An ideal I of the triple system A is *trivial* if $<IAI> = 0$. The triple system A is called *semiprime* (resp. *prime*) if contains no nonzero trivial ideals (resp. $<IAJ> = 0$ implies $I = 0$ or $J = 0$; I, J ideals of A). Clearly every prime triple system is semiprime. As in the binary case (see for instance [12, 8.2]) it can be shown that a semiprime triple system is a subdirect sum of prime triple systems. A triple system A is said to be *simple* if

* This work has been partially supported by the "Plan Andaluz de Investigación y desarrollo tecnológico" and the DGICYT with project no. PS89-0119

S. González (ed.), Non-Associative Algebra and Its Applications, 73–79.

$<AAA>\neq 0$ and its only ideals are 0 and A. We define the *centroid* $Z(A)$ and *metacentroid* $C(A)$ of a triple system A in the usual way

$$Z(A) = \{T \in End_R(A) : [T, X] = 0,\ X \in \{L, M, R\}\}.$$

$$C(A) = \{T \in End_R(A) : [T, X] = 0,\ X \in \{L, R\}\},$$

As in [6, Theorem 1.1] it can be shown that $Z(A)$ is a commutative integral domain with 1 if A is prime in which case A is a torsion free $Z(A)$–module. Moreover $Z(A)$ is independent of the ring of scalars. The triple system A will be called *central* over R in case $R \cong Z(A)$. Let A a be prime triple system over R. A *partially defined centralizer* of A will be a pair (T, U) where U is a nonzero ideal of A and T is a linear mapping from U into A satistying $[T, X] = 0$ for $X \in \{L, M, R\}$. For any partially defined centralizer (T, U) on A we have that $Ker\ T$ and $T(A)$ are ideals of A with $<ker\ T, A, T(A)> = 0$. Hence either $T = 0$ or T is an injection. As in the binary case, we define the equivalence relation $(T, U) \sim (S, V)$ if and only if $T = S$ in $U \cap V$. We note that the primeness of A gives $0 \neq <UAV> \subset U \cap V$. We denote by $\overline{(T, U)}$ to the equivalence class of (T, U). The set $E_z(A)$ of all equivalence classes of the partially defined centralizers is called the *extended centroid* of A. Addition and product in $E_z(A)$ are defined by

$$\overline{(T,U)} + \overline{(S,V)} = \overline{(T+S, U \cap V)},$$

$$\overline{(T,U)}\ \overline{(S,V)} = \overline{(T(S), V \cap S^{-1}(U))}.$$

This last operation is well defined (see [6, II]). Argueing as in [6, Theorem 2.1] we can prove that $E_z(A)$ is a field. Obviuosly $Z(A) \subset E_z(A)$. A element e of an alternative triple system A is said to be a *tripotent* if $<eee> = e$. If e is a tripotent of the alternative triple system A, then A splits in the Peirce decomposition

$$A = A_{11} \oplus A_{10} \oplus A_{01} \oplus A_{00},$$

where $A_{ij} = \{x \in A : L(e,e)x = ix, R(e,e)x = jx\}$. A tripotent e is said to be *maximal* if $A_{00} = 0$.

Lemma 1 ([9, Lemma 9.6, 11.1], [10, p. 203-204]) *Let e be a maximal tripotent of an alternative triple system A. Then the following composition rules hold*

(1) $<A_{11}A_{11}A_{11}> \subset A_{11}$	(8) $<A_{01}A_{11}A_{01}> \subset A_{10}$
(2) $<A_{11}A_{11}A_{10}> \subset A_{10}$	(9) $<A_{11}A_{01}A_{10}> \subset A_{01}$
(3) $<A_{01}A_{11}A_{11}> \subset A_{01}$	(10) $<A_{10}A_{01}A_{11}> \subset A_{01}$
(4) $<A_{10}A_{10}A_{11}> \subset A_{11}$	(11) $<A_{01}A_{10}A_{01}> \subset A_{11}$
(5) $<A_{11}A_{01}A_{01}> \subset A_{11}$	(12) $<A_{01}A_{10}A_{10}> \subset A_{01}$
(6) $<A_{10}A_{10}A_{10}> \subset A_{10}$	(13) $<A_{10}A_{01}A_{01}> \subset A_{10}$
(7) $<A_{01}A_{01}A_{01}> \subset A_{01}$	

while all other compositions are zero. Moreover $\mathcal{R} = A_{11}$ is an alternative algebra with product $a \cdot b = <aeb>$, unit element e and involution $a \mapsto a^* = <eae>$, and *for any* $x, y \in A$; $a, b \in A_{11}$; $f, g, h \in A_{10}$; $u, v, w \in A_{01}$, *the following formulas are satisfied:*

(14)
$$<xay>= (x \cdot a^*) \cdot y$$

(15)
$$a \cdot (b \cdot f) = (a \cdot b) \cdot f, \quad (u \cdot a) \cdot b = u \cdot (a \cdot b)$$

(16)
$$\left\{ \begin{array}{ccc} <fge>^* & = & <gfe> \\ a\cdot <fge> & = & <(a \cdot f)ge> \\ <fgb> \cdot a & = & <fg(b \cdot a)> \\ <fga> & = & <f(a^* \cdot g)e>. \end{array} \right\}$$

(17)
$$\left\{ \begin{array}{ccc} <euv>^* & = & <evu> \\ <buv> \cdot a & = & <bu(v \cdot a)> \\ a\cdot <buv> & = & <(a \cdot b)uv> \\ <auv> & = & <e(u \cdot a^*)v> \end{array} \right\}$$

(18)
$$\left\{ \begin{array}{cccc} <fua> & = & - <auf> & \\ < fua > \cdot b & = & < fu(a \cdot b) > & = \\ < (a \cdot f)ub > & = & < f(u \cdot a^*)b > & \end{array} \right\}$$

(19)
$$u \cdot u = 0, \; a \cdot (u \cdot v) = (u \cdot a) \cdot v = u \cdot (v \cdot a) =<ua^* v>$$

(20)
$$<fgh>=<fge> \cdot h$$

(21)
$$<uvw>= u\cdot <evw> + <ev(u \cdot w)>$$

(22)
$$<ufv>= - <(u \cdot v)fe>$$

(23)
$$<ufg>= u\cdot <fge>$$

(24)
$$<fuv>=<fue> \cdot v.$$

Next we give several examples of alternative triple systems. If A is an alternative R–algebra with involution $*$, the R–module A endowed with the triple product given by $<xyz>= (xy^*)z$ is an alternative triple system over R called the **-isotope* of A and will be denoted by $A^{(*)}$. The triple system $A^{(*)}$ is simple if A is simple. Let X be a free C–module over a commutative unital ring C with involution such that the subring K of symmetric elements of C is a field, $\phi : X \times X \to C$ a nondegenerate sesquilinear form and $J : X \to X$ a semilinear map satisfying $J^2 = -\gamma 1, 0 \neq \gamma \in K$, and such that the bilinear form α given by $\phi(x, y) + \alpha(Jx, y)) = 0$ is a nondegenerate alternate form. Then X becomes a simple alternative (non associative if $dim_K(X) \geq 4$) triple system over K with the triple product

$$<xyz>= \phi(y, z)x + \alpha(x, z)Jy.$$

2. Prime alternative triple systems

Throughout this paragraph we denote by A a prime alternative triple system over R with maximal tripotent e and by $\mathcal{R}$ the alternative R-algebra given by the lemma 1.

Lemma 2 *Let I be an ideal of $\mathcal{R}$. Then $<IA_{01}A_{01}> + <A_{10}A_{10}I> \subset I$.*

Let $f, g \in A_{01}$, $a \in I$. Using (16) we obtain $<fga>=<fg(e \cdot a)>=<fge> \cdot a \in I$. Analogously (17) proves $<IA_{01}A_{01}> \subset I$.

Lemma 3 *$\mathcal{R}$ is a $*$-prime algebra and therefore semiprime.*

Let I, J be $*$-ideals of $\mathcal{R}$, that is, $I^* \subset I$, $J^* \subset J$, with $I \cdot J = 0$. This implies that $J \cdot I = 0$. By [3, Corollary 1 p. 212],

$$\tilde{I} = I \oplus I \cdot A_{10} \oplus A_{01} \cdot I,$$

$$\tilde{J} = J \oplus J \cdot A_{10} \oplus A_{01} \cdot J,$$

are ideals of A. The lemma follows from the fact that $<\tilde{I}A\tilde{J}>= 0$. This last result is obtained from a extensive use of lemma 1, the Peirce relations (1)-(13) and formulas (14)-(24). Thus $\mathcal{R}$ is semiprime (see for instance [1]).

Lemma 4 *If either $A_{10} \neq 0$ or $A_{01} \neq 0$ and A is nondegenerate, that is, $M(x, x) = 0$ implies $x = 0$, then $\mathcal{R}$ is associative.*

Let $A_{10} \neq 0$. As in [8, 11.7] we consider the map $F : \mathcal{R} \to End_R(A_{10})$ such that to each $a \in \mathcal{R}$ assigns the endomorphism $L(a, e)$ of A_{10}. It follows from (13) that F is a morphism of R-algebras and therefore $I = ker\ F$ is an ideal of $\mathcal{R}$. On the other hand (16) proves that $< A_{10}A_{10}e >=< A_{10}A_{10}A_{11} >$ is a $*$-ideal of $\mathcal{R}$. Let a be an element of I. Again by (16) we have $a \cdot < fge >=< (a \cdot f)ge >= 0$. So $I \cdot <A_{10}A_{10}A_{11}>= 0$. Now consider the morphism of algebras with involution $G : \mathcal{R} \to End_R(A_{10}) \oplus End_R(A_{10})^{op}$ given by $G(a) = (L(a, e), L(a^*, e))$. Is obvious that $Ker\ G$ is a $*$-ideal of $\mathcal{R}$ contained in I. Since $ker\ G \cdot <A_{10}A_{10}A_{11}>= 0$, either $Ker\ G = 0$ or $< A_{10}A_{10}A_{11} >= 0$. In this last case, lef f be a nonzero element of A_{10}. The Peirce relations (1)-(13) and formula (20) show that $M(f, f) = 0$. Contradiction. So $Ker\ G = 0$ and $\mathcal{R}$ is embedded in an associative algebra. Suppose now that $A_{01} \neq 0$. As in the above case, we obtain $<A_{11}A_{01}A_{01}> \cdot Ker\ H = 0$ for convenient morphism H of associative algebras being $< A_{11}A_{01}A_{01} >$ and $Ker\ H$ $*$-ideals of $\mathcal{R}$. If $<A_{11}A_{01}A_{01}>= 0$ and $0 \neq u \in A_{01}$, the Peirce relations (1)-(13) and formulas (19) and (22) prove that $M(u, u) = 0$ and the proof is finished.

Lemma 5 *A_{10}, A_{01} are $\mathcal{R}$-modules. Moreover if A is either associative or nondegenerate, then the maps*

$$\begin{array}{ll} \Phi : A_{10} \times A_{10} \to \mathcal{R} & \Psi : A_{01} \times A_{01} \to \mathcal{R} \\ (f, g) \mapsto <fge> & (u, v) \mapsto <evu> \end{array}$$

are nondegenerate sesquilinear forms.

The formulas (15) allow considerer to A_{10} and A_{01} as $\mathcal{R}$–modules. The sesquilinearity of Φ and Ψ is obtained from (16) and (17). Let f be an element of radical of Φ, that is, $<fA_{10}e>= 0$. By (20) we obtain $<fA_{10}A_{10}>= 0$, which gives $M(f,f) = 0$ and $f = 0$ if A is nondegenerate. If A is associative aplying [5, Paragraph 3.2] or [7, 1.9.3] we obtain also $f = 0$ and Φ is nondegenerate. Finally let $u \in A_{01}$ with $<eA_{01}u>= 0$. Taking into account (19), (21), (22) and the fact that formulas (8)-(13) are identically zero when A is associative, we obtain again $M(u,u) = 0$ and the proof finishes as in the above case.

Lemma 6 *The triple system A is associative if and only if $\mathcal{R}$ is associative and either A_{10} or A_{01} are zero.*

If $\mathcal{R}$ is associative and either A_{10} or A_{01} are zero, the left hand sides of relations (8)-(13) vanish and A is associative by aplication of [10, Theorem 1 p. 210]. Conversely, assume A associative. Obviously $\mathcal{R}$ is associative. On the other hand $I =<A_{10}A_{10}e>=<A_{10}A_{10}A_{11}>$ and $J =<eA_{01}A_{01}>=<A_{11}A_{01}A_{01}>$ are $*$–ideals of $\mathcal{R}$ (see proof of lemma 4). As in the proof of [9, 11.8], we have $I \cdot J = 0$ which implies, by lemma 3, that either I or J vanish. It follows from lemma 5 that either A_{10} or A_{01} are zero and this lemma is proved. It is easy and wellknown that every nonzero $*$–ideal I of a $*$–prime (associative) ring R contains a nonzero symmetric element. For completeness we disgress to prove this. If $x + x^* = x \cdot x^* = 0$ for any $x \in I$, I is a skewcommutative semiprime ring with $x^2 = 0$ for any $x \in I$ which is imposible. The known structure of semiprime rings with involution shows also that if $a \cdot b = 0$ implies either $a = 0$ or $b = 0$ for any symmetric elements of R, then $Sym(R,*)$ contains no nonzero zerodivisors. Both of the above results will be used in the following without explicit mention.

Lemma 7 *If $A_{10} \neq 0 \neq A_{01}$, and A is nondegenerate, then $\mathcal{R}$ is commutative.*

Let $I = \{a \in \mathcal{R} : a \cdot \mathcal{U} = a^* \cdot \mathcal{U} = 0\}$, with $\mathcal{U} = A_{01} \cdot A_{01} \subset A_{10}$. It is proof in [10, Lemma 3 p.215] (or [9, 11.9]) that $\mathcal{U} \neq 0$ and I is a proper $*$–ideal of $\mathcal{R}$ containing all the commuters ($[\mathcal{R},\mathcal{R}] \subset I$). As in the proof of lemma 4, we have $I \cdot <\mathcal{U}A_{10}e>= 0$. If $I \neq 0$, some nonzero symmetric element of $\mathcal{R}$ lies in I and therefore $<\mathcal{U}A_{10}e>= 0$. Let $u,v \in A_{01}$, $f,g \in A_{10}$. By (20) we have $<(u \cdot v)fg>=<(u \cdot v)fe> \cdot g = 0$ and $M(u \cdot v, u \cdot v) = 0$. Since $\mathcal{U} \neq 0$, we obtain a contradiction. Then $I = 0$ and $\mathcal{R}$ is commutative. Combining the above lemmas, we can give a classification of nondegenerate prime triple systems with maximal tripotent attending to extinction of the Peirce subspaces A_{10} and A_{01}. So if A is a nondegenerate prime alternative triple system with a maximal tripotent, then A is of one and only one of the following types

- Type I) Only one of the A_{10}, A_{01} is zero, in which case the triple system A is associative.
- Type II) $A_{10} = A_{01} = 0$, in which case A is the $*$–isotope of a $*$–prime nondegenerate alternative non associative algebra with unity and involution.
- Type III) $A_{10} \neq 0 \neq A_{01}$. In which case $\mathcal{R}$ is associative commutative with unity.

Types II and III are *properly alternative*. In order to obtain more information of type III we will study a scalar extension of A.

Theorem 1 ([3, Theorem 4 p. 221]) *If Z is the center of $\mathcal{R}$ (as algebra) and $Z^+ = \{a \in Z : a^* = a\}$, then the maps $\eta : C(A) \to Z$, $\zeta : Z \to C(A)$ given by $\eta(T) = T(e)$, and $\zeta(a) = L(a,e)$ in $A_{11} \oplus A_{10}$ and $R(a,e)$ in A_{01}, are isomorphisms of algebras with $\eta^{-1} = \zeta$ and $\eta(Z(A)) = Z^+$.*

The above theorem allows to define a involution $T \mapsto T^*$ in the metacentroid $C(A)$ of A such that $Z(A) = Sym(C(A), *)$. Let Q be the field of fractions of $Z(A)$. In an annalogous way to [6, Theorem 1.3] it can be proved that if A is a prime nondegenerate alternative triple system with, maximal tripotent e, then $A \otimes_{Z(A)} Q$ is a central prime nondegenerate alternative triple system where $e \otimes 1$ is a maximal tripotent. Analogously it can be proved that $C(A \otimes Q) \cong Z^{-1}C(A)$. Next we prove that $A \otimes Q$ is simple if every nonzero ideal I of $\mathcal{R}$ satisfies $I \cap Z^+ \neq 0$. In fact, if I is a nonzero Q–ideal of $A \otimes Q$, I contains a certain $a \otimes T^{-1}$ and therefore some $b \otimes 1$ lies in I. Then $J = I \cap \bar{A}$ is a nonzero ideal of $\bar{A}$ ($\bar{A} = A \otimes 1 \cong A$). Put $J = J_{11} \oplus J_{10} \oplus J_{01}$ with $J_{ij} = \bar{A}_{i,j} \cap I$. Let $0 \neq x_{11} + x_{10} + x_{01} \in J$ where $x_{ij} \in J_{ij}$. It follows from lemma 5 that $J_{11} \neq 0$ and by [10, Lemma 2 p.211] J_{11} is a nonzero $*$–ideal of $\mathcal{R}$. Since J contains some nonzero symmetric element, $I = A \otimes Q$. The condition that any nonzero ideal of $\mathcal{R}$ has nonzero elements in Z^+ is clearly satisfied in each triple system of type III while some additional conditions are necessary if the triple system is of type II (see for instance [12, Corollary 7.7]). As in [6, Theorem 2.1, Lemma 2.2] it can be proved that the extended centroid $E_z(A)$ of A is a field and

$$I = \{\sum_i [u_i \otimes \lambda_i \rho_i - T_i(u_i) \odot \rho_i : \lambda_i, \rho_i \in E_z(A), (T_i, U_i) \in \lambda_i, u_i \in U_i\}$$

is an ideal of $A \otimes E_z(A)$ with $I \cap \bar{A} = 0$. Can be considerer the maximal ideal M of $A \otimes Ez(A)$ respect to this property and to define the *closure central* of a prime triple system A as the quotient $A \otimes E_z(A)/M$. However, in our case, if (T, U) is a partially defined centralicer and $U \cap Z \neq 0$, then $U = A$ and $T \in Z(A)$, that is

Theorem 2 *The central closure of an alternative triple system of type III is a simple alternative triple system with maximal tripotent and its agrees with the scalar extension by the field of fractions of the centroid.*

References

1. Baxter W.E. and Martindale 3rd W.S.:1992, 'Extended centroid in *–prime rings', *Comm. in Algebra* Vol. no. **10** (8), pp. 847–874
2. Castellón A. and Cuenca J.A.:1992, 'Associative H^*-triple systems, *In Nonassociative Algebraic Models, (eds. González S. and Myung H.C.), Nova Science Publ., New York*, pp. 45–67
3. Castellón A. and Cuenca J.A.:1992, 'Alternative H^*-triple systems, *Comm. in Algebra* Vol. no. **20** (11), pp. 3191–3206
4. Castellón A. and Martín C. 'On the socle of alternative triple systems, *preprint*.

5. Cuenca J.A., García A. and Martín C.:1989 'Jacobson density for associative pairs and its applications, *Comm. in Algebra* Vol. no. **17** (10), pp. 2595–2610
6. Erickson T.S., Martindale 3rd W.S. and Osborn J.M.:1975, 'Prime nonassociative algebras, *Pac. J. Math.*, Vol. no. **60**, pp. 49-63
7. Fernández A. and García E.:1990, 'Prime associative triple systems with nonzero socle, *Comm. in Algebra* Vol. no. **18** (1), pp. 1–13
8. Loos O.:1972, 'Alternative triplesysteme, *Math. Ann.* Vol. no. **198**, pp. 205–238
9. Loos O.:1975 'Jordan pairs, *Lecture Notes in Mathematics Springer-Verlag, Berlin-Heidelberg-New York*, Vol. no. **460**
10. Meyberg K.:1972, 'Lectures on algebras and triple systems, *Lecture notes, The University of Virginia, Charlottesville*
11. Slater M.:1968, 'Ideals in semiprime alternative rings, *Journal of Algebra*, Vol. no. **8**, pp. 60–76
12. Zhevlakov K.A., Slin'ko A.M., Shestakov I.P. and Shirshov A.I.:1982 'Rings that are nearly associative, Academic Press, New York-London

M-IDEALS OF SCHREIER TYPE AND THE DUNFORD-PETTIS PROPERTY

JESÚS M.F. CASTILLO and FERNANDO SÁNCHEZ
Departamento de Matemáticas, Facultad de Ciencias, E-06071 Badajoz (Spain)
e-mail: castillo@ba.unex.es

and

MANUEL GONZÁLEZ*
Departamento de Matemáticas, Facultad de Ciencias, E-39071 Santander (Spain)
e-mail: mgonzalez@ccucvx.unican.es

Abstract. Given a compact subspace $\mathcal{F}$ of $\{0,1\}^{\mathbf{N}}$, i.e., a family of subsets of positive integers that is compact under the topology of pointwise convergence, the M-ideal generated by $\mathcal{F}$ is studied. Moreover, we prove that for those M-ideals, the Dunford-Pettis and the hereditary Dunford-Pettis properties are equivalent.

Key words: M-ideal, Dunford-Pettis property, Banach space

1. Introduction

In what follows $P_\infty(\mathbf{N})$ denote the set of all infinite subsets of $\mathbf{N}$, the positive integers. If E is a subset of $\mathbf{N}$ and x is a sequence, then Ex denotes the product sequence of x by the characteristic function of E. If A and B are sets, $A < B$ means that $maxA < minB$.

Proposition 1.1 *Let $\mathcal{F}$ be a family of finite subsets of $\mathbf{N}$ such that*
i) if $G \subset F \in \mathcal{F}$, then $G \in \mathcal{F}$,
ii) $\{n\} \in \mathcal{F}$ for all $n \in \mathbf{N}$, and
iii) for all $Z \in P_\infty(\mathbf{N})$ there is $B \subset Z$ finite such that $B \notin \mathcal{F}$.
Then $\mathcal{F}$ is a countable compact metric space, viewed as the space $\{1_F : F \in \mathcal{F}\}$ under the topology of pointwise convergence. The sequence (x_n) defined by $x_n(F) = 1_F(n)$ converges pointwise to 0 on $\mathcal{F}$, and each x_n is continuous on $\mathcal{F}$. Therefore, it is a weakly null basic sequence in $C(\mathcal{F})$.

A family $\mathcal{F}$ satisfying the conditions of Proposition 1.1 shall be termed *adequate.* Now, given an adequate family $\mathcal{F}$ of subsets of $\mathbf{N}$, a Banach space $S_\mathcal{F}$ can be constructed as the completion of the space of finite sequences with respect to the norm

$$||x|| = sup\{||Ex||_1 := \sum_{n\in E} |x_n| : E \in \mathcal{F}\},$$

* Supported in part by DGICYT Grant PB91-0307 (Spain)

S. González (ed.), Non-Associative Algebra and Its Applications, 80–85.

for which the standard basis (e_n) is a weakly null unconditional Schauder basis. It is not hard to see that the space $S_{\mathcal{F}}$ is a subspace of a $C(K)$–space with K countable. Therefore, every closed subspace of $S_{\mathcal{F}}$ contains a copy of c_0.

If A is a subset of $\mathbf{N}$, the space $S_{\mathcal{F}}(A)$ is defined as the space of sequences (x_n) in $S_{\mathcal{F}}$ such that $x_n = 0$ if $n \notin A$.

Examples 1.2 *A.* If $\mathbf{N}$ is partitioned into a countable family of consecutive finite sets (A_n) with $card(A_n) = f(n)$, and a family $\mathcal{F}$ is defined by: $A \in \mathcal{F}$ if and only if $A \subset A_n$ for some $n \in \mathbf{N}$; then it is not difficult to see that the space $S_{\mathcal{F}}$ is $c_0(\ell_1^{f(n)})$.

B. The Schreier space [8] is obtained taking as $\mathcal{F}$ the family of admissible sets. A finite subset E of $\mathbf{N}$ will be called *admissible* if it is $\emptyset$ or $cardE \leq minE$. The Schreier sequence, i.e., the standard basis of the Schreier space, was introduced in [8] as the first example of a weakly null sequence without Banach-Saks subsequences. Recall that a sequence (x_n) in a Banach space is said to be a *Banach-Saks sequence* if it has norm convergent arithmetic means; i.e., if the sequence $(x_1 + \cdots + x_n)/n$ converges.

C. The space of Schachermayer [7] is obtained choosing as $\mathcal{F}$ the family of totally admissible sets: For each $n \in \mathbf{N}$ we write $t(n) = v/2u \in [0,1]$ if $n = 2^u + v$ $(0 \leq v \leq 2^u - 1)$. A finite subset $\{n_1, \ldots, n_l\}$ of $\mathbf{N}$ will be called *totally admissible* if it is $\emptyset$ or:

1) $l \leq n_1$, and

2) Let p be defined by $2^{p-1} < n_1 \leq 2^p$. For every $0 \leq j < 2^p$ there is at most one i such that $t(n_i) \in [j/2^p, (j+1)/2^p)$.

The space of Schachermayer was introduced to show that the Banach-Saks property (i.e., the property that every bounded sequence admits a Banach-Saks subsequence) is not L_2–hereditary.

D. The lunatic space [2]. This space is obtained taking as $\mathcal{F}$ the family of lunatic sets. The *range* of a finite set $A \subset \mathbf{N}$ is defined by

$$r(A) = inf\{b - a : a, b \in A; a < b\}.$$

A finite set $A \subset \mathbf{N}$ is said to be *lunatic* if $1 + r(A) \geq card(A)$.

Definition 1.3 *A Banach space X is said to have the* Dunford-Pettis property *(DPP) if any weakly compact operator $T : X \to Y$ transforms weakly compact sets of X into relatively compact sets of Y. Equivalently, given weakly null sequences (x_n) in X and (x_n^*) in X^*, we have $lim < x_n^*, x_n >= 0$.*

A Banach space X is said to have the hereditary Dunford-Pettis property *(HDPP) if every subspace of X has the DPP.*

The spaces L_1 and $C(K)$ are examples of spaces with the DPP, and the spaces ℓ_1 and c_0 are examples of spaces having the HDPP.

A profound characterization, due to Elton (see [5]), of the HDPP is: *every normalized weakly null sequence admits a subsequence equivalent to the canonical basis of c_0.* Another useful characterization for this property is in [4]: every weakly null sequence contains a subsequence such that

$$\left\| \sum_{k=1}^{n} x_k \right\| \leq C$$

for some $C > 0$ and every $n \in \mathbf{N}$.

2. M-ideals and the Dunford-Pettis property

A Banach space X is said to be an *M-ideal* (or an M-ideal in its bidual) if the following equality holds isometrically:

$$X^{***} = X^* \oplus_1 (X^{**}/X)^*.$$

In [9], the following equivalent condition is given: There is a dense subset Δ of $Ball(X)$, the unit ball of X, such that

$$\forall x^{**} \in Ball(X^{**}), \forall x \in \Delta, \forall \epsilon > 0, \exists y \in X : \|x^{**} \pm x - y\| \leq 1 + \epsilon.$$

In our case, we shall take as Δ the set of all finite sequences of $Ball(S_{\mathcal{F}})$. Therefore, the space $S_{\mathcal{F}}$ is and M-ideal if and only if

$$lim_{N\to\infty}\|x + [N,\infty)x^{**}\| = 1 \qquad (\dagger)$$

for all $x^{**} \in Ball(S_{\mathcal{F}}^{**})$ and all $x \in \Delta$.

To characterize the spaces $S_{\mathcal{F}}$ which are M-ideals we introduce a hierarchy of functions.

Definition 2.1 *Let $k \in \mathbf{N}$. The function $g_k : \mathbf{N^k} \to \mathbf{N} \cup \{\infty\}$ is defined by*

$$g_k(n_1, \ldots, n_k) := sup\{cardA : \{n_1, \ldots, n_k\} \subset A \in \mathcal{F}\}.$$

We shall say that the function g_k is finite *if it always takes finite values.*

Proposition 2.2 *If $(S_{\mathcal{F}})^{**} \subset c_0$ and the function g_1 is finite, then $S_{\mathcal{F}}$ is an M-ideal.*

Proof. We proceed as in [9, Prop. 2.1]. Let $x^{**} = (a_n)$ be an element in the unit ball of $(S_{\mathcal{F}})^{**}$, $x \in \Delta$ and $\epsilon > 0$. Consider an integer $N \in \mathbf{N}$ such that $x_n = 0$ if $n \geq N$, and then choose an integer $M \geq N$ such that $g_1(k)a_n \leq \epsilon$ if $1 \leq k \leq N$ and $n \geq M$. Then, the sequence $y = [1, M]x^{**}$ belongs to $S_{\mathcal{F}}$.

If $E \in \mathcal{F}$ and $\{1, \ldots, N\} \cap E = \emptyset$ then $\sum_{n\in E} |a_n \pm x_n - y_n| \leq \|x^{**}\| \leq 1$. Finally, if $E \in \mathcal{F}$ and $\{1, \ldots, N\} \cap E \neq \emptyset$ then $card(E) \leq g_1(k)$ for some $k \leq N$ and therefore $\sum_{n\in E} |a_n \pm x_n - y_n| \leq \|x\| + \epsilon \leq 1 + \epsilon$. □

Remark 2.3 There is a simple form to verify the first hypothesis: if for every $Z \in P_\infty(\mathbf{N})$ there is $Z' \subset Z$ infinite such that $sup\{card(A \cap Z') : A \in \mathcal{F}\} = \infty$, then $(S_{\mathcal{F}})^{**} \subset c_0$.

Proposition 2.4 *If $S_{\mathcal{F}}$ is an M-ideal then either the function g_1 is finite or, for all $k \in \mathbf{N}$, the function g_k is not finite.*

Proof. Assume that for some k the function g_k is finite and that, moreover, the function g_1 is not finite. There is no loss of generality assuming that $g_1(1) = \infty$ and that for some sequence (B_n) of consecutive finite subsets of $\mathbf{N}$ such that $card(B_n) \to \infty$, the set $\{1\} \cup B_n \in \mathcal{F}$. We shall restrict ourselves in what follows to work into the space $S_{\mathcal{F}}(\cup B_n)$.

Let $N_1 = sup\{g_k(n_1, \ldots, n_k) : n_1, \ldots, n_k \in B_1\}$, and pick some set $C_1 \in \mathcal{F}$ where that supremum is attained. Find some $B_{j(2)} > C_1$, obtaining thus some N_2 and C_2, and repeat inductively this process. It can be readily assumed that $N_n \to \infty$ and thus that $2N_n < N_{n+1}$.

Let us prove that the element $x^{**} = (x_n)$ defined by $x_n = 1/N_j$ when $n \in C_j$, and 0 otherwise, belongs to $(S_{\mathcal{F}})^{**} \setminus S_{\mathcal{F}}$. Recall that $x^{**} \in (S_{\mathcal{F}})^{**}$ if and only if $\|x^{**}\| < \infty$. Let $A \in \mathcal{F}$, and let $d_i = card(A \cap C_i)$, $i = 1, \ldots, m$. If $d_i \leq k$ for all i, then $\|Ax^{**}\|_1 \leq 2k/N_1 \leq 2k$. Otherwise, let j be the first index such that $d_j > k$, which implies $cardA \leq N_j$ and thus $\|Ax^{**}\|_1 \leq 2k + (d_j + \ldots + d_m)/N_j \leq 2k + 1$.

This choice of x^{**} together with $x = e_1$ in (†) shows that the space $S_{\mathcal{F}}$ is not an M-ideal. □

Example 2.5 Let P_n denote the set $\{2^n, \ldots, 2^{n+1} - 1\}$. Let $\mathbf{N} = \cup A_k$ be a disjoint decomposition of $\mathbf{N}$ into infinite sets. Define a family $\mathcal{F}$ by means of: A set $A \in \mathcal{F}$ if and only if A has only one point or, for some k and j, $A \subset P_k \cup P_j$, where $j \in A_k$. The space $S_{\mathcal{F}}$ is not an M-ideal although all the functions g_k are infinite.

Example 2.6 Define a family $\mathcal{F}$ by means of: A set $A \in \mathcal{F}$ if and only if A has only one point or, $A \subset \cup_{i \in I} P_i$, with I admissible. The space $S_{\mathcal{F}}$ has the property of Remark 2.3, is an M-ideal and all the functions g_k are infinite.

The following result characterizes the spaces $S_{\mathcal{F}}$ having the DPP, when they are M-ideals, in terms of which could be understood as "the g_0 function".

Proposition 2.7 *Let $\mathcal{F}$ be an adequate family such that $S_{\mathcal{F}}$ is an M-ideal. The following are equivalent:*
i) $S_{\mathcal{F}}$ has DPP, and
ii) $S_{\mathcal{F}}$ has HDPP.

Proof. Let us define the function $g_0(n) = sup\{max(A) : n \in A, A \in \mathcal{F}\}$. We shall show that i) and ii) are equivalent to the condition

iii) $g_0(n) < \infty$ for all $n \in \mathbf{N}$.

The implication *iii)* $\Rightarrow$ *ii)* is as follows. Let (x_n) be a normalized weakly null sequence in $S_{\mathcal{F}}$. By an standard perturbation argument, we can suppose that that sequence is formed by certain blocks of the canonical basis (e_n) of $S_{\mathcal{F}}$. Let B_n be the support of x_n. Pick $y_1 = x_1$.

We inductively define a subsequence $(x_{j(n)})$ as follows: $j(1) = 1$, and $j(n+1)$ is the first index j such that $g_0(i) \leq min(B_j)$ for all $i \in B_{j(n)}$. Taking the subsequence $y_n = x_{j(n)}$ one has that

$$\left\| \sum_{m=1}^{N} y_m \right\| \leq 1$$

for all $N \in \mathbf{N}$. This implies that $S_{\mathcal{F}}$ has the HDPP.

To prove the implication $i) \Rightarrow iii)$, suppose that $g_0(k) = \infty$ for some $k \in \mathbf{N}$. Then, there is $A \in P_\infty(\mathbf{N})$ such that $\{k, m\} \in \mathcal{F}$ for all $m \in A$. The projection in $S_{\mathcal{F}}$ onto $S_{\mathcal{F}}(A)$ is continuous and then $S_{\mathcal{F}}(A)$ has the DPP. Consequently, $(S_{\mathcal{F}}(A))^{**}$ is not contained in c_0. It is evident that, in this case, there is $B \subset A$ infinite, $\delta > 0$ and $x^{**} = (x_n)$ in the unit ball of $(S_{\mathcal{F}}(A))^{**}$ such that $x_n > \delta$ for all $n \in B$. We have $||x^{**} \pm e_k - y|| \geq 1 + \delta$ for all $y \in S_{\mathcal{F}}$, which is impossible from hypothesis. □

Examples 2.8 *A. Renormings of* c_0. The space $c_0(n)$ of null sequences (x_n) with the norm

$$||(x_n)|| = sup_{i_1 < \ldots < i_n} \frac{1}{n} \sum_{j=1}^{n} |x_{i_j}|,$$

is an M-ideal if and only if $n = 1$ (the only case in which the function g_0 is finite). Analogously, given any function $\eta : \mathbf{N} \to \mathbf{N}$, the space $S_{\mathcal{F}}$ defined by the family

$$A \in \mathcal{F} \Leftrightarrow max(A) < sup\{\eta(j) : j \in A\}$$

is an M-ideal.

B. The Schreier's space is an M-ideal [9]. In fact $g_1(n) = n$. Analogously, given any function ϕ, the space $S_{\mathcal{F}}$ defined by the family

$$A \in \mathcal{F} \Leftrightarrow card(A) < sup\{\phi(j) : j \in A\}$$

is an M-ideal.

C. The space of Schachermayer is not an M-ideal: it has the DPP (see [3]) but the function g_0 is not finite. The function g_1 is finite, which shows that Proposition 2.2 is not an equivalence.

D. The lunatic space is not an M-ideal. The function g_1 is infinite but the function g_2 is not.

E. Leung's space [6] is an M-ideal. A finite subset $A \subset \mathbf{N}$ is said to be *Leung-admissible* ($A \in \mathcal{L}$) if it is admissible and $card(A) = 2^i$ for some $i \geq 0$. The space L is defined as the completion the space of finite sequences (x_n) under the norm

$$||(x_n)|| = sup\{||A(x_n)||_{\sqrt{i}} : i \in \mathbf{N}, A \in \mathcal{L}, card(A) = 2^i\}$$

where $A(x_n)$ is the sequence equal to (x_n) if $n \in A$ and equal to zero otherwise, $||\cdot||_p$ is the ℓ_p−norm and the supremum is taken over all Leung-admissible sets.

A standard computation shows that $L^{**} \subset c_0$. Let Δ be the set of sequences with finite support contained in the unit ball of L. Let $x^{**} = (a_n)$ in the unit ball of L^{**}, $x \in \Delta$, and $\epsilon > 0$. Consider an integer $N \in \mathbf{N}$ such that $x_n = 0$ if $n \geq N$, and choose an integer $M \geq N$ such that $a_n \leq \epsilon/N$ if $n \geq M$. Now let $y = [1, M]x^{**}$. Then $y \in S_{\mathcal{F}}$. Next, let A be a Leung-admissible set. If $min(A) \leq N$ then $card(A) \leq N$. On the other hand, if $min(A) > N$ then $\{1, \ldots, N\} \cap A = \emptyset$. Thus, we have the inequality $||x^{**} \pm x - y|| \leq 1 + \epsilon$.

Comment. The preceding results seem to indicate that, among spaces of Schreier type $S_{\mathcal{F}}$, there are essentially two types of M-ideals: the type of c_0; i.e., the $S_{\mathcal{F}}$ space defined by some function g_0, which corresponds to those having DPP; and the type of Schreier space; i.e., the $S_{\mathcal{F}}$ space defined by some function g_1, which corresponds to those having not DPP.

References

1. D.E. Alspach and S. Argyros. Complexity of weakly null sequences, Dissertationes Math. CCCXXI (1993).
2. J.M.F. Castillo. A variation on Schreier's space, Riv. Mat. Univ. Parma (to appear).
3. J.M.F. Castillo and M. González. New results on the Dunford-Pettis property, Preprint.
4. J.M.F. Castillo and M. González. The Dunford-Pettis property is not a three space property. Israel J. of Math. 81 (1993) 297-299.
5. J. Diestel. A survey of results related to the Dunford-Pettis property, Contemporary Math. vol 2, Amer. Math. Soc. (1979) 15-60.
6. D. Leung. Uniform convergence of operators and Grothendieck spaces with the Dunford-Pettis property, Math. Z. 197 (1988) 21-32.
7. W. Schachermayer. The Banach-Saks property is not L_2-hereditary, Israel J. Math. 40 (1981) 340-344.
8. J. Schreier. Ein Gegenbeispiel zur schwachen Konvergenz, Studia Math. 2 (1930) 58-62.
9. D. Werner. New classes of Banach spaces which are M-ideals in their biduals, Math. Proc. Cambridge Philos. Soc. 111 (1992) 337-354.

SPECTRA OF ELEMENTS OF A NONASSOCIATIVE ALGEBRA

ANTON CEDILNIK
Biotehniška fakulteta, University of Ljubljana
Večna pot 83, 61000 Ljubljana, Slovenia

Abstract. In the contribution we generalize the notion of spectrum to any (nonassociative) algebra. The spectra of elements of a given algebra are such sets of elements from some other algebra that the Spectrum mapping theorem is valid for polynomials. If the two algebras are topological, we extend the validity of Spectral mapping th. to functions which are limits of sequences of polynomials. In this case the spectral radius keeps its meaning almost unchanged.

Key words: nonassociative algebra, spectrum, local homomorphism, spectral radius.

The problem of generalizing the notion of spectrum to any nonassociative algebra is above all searching for those properties of spectrum, which give to it the importance and have interesting consequences. Let's write down some most important theorems on spectra.

1. Classical theory! The spectrum of a square matrix C is a set of zeros of its characteristic or minimal polynomial.
2. The spectrum $\mathrm{Sp}(x)$ of an x of a noncommutative Jordan Banach algebra is nonvoid compact set of complex numbers. The map $x \mapsto \mathrm{Sp}(x)$ is upper semi-continuous ([1], [5]).
3. Spectral mapping theorem! In the associative Banach algebra A it holds $\mathrm{Sp}(f(x)) = f(\mathrm{Sp}(x))$ for any $x \in A$ and any power series f with radius of convergence $R > r(x)$, where $r(x) = \lim_{n\to\infty} ||x^n||^{1/n} = \max|\,\mathrm{Sp}(x)|$.
4. Gelfand theory! In a commutative complex Banach algebra A we have for $x \in A: \quad \mathrm{Sp}(x) = \{\varphi(x) |\ \varphi \in \mathrm{Hom}(A, \mathbb{C})\}$.
5. Vidav's lemma! In a Jordan Banach algebra A it holds for $a \in \mathrm{Her}(A)$: $V(a) = \mathrm{co}\ \mathrm{Sp}(a)$, $r(a) = v(a) = ||a||$, where $V(a)$ is the numerical range and $v(a)$ the numerical radius ([6]).

It seems that the common idea of these theorems is a similarity between the element of algebra and the elements of its spectrum. *We claim it is the Spectral mapping theorem for polynomials that most emphasizes algebraic similarity.* It seems a good choice because this theorem can be easily formulated in any nonassociative algebra; for the notion of inverse element it is not the case.

Another generalization follows from such observations as the next three.

a) A spectrum of an element of a real or complex associative algebra is a subset of one- or twodimensional algebra $\mathbb{C}$.

b) Let B be a subalgebra of an associative complex algebra A with unit 1, and $1 \in B$. Then we can define for $x \in A : \mathrm{Sp}(x) := \{t \in B |\ \exists (t - x)^{-1}\}$. Such a spectrum has some of nice properties of the usual spectrum.

S. González (ed.), Non-Associative Algebra and Its Applications, 86–92.

c) The spectrum of a quantum operator should be a set of values of certain physical quantity, but such values are besides scalars also vectors, tensors and tensor–like objects, elements of subalgebras and quotient algebras of the tensor algebra.

Therefore we'll define a spectrum of an element of given algebra as a subset of some another algebra.

A spectrum which suits topological structure of algebra, will be defined in the similar way: the Spectral mapping theorem must be true for functions which are limits of generalized sequences of polynomials. We shall find out that in this case the spectral radius more or less keeps its role.

1. Algebraic spectra

Let $\mathbb{F}$ be commutative field and $\mathbb{F}[X]$ the set of polynomials with coefficients from $\mathbb{F}$; polynomial is a linear combination of formal nonassociative powers: $X, X^2, XX^2, X^2X, \ldots$ [1]. $\mathbb{F}^*[X] := \mathbb{F}[X]\backslash\{0\}$. Let C be a subset of an algebra A; the symbol $\operatorname{gen} C$ will designate the smallest subalgebra including C. Specially: $\operatorname{gen}\{a\} = \{M(a) |\ M(X) \in \mathbb{F}[X]\}$. Additional symbol: $\operatorname{gen}^*\{a\} := \{M(a) |\ M(X) \in \mathbb{F}^*[X]\}$. Of course, $\operatorname{gen}^*\{a\} \cup \{0\} = \operatorname{gen}\{a\}$. If $\operatorname{gen}^*\{a\} \neq \operatorname{gen}\{a\}$, then $\dim \operatorname{gen}\{a\} = \infty$ and a is **transcendental**. If $\operatorname{gen}^*\{a\} = \operatorname{gen}\{a\}$, we'll call a **semialgebraic**. And if $\dim \operatorname{gen}\{a\} < \infty$, a is **algebraic**. In this section A and B will always be two $\mathbb{F}$–algebras.

Definition 1. *(A,B)–***spectrum** *is a map $s(.,A,B) : A \to 2^B$ with properties*
(AS 1) $s(0,A,B); = \{0\}$
(AS 2) $\forall M(X) \in \mathbb{F}^*[X]$, $\forall x \in A$:

$$s(M(a),A,B) := \{M(x) |\ x \in s(a,A,B)\}. \tag{1}$$

The set $s(a,A,B)$ is (A,B) –**spectrum of** *a.*

(1) will be written simplier

$$s(M(a),A,B) = M(s(a,A,B)). \tag{2}$$

A special case: $\forall a \in A$, $\forall \lambda \in \mathbb{F}\backslash\{0\}$:

$$s(\lambda a,A,B) = \lambda s(a,A,B). \tag{3}$$

Proposition 2. *If $\{s_i(.,A,B) |\ i \in I\}$ is a family of spectra, the map $a \mapsto \bigcup_{i\in I} s_i(a,A,B)$ is again an (A,B)-spectrum.*

Zero spectrum: $s^0(a,A,B) = \{0\}$ for all $a \in A$. **Maximal spectrum** $s^m(a,A,B)$ is a union of all (A,B)–spectra. If A has the unit 1_A and B 1_B and if for a spectrum $s(.,A,B)$ holds $s(1_A,A,B) = \{1_B\}$, then s will be named **unital spectrum**; the union of all unital spectra $s^u(.,A,B)$ is **maximal unital spectrum**. Unital spectra do not exist allways; it can be proved that the algebra of square matrices $M_n(\mathbb{R})$ $(n > 1)$ has only $s^0(.,M_n(\mathbb{R}),\mathbb{R})$.

Proposition 3. *Let $M(X) \in \mathbb{F}[X]$ and $a \in A$ such that $M(a) = 0$.*

[1] In the case we work only in some concrete variety of algebras, it is sensible not to distinguish polynomials which are identically equal.

Then $s^m(a, A, B) \subset \{x \in B |\, M(x) = 0\}$.

Example: Let $\mathbb{F}$ be a subfield of algebraically closed field $\mathbb{E}$. Then the map

$$a \mapsto s(a) := \begin{cases} \{0\} & \text{(for semialgebraic } a) \\ \mathbb{E} & \text{(for transcendental } a) \end{cases}$$

is an $(A, \mathbb{E})$–spectrum. Maximal spectrum is sometimes too big to be interesting. □

Proposition 4. *Let $C \subset A$ be a subalgebra; then the restriction of an (A, B)–spectrum s to C is a (C, B)–spectrum. Let $D \subset B$ be a subalgebra; then any (A, D)–spectrum is also an (A, B)–spectrum.*

An important consequence of Proposition 4: if A and B have units and if there exists $s^u(., A, \mathbb{F})$, there exists $s^u(., A, B)$ as well.
We shall designate the set of homomorphisms from A to B with $\mathrm{Hom}(A, B)$.

Proposition 5. *The map $a \mapsto s(a) := \{\varphi(a)|\, \varphi \in \mathrm{Hom}(A, B)\}$ is an (A, B)–spectrum. If $A = \mathrm{gen}\{c\}$, this spectrum is maximal.*

Proposition 6. *Let A, B, C, D be four $\mathbb{F}$–algebras, $\varphi \in \mathrm{Hom}(C, A)$, $\psi \in \mathrm{Hom}(B, D)$ and $s(., A, B)$ a spectrum. Then the map $c \mapsto s_1(c) := \psi(s(\varphi(c), A, B))$ is a (C, D)–spectrum.*

Definition 7. *To any $a \in A$ we associate a set of local homomorphisms $h(a) \subset \mathrm{Hom}(\mathrm{gen}\{a\}, B)$ such that*
(SC 1) $h(0) = \{0\}$;
(SC 2) if $a, b \in A$ and $c \in \mathrm{gen}^*\{a\} \cap \mathrm{gen}^*\{b\}$, for any $\varphi \in h(a)$ there exists such $\psi \in h(b)$ that $\varphi(c) = \psi(c)$.
*The family $\{h(a)|\, a \in A\}$ will be called (A,B)–***spectral covering.**

Proposition 8. *If $\{h(a)|\, a \in A\}$ is an (A, B)–spectral covering, $a \in A$ and $c \in \mathrm{gen}^*\{a\}$, then*

$$h(c) = \{\varphi|_{\mathrm{gen}(x)}|\, \varphi \in h(a)\}. \tag{4}$$

Proof. Use Definition 7 for $c \in \mathrm{gen}^*\{a\} \cap \mathrm{gen}^*\{c\} = \mathrm{gen}^*\{c\} \cap \mathrm{gen}^*\{a\}$. In fact, Proposition is equivalent to Definition 7. □

Theorem 9. *Any (A, B)–spectral covering $\{h(a)|\, a \in A\}$ determines some (A, B)–spectrum $s(., A, B)$ so that for each $a \in A$:*

$$s(a, A, B) = \{\varphi(a)|\, \varphi \in h(a)\}. \tag{5}$$

To any (A, B)–spectrum there belongs one and only one (A, B)–spectral covering so that (5) holds.

Proof. The first part of Theorem follows from (4) and (2). Now, let's have $a \in A$ and $x \in s(a, A, B)$ and define a map $\varphi_{a,x} : \mathrm{gen}\{a\} \to B$ with the rule: $\varphi_{a,x}(M(a)) = M(x)$ for all $M(X) \in \mathbb{F}[X]$. $\varphi_{a,x}$ is well defined local homomorphism. The set of such maps is, say, $h(a)$. Then the family $\{h(a)|\, a \in A\}$ is an (A, B)–spectral covering, which can be shown by a straightforward verification. □

Theorem 10. *Let 1_A and 1_B be units of A and B respectively, and $\mathbb{F}$ algebraically closed. If there exists $s^u(., A, B)$ then for any $\lambda \in \mathbb{F}$ and $a \in A$ it is*

$$s^u(\lambda 1_A + a, A, B) = \lambda 1_B + s^u(a, A, B). \tag{6}$$

Therefore the spectral mapping th. is valid also for polynomials with the constant term.

Proof. For $a \in A$ we define

$$s(a) := \bigcup_{\lambda \in \mathbb{F}} \{x + \lambda 1_B | \; x \in s^u(a - \lambda 1_A, A, B)\}. \tag{7}$$

For $M(X) \in \mathbb{F}^*[X]$ and $\mu \in \mathbb{F}$ there exists such $N_\mu(X) \in \mathbb{F}[X]$ that $M(c + \mu 1_A) = N_\mu(c) + M(\mu)1_A$.

$$\begin{aligned} s(M(a)) &= \bigcup_\lambda \{x + \lambda 1_B | \; x \in s^u(M(a) - \lambda 1_A, A, B)\} \\ &= \bigcup_\mu \{x + M(\mu)1_B | \; x \in s^u(M(a) - M(\mu)1_A, A, B)\} \\ &= \bigcup_\mu \{x + M(\mu)1_B | \; x \in s^u(N_\mu(a - \mu 1_A), A, B)\} \\ &= \bigcup_\mu \{N_\mu(y) + M(\mu)1_B | \; y \in s^u(a - \mu 1_A, A, B)\} = \{M(z) | \; z \in s(a)\} \end{aligned}$$

$s(0) = \{0\}$, $s(1_A) = \{1_B\}$, therefore s is unital spectrum and is included in s^u:

$$s(a) = s^u(a - \lambda 1_A, A, B) + \lambda 1_B \subset s^u(a, A, B).$$

This is alredy (6). □

Proposition 11. *Let's have three* $\mathbf{F}$*-algebras* A, B, C *and spectra* $s_1(., A, B)$ *and* $s_2(., B, C)$. *Then the map* $a \mapsto \bigcup\{s_2(x, B, C) | \; x \in s_1(a, A, B)\}$ *is an* (A, C) *- spectrum.*

Example. The map $a \mapsto \{a\}$ $(a \in A)$ is an (A, A)–spectrum which is unital iff A has a unit. □

Theorem 12. *Let* $\mathbf{E}$ *be an algebraically closed field and* $\mathbf{F} \subset \mathbb{E}$, *and* A *power–associative with unit* 1. *Then* $s^u(., A, \mathbf{E})$ *exists and for* $a \in A$ *it is*

$$s^u(a, A, \mathbb{E}) = \mathrm{Sp}(a, \mathrm{gen}\{a, 1\} \otimes \mathbb{E}), \tag{8}$$

where Sp *is the usual spectrum (defined with inverse) on the algebra* $\mathrm{gen}\{a, 1\} \otimes \mathbf{E}$. *There also holds:*

$$s^m(a, A, \mathbf{E}) = s^u(a, A, \mathbb{E}) \cup \{0\}. \tag{9}$$

Proof. Inclusion "$\supset$" in (8) and (9) follows from known properties of the usual spectrum, defined with inverse and quasiinverse in an associative algebra. For non-algebraic a the inclusion "$\subset$" holds because of $\mathrm{Sp}(a, \mathrm{gen}\{a, 1\} \otimes \mathbb{E}) = \mathbb{E}$. Algebraic a corresponds to some polynomial equation and Proposition 3 gives the inclusion "$\subset$" again. □

2. Topological spectrum

Let $\mathbb{F}$ be $\mathbf{R}$ or $\mathbf{C}$ and A, B be two Hausdorff topological $\mathbf{F}$–algebras (multiplication is jointly continuous). We shall define the topological spectrum by extending the validity of Spectral mapping theorem. Let $\{M_i(X)|\, i \in I\}$ be a generalized sequence of polynomials from $\mathbb{F}[X]$ and define $F(X) := \lim_{i\in I} M_i(X)$. Such formal function will be called **limit of polynomial sequence** (abbreviated LPS). The **domain** $\mathcal{D}_C(F)$ of F in the algebra $C(= A$ or $B)$ is $\mathcal{D}_C(F) := \{a \in C|\, \exists \lim M_i(a) =: F(a)\}$. $\mathcal{D}_C(F)$ is always nonvoid since it contains 0.

Definition 13. Topological (A, B)**–spectrum** *is a map* $\sigma(.,A,B) : A \to 2^B$ *with properties:*
(TS 1) $\sigma(0, A, B) = \{0\}$,
(TS 2) *for each LPS F and $a \in A$, the next implication holds:*

$$a \in \mathcal{D}_A(F) \Rightarrow \sigma(a, A, B) \subset \mathcal{D}_B(F) \;\wedge\; \sigma(F(a), A, B) = F(\sigma(a, A, B)).$$

Any topological (A, B)–spectrum is also (A, B)–spectrum in the algebraic sense. Topological spectrum is always nonvoid since for LPS $F(X) = \lim_{n\to\infty} \frac{1}{n}X = 0$ it is $\{0\} = \sigma(0) = \sigma(F(a)) = F(\sigma(a)) \neq \phi$. The simpliest case of a topological spectrum is the zero spectrum. Therefore topological spectra always exist and, since the union of topological spectra is topological spectrum again, there exists **maximal topological spectrum** $\sigma^m(.,A,B)$. The union of unital topological spectra (if they exist) is the **maximal unital topological spectrum** $\sigma^u(.,A,B)$. It is also obvious that if $\dim \operatorname{gen}\{a\} < \infty$ for any $a \in A$, every spectrum is topological.

Proposition 14. *Let $a \in A$; for any $b \in \overline{\operatorname{gen}\{a\}}$ define*

$$\sigma(b) := \{\varphi(b)|\, \varphi \in \operatorname{Hom}(\overline{\operatorname{gen}\{a\}}, B) \wedge \varphi \text{ continuous}\}. \tag{10}$$

Then σ is a topological $(\overline{\operatorname{gen}\{a\}}, B)$–spectrum.

Let A have a unit 1_A; then we introduce the symbol $G(a) := \overline{\operatorname{gen}\{a, 1_A\}}$.

Proposition 15. *Let A and B have units 1_A and 1_B respectively and define*

$$\sigma(b) := \{\varphi(b)|\, \varphi \in \operatorname{Hom}(G(a), B) \;\wedge\; \varphi \text{ continuous} \;\wedge\; \varphi(1_A) = 1_B\}. \tag{11}$$

Then σ is a unital topological $(G(a), B)$–spectrum.

Of course, Proposition 15 does not ensure the existence of the unital spectrum.

Theorem 16. *Let $\sigma(.,A,B)$ be a topological spectrum and $\{h(a)\}$ the corresponding spectral covering. Then all local homomorphisms from the covering are continuous and if $c \in \overline{\operatorname{gen}\{a\}} \cap \overline{\operatorname{gen}\{b\}}$ for some $a, b \in A$, for every $\varphi \in h(a)$ there exists such $\psi \in h(b)$ that $\overline{\varphi}(c) = \overline{\psi}(c)$, where $\overline{\varphi}$ and $\overline{\psi}$ are continuous extensions of φ and ψ to $\overline{\operatorname{gen}\{a\}}$ and $\overline{\operatorname{gen}\{b\}}$ respectively.*

Proof. Take $a \in A$ and $x \in \sigma(a, A, B)$ and define a map $\varphi_{a,x} : \overline{\operatorname{gen}\{a\}} \to B$, $\varphi_{a,x}(F(a)) := F(x)$ for each LPS F, for which $a \in \mathcal{D}_A(F)$. Easy verification shows that $\varphi_{a,x}$ is a well defined map which is continuous since it maps a limit of a convergent sequence into the limit of the sequence of images. It is also obvious that $\varphi_{a,x} \in \operatorname{Hom}(\overline{\operatorname{gen}\{a\}}, B)$.

For $a, b \in A$ and $c \in \overline{\operatorname{gen}\{a\}} \cap \overline{\operatorname{gen}\{b\}}$ there exist such LPS F_1 and F_2 that $c = F_1(a) = F_2(b)$.

$$\sigma(c, A, B) = \sigma((F_1(a), A, B) = F_1(\sigma(a, A, B)) = \sigma((F_2(b), A, B) = F_2(\sigma(b, A, B)).$$

For some $x \in \sigma(a, A, B)$ there exists such $y \in \sigma(b, A, B)$ that

$$z = F_1(x) = \varphi_{a,x}(F_1(a)) = \varphi_{a,x}(c) = F_2(y) = \varphi_{b,y}(F_2(b)) = \varphi_{b,y}(c).$$

$\{\{\varphi_{a,x}|_{\text{gen}\{a\}}|\ x \in \sigma(a, A, B)\}|\ a \in A\}$ is then a spectral covering corresponding to the spectrum $\sigma(., A, B)$. Since by Theorem 9 the spectral covering is unique, the proof is complete. □

The product of n factors a in a certain arrangement will be called **pure power**. There are (in a concrete algebra at most) $\frac{1}{n}\binom{2n-2}{n-1}$ pure n–th powers of an a and we shall denote any of them $p^n(a)$: $p^1(a) := a$, $p^2(a) := a^2$, $p^3(a) := aa^2$ or a^2a, etc. Two special pure powers:

$$p_L^n(a) := L_a^{n-1}(a) = a(a(\ldots(aa^2)\ldots)),\ p_R^n(a) := R_a^{n-1}(a) = ((\ldots(a^2a)\ldots)a)a.$$

A convex combination of pure n–th powers of an a is **mixed power** $P^n(a)$: $P^3(a) := \lambda aa^2 + (1-\lambda)a^2a \quad (0 \le \lambda \le 1)$, etc.

Up to the end let $(A, \|.\|)$ and $(B, \|.\|)$ be normed algebras and suppose that the norms are algebraic: $\|ab\| \le \|a\|\|b\|$, $\|xy\| \le \|x\|\|y\|$ $(a, b \in A,\ x, y \in B)$.

(Lower) spectral radius of an $a \in A$:

$$r(a) := \liminf_{n\to\infty} \inf\{\|P^n(a)\|^{1/n} |\ P^n \text{ mixed } n - \text{th power}\}.$$

Proposition 17.

$$r(a) = \inf_n \inf_P\{\|P^n(a)\|^{1/n}\} = \lim_{n\to\infty} \inf_P\{\|P^n(a)\|^{1/n}\} \le \|a\|. \tag{12}$$

Proof is quite similar to the classical one for associative algebras. □

Proposition 18. *The next statements are equivalent for $a \in A$:*

(i) $r(a) = \|a\|$;

(ii) $\|P^n(a)\| = \|a\|^n \quad (\forall P^n)$.

Theorem 19. *Let $\sigma(., A, B)$ be a topological (A, B)–spectrum, $a \in A$ and $x \in \sigma(a, A, B)$. Then: $r(x) \le r(a)$.*

Proof. Let $\{h(c)|\ c \in A\}$ be the corresponding spectral covering and $\varphi \in h(a)$ such the $\varphi(a) = x$.

$$\|P^n(x)\| = \|\varphi(P^n(a))\| \le \|\varphi\|\|P^n(a)\|.$$

$\{P^n(x)|\ P^n$ mixed n–th power $\}$ is compact, therefore there exists $P_0^n(x)$ with minimal norm.

$$\|P_0^n(x)\| \le \|\varphi\|\|P_0^n(a)\| = \|\varphi\| \inf_P \|P^n(a)\|.$$

$$\inf_P \|P^n(x)\| \le \|\varphi\| \inf_P \|P^n(a)\|,$$

$$\inf_P \|P^n(x)\|^{1/n} \le \|\varphi\|^{1/n} \inf_P \|P^n(a)\|^{1/n}.$$

Letting $n \to \infty$ we get Proposition. □

Theorem 20. *Let A have a unit 1_A and B a unit 1_B and $\sigma(., A, B)$ be a unital topological spectrum. Then for $\alpha \in \mathbb{F}$, $a \in A$:*

$$\sigma(\alpha 1_A + a, A, B) = \alpha 1_B + \sigma(a, A, B). \tag{13}$$

Proof. For $\lambda > r(a)$ it holds:

$$\overline{\text{gen}\{\lambda 1_A - a\}} = \overline{\text{gen}\{a, 1_A\}}. \tag{14}$$

This will be proven in the following way. For $b := \frac{1}{\lambda}a$ and $\varepsilon := \lambda - r(a)$ we get by induction:

$$[1_A - P^m(b)] + [1_A - P^n(b)] - [1_A - P^m(b)][1_A - P^n(b)] = 1_A - P^{m+n}(b) \in \overline{\text{gen}\{\lambda 1_A - a\}}.$$

For $k = m + n$ big enough, and suitable power P_0^k it holds:

$$\|P_0^k(b)\| \leq \left(r(b) + \frac{\varepsilon}{2\lambda}\right)^k = \left(1 - \frac{\varepsilon}{2\lambda}\right)^k.$$

Therefore, $\lim_{k\to\infty}[1_A - P_0^k(b)] = 1_A$, and (14) is confirmed. Hence, if $\varphi \in h(\lambda 1_A - a)$, a local homomorphism from the spectral covering, according to Theorem 16 it is: $\varphi(1_A) = 1_B$ and $\varphi(a) \subset \sigma(a, A, B)$. Then $\alpha 1_B + \varphi(a) \subset \sigma(\alpha 1_A + a, A, B)$ for any α and the proof is complete. □

References

1. M. Benslimane, A. Rodríguez Palacios, Caractérisation spectrale des algèbres de Jordan Banach non commutatives complexes modulaires annihilatrices, J. Algebra, **140** (1991), 344-354.
2. A. Cedilnik, Spektri in zaloge vrednosti v nekaterih neasociativnih Banachovih modulih, Research report, IMFM, Ljubljana 1991.
3. R. E. Harte, Spectral mapping theorems, Proc. Royal Irish Acad. **72** (1972), Sec. A, No. 7, 89 - 107.
4. T. J. Ransford, Generalized spectra and analytic multivalued functions, J. London Math. Soc. (2), **29** (1984), 306 - 322.
5. C. Viola Devapakkiam, Jordan algebras with continuous inverse, Math. Japon., **16** (1971), 115 - 125.
6. M. A. Youngson, A Vidav theorem for Banach Jordan algebras, Math. Proc. Camb. Phil. Soc., **84** (1978), 263 - 272.

RANGES OF ELEMENTS OF A NONASSOCIATIVE ALGEBRA

ANTON CEDILNIK
Biotehniška fakulteta, University of Ljubljana
Večna pot 83, 61000 Ljubljana, Slovenia

Abstract. We generalize the notion of the numerical range in such a way that the ranges of elements of a given normed algebra are sets of elements of some other normed algebra, and that the states are linear operators. We show that many properties of numerical range rest unchanged, including the connection between range and spectrum. For Hermitian elements we deduce Vidav's lemma and associator identities $[x,x,x] = [x,x^2,x] = 0$.

Key words: nonassociative normed algebra, numerical range, algebraic range, exponential function, Hermitian element, Vidav's lemma.

Similar reasons due to which we generalized the notion of spectrum of element of algebra in [3] in such a way, that the points of a spectrum are elements of some other algebra, suggest we generalize also the numerical range. Some of it has already been done in special cases (for instance in [4]), but here we shall do this in a general sense so that we shall define states as linear operators. Much trouble is caused by the fact that it is not possible to use Hahn – Banach theorem and that instead of the function $\Re(.)$ we have to use logaritmic norm; nevertheless it is possible to regenerate almost entire theory of numerical range. We have problems also with the exponential function which is defined with mixed powers. The nonassociativity causes some estimates to be weaker then in the associative case.
Hermitian elements keep their central role in this theory, which is evident from very strong conclusions in 2. section. The validity of Vidav's lemma again confirms suitability of the definition of spectrum in [3]. The use of spectrum also reveals a surprising fact that the subalgebra generated by a Hermitian elements is almost associative.
In this study we use the symbols from [3]: $\operatorname{gen} C$, $G(a)$, p^n, p_L^n, p_R^n, σ, σ^u, $r(.)$.

1. Algebraic ranges

Let $(A, ||.||)$ and $(B, \|.\|)$ be normed algebras over $\mathbb{F} = \mathbb{R}$ or $\mathbb{C}$ with units 1_A and 1_B. For the norms we shall presume, that they are algebraic and $||1_A|| = \|1_B\| = 1$; if those were not true, there exist equivalent norms in A and B which have these properties.
Suppose that for $a \in A$ its $G(a)$ is Banach algebra. Then we can define **exponential functions**:

$$\operatorname{Exp}(a) := 1_A + \sum_{n=1}^{\infty} \frac{1}{n!} P^n(a),$$

S. González (ed.), Non-Associative Algebra and Its Applications, 93–98.

where $\{P^n | n \in \mathbb{Z}^+\}$ is a sequence of mixed powers. Special cases:

$$\operatorname{Exp}_L(a) := 1_A + \sum \frac{1}{n!} p_L^n(a)$$

and similary $\operatorname{Exp}_R(a)$.

The following Proposition will contain several important properties of exponential functions. The proofs are not difficult but sometimes pretty long. For any $a \in A$ in the Proposition, $G(a)$ is supposed to be complete.

Proposition 1.

1. *For $a \in A$ and any* Exp *it holds:*

$$\|\operatorname{Exp}(a)\| \leq e^{\|a\|}. \tag{1}$$

2. *For* Exp_1 *and* Exp_2 *and* $0 \leq \lambda \leq 1$ *there exists* Exp_3 *such that for any $a \in A$:*

$$\lambda \operatorname{Exp}_1(a) + (1 - \lambda) \operatorname{Exp}_2(a) = \operatorname{Exp}_3(a). \tag{2}$$

3. *For* Exp_1 *and* $\lambda \geq 0$ *there exists* Exp_2 *such that for any $a \in A$:*

$$\operatorname{Exp}_1(\lambda 1_A + a) = e^{\lambda} \operatorname{Exp}_2(a). \tag{3}$$

4. *For* Exp_1 *and* Exp_2 *and* $\lambda \geq 0$ *there exists* Exp_3 *such that for any $a \in A$ and $\alpha, \beta \in \mathbb{F}$, $\alpha/\beta = \lambda$:*

$$\operatorname{Exp}_1(\alpha a) \operatorname{Exp}_2(\beta a) = \operatorname{Exp}_3((\alpha + \beta)a). \tag{4}$$

5. *For* Exp_1 *and* P^n *there exists* Exp_2 *such that for any $a \in A$:*

$$P^n(\operatorname{Exp}_1(a)) = \operatorname{Exp}_2(na). \tag{5}$$

6. *Let $\|a\| \leq \frac{2}{3}$; then for any* Exp *it is:* $\operatorname{Exp}(a) \operatorname{Exp}(-a) \neq 0$.
7. *For $a \in A$ we have:*

$$\operatorname{Exp}_L(a) = \operatorname{Exp}(L_a)(1_A) \neq 0, \tag{6}$$

and similarly for Exp_R.

8. *For $a \in A$ and $\lambda \in \mathbb{F}$:*

$$\operatorname{Exp}_L(\lambda 1_A + a) = e^{\lambda} \operatorname{Exp}_L(a), \tag{7}$$

and similarly for Exp_R.

9. *Let $f : A \times \mathbb{Z}^+ \to A$ be such a map that for any $a \in A$ it holds: $\lim_{n\to\infty} f(a, n) = a$. Then:*

$$\lim_{n\to\infty} p_L^n(1_A + \frac{1}{n} f(a, n)) = \operatorname{Exp}_L(a), \tag{8}$$

and similarly for p_R and Exp_R.

10. *For $a \in A$ the set* $\{\operatorname{Exp}(a)\}$ *of values of all exponential functions is convex and compact.*
11. *For $a \in A$ and any exponential function* Exp_0 *it holds:*

$$r(\operatorname{Exp}_0(a)) \geq \lim_{n\to\infty} \inf_{\operatorname{Exp}} \{\|\operatorname{Exp}(na)\|^{1/n}\}. \tag{9}$$

Definition 2. *The set of* (A,B)**–states**:

$$\mathcal{D}(A,B) := \{\varphi \in \mathcal{B}(A,B) |\ \varphi(1_A) = 1_B \wedge \|\varphi\| = 1\}.$$

B**–range** *of an* $a \in A$: $V(a,A,B) := \{\varphi(a) |\ \varphi \in \mathcal{D}(A,B)\}$.
B**–radius** *of* a: $v(a,A,B) := \sup \|V(a,A,B)\|$.

Although these definitions are rather wider in comparison to the ones in the usual theory of numerical range, many properties rest almost unchanged, together with their proofs.

$\mathcal{D}(A,B)$ is a nonvoid closed convex subset of $\mathcal{B}(A,B)$.
$V(a,A,B)$ $(a \in A)$ is a nonvoid bounded convex subset of B.

$$\forall a \in A,\ \forall \alpha,\beta \in \mathbb{F} \ :\ V(\alpha 1_A + \beta a, A, B) = \alpha 1_B + \beta V(a,A,B). \tag{10}$$

$$\forall a,b \in A \ :\ V(a+b,A,B) \subset V(a,A,B) + V(b,A,B). \tag{11}$$

$$C \subset A \text{ subalgebra} \Rightarrow \forall a \in C :\ V(a,A,B) \subset V(a,C,B).$$

$$D \subset B \text{ subalgebra} \Rightarrow \forall a \in A :\ V(a,A,D) \subset V(a,A,B).$$

$$\forall a \in A \ :\ V(a,A,\mathbb{F})1_B \subset V(a,A,B). \tag{12}$$

$$V(a,A,B) = V(a^2,A,B) = \{0\} \Rightarrow a = 0.$$

$$V(a,A,B) = \{0\} \wedge \mathbb{F} = \mathbb{C} \Rightarrow a = 0.$$

$$\forall T \in \mathcal{B}(A) \ :\ V(T(1_A),A,B) \subset V(T,\mathcal{B}(A),B). \tag{13}$$

$$\forall a \in A :\ V(a,A,B) = V(L_a,\mathcal{B}(A),B) = V(R_a,\mathcal{B}(A),B). \tag{14}$$

It follows directly from Hahn – Banach theorem that if $\sigma(.,A,\mathbb{F})$ is a unital topological spectrum then for $a \in A$:

$$\sigma(a,A,\mathbb{F}) \subset V(a,A,\mathbb{F}), \tag{15}$$

$$r(a) \le v(a,A,\mathbf{F}), \tag{16}$$

since $r(a) \le \liminf_{n\to\infty} \|L_a^n 1_A\|^{1/n} \le r(L_a) \le v(L_a,\mathcal{B}(A),\mathbb{F}) = v(a,A,\mathbb{F})$.

$$\begin{aligned} V(a,A,B) &= \bigcup_{c\in A} \left\{\varphi(ca) |\ \varphi(c) = 1_B \wedge \|\varphi\| = \frac{1}{\|c\|}\right\} \\ &= \bigcup_{c\in A} \left\{\varphi(ac) |\ \varphi(c) = 1_B \wedge \|\varphi\| = \frac{1}{\|c\|}\right\}. \end{aligned} \tag{17}$$

$$V(a,A,B) \subset \bigcap_{\zeta\in\mathbb{F}} \{z \in B |\ \|\zeta 1_B - z\| \le \|\zeta 1_A - a\|\}. \tag{18}$$

$$\mu_A^+ : A \to \mathbb{R},\quad \mu_A^+(a) := \lim_{\tau\searrow 0} \frac{1}{\tau}(\|1_A + \tau a\| - 1)$$

is the logaritmic norm on A, and similarly μ_B^+ on B. For any $a \in A$:

$$\Re(V(a,A,\mathbb{F})) = \mu_B^+(V(a,A,B)) = [-\mu_A^+(-a), \mu_A^+(a)], \tag{19}$$

$$\mu_A^+(a) \le v(a,A,\mathbf{F}) \le v(a,A,B) \le \|a\|. \tag{20}$$

If A is Banach algebra and $a \in A$, then

$$\forall \lambda \geq 0 : \|\mathrm{Exp}_L(\lambda a)\| \leq e^{\lambda \mu_A^+(a)}; \tag{21}$$
$$\forall \lambda \leq 0 : \|\mathrm{Exp}_L(\lambda a)\| \leq e^{-\lambda \mu_A^+(-a)}; \tag{22}$$
$$\forall \zeta \in \mathbb{F} : \|\mathrm{Exp}_L(\zeta a)\| \leq e^{|\zeta| v(a,A,B)}. \tag{23}$$

(21) - (23) hold also for $\mathrm{Exp}_R(c)$, and if they hold for Exp_1 and Exp_2, they are true also for Exp_3, where $\mathrm{Exp}_3(c) = \varepsilon \,\mathrm{Exp}_1(c) + (1-\varepsilon)\,\mathrm{Exp}_2(c)$ or $\mathrm{Exp}_3(c) = \mathrm{Exp}_1(\varepsilon c)\,\mathrm{Exp}_2((1-\varepsilon)c) \quad (0 \leq \varepsilon \leq 1)$.

2. Hermitian elements

Let $(A, \|.\|)$ and $(B, \|.\|)$ be complex Banach algebras with algebraic norms and with the units 1_A and 1_B, $\|1_A\| = \|1_B\| = 1$. As usualy, $a \in A$ is **Hermitian** if $V(a, A, \mathbb{C}) \subset \mathbb{R}$, and **positive** if $V(a, A, \mathbb{C}) \subset [0, \infty)$. $\mathrm{Her}(A)$ is the set of Hermitian elements from A, and $\mathrm{Pos}(A)$ is the set of positive ones. Equivalent definitions:

$$a \in \mathrm{Her}(A) \iff \mu_A^+(\pm ia) = 0.$$
$$a \in \mathrm{Pos}(A) \iff a \in \mathrm{Her}(A) \wedge \mu^+(-a) \leq 0.$$

From (15) it follows that

$$a \in \mathrm{Her}(A) \implies \sigma(a, A, \mathbb{C}) \subset \mathbb{R},$$
$$a \in \mathrm{Pos}(A) \implies \sigma(a, A, \mathbb{C}) \subset [0, \infty),$$

for any unital topological $(A, \mathbb{C})$–spectrum σ.
A consequences of (16):

$$a \in \mathrm{Her}(A) \implies r(a) \leq \sup |\mu_A^+(\pm a)| = v(a, A, \mathbb{C});$$
$$a \in \mathrm{Pos}(A) \implies r(a) \leq \mu_A^+(a).$$

Sinclair's theorem ([6]) for $a \in \mathrm{Her}(A)$ and $\lambda, \mu \in \mathbb{C}$:

$$v(\lambda 1_A + \mu a, A, \mathbb{C}) = v(\lambda 1_A + \mu a, A, B) = \|\lambda 1_A + \mu a\|. \tag{24}$$

(24) suggests that for $a \in \mathrm{Her}(A)$, $V(a, A, B)$ is not much "bigger" then $V(a, A, \mathbb{C})$; actually, it can be proved ([2]) that

$$V(a, A, B) = \{x \in B |\ V(x, B, \mathbb{C}) \subset V(a, A, \mathbb{C})\}. \tag{25}$$

$\mathrm{Her}(A)$ is closed for the product $a, b \mapsto i(ab - ba) = i[a, b]$.
$a \in \mathrm{Her}(a) \implies L_a, R_a \in \mathrm{Her}(\mathcal{B}(A))$.
Two useful lemmas ([5]):
(i) $a \in \mathrm{Her}(A)$, $T \in \mathrm{Her}(\mathcal{B}(A))$, $T(1_A) = 0 \implies iT(a) \in \mathrm{Her}(A)$;
(ii) $a, b, c \in \mathrm{Her}(A) \implies [a, b, c] = (ab)c - a(bc) \in \mathrm{Her}(A)$.

Theorem 3. *Let $a \in A$ and $\lambda \in V(a, A, \mathbb{C})$ be a vertex of the numerical range with its inner angle $\alpha < \pi/2$. Then*

$$\lambda \in \sigma^u(a, G(a), \mathbb{C}). \tag{26}$$

Proof. We will follow the idea of B. Schmidt and A. M. Sinclair, who proved Theorem (independently of each other) for associative algebras.
Replace a with $b := e^{i\beta}(a - \lambda 1_A)$, where β is chosen in such a way that $V(b, A, \mathbb{C})$ is inside the angle $\varphi = \pi \pm \frac{\alpha}{2}$ in a complex plane. Since $\max \Re(V(e^{i\gamma}b, A, C)) = 0 = \mu_A^+(e^{i\gamma}b)$ for any $\gamma : |\gamma| < (\pi - \alpha)/2$, it follows from (21):

$$\| \operatorname{Exp}(\tau e^{i\gamma}b)\| \leq 1 \tag{27}$$

($\tau \geq 0$), where Exp is a suitable exponential function (see the end of the third section). Because of $0 \in V(b, A, \mathbb{C})$, there exists $\varphi \in D(A, \mathbb{C})$, such that $\varphi(b) = 0$. We will prove by induction that $\varphi(P^n(b)) = 0$ for any mixed power; of course it is enough to prove it for pure powers only.
Suppose that $\varphi(p^k(b)) = 0$ for $k < n$ and consider some p_0^n.

$$\operatorname{Exp}_\omega(b) := \operatorname{Exp}_L(\omega_1 b)\operatorname{Exp}_L(\omega_2 b)\ldots\operatorname{Exp}_L(\omega_n b),$$

where $\sum \omega_i = 1$, $\omega_i \geq 0$, and the product has the same arrangement of Exp–s as $p_0^n(b)$ of $b - s$. Then $\operatorname{Exp}_\omega$ fulfills (27) and a part of its n–th term $P_\omega^n(b)$ is $\omega_1 \ldots \omega_n p_0^n(b)$.

$$\begin{aligned} 1 \geq |\varphi(\operatorname{Exp}_\omega(\tau e^{i\gamma}b))| &= |1 + \frac{\tau^n}{n!}e^{ni\gamma}\varphi(P_\omega^n(b)) + \cdots| \\ &= |1 + \frac{\tau^n}{n!}e^{ni\gamma}\varphi(P_\omega^n(b))| + o(\tau^n) = 1 + \frac{\tau^n}{n!}\Re(e^{ni\gamma}\varphi(P_\omega^n(b)) + o(\tau^n). \end{aligned}$$

Therefore: $\Re(e^{ni\gamma}\varphi(P_\omega^n(B))) \leq 0$. But because of the variability of γ we get $\varphi(P_\omega^n(b)) = 0$ for any set of ω-s; so, $\varphi(p_0^n(b)) = 0$ too. Therefore φ is a local homomorphism on $G(a)$ and, since $\varphi(a) = \lambda$, Proposition 15 from [3] gives (26). □

If we apply this Theorem to an $a \in \operatorname{Her}(A)$ and $\lambda = \mu_A^+(a)$ or $\lambda = -\mu_A^+(-a)$, we get **Vidav's lemma**.

Theorem 4. *For $a \in \operatorname{Her}(A)$ it holds:*

$$\pm \mu_A^+(\pm a) \in \sigma^u(a, G(a), \mathbb{C}), \tag{28}$$

$$\operatorname{co} \sigma^u(a, G(a), \mathbb{C}) = V(a, A, \mathbb{C}), \tag{29}$$

$$\max |\sigma^u(a, G(a), \mathbb{C})| = r(a) = v(a, A, B) = \|a\|. \tag{30}$$

Since $\sigma(P^n(a), G(a), \mathbb{C}) = \sigma(a, G(a), \mathbb{C})^n$, we get:

$$a \in \operatorname{Her}(A) \Longrightarrow \|a\|^n = r(a)^n \leq r(P^n(a)) \leq \|P^n(a)\| \leq \|a\|^n.$$

If we consider (16) too, we find out:

$$r(P^n(a)) = v(P^n(a), A, B) = \|P^n(a)\| = \|a\|^n. \tag{31}$$

Corollary 5. $a \in \operatorname{Pos}(A)$, $\{\alpha_i | \ i \in \mathbb{N}\} \subset [0, \infty)$; *if* $\sum_{n=1}^{\infty} \alpha_n \|a\|^n < \infty$ *then*

$$\|\alpha_0 1_A + \sum_{n=1}^{\infty} \alpha_n P^n(a)\| = \alpha_0 + \sum_{n=1}^{\infty} \alpha_n \|a\|^n \tag{32}$$

for any sequence of powers P^n.

Proof. From (29) it follows that $\mu_A^+(a) = ||a||$. Let φ be the local homomorphism from the proof of the Theorem 3. Then:

$$\alpha_0+\sum \alpha_n||a||^n = \varphi(\alpha_0 1_A+\sum \alpha_n P^n(a)) \le ||\alpha_0 1_A+\sum \alpha_n P^n(a)|| \le \alpha_0+\sum \alpha_n||a||^n$$

□

Corollary 6. *For* $a \in \mathrm{Her}(A)$ *and* $\tau \in \mathbf{R}$ *there is*

$$||\,\mathrm{Exp}(i\tau a)|| \ge 1. \tag{33}$$

If Exp *is such an exponential function that (21) holds, then there is "=" in (33).*

Proof. Let $J \subset G(a)$ be the closed twosided ideal generated by the set of associators. Because of the existence of unital $(A,\mathbb{C})$–spectrum, $J \ne G(a)$. Let $H := G(a)/J$ with the quotient norm $|||.|||$, and $G(a) \xrightarrow{\wedge} H$ the quotient map. Then, considering the associativity of H,

$$1 = |||\,\mathrm{Exp}(i\tau\widehat{a})||| = |||(\mathrm{Exp}(i\tau a))^\wedge||| \le ||\,\mathrm{Exp}(i\tau a)||.$$

The fact that $\widehat{a}$ is Hermitian is easily verified from the definition. □

Proposition 7. $a \in \mathrm{Her}(A) \Longrightarrow [a,a,a] = 0$.

Proof. $i[L_a,R_a] \in \mathrm{Her}(\mathcal{B}(A))$ and $i[L_a,R_a](1_A) = 0 \Longrightarrow i[L_a,R_a](a) = i(aa^2 - a^2a) \in \mathrm{Her}(A)$.
$\sigma^u(i[a,a,a],A,\mathbb{C}) = \{0\} \Longrightarrow V(i[a,a,a],A,\mathbb{C}) = \{0\} \Longrightarrow i[a,a,a] = 0$.□

Proposition 8. $a \in \mathrm{Her}(A) \Longrightarrow [a,a^2,a] = 0$.

Proof. The same as the previous proof, but with operator $[L_a,[L_a,R_a]]$.□

We can continue in such a way with the operators $i[L_a,[L_a,[L_a,R_a]]]$, $[[L_a,[L_a,R_a]],[L_a,R_a]]$, etc., but there still remains the question of the validity of $[a,a,a^2] = 0$, which would be enough for the associativity of $G(a)$.

References

1. A. Bensebah, JV–algèbres et JH*–algebres, Canad. Math. Bull., **34** (4) (1991), 447 – 455.
2. A. Cedilnik, Spektri in zaloge vrednosti v nekaterih neasociativnih Banachovih modulih, Research report, IMFM, Ljubljana 1991.
3. A. Cedilnik, Spectra of elements of a nonassociative algebra, These Proceedings.
4. D. R. Farenick, Matricial Extensions on the Numerical Range: A Brief Survey, Linear and Multilinear Algebra, **34** (1993), 197 – 211.
5. J. Martínez – Moreno, A. Mojtar – Kaidi, A. Rodríguez – Palacios, On a nonassociative Vidav – Palmer theorem, Quart. J. Math. Oxford (2), **32** (1981), 435 – 442.
6. A. Rodríguez – Palacios, Non–associative normed algebras spanned by Hermitian elements, Proc. London Math. Soc. (3), **47** (1983), 258 – 274.
7. M. A. Youngson, A. Vidav theorem for Banach Jordan algebras, Math. Proc. Camb. Phil. Soc., **84** (1978), 263 – 272.

COORDINATIZATION OF JORDAN ALGEBRAS OVER LOCALLY RINGED SPACES

JOHANNES CLEVEN
Fachbereich Mathematik, FernUniversität - Gesamthochschule -
Lützowstraße 125
D-58084 Hagen, Germany

Abstract. The classical theory of reduced Jordan algebras is rephrased in the setting of locally ringed spaces. Coordinatization of such algebras leads to the concept of composition triples, i.e. a generalization of composition algebras, over locally ringed spaces. Over a fixed locally ringed space composition triples define a category as well as reduced Jordan algebras do. A natural equivalence between these categories is shown. A generalization of the classical Cayley-Dickson-Doubling Process allows the classification of composition triples over the projective line. By taking global sections composition triples of rank 8 induce Jordan algebras over the base field with pretty big radicals.

Key words: Jordan algebra, composition triple, quadratic space (Mathematics Subject Classification (1991): 17C50)

Introduction

In 1954 N. Jacobson [J1] proved the following coordinatization theorem. Let J be a Jordan algebra over a field of characteristic $\neq 2$ satisfying the following conditions: (1) J has an identity $1 = e_1 + \ldots + e_n$ where the e_i are nonzero orthogonal idempotents and $n \geq 3$, (2) J contains $n-1$ elements $u_{1j}, j = 2, \ldots, n$, such that $e_1 u_{1j} = e_j u_{1j} = \frac{1}{2} u_{1j}$ and $u_{1j}^2 = e_1 + e_j$. Then there exists an algebra D with an identity and an involution which is associative if $n \geq 4$ and alternative with its self-adjoint elements in the nucleus if $n = 3$, and an isomorphism $\varphi : J \to H(D_n)$ (the Jordan algebra of hermitian matrices with entries in D).

In case the Jordan algebra J is reduced with respect to nonzero orthogonal idempotents $e_1, \ldots, e_n$ $(n \geq 3)$, and $e_1 + \ldots + e_n = 1$, and J is central simple N. Jacobson [J2] showed that J is isomorphic to $H(D_n, j)$ where D is a composition algebra over k and j is an involution on D_n. For $n \geq 4$ D is associative and for $n = 3$ D is alternative.

In this paper we consider Jordan algebras over a locally ringed space X which are reduced with respect to a triple of nonzero orthogonal idempotents. This triple of idempotents allows a Peirce decomposition of J. As the off-diagonal Peirce components in general do not contain global sections u_{ij} with $u_{ij}^2 = \lambda(e_i + e_j)$ where λ is an invertible global section in $\mathcal{O}_X$, we cannot expect to get a version of the coordinatization theorem which involves one single algebra D over X. What we get is a triple of $\mathcal{O}_X$-modules equipped with quadratic forms and multiplications between

S. González (ed.), Non-Associative Algebra and Its Applications, 99–105.

the components of such a triple. By such a so-called composition triple our Jordan algebra J is completely determined.

The terminology adopted in this paper is the standard one, cf. [J2], [P1], concepts from algebraic geometry not explained in the text are to be understood in the sense of Hartshorne [H]. Due to the lack of space proofs are omitted and will appear elsewhere.

1. A categorical equivalence

1.1. Let X be a locally ringed space, $\mathcal{O}_X$ its structure sheaf, where $\frac{1}{2} \in \Gamma(X, \mathcal{O}_X)$, i.e. the global section $2 \in \Gamma(X, \mathcal{O}_X)$ is invertible. We call a locally free $\mathcal{O}_X$-module $\mathcal{A}$ of finite rank an algebra over X iff

(i) there is a multiplication ($\mathcal{O}_X$-bilinear mapping) $\mathcal{A} \times \mathcal{A} \to \mathcal{A}$ denoted by $(u, v) \mapsto uv$ for all local sections u, v in $\mathcal{A}$;

(ii) there is a global section $1 \in \Gamma(X, \mathcal{O}_X)$ such that $u1|_U = u = 1|_U u$ for all $u \in \Gamma(U, \mathcal{A}), U \subset X$ open. We call 1 the unity of $\mathcal{A}$.

1.2. Let $\mathcal{J}(X)$ be the category of reduced Jordan-Azumaya algebras over X, consisting of pairs $(\mathcal{J}, (e_1, e_2, e_3))$ where $\mathcal{J}$ is a commutative algebra over X, and (e_1, e_2, e_3) is a triple of orthogonal idempotents $e_i \in \Gamma(X, \mathcal{J})$ with $e_1 + e_2 + e_3 = 1$ (the unity of $\mathcal{J}$) and $e_i \neq 0, 1$ such that:

(i) $u^2(uv) = u(u^2v)$ for all local sections u, v in $\mathcal{J}$;

(ii) $\mathcal{J}(p) := \mathcal{J}_p \otimes_{\mathcal{O}_{X,p}} \kappa(p)$ is a central simple Jordan algebra over $\kappa(p) := \mathcal{O}_{X,p}/m_{X,p}$, the residue field over $p \in X$ $(\forall p \in X)$;

(iii) $\mathcal{J}_{ii} \cong \mathcal{O}_X e_i$ as algebras over X for all i, where $\mathcal{J}_{ij}$ denotes the ij-th Peirce component of the Peirce decomposition of $\mathcal{J}$ with respect to (e_1, e_2, e_3).

The Peirce decomposition of Jordan algebras in the setting of locally ringed spaces can be defined in a similar way as in the classical situation of Jordan algebras over a field. The Peirce components are locally free $\mathcal{O}_X$-modules.

The morphisms $\varphi : (\mathcal{J}, (e_1, e_2, e_3)) \to (\mathcal{J}', (e_1', e_2', e_3'))$ in the category $\mathcal{J}(X)$ are algebra homomorphisms $\varphi : \mathcal{J} \to \mathcal{J}'$ with $\varphi(e_i) = e_i', i \in \{1, 2, 3\}$.

A standard consequence of the Peirce decomposition is the following Lemma. For the properties of the Peirce decomposition confer [J2] or [P1].

Lemma 1.3. Let $(\mathcal{J}, (e_1, e_2, e_3))$ be an object in $\mathcal{J}(X)$. Then the Peirce decomposition of $\mathcal{J}$ with respect to (e_1, e_2, e_3) defines a triple

$$((\mathcal{J}_{12}, N_{12}), (\mathcal{J}_{13}, N_{13}), (\mathcal{J}_{23}, N_{23}))$$

where $N_{ij} : \mathcal{J}_{ij} \to \mathcal{O}_X$ are quadratic forms, defined on the off-diagonal Peirce components $\mathcal{J}_{ij}$ by

$$4u_{ij}^2 = N_{ij}(u_{ij})(e_i + e_j)$$

for all local sections u_{ij} in $\mathcal{J}_{ij}$, $i, j \in \{1, 2, 3\}$, $i \neq j$.
The quadratic form N_{ij} is regular, i.e. the polar of N_{ij} induces an $\mathcal{O}_X$-module isomorphism

$$N_{ij} : \mathcal{J}_{ij} \to \check{\mathcal{J}}_{ij} = \mathcal{H}om_{\mathcal{O}_X}(\mathcal{J}, \mathcal{O}_X).$$

Further the quadratic forms N_{ij} allow composition, i.e. we get the following identity:

$$N_{ik}(u_{ij}u_{jk}) = N_{ij}(u_{ij})N_{jk}(u_{jk})$$

for all local sections $u_{\mu\nu}$ in $\mathcal{J}_{\mu\nu}$, where $\{i, j, k\} = \{1, 2, 3\}$.

Remark 1.4. The regularity of the quadratic forms N_{ij} is guaranteed by the fact that $\mathcal{J}$ is a Jordan-Azumaya algebra, in particular that $\mathcal{J}(p)$ is central simple and thus isomorphic to $H(D_3, j)$, where D is a composition algebra over $\kappa(p)$. N_{ij} induces a quadratic form on $\mathcal{J}(p)_{ij}$ denoted by $N_{ij} \otimes \kappa(p)$. We can find a scalar $\lambda \in \kappa(p)\,(\lambda \neq 0)$ such that $N_{ij} \otimes \kappa(p) = \lambda n$ where n is the regular quadratic form on D.

Our Lemma leads us to the following definition.

Definition 1.5. Let $\mathcal{C}(X)$ the category of composition triples $\mathcal{C} = (\mathcal{C}_1, \mathcal{C}_2, \mathcal{C}_3)$ over X, which are characterized by following data:

(i) $\mathcal{C}_i = (\mathcal{C}_i, N_i)$ (with a certain abuse of notation) are locally free $\mathcal{O}_X$-modules of identical (constant) ranks equipped with regular quadratic forms $N_i : \mathcal{C}_i \to \mathcal{O}_X$,

(ii) there is a multiplication (an $\mathcal{O}_X$-bilinear mapping) $\mathcal{C}_1 \times \mathcal{C}_2 \to \mathcal{C}_3$ denoted by $(u_1, u_2) \mapsto u_1 u_2$

subject to the condition

$$N_3(u_1 u_2) = N_1(u_1)N_2(u_2)$$

for all local sections u_i in $\mathcal{C}_i$ (the quadrataic forms allow composition). Morphisms $\varphi : \mathcal{C} \to \mathcal{C}'$ are triples $\varphi = (\varphi_1, \varphi_2, \varphi_3)$ of $\mathcal{O}_X$-module homomorphisms such that
(a) the diagram

$$\begin{array}{ccc} \mathcal{C}_1 \times \mathcal{C}_2 & \longrightarrow & \mathcal{C}_3 \\ \downarrow \varphi_1 \times \varphi_2 & & \downarrow \varphi_3 \\ \mathcal{C}'_1 \times \mathcal{C}'_2 & \longrightarrow & \mathcal{C}'_3 \end{array}$$

is commutative, where the horizontal arrows denote the multiplication on $\mathcal{C}$ and $\mathcal{C}'$, respectively;
(b) $N'_i \circ \varphi_i = N_i$, i.e. φ_i is an isometry.

Remarks 1.6. **(i)** A composition algebra over X (cf. [P2]) is a composition triple over X.
(ii) Due to the regularity of all $N_i : \mathcal{C}_i \to \mathcal{O}_X$ we obtain multiplications $\mathcal{C}_i \times \mathcal{C}_j \to \mathcal{C}_k$ for all cyclic permutations (i,j,k) of $(1,2,3)$ such that

$$\begin{array}{ll} \textbf{(a)}\ N_k(u_iu_j, u_k) = N_i(u_i, u_ju_k) & \textbf{(b)}\ N_k(u_iu_j) = N_i(u_i)N_j(u_j) \\ \textbf{(c)}\ (u_iu_j)u_i = N_i(u_i)u_j & \textbf{(d)}\ u_i(u_ku_i) = N_i(u_i)u_k \\ \textbf{(e)}\ N_k(u_iu_j, u_k)u_i = (u_ku_i)(u_iu_j) + N_i(u_i)u_ju_k & \end{array}$$

for all local section u_μ in $\mathcal{C}\mu$.
(iii) Locally a composition triple induces a composition algebra. To be more precise: Let $p \in X$, then there is an open subset $U_p \subset X$ such that we can find local sections s_i in $\mathcal{C}_i$ with $N_i(s_i) \in \Gamma(U_p, \mathcal{O}_X)$ is invertible. This is implied by the regularity of N_i. On $\mathcal{C}_{3|U_p}$ we define a multiplication as follows:

$$x \circ y := (s_2x)(ys_1) \text{ for all local sections } x, y \text{ in } \mathcal{C}_{3|U_p}.$$

Then it can be shown that $(\mathcal{C}_{3|U_p}, \circ)$ is a composition algebra over U_p.
(iv) As consequence of this construction we see that composition triples occur only in ranks 1, 2, 4, and 8.

Theorem 1.7. The Peirce decomposition defines a functor $F : \mathcal{J}(X) \to \mathcal{C}(X)$ by $(\mathcal{J}, (e_1, e_2, e_3)) \mapsto ((\mathcal{J}_{12}, N_{12}), (\mathcal{J}_{13}, N_{13}), (\mathcal{J}_{23}, N_{23}))$. F is an isomorphism, i.e. there is a functor $G : \mathcal{C}(X) \to \mathcal{J}(X)$ such that we get natural equivalences

$$\text{(a) } F \circ G \cong \mathrm{id}_{\mathcal{C}(X)} \text{ and (b) } G \circ F \cong \mathrm{id}_{\mathcal{J}(X)}.$$

Remark about the proof: We obtain G by means of McCrimmon's "General construction of quadratic Jordan algebras" [M]. $G(B)$ carries a so-called cubic form structure.

2. The Cayley-Dickson-Doubling Process in the category of composition triples

By theorem 1.7 we have classified reduced Jordan-Azumaya algebras over X up to isomorphism when we have classified the composition triples over X up to isomorphism. A useful implement concerning the classification of composition triples is the so-called Cayley-Dickson-Doubling Process.

Theorem 2.1 (Cayley-Dickson-Doubling). **(i)** Let $\mathcal{D}$ be a composition triple over X. Let $\mathcal{P} = (\mathcal{P}_1, \mathcal{P}_2, \mathcal{P}_3)$ a triple consisting of regular quadratic spaces over X of identical ranks equal to rank $\mathcal{D}$ such that $(\mathcal{P}_1, \mathcal{P}_2, \mathcal{D}_3), (\mathcal{D}_1, \mathcal{P}_2, \mathcal{P}_3)$ and $(\mathcal{P}_1, \mathcal{D}_2, \mathcal{P}_3)$ are composition triples subject to the condition

$$(*) \qquad N_{\mathcal{D}_3}(d_1d_2, p_1p_2) + N_{\mathcal{P}_3}(d_1p_2, p_1d_2) = 0$$

for all local sections d_i in $\mathcal{D}_i, p_i$ in $\mathcal{P}_i$, where $N_{\mathcal{D}_3}(-,-)$ and $N_{\mathcal{P}_3}(-,-)$ is the polar of the quadratic form $N_{\mathcal{D}_3} : \mathcal{D}_3 \to \mathcal{O}_X$ and $N_{\mathcal{P}_3} : \mathcal{P}_3 \to \mathcal{O}_X$, respectively.
Define

$$\mathcal{C}_i := \mathcal{D}_i \oplus \mathcal{P}_i, \; N_{\mathcal{C}_i} = N_{\mathcal{D}_i} \oplus N_{\mathcal{P}_i}$$

and define a multiplication

$$\circ : \mathcal{C}_1 \times \mathcal{C}_2 \to \mathcal{C}_3$$

by

$$(d_1, p_1) \circ (d_2, p_2) := (d_1 d_2 + p_1 p_2, p_1 d_2 + d_1 p_2).$$

Then Cay $(\mathcal{D}, \mathcal{P}) := (\mathcal{C}_1, \mathcal{C}_2, \mathcal{C}_3)$ equipped with the multiplication $\circ$ is a composition triple over X.

(ii) Let $\mathcal{D}$ be a composition triple over X, let $\mathcal{D} \hookrightarrow \mathcal{C}$ be a composition subtriple of $\mathcal{C}$ with 2 rank $\mathcal{D} =$ rank $\mathcal{C}$. Then exists a triple $\mathcal{P} = (\mathcal{P}_1, \mathcal{P}_2, \mathcal{P}_3)$ of regular quadratic spaces over X such that $\mathcal{D}$ and $\mathcal{P}$ suffice the conditions of (i). The identity of $\mathcal{D}$ extends to an isomorphism φ : Cay $(\mathcal{D}, \mathcal{P}) \to \mathcal{C}$.

Remarks 2.2. **(i)** This Cayley-Dickson Process generalizes the classical Cayley-Dickson Process of composition algebras over fields, and H. P. Petersson's Cayley-Dickson Process formulated in the setting of composition algebras over locally ringed spaces [P2].

(ii) In the classical setting of composition algebras over fields the condition $(*)$ implies the associativity of $\mathcal{D}$.

3. Composition triples over the projective line

Let $\mathbb{P}^1_k$ be the projective line over the field k (char $k \neq 2$).

Each composition triple consists of regular quadratic spaces. Due to the assumption $\frac{1}{2} \in \Gamma(X, \mathcal{O}_X)$ the category of regular quadratic spaces corresponds bijectively to the category of regular symmetric bilinear spaces over X. A theorem of Knebusch [K] tells us what regular symmetric bilinear spaces over the projective line look like.

Theorem 3.1 (Knebusch). Let $\mathcal{E}$ be a regular symmetric bilinear space over $\mathbb{P}^1_k$, then $\mathcal{E}$ decomposes into an orthogonal sum of hyperbolic planes $h(\mu_i) := \mathcal{O}_X(\mu_i) \oplus \mathcal{O}_X(-\mu_i)$ equipped with the canonical hyperbolic form $N_{h(\mu)}((t, \check{t})) =< t, \check{t} >$, and a regular symmetric bilinear space $\sigma^*(V)$ defined over k, i.e. V is a regular symmetric bilinear space over k, where $\sigma : \mathbb{P}^1_k \to k$ is the so-called structure morphism. We have $\mathcal{E} \cong h(\mu_1) \perp \ldots \perp h(\mu_n) \perp \sigma^*(V)$.

Definition 3.2. **(i)** We call a composition triple $\mathcal{C}$ over $\mathbb{P}^1_k$ not defined over k, if a least one component of $\mathcal{C}$ contains a hyperbolic plane $h(\mu)$ with $\mu > 0$ as an orthogonal summand.

(ii) We call a composition triple over $\mathbb{P}^1_k$ of rank 2 a torus, if each component of $\mathcal{C}$ is a hyperbolic plane.

The following Lemma classifies the tori over $\mathbb{P}^1_k$.

Lemma 3.3. Let $\mathcal{C}$ be a torus over $\mathbb{P}^1_k$. Then $\mathcal{C}$ is represented up to isomorphism and cyclic permutation of $\mathcal{C}'s$ components by the following torus $\tilde{\mathcal{C}}$ over $\mathbb{P}^1_k$. Define

$$\tilde{\mathcal{C}} := (h(l_1), h(l_2), h(l_3))$$

equipped with the multiplication

$$h(l_1) \times h(l_2) \rightarrow h(l_3)$$

defined by

$$(t_1, \check{t}_1) \circ (t_2, \check{t}_2) = (t_1 \otimes t_2, \check{t}_1 \otimes \check{t}_2)$$

where $l_3 = l_1 + l_2 \geq 0$, $l_1, l_2 \in \mathbb{N}_0$; where we have identified $\mathcal{O}_X(l_1 + l_2) = \mathcal{O}_X(l_1) \otimes_{\mathcal{O}_X} \mathcal{O}_X(l_2)$.

A theorem of H. P. Petersson [P2] shows that a composition algebra over $\mathbb{P}^1_k$ contains a torus that is defined over k. By means of Knebusch's theorem one can prove the following theorems.

Theorem 3.4. Let $\mathcal{C}$ be a composition triple over $\mathbb{P}^1_k$ that is not defined over k. Then $\mathcal{C}$ contains a torus $\mathcal{T}$, i.e. a composition subtriple of rank 2 that is a torus.

Theorem 3.5. Let $\mathcal{C}$ be a composition triple of rank 8 over $\mathbb{P}^1_k$.
Then $\mathcal{C}$ contains a composition subtriple $\mathcal{D}$ of rank 4.

In case $\mathcal{C}$ contains two components in which we can find global sections with invertible norm, then is $\mathcal{C}$ induced by a composition algebra over $\mathbb{P}^1_k$. Due to the classification of composition algebras over $\mathbb{P}^1_k$, carried out by H. P. Petersson, the statements of 3.5 and 3.6 are quite trivial. For the remaining case analyse the multiplication $\mathcal{C}_1 \times \mathcal{C}_2 \rightarrow \mathcal{C}_3$ after restriction on hyperbolic planes $h(\mu_1) \times h(\mu_2) \rightarrow \mathcal{C}_3$ on local sections in $h(\mu_1)$ and $h(\mu_2)$. By means of Knebusch's theorem and Krull-Schmidt we can construct a torus and furthermore a subtriple of rank 4.

An obvious consequence of 3.5 and 3.6 is the following corollary.

Corollary 3.6. Every composition triple $\mathcal{C}$ of rank ≥ 4 over $\mathbb{P}^1_k$ is obtained by iterating the Cayley-Dickson process. The components of $\mathcal{C}$ decompose completely into orthogonal sums of hyperbolic planes.

Instead of writing down the series which classify composition triples over $\mathbb{P}^1_k$, we consider now the global sections of Jordan algebras induced by composition triples of rank 8 over $\mathbb{P}^1_k$.
Let $\mathcal{C}$ be a composition triple of rank 8 over $\mathbb{P}^1_k$, let $\mathcal{J}_\Gamma := \Gamma(\mathbb{P}^1_k, G(\mathcal{C}))$ the global sections of $G(\mathcal{C})$ the Jordan algebra induced by $\mathcal{C}$.

There is one series of rank 8 composition triples $\mathcal{C}$ over $\mathbb{P}^1_k$ which has the following properties in case the components of these composition triples do not contain an orthogonal summand that is defined over k.

(i) each element of the off-diagonal part of $\mathcal{J}_\Gamma$ is nilpotent of index 3,

(ii) the off-diagonal part of $\mathcal{J}_\Gamma$ is a nilpotent algebra of index 6,

(iii) $\dim_k \mathcal{J}_\Gamma = 15 + 10\alpha + 6\beta + 6\gamma + 6\delta \geq 43$

(iv) $\dim_k \operatorname{rad} \mathcal{J}_\Gamma = 12 + 10\alpha + 6\beta + 6\gamma + 6\delta \geq 40$

where $\alpha, \beta, \gamma, \delta \in \mathbb{N}$ are certain parameters which determine the isomorphism class of composition triples within a series.

Among the rank 8 composition triples over $\mathbb{P}^1_k$ we find the octonion algebras over $\mathbb{P}^1_k$. It is quite easy to show that $\mathcal{J}_\Gamma$ is an exceptional Jordan algebra over k in case C is an octonion algebra over $\mathbb{P}^1_k$. We can use Glennie's s-identity of degree 8 $G_8(x, y, z)$ and put in for x, y, z the usual elements on $\mathcal{J}_\Gamma[G]$. As the structure of rank 8 composition triples are similar to a certain extend one might expect that $\mathcal{J}_\Gamma$ is always an exceptional Jordan algebra. This is now an open problem again.

References

[G] C. M. Glennie. *Some identities valid in special Jordan algebras but not valid in all Jordan algebras.* Pacific J. Math. 16 (1966), no. 1, 47 - 59.

[H] R. Hartshorne. *Algebraic geometry.* Graduate Texts in Math., vol. 52, Springer-Verlag, New York, Heidelberg und Berlin, 1977.

[J1] N. Jacobson. *Structure of alternative and Jordan bimodules.* Osaka Math. J. 6 (1954), 1 - 71.

[J2] - *Structure and Representations of Jordan Algebras.* Amer. Math. Soc. Colloq. Publ., Vol. 39, Providence, Rhode Island, 1968.

[K] M. Knebusch. *Grothendieck- und Wittringe von nicht-ausgearteten symmetrischen Bilinearformen.* Sitzungsber. Heidelb. Akad. Wiss. Math.-Natur. Kl., Springer-Verlag, Berlin, Heidelberg und New York, 1970.

[M] K. McCrimmon. *The Freudenthal-Springer-Tits constructions of exceptional Jordan algebras.* Trans. Amer. Math. Soc. 139 (1969), 495 - 510.

[P1] H. P. Petersson. *Einführung in die Theorie der Jordan-Algebren.* Kurs 1317 der FernUniversität, Fachbereich Mathematik, Hagen, 1980.

[P2] - *Composition algebras over algebraic curves of genus zero.* Trans. Amer. Math. Soc. 337 (1993) 1, 473 - 493.

SUR LA DÉCOMPOSITION DE PEIRCE

JOÃO CARLOS DA MOTTA FERREIRA*
Universidade Federal de Mato Grosso do Sul, Departamento de Matemática, Caixa Postal 649, 79070-900 Campo Grande,MS, Brsil

and

ARTIBANO MICALI
Département des Sciences Mathématiques, Université Montpellier II, Place Eugène Bataillon, 34095 Montpellier cedex 5, France

Abstract. The aim of this paper is to give a general theory of Peirce decomposition in nonassociative algebras. This theory generalizes the theory of associative and alternative algebras and permit also to us, among other results, the characterization of the nilradical of a finite dimensional algebra which is flexible and power-associative.

Key words: AMS Subject Classification 17 A 01, 17 A 05, 17 A 20, 17 A 60.

1. Historique

Nous rappelons ici que les notions d'éléments idempotents et nilpotents ont été introduites par le mathématicien Benjamin Peirce (1809-1880) à partir de 1870 dans ses recherches concernant la structure générale des algèbres réelles ou complexes associatives de dimension finie et il démontre qu'une algèbre possédant au moins un élément non nilpotent, admet un idempotent non nul; c'est encore lui qui écrit l'identité

$$x = exe + (xe - exe) + (ex - exe) + (x - xe - ex + exe)$$

valable dans toute algèbre alternative et pour tout idempotent e dans cette algèbre (cf. [5]) sans toutefois aboutir effectivement à ce que nous appelons aujourd'hui la décomposition de Peirce (cf. [3], Note Historique). La suite a été accomplie par A. Adrian Albert et son Ecole (cf. [2]).

2. Préliminaires

Par la suite F désignera un corps commutatif pour lequel, chaque fois que la caractéristique intervient nous le mentionnerons. Pour toute F-algèbre U, le *commutateur* de U, défini par $[x, y] = xy - yx$ pour x et y parcourant U, mesure

* Supported by CAPES, Brazil, Processo n° 1116/91-13.

S. González (ed.), Non-Associative Algebra and Its Applications, 106–113.

le *degré de non commutativité* de l'algèbre U et l'*associateur* de U, défini par $(x,y,z) = (xy)z - x(yz)$ pour x, y et z parcourant U, mesure le *degré de non associativité* de l'algèbre U. Rappelons que l'associateur d'une F-algèbre U est une application F-linéaire dans chaque variable et qu'il vérifie l'*identité de Teichmüller*, à savoir, $x(y,z,t) + (x,y,z)t = (xy,z,t) - (x,yz,t) + (x,y,zt)$, pour x, y, z et t parcourant U.

Nous dirons qu'une F-algèbre U admet une *décomposition de Peirce relative à un ensemble d'idempotents* $\{e_1,\ldots,e_n\}$ (de U), deux à deux orthogonaux, si les *espaces de Peirce* de U définis par

$$U_{ij} = \{x \mid x \in U, e_k x = \delta_{ki} x, x e_k = \delta_{jk} x\},$$

$(i,j = 0,1,\ldots,n)$, où les δ_{ij}, $(i,j = 0,1,\ldots,n)$, sont les *deltas de Kronecker*, vérifient la condition $U = \bigoplus_{i,j=0}^{n} U_{ij}$.

Nous dirons qu'une F-algèbre U est une *algèbre avec décomposition de Peirce* si U possède décomposition de Peirce relative à tous les ensembles d'idempotents de U, deux à deux orthogonaux.

Le lemme qui suit généralise le lemme 2 de [7] sur la décomposition de Peirce relative au cas d'un seul idempotent, pour un ensemble fini d'idempotents deux à deux orthogonaux.

2.1. LEMME

Soient U une F-algèbre et $\{e_1,\ldots,e_n\}$ un ensemble d'idempotents de U, deux à deux orthogonaux. Les assertions suivantes sont équivalentes: (i) l'algèbre U possède décomposition de Peirce relative à l'ensemble d'idempotents $\{e_1,\ldots,e_n\}$, vérifiant les conditions $U_{ij}U_{jl} \subset U_{il}$, $(i,j,l = 0,1,\ldots,n)$, $U_{ij}U_{ij} \subset U_{ji}$, $(i,j = 0,1,\ldots,n)$, et $U_{ij}U_{kl} = 0$ si $j \neq k$ et $(i,j) \neq (k,l)$, $(i,j,k,l = 0,1,\ldots,n)$; (ii) les idempotents $e_1,\ldots,e_n$ vérifient les conditions $(e_k,x,y) = -(x,e_k,y) = (x,y,e_k)$, $(k = 1,\ldots,n)$, quels que soient les éléments x et y dans U.

Dans la suite de cet article nous supposerons que toute décomposition de Peirce vérifie les conditions du lemme ci-dessus.

3. Idempotents primitifs et principaux

On montre, dans ce paragraphe, qu'à partir de la décomposition de Peirce on peut généraliser, de façon analogue, les résultats concernant les idempotents primitifs et principaux des algèbres alternatives.

Soient U une F-algèbre et e un idempotent de U. On dit que e est un *idempotent principal* de U si U ne possède aucun idempotent non nul u orthogonal à e, ou encore, tel que $ue = eu = 0$.

Les démonstrations des deux théorèmes qui suivent sont des conséquences des décompositions de Peirce et se font comme dans le cas alternatif.

3.1. THÉORÈME

Soit U une F-algèbre de dimension finie ayant au moins un idempotent non nul. Si U est une algèbre avec décomposition de Peirce, alors U possède un idempotent principal.

Soient U une F-algèbre et e un idempotent non nul de U. On dit que e est un ***idempotent primitif*** de U, s'il n'existe pas des idempotents non nuls u et v de U tels que $e = u + v$ et $uv = vu = 0$.

3.2. THÉORÈME

Soit U une F-algèbre de dimension finie ayant des idempotents non nuls. Si U est une algèbre avec décomposition de Peirce, alors pour tout idempotent non nul u de U, il existe des idempotents primitifs $e_1, \ldots, e_r, e_{r+1}, \ldots, e_n$, $(1 \leq r \leq n)$, deux à deux orthogonaux, tels que $u = e_1 + \cdots + e_r$ et $e = e_1 + \cdots + e_n$ est un idempotent principal de U.

4. Algèbres flexibles à puissances associatives avec décomposition de Peirce

Le but de ce paragraphe est de généraliser, aux algèbres flexibles et à puissances associatives avec décomposition de Peirce, la théorie connue pour les algèbres alternatives. Les propriétés (3.20) à (3.31) de la proposition 3.4. de [6], dans le cas alternatif, sont encore valables pour toute décomposition de Peirce, relative à un ensemble d'idempotents deux à deux orthogonaux, dans les algèbres flexibles et à puissances associatives. Les propriétés (3.20) à (3.28) de la proposition 3.4. de [6] sont une conséquence directe de la condition (i) du lemme 2.1., de l'identité de Teichmüller et de la flexibilité tandis que les propriétés (3.29) à (3.31) de la proposition 3.4. de [6] sont une conséquence de la condition (i) du lemme 2.1., de la flexibilité et de l'associativité des puissances. Par la suite, on caractérisera le nilradical (nilidéal maximal) de telles algèbres comme étant l'ensemble de ses éléments proprement nilpotents, résultat analogue à celui des algèbres alternatives.

Si U est une F-algèbre à puissances associatives, on dira qu'un élément x de U est ***proprement nilpotent*** dans U si pour tout élément a dans U, ax et xa sont des éléments nilpotentes dans U.

Le lemme suivant est une conséquence de la décomposition de Peirce, du fait que l'on est sur des algèbres à puissances associatives et du *théorème d'Albert* (cf. [6], proposition 3.3):

4.1. LEMME

Soit U une F-algèbre de dimension finie et à puissances associatives ayant des idempotents non nuls. Si U est une algèbre avec décomposition de Peirce, alors un idempotent e de U est principal si, et seulement si, la composante U_{00} relative à e est une nilalgèbre.

Le lemme suivant est une conséquence de la décomposition de Peirce et du fait que l'algèbre soit flexible et à puissances associatives:

4.2. Lemme

Soit U une F-algèbre flexible et à puissances associatives. Si U possède décomposition de Peirce relative à un ensempble d'idempotents deux à deux orthogonaux, les propriétés suivantes sont vérifiées: (a) $(xy)^m = x((yx)^{m-1}y) = (x(yx)^{m-1})y$, quels que soient x dans U_{ij} et y dans U_{ji}, $(i,j = 0, 1, \ldots, n; i \neq j)$ et pour tout entier $m \geq 2$; (b) xy est nilpotent si, et seulement si, yx est nilpotent, quels que soient x dans U_{ij} et y dans U_{ji}, $(i,j = 0, 1, \ldots, n; i \neq j)$.

4.3. Lemme

Soit U une F-algèbre de dimension finie à puissances associatives et à élément unité dans laquelle l'élément unité, noté 1, est son unique idempotent non nul. Les assertions suivantes sont alors équivalentes: (1) un élément x de U n'est pas nilpotent; (2) il existe un élément y dans $< 1, x >$, la sous-F-algèbre de U commutative et associative engendrée par les éléments 1 et x, tel que $xy = yx = 1$.

(1) $\Rightarrow$ (2). Par hypothèse, la sous-F-algèbre $< x >$ de U est commutative, associative et elle n'est pas une nilalgèbre. D'après le théorème d'Albert, cette sous-algèbre contient un idempotent non nul lequel est nécessairement l'élément unité. Donc, il existe un entier $n \geq 1$ et des scalaires $\alpha_1, \ldots, \alpha_n$ dans F tels que $1 = \alpha_1 x + \cdots + \alpha_n x^n$ et, par suite, $< 1, x > = < x >$. L'élément $y = \alpha_1 1 + \cdots + \alpha_n x^{n-1}$ vérifie les conditions souhaitées.

(2) $\Rightarrow$ (1). Tout d'abord, observons que pour tout entier $m \geq 1$, $x^m y = x^{m-1}$, où l'on pose $x^0 = 1$. Si x était nilpotent, il existerait un entier $r \geq 1$ tel que $x^r \neq 0$ et $x^{r+1} = 0$ d'où $0 = x^{r+1} y = x^r$, ce qui est absurde.

4.4. Théorème

Soit U une F-algèbre de dimension finie, flexible et à puissances associatives, ayant des idempotents non nuls. Supposons que U soit une algèbre avec décomposition de Peirce et que pour toute sous-F-algèbre T de U à élément unité dans laquelle l'unité est l'unique idempotent non nul, l'ensemble des éléments nilpotents de T est un nilidéal de T. Alors, le nilradical (nilidéal maximal) de U est l'ensemble de tous ses éléments proprement nilpotents.

Considérons la décomposition de Peirce de U relative à un ensemble $\{e_1, \ldots, e_n\}$ d'idempotents primitifs de U, deux à deux orthogonaux, tel que $e = e_1 + \cdots + e_n$ soit un idempotent principal de U. Alors, U_{00} est une sous-nilalgèbre de U et e_i, l'élément unité de $U_{ii}, (i = 1, \ldots, n)$, est l'unique idempotent de $U_{ii}, (i = 1, \ldots, n)$. L'hypothèse du théorème peut donc être appliquée à chaque U_{ii}, $(i = 1, \ldots, n)$. Ainsi, tout élément nilpotent de $U_{ii}, (i = 0, 1, \ldots, n)$, est proprement nilpotent dans U. Considérons donc les ensembles

$$W_{ij} = \{s \mid s \in U_{ij}, \text{tous les éléments de } sU_{ji} \text{ sont nilpotents}\}, (i,j = 0, 1 \ldots, n).$$

Chaque ensemble $W_{ij}, (i,j = 0, 1, \ldots, n)$, est un sous-espace vectoriel de U et $W_{ij} \subset R, (i,j = 0, 1 \ldots, n)$, R étant l'ensemble des éléments proprement nilpotents de U (cf. lemme 4.2.,(b) et les propriétés (3.29) et (3.30) de la proposition 3.4. de [6]). Soit $W = \sum_{i,j} W_{ij}$, notons $< R >$ le sous-F-espace vectoriel de U engendré par

R et montrons que $W =< R >$ est un idéal bilatère de U. En effet, on sait que $W \subset< R >$ et supposons qu'il existe un élément $x \in R$ tel que $x \notin W$. Alors, pour $x = \sum_{i,j} x_{ij}$ avec $x_{ij} \in U_{ij}$, il existe un couple d'indices $(i,j), (i,j = 1, \ldots, n)$, tel que $x_{ij} \notin W_{ij}$. D'après les lemmes 4.2. et 4.3., il existe un élément $b_{ji} \in U_{ji}$ tel que $b_{ji}x_{ij} = e_j$ et la proposition 3.4., (3.31), de [6] nous dit alors que $b_{ji}x$ n'est pas nilpotent dans U, ce qui est absurde. On a ainsi montré que $R \subset W$ donc $W =< R >$. Pour démontrer que W est un idéal à droite de U observons, tout d'abord, que (a) $W_{ij}U_{jk} \subset W_{ik}, (i,j,k = 0,1,\ldots,n)$, et (b) $W_{ij}U_{ij} \subset W_{ji}, (i,j = 0,1,\ldots,n; i \neq j)$, d'après les hypothèses du théorème et les propriétés (3.22) et (3.24) de [6]. Ainsi,

$$WU = (\sum_{i,j} W_{ij})(\sum_{k,l} U_{kl}) = \sum_{i,j,l} W_{ij}U_{jl} + \sum_{i \neq j} W_{ij}U_{ij} \subset W,$$

c'est à dire que W est un idéal à droite de U. De même, on montre que W est un idéal à gauche de U et, par suite, $W =< R >$ est un idéal bilatère de U et il est clair que cet idéal ne dépend pas du choix des idempotents utilisés dans sa construction. Si l'on suppose que l'idéal $< R >$ ne soit pas un nilidéal, d'après la proposition 3.3. de [6], il possède un idempotent $u \neq 0$ et d'après le théorème 3.2., il existe, dans U, des idempotents primitifs $u_1, \ldots, u_n$, deux à deux orthogonaux, tels que $u = u_1 + \cdots + u_r (r \leq n)$ et $v = u_1 + \cdots + u_n$ et un idempotent principal. Alors, pour $W = \sum_{i,j} W_{ij}$ et si l'on écrit $u = \sum_{i,j} s_{ij}$ où $s_{ij} \in W_{ij} (i,j = 0,1,\ldots,n)$, on a $u_1 \in W_{11}$ et donc u_1 est un élément nilpotent, ce qui est absurde. On a ainsi montré que $R =< R >$ est un nilidéal maximal de U ou encore, R est le nilradical de U.

Comme conséquence immédiate du théorème ci-dessus on a les deux résultats suivants, les hypothèses étant celles du théorème:

4.5. Corollaire

Soit $U = U_{11} + U_{10} + U_{01} + U_{00}$ la décomposition de Peirce de U relative à un idempotent non nul e de U. Alors, le nilradical de U_{ii} est $R \cap U_{ii}$, $(i = 0,1)$, où R est le nilradical de U.

4.6. Corollaire

Si $U = U_{11} + U_{10} + U_{01} + U_{00}$ est la décomposition de Peirce de U relative à un idempotent principal e de U, alors $U_{10} + U_{01} + U_{00} \subset R$.

Sous les hypothèses du théorème 4.4. et pour tout corps commutatif F, on a le résultat suivant:

4.7. Théorème

Soit U une F-algèbre non nulle de dimension finie, flexible, à puissances associatives et semi-simple. Supposons que U soit une algèbre avec décomposition de Peirce et que pour toute sous-F-algèbre T de U à élément unité dans laquelle l'unité est l'unique idempotent non nul, l'ensemble des éléments nilpotents de T est un nilidéal de T. Alors U est une algèbre à élément unité.

En effet, soit $U = U_{11} + U_{10} + U_{01} + U_{00}$ la décomposition de Peirce de U relative à un idempotent principal e de U. D'après le corollaire 4.6., $U_{10} + U_{01} + U_{00} \subset R$ et comme $R = 0$, nécessairement $U = U_{11}$ et e est l'élément unité de U.

5. Décomposition de Wedderburn et les algèbres flexibles à puissances associatives avec décomposition de Peirce

Soit U une F-algèbre flexible et à puissances associatives. On dit que U possède une *décomposition de Wedderburn* s'il existe une sous-F-algèbre σ de U qui vérifie les conditions suivantes: (1) $U = R \oplus \sigma$, où R est le nilradical de U; (2) l'algèbre σ est isomorphe à l'algèbre quotient U/R.

Dans la suite de ce paragraphe, les hypothèses sont celles du théorème 4.4. On a ainsi le lemme suivant:

5.1. LEMME

Soit U une F-algèbre. Pour tout idempotent principal e de U, le nilradical R de U s'écrit $R = R_1 \oplus U_{10} \oplus U_{01} \oplus U_{00}$, où R_1 est le nilradical de la sous-algèbre U_{11} de U.

5.2. THÉORÈME

Soit U une F-algèbre. Pour toute décomposition de Peirce de U relative à un idempotent principal e, les assertions suivantes sont vérifiées: (1) l'algèbre quotient U/R est isomorphe à l'algèbre quotient U_{11}/R_1; (2) si U_{11} possède une décomposition de Wedderburn, alors l'algèbre U possède aussi une décomposition de Wedderburn.

En effet, $U = R + U_{11}$ et comme R est un idéal de U, $R_1 = R \cap U_{11}$ est un idéal de U_{11} d'où l'isomorphisme de F-algèbres $U/R \simeq U_{11}/R_1$. Cela démontre l'assertion (1). Pour ce qui est de (2), si U_{11} possède une décompsition de Wedderburn, il existe une sous-algèbre σ_1 de U_{11} vérifiant $U_{11} = R_1 \oplus \sigma_1$ et σ_1 est isomorphe, en tant que F-algèbre, à l'algèbre quotient U_{11}/R_1. Il s'ensuit que $U = U_{11} \oplus U_{10} \oplus U_{01} \oplus U_{00} = R \oplus \sigma_1$ (cf. lemme 5.1.), d'où le résultat voulu.

6. Applications aux algèbres de Moufang

Une généralisation naturelle d'algèbre alternative est celle d'algèbre de Moufang (cf. [4]), à savoir, si F est un corps commutatif et U une F-algèbre flexible, on dit que U est une *algèbre de Moufang* si elle vérifie les *identités de Moufang*: M1) $((zx)y)x = z(xyx)$ (*identité de Moufang à droite*); M2) $x(y(xz)) = (xyx)z$ (*identité de Moufang à gauche*); M3) $(xz)(yx) = x(zy)x$ (*identité de Moufang au milieu*). Les auteurs ont donné un exemple d'une algèbre de Moufang qui n'est pas alternative et, en utilisant des propriétés élémentaires, ils ont établi des relations entre les algèbres de Moufang et les algèbres de Jordan non commutatives, les algèbres de Malcev et les algèbres de Lie.

L'étude des algèbres de Moufang a été entrepris dans [1] où les auteurs ont démontré des propriétés analogues à celles des algèbres alternatives et ont caractérisé, en dimension finie, les algèbres de Moufang simples et semi-simples en montrant qu'elles coïncident exactement avec les algèbres alternatives simples et semi-simples, respectivement.

Dans [4], il est aussi fait mention de classes d'algèbres plus générales que celle des algèbres de Moufang, à savoir, les algèbres de Moufang à droite et les algèbres de Moufang au milieu, mais dans ces classes d'algèbres, aucune étude approfondue n'est, à notre connaissance, faite jusqu'au moment. La théorie développée dans cet article nous permet d'éclairer les structures d'algèbres de Moufang à droite et au milieu. C'est ce que nous ferons par la suite.

Une F-algèbre flexible U est dite *algèbre de Moufang à droite* (resp. *algèbre de Moufang au milieu*), si elle vérifie l'identité de Moufang à droite (resp. l'identité de Moufang au milieu).

On sait (cf. [4]) que toute algèbre de Moufang ainsi que toute algèbre de Moufang au milieu est à puissances associatives. Par contre, il existe des algèbres de Moufang à droite qui ne sont pas à puissances associatives.

6.1. Exemple

Soit U une F-algèbre de dimension 4 dont la table de multiplication relative à une base $\{e_1, \ldots, e_4\}$ s'écrit $e_2^2 = e_3$ et $e_3^2 = e_4$, tous les autres produits étant nuls. On vérifie que U est une F-algèbre commutative, qui vérifie l'identité de Moufang à droite et donc à gauche aussi; par contre, l'algèbre U n'est pas de Jordan ni à puissances associatives car $(e_2, e_2, e_2^2) = e_2^2 e_2^2 - e_2(e_2 e_2^2) = e_4$. Cet exemple nous montre que la classe des algèbres flexibles qui satisfont l'identitè de Moufang à droite n'est pas une sous-classe de celle des algèbres à puissances associatives ou des algèbres de Jordan.

En termes d'associateurs, la flexibilité d'une algèbre U s'écrit: (1) $(x, y, x) = 0$ et (2) $(x, y, z) + (z, y, x) = 0$; la structure de Moufang à droite s'écrit: (3) $(yz, x, t) + (yt, x, z) + (y, z, xt) + (y, t, xz) = 0$ et (4) $(y, zx, t) + (y, tx, z) + (y, z, x)t + (y, t, x)z = 0$; la structure de Moufang au milieu s'écrit: (5) $(y, z, xt) + (t, z, xy) + y(z, x, t) + t(z, x, y) = 0$ et (6) $(yz, x, t) + (tz, x, y) + (y, z, x)t + (t, z, x)y = 0$. Ainsi, toute algèbre de Moufang à droite vérifie les identités (1), (2), (3) et (4) et toute algèbre de Moufang au milieu vérifie les identités (1), (2), (5) et (6).

6.2. Théorème

Soient F un corps commutatif de caractéristique différente de 2 et U une F-algèbre de Moufang à droite. Alors, U possède décomposition de Peirce relative à tout ensemble d'idempotents deux à deux orthogonaux.

En effet, il suffit de voir que tout ensemble d'idempotents de U, deux à deux orthogonaux, vérifie les conditions du lemme 2.1.

6.3. Théorème

Soit F un corps commutatif de caractéristique différente de 2. Toute F-algèbre de Moufang à droite simples ayant des idempotents non nuls est alternative.

On choisit un idempotent $e \neq 0$ de U; si e est l'élément unité de U, l' algèbre est nécessairement alternative. Sinon, le théorème 2 de [7] nous montre que U est encore alternative.

6.4. Théorème

Soient F un corps commutatif de caractéristique différente de 2 et U une F-algèbre de Moufang à droite de dimension finie et à puissances associatives. Alors: (1) le nilradical (nilidéal maximal) R de U est l'ensemble de ses éléments proprement nilpotents; (2) si $U \neq 0$ est semi-simple, alors U est une algèbre alternative à élément unité; (3) si U/R est séparable, alors $U = R \oplus \sigma$, où σ est une sous-algèbre de U isomorphe à l'algèbre quotient U/R.

En effet, l'assertion (1) est une conséquence du théorème 4.4.; l'assertion (2) résulte du théorème 4.7. puisque U est une algèbre à élément unité. L'assertion (3) découle du théorème 5.2. car l'algèbre U_{11} étant à élément unité, elle est alternative et, par suite, la décomposition de Wedderburn y est vérifiée. Donc, la décomposition de Wedderburn de U s'ensuit.

Des résultats analogues à ceux établis ci-dessus peuvent l'être pour des algèbres de Moufang au milieu, sans aucune hypothèse sur la caractéristique du corps de base F, tout en rappelant que les algèbres de Moufang au milieu sont déjà à puissances associatives.

6.5. Exemple

Soit U une F-algèbre de dimension 6 dont la table de multiplication relative à une base $\{e_1, \ldots, e_6\}$ s'écrit $e_1e_2 = -e_2e_1 = e_3$, $e_2e_4 = -e_4e_2 = e_5$ et $e_3e_5 = -e_5e_3 = e_6$, tous les autres produits étant nuls. On vérifie que U est une algèbre flexible à puissances associatives qui vérifie les identités de Moufang à droite et à gauche ainsi que l'identité de Jordan mais qui n'est pas une algèbre de Moufang.

Bibliographie

1. L. Acosta, L.A. de Capua, J.C. da Motta Ferreira et A. Micali, Sur les algèbres de Moufang, Université de Montpellier, preprint 1992.

2. A.A. Albert, Structure of Algebras, Amer. Math. Soc., Colloquium Publications, Volume 24, New York 1939.

3. N. Bourbaki, Algèbre, Chapitre 8, Hermann, Paris 1958.

4. L.A. de Capua, A. Koulibaly et A. Micali, Autour des algèbres de Malcev, Rivista di Matematica Pura ed Applicata 8 (1991), 29-40.

5. B. Peirce, Linear Associative Algebra, Amer. J. Math. 4 (1881), 97-221.

6. R.D. Schafer, An introduction to nonassociative algebras, Academic Press, New York 1966.

7. M. Rich, Rings with idempotents in their nuclei, Trans. Amer. Math. Soc. 208 (1975), 81-90.

ON TRANSITIVE LEFT–SYMMETRIC ALGEBRAS

ALBERTO ELDUQUE*
Departamento de Matemáticas, Universidad de Zaragoza
50009 Zaragoza, Spain

and

HYO CHUL MYUNG
Department of Mathematics, University of Northern Iowa
Cedar Falls, Iowa 50614, USA

Abstract. The relationships of transitive left–symmetric algebras with certain solvable Lie algebras are studied and new proofs are given of the following results: i) the associated Lie algebra of a transitive left symmetric algebra is solvable, and ii) a left–symmetric algebra is transitive if and only if the right multiplication operators are nilpotent.

Let L be a finite-dimensional Lie algebra over a field of characteristic 0 and denote by $(L, *)$ an algebra with multiplication $*$ defined on L. Then, $(L, *)$ is called a transitive left-symmetric algebra if it satisfies the identity $(x, y, z)^* = (y, x, z)^*$ with $(L, *)^- = L$ and $1 + \rho_x : L \to L : y \to y + y * x$ is bijective for all $x \in L$. The main concern in this work is to investigate the relationships of transitive left-symmetric algebras with certain solvable Lie algebras without nonzero semisimple elements, and to present alternative proofs to the known results on left-symmetric algebras $(L, *)$: (1) If $(L, *)$ is transitive, then L is solvable and (2) $(L, *)$ is transitive if and only if ρ_x is nilpotent for all $x \in L$. These algebras interplay with a geometric problem as follows: If G is a simply connected real Lie group with Lie algebra L, then there is a bijective correspondence between the set of affine structures on G and the set of transitive left-symmetric algebras $(L, *)$ on L.

1. Left–symmetric algebras

Let A be a nonassociative algebra with multiplication denoted by $x * y$ over a field F. Let $(x, y, z)^* = (x * y) * z - x * (y * z)$ denote the associator in A. Then, A is called *left–symmetric* if it satisfies the symmetric identity

$$(x, y, z)^* = (y, x, z)^* \tag{1.1}$$

for all $x, y, z \in A$. A right–symmetric algebra is similarly defined and its opposite algebra becomes left–symmetric [5]. Left–symmetric algebras arise in the study of affine structures on certain Lie groups [4],[1],[3],[9]. Left–symmetric algebras satisfying the identity $(x * y) * z = (x * z) * y$ have been used for the study of operators in the formal calculus of variations (see [7]). Such algebras were called Novikov algebras in [6].

* Supported by the DGICYT, Ps. 90–0129 and by the DGA (PCB–6/91)

S. González (ed.), *Non-Associative Algebra and Its Applications*, 114–121.

Let A^- denote the algebra with multiplication $[x,y]^* = x*y - y*x$ defined on A. If A^- is a Lie algebra, then A is said to be *Lie-admissible*. Since any algebra satisfies the identity $[[x,y]^*,z]^* + [[y,z]^*,x]^* + [[z,x]^*,y]^* = (x,y,z)^* + (y,z,x)^* + (z,x,y)^* - (x,z,y)^* - (z,y,x)^* - (y,x,z)^*$, it follows from (1.1) that any left-symmetric algebra is Lie-admissible.

Left-symmetric algebras to be considered in our discussion are ones that interplay closely with affine structures on a certain real Lie group G whose minus algebras are the Lie algebra of G. Thus, let L be a Lie algebra with multiplication $[x,y]$ over F, and denote by $(L,*)$ an algebra with multiplication $x*y$ defined on L. If $(L,*)^- = L$, i.e., $[x,y]^* = [x,y]$ for all $x,y \in L$, then $(L,*)$ is said to be *compatible* with L [4]. For a compatible $(L,*)$, we have:

Lemma 1.1. *Let L be a Lie algebra over F and let $(L,*)$ be compatible with L. Then, $(L,*)$ is left-symmetric if and only if it satisfies the identity*

$$[x,y]*z = x*(y*z) - y*(x*z)\ . \tag{1.2}$$

Proof. This follows from expanding (1.1). □

Let λ_x and ρ_x denote the left and right multiplications in $(L,*)$ by x. Identity (1.2) means that the map

$$\lambda : L \longrightarrow End_F L \equiv E(L)\ :\ x \mapsto \lambda_x \tag{1.3}$$

is a representation of L acting on L. In [1], a representation $\sigma : L \to E(L)$ satisfying $\sigma(x)y - \sigma(y)x = [x,y]$ for all $x,y \in L$ is called a *Koszul–Vinberg structure* on L. It is immediate from (1.2) that there is a bijective correspondence between the set of Koszul–Vinberg structures σ on L and the set of compatible left-symmetric algebras $(L,*)$, under the map $\sigma \mapsto x*y = \sigma(x)y$ $(x,y \in L)$.

Using differential geometric jargons [1],[4],[9], we define:

Definition 1.1. Let L be any Lie algebra over F. A left-symmetric algebra $(L,*)$ compatible with L is called *transitive* if the linear map

$$1 + \rho_x : L \longrightarrow L\ :\ y \mapsto y + y*x \tag{1.4}$$

is bijective for all $x,y \in L$. □

Definition 1.2. Let V be a vector space over F and L be a Lie subalgebra of $(End_F V)^- \equiv E(V)^-$. Then, L is said to be *complete* if there exists a subspace W of V of codimension one such that $LV = W$ and, for any $a \in V - W$, the linear map δ_a given by

$$\delta_a : L \longrightarrow W\ :\ x \mapsto xa \tag{1.5}$$

is bijective. □

The following result gives a relationship between transitive left-symmetric algebras and complete Lie algebras.

Theorem 1.2. *Let L be an arbitrary Lie algebra over a field F and $(L,*)$ be a left–symmetric algebra compatible with L. Let $V = L \oplus F$ with $W = L$. Define the map*

$$\psi : L \longrightarrow E(V)^- \ : \ \psi(x)(y,\alpha) = (\alpha x + \rho_y x, 0)$$

for $x, y \in L$ and $\alpha \in F$. Then, ψ is a faithful representation of L such that $\psi(L)V = W$. Moreover, $(L,*)$ is transitive if and only if $\psi(L)$ is complete.

Conversely, let L be a Lie subalgebra of $E(V)^-$ with the property that there is a subspace W of V of codimension one with $LV = W$, so that $V = W \oplus Fa$. Assume that the map $\delta = \delta_a$ defined by (1.5) is bijective. Then, the algebra $(L,*)$ with multiplication $x * y$ given by

$$x * y = \delta^{-1}(x\delta(y)) = \delta^{-1}((xy)a) \tag{1.6}$$

is left–symmetric and compatible with L, and $(L,*)$ is transitive if and only if L is complete.

Proof. For the first part, if $x, y, z \in L$, then one has

$$\begin{aligned} [\psi(x), \psi(y)](z,\alpha) &= (x*(y*z) - y*(x*z) + \alpha[x,y], 0) \\ &= ([x,y]*z + \alpha[x,y], 0) \\ &= \psi([x,y])(z,\alpha), \end{aligned}$$

using (1.2). Hence, ψ is a representation of L. Clearly, ψ is faithful and $\psi(L)V = W$. For any $b \in V - W$, we can let $b = (x,1)$ and identify $y = (y,0)$ for $y \in L = W$. If $\delta_b : \psi(L) \to W$ is given by (1.5), then it follows easily that $\delta_b \circ \psi = 1 + \rho_x$ (see (1.4)), and hence δ_b is bijective if and only if $1 + \rho_x$ is bijective.

For the converse, let $x, y, z \in L$ be any elements. Clearly, $x * y - y * x = \delta^{-1}([x,y]a) = [x,y]$ by (1.6), hence $(L,*)$ is compatible with L. Since $x*(y*z) = \delta^{-1}((xyz)a)$,

$$\begin{aligned} [x,y]*z = \delta^{-1}([x,y]\delta(z)) &= \delta^{-1}((xyz)a) - \delta^{-1}((yxz)a) \\ &= x*(y*z) - y*(x*z) \ . \end{aligned}$$

Thus, by Lemma 1.1, $(L,*)$ is left–symmetric. If $b \in V - W$, then we can let $b = \delta(x) + a$ for some $x \in L$. It is easy to see that $\delta_b = \delta \circ (1 + \rho_x)$, and hence $(L,*)$ is transitive if and only if L is complete. □

A version of Theorem 1.2 for the finite–dimensional case is also given in [9]. In this case, one has $\dim L = \dim W$. The following result is immediate from Definition 1.2.

Proposition 1.3. *Let V be a finite–dimensional vector space over F and let L be a Lie subalgebra of $E(V)^-$. Assume that V contains a subspace W of codimension one such that $LV = W$. Then, L is complete if and only if $\dim L = \dim W$ and $xa \neq 0$ for all $0 \neq x \in L$ and all $a \in V - W$.* □

Henceforth, all Lie algebras and modules considered are assumed to be finite–dimensional over a field F of characteristic 0. The primary concern of this paper is to investigate the relationship of transitive left–symmetric algebras with solvable Lie subalgebras of $E(V)^-$ without nonzero semisimple elements. The method employed here enables us to give a new proof to the known result that a left-symmetric algebra $(L,*)$ is transitive if and only if ρ_x is nilpotent for all $x \in L$ [9]. Our proof of this result is based on the conjugacy theorem of Cartan subalgebras in a Lie algebra and on the algebraic hull of a solvable Lie subalgebra in $E(V)^-$. On the other hand, the proof presented in [9] utilizes a conjugacy theorem in Lie groups as well as results on unipotent subgroups. Thus, our discussion is purely algebraic.

Geometric origin and background of left-symmetric algebras may be found in [1],[4],[9] and references therein. Transitivity defined above has been termed completeness in [1],[9]. An affine structure on a Lie group G is a left G-invariant affine connection ∇ having zero torsion and zero curvature, which is also complete [1],[4]. Here, ∇ is defined to be complete if any geodesic of ∇ can be defined on all time intervals. On the other hand, it is known that if G is simply connected, then the connection ∇ above is complete if and only if its associated left-symmetric algebra $(L,*)$ is transitive in the sense of Definition 1.1 [4], where L is the Lie algebra of G. The classification of transitive left-symmetric algebras $(L,*)$ is known for dimension ≤ 3. For the case of $\dim L = 4$, it is known when L is nilpotent and $\ker \lambda = 0$ for the map λ given by (1.3) (i.e., without translations) [4]. For the general case of dimension 4, the classification is an open problem. A recipe for constructing all transitive left-symmetric algebras, once the same problem with L nilpotent is solved, is given in [9]. As we show in the next section, if $(L,*)$ is transitive, then L is necessarily solvable.

2. Solvable Lie algebras without semisimple elements

In what follows, V denotes a finite-dimensional vector space over an arbitrary field F of characteristic 0. In this section, we investigate the structure of certain solvable Lie subalgebras of $E(V)^-$ without nonzero semisimple elements. The results obtained in this section are also instrumental for the proof of our main result in the next section.

For $x \in E(V)$, let $x = x_n + x_s$ be the Jordan-Chevalley decomposition of x, where x_n and x_s are the nilpotent and semisimple parts of x. Thus, if $x = x_s$, then x is semisimple, i.e., x is diagonalizable after scalar extension. For $L \subseteq E(V)$, denote

$$L_n = \{x_n : x \in L\}, \quad L_s = \{x_s : x \in L\}.$$

Hence, a Lie subalgebra L of $E(V)^-$ contains no nonzero semisimple elements if and only if $L \cap L_s = 0$.

Lemma 2.1. *Let L be a Lie subalgebra of $E(V)^-$, and assume that there is a subspace W of V of codimension one such that $LV \subseteq W$ and $xv \neq 0$ for all $0 \neq x \in L$ and all $0 \neq v \in V - W$. Then, L is solvable and $L \cap L_s = 0$.*

Proof. We employ a similar argument as the one given in [3, p.30]. Assume that L is not solvable. Then, L contains a nonzero semisimple subalgebra S. Since V is a completely reducible S-module, there is an S-submodule Fa of V such that $V = W \oplus Fa$ for some $a \in V - W$. But, $Sa = [S,S]a = 0$, a contradiction. If $x \in L \cap L_s$ and $S = Fx$, then V is again a completely reducible S-module. As above, we have $V = W \oplus Fa$ for some $a \in V - W$ and $xa \in LV \cap Fa \subseteq W \cap Fa = 0$. Hence, it must be that $x = 0$. □

Corollary 2.2. *Let L be a Lie subalgebra of $E(V)^-$.*

(i) *If L is complete, then L is solvable and $L \cap L_s = 0$.*

(ii) *If $(L, *)$ is a transitive left-symmetric algebra compatible with L, then L is solvable.*

Proof. Part (i) is an immediate consequence of Lemma 2.1. For part (ii), we notice that L is isomorphic to the Lie algebra $\psi(L)$ which is complete by Theorem 1.2. Hence, by part (i), L is solvable. □

The following result shows that if L ($\subseteq E(V)^-$) is solvable, then $L \cap L_s = 0$ if and only if $L \cap H_s = 0$ for any Cartan subalgebra H of L, which also remains unchanged under any scalar extension. The proof of this utilizes the conjugacy theorem of Cartan subalgebras.

Theorem 2.3. *Let L be a solvable Lie subalgebra of $E(V)^-$ and H be any Cartan subalgebra of L. Let K denote the algebraic closure of F. Then, the following are equivalent:*

(i) $L \cap L_s = 0$;

(ii) $L \cap H_s = 0$;

(iii) $L_K \cap (L_K)_s = 0$, *where* $L_K = K \otimes_F L \subseteq E(V_K)^-$.

Proof. It is clear that (iii) ⇒ (i) ⇒ (ii). To verify (ii) ⇒ (iii), let $x \in L_K \cap (L_K)_s$ and let

$$L_K = L_0(x) \oplus \sum_{\alpha \neq 0} \oplus L_\alpha(x)$$

be the root space decomposition of L_K relative to $ad\, x$. Assume that H' is a Cartan subalgebra of $L_0(x)$. Since $[x, L_0(x)] = 0$, x is in the normalizer of H' in $L_0(x)$, which equals H'. It follows from this that H' is also its own normalizer in L_K, and hence is a Cartan subalgebra of L_K. Therefore, there is an inner automorphism τ of L_K such that $\tau(H') = H_K = K \otimes_F H$. Since $\tau(x)$ and x are similar endomorphisms of V_K [8, p. 159], $\tau(x)$ is also semisimple and $\tau(x) \in L_K \cap (H_K)_s = (L \cap H_s)_K = 0$. Thus, $x = 0$ and $L_K \cap (L_K)_s = 0$. □

Recall that a subalgebra L of $E(V)^-$ is called *algebraic* if it is a Lie algebra of an algebraic group $G \subseteq E(V)$. For any subalgebra L of $E(V)^-$, the algebraic hull of L, denoted by $\tilde{L}$, is the smallest algebraic Lie subalgebra of $E(V)^-$ containing L (i.e., the intersection of all algebraic Lie subalgebras of $E(V)^-$ containing L [2, p. 173]). Thus, $\tilde{L}$ contains L_n and L_s as well as all replicas of each $x \in L$ (i.e., the elements in $\widetilde{Fx}$) [2, p. 180; 10, p.124]. Also, any ideal of L is an ideal of $\tilde{L}$ and $[L, L] = [\tilde{L}, \tilde{L}]$ (see [2, p.173]). We notice that this implies that if L is solvable, then so is $\tilde{L}$ [2, p.309].

Assume now that L is a solvable subalgebra of $E(V)^-$ and H is a Cartan subalgebra of L. Let $\nu(L)$ denote the nilradical of L. Then, $L = \nu(L) + H$ and $[L, L] = [\tilde{L}, \tilde{L}] \subseteq \nu(L)$, and hence $\nu(L) + H_n$ is a nilpotent ideal of $\tilde{L}$. Since $\tilde{L}$ equals the algebraic hull of $\nu(L) + H_n + H_s$, one has $\tilde{L} = \nu(L) + H_n + \widetilde{H_s}$. Since $\tilde{L}$ is solvable, $\nu(L) + H_n$ acts nilpotently on V and $\widetilde{H_s}$ semisimply on V. Therefore,

$$\nu(\tilde{L}) = \nu(L) + H_n\ , \quad \tilde{L} = \nu(\tilde{L}) \oplus \widetilde{H_s}\ .$$

Lemma 2.4. *If L is a solvable subalgebra of $E(V)^-$, then $L_n \subseteq \nu(\tilde{L})$.*

Proof. For any $x \in \tilde{L}$, let $x = n + s$ for $n \in \nu(\tilde{L})$ and $s \in \widetilde{H_s}$. Since $\nu(\tilde{L})$ is nilpotent, $\nu(\tilde{L})^r = 0$ on V but $\nu(\tilde{L})^{r-1} \neq 0$ for some $r > 0$. If we let $V^i = \nu(\tilde{L})^i V$, $i = 0, 1, \ldots, r$, then $V \supset V^1 \supset \cdots \supset V^r = 0$, and it is easily seen that each V^i is an $\tilde{L}$–submodule of V such that $nV^i \subseteq V^{i+1}$. Thus, x is nilpotent on V if and only if x is nilpotent on each factor V^i/V^{i+1} if and only if s is nilpotent on each V^i/V^{i+1}. This implies that if x acts nilpotently on V, then so does s on V and hence $s = 0$ since $\widetilde{H_s}$ consists of semisimple elements. Thus, the Lemma follows. □

Lemma 2.5. *If L is a solvable subalgebra of $E(V)^-$, then $L \cap L_s = 0$ if and only if $\dim \nu(\tilde{L}) = \dim L$.*

Proof. Let H be a Cartan subalgebra of L. Since $\nu(\tilde{L}) = \nu(L) + H_n \subseteq \nu(L) + H + H_s = L + H_s$, we have

$$\tilde{L} = \nu(\tilde{L}) \oplus \widetilde{H_s} = L + \widetilde{H_s}\ .$$

From Theorem 2.3, it follows that $L \cap \widetilde{L_s} = 0$ if and only if $L \cap H_s = 0$ if and only if $L \cap \widetilde{H_s} = 0$, since $H_s \subseteq \widetilde{H_s}$ and $\widetilde{H_s}$ consists of semisimple elements. Thus, $\tilde{L} = L \oplus \widetilde{H_s}$ if and only if $L \cap L_s = 0$ if and only if $\dim \nu(\tilde{L}) = \dim L$. □

3. Transitive left–symmetric algebras

As before, V denotes a finite–dimensional vector space over F of characteristic 0. All left–symmetric algebras $(L, *)$ are assumed to be compatible with L. The primary objective is to give an alternative proof of the following result (also see [9]):

Theorem 3.1. *Let L be an arbitrary finite–dimensional Lie algebra over F and let $(L, *)$ be any left–symmetric algebra defined on L. Then, $(L, *)$ is transitive if and only if the right multiplication ρ_x in $(L, *)$ is nilpotent for all $x \in L$.* □

For the proof of Theorem 3.1, we first verify:

Theorem 3.2. *Let L be a Lie subalgebra of $E(V)^-$ with $L \cap L_s = 0$. Assume that there exists a subspace W of V of codimension one such that $LV \subseteq W$ and $\dim W = \dim L$. Let H be any Cartan subalgebra of L. Then, $H_s a = 0$ for some $a \in V - W$, and moreover, the following are equivalent:*

(i) *L is complete;*

(ii) *$\delta_a|_{\nu(\tilde{L})}$ is bijective, where $\delta_a : \tilde{L} \to W$ is given by (1.5);*

(iii) *There exists an element $b \in V - W$ such that $\delta_b|_{\nu(\tilde{L})}$ is bijective;*

(iv) *$\nu(\tilde{L})$ is complete.*

Proof. Since H_s acts semisimply on V and $LV \subseteq W$, we find an element $a \in V - W$ such that $H_s a = 0$. Notice also that $\tilde{L}V \subseteq W$.

(i) $\Rightarrow$ (ii). Assume that L is complete. Thus, by Corollary 2.2, L is solvable with $L \cap L_s = 0$, and by Lemma 2.5 $\dim L = \dim \nu(\tilde{L})$. Let $x = n + h_n \in \nu(\tilde{L}) = \nu(L) + H_n$ with $n \in \nu(L)$ and $h \in H$ such that $\delta_a(x) = 0$. Since $H_s a = 0$ and L is complete, $0 = \delta_a(x) = xa = (n + h_n + h_s)a$, so $n + h = 0$ ($h = h_n + h_s$). Hence, $h_s = -(n + h_n) \in \nu(\tilde{L}) \cap H_s = 0$, so that $x = 0$ and δ_a is bijective on $\nu(\tilde{L})$.

(ii) $\Rightarrow$ (iii) is clear.

(iii) $\Rightarrow$ (iv). Assume that $\delta = \delta_b|_{\nu(\tilde{L})} : \nu(\tilde{L}) \to W$ is bijective. Thus, by Theorem 1.2, the algebra $(\nu(\tilde{L}), *)$ defined by

$$x * y = \delta^{-1}(x\delta(y)) = \delta^{-1}((xy)b)$$

is left-symmetric. Since $\nu(\tilde{L})$ acts nilpotently on V,

$$\lambda_x^r(y) = x * (x * (\cdots (x * y) \cdots) = \delta^{-1}(x^r \delta(y)) = 0$$

for some $r > 0$; that is, the left multiplication λ_x in $(\nu(\tilde{L}), *)$ is nilpotent for all $x \in \nu(\tilde{L})$. It follows from this and [4, Theorem 2.2] that all right multiplications ρ_x in $(\nu(\tilde{L}), *)$ are nilpotent. This shows that $(\nu(\tilde{L}), *)$ is transitive and hence $\nu(\tilde{L})$ is complete by the second part of Theorem 1.2.

(iv) $\Rightarrow$ (i). For any $b \in V - W$, if $\delta_b(x) = xb = 0$ for some $x \in L$, then $x_n b = 0 = x_s b$ (x_n and x_s are polynomials in x without constant terms). But, by Lemma 2.4, $x_n \in \nu(\tilde{L})$, and since $\nu(\tilde{L})$ is complete, $x_n = 0$. Hence, $x = x_s$ is semisimple, and $x = 0$ by hypothesis. Therefore, L is complete. □

Theorem 3.3. *Let $(L, *)$ be a left-symmetric algebra over F and let K be the algebraic closure of F. Then, $(L, *)$ is transitive if and only if $(L_K, *)$ is transitive.*

Proof. Let $\psi(L) \subseteq E(V)^-$ be the Lie algebra as in Theorem 1.2, where $V = L \oplus F$. Now, by Theorem 1.2, $(L, *)$ is transitive if and only if $\psi(L)$ is complete, but by Corollary 2.2 and Theorem 3.2, the latter is equivalent to $L \cap L_s = 0$ and the condition (ii) of Theorem 3.2. Since the last two properties are independent of scalar extensions, the result follows. □

Now, Theorem 3.1 can be easily proven:

Proof of Theorem 3.1. If ρ_x is nilpotent for all $x \in L$, then $1 + \rho_x$ is clearly bijective and hence $(L, *)$ is transitive. Assume that $(L, *)$ is transitive. Then, by Theorem 3.3, we may assume that F is algebraically closed. But, we find that $1 + \rho_x$ is invertible for all $x \in L$ if and only if $\alpha + \rho_x$ is invertible for all $0 \neq \alpha \in F$ and all $x \in L$ if and only if the only eigenvalue of ρ_x is 0 for all $x \in L$. Since F is algebraically closed, the last statement is equivalent to the nilpotence of ρ_x for all $x \in L$. □

References

1. N.B. Boyom, The lifting problem for affine structures in nilpotent Lie groups, Trans. Amer. Math. Soc. **313** (1989), 347–379.

2. C. Chevalley, *Théorie des Groupes de Lie.* vol. II: *Groupes Algébriques*, Actualités Sci. Indust., no. 1152, Hermann, Paris, 1951.
3. J. Helmstetter, Radical d'une algèbre symmetrique à gauche, Ann. Inst. Fourier, Grenoble, **29** (1979), 17–35.
4. H. Kim, Complete left-invariant affine structures on nilpotent Lie groups, J. Differential Geom. **24** (1986), 373–394.
5. E. Kleinfeld, On rings satisfying $(x, y, z) = (x, z, y)$, Algebras Groups Geom. **4** (1987), 129–138.
6. J.M. Osborn, Novikov algebras, Nova J. Algebra Geom. **1** (1992), 1–14.
7. J.M. Osborn and E. Zelmanov, Nonassociative algebras related to Hamiltonian operators in the formal calculus of variations, to appear.
8. A.A. Sagle and R.E. Walde, *Introduction to Lie Groups and Lie Algebras*, Academic Press, New York, 1973.
9. D. Segal, The structure of complete left-symmetric algebras, Math. Ann. **293** (1992), 569–578.
10. G.B. Seligman, *Algebraic Groups*, Lecture Notes, Yale University, 1964.

SPECTRAL STUDY OF SOME TOPOLOGICAL JORDAN ALGEBRAS

N. EL YACOUBI
Dep. of Math. and Inf. B.P. 1014 Rabat Morocco
Fax 212 7 77 54 71

Many fundamental results of the spectral theory of Banach algebras have been generalized, on one hand, to some more general associative topological algebras, for example, locally multiplicatively convex algebras (Michael 1952), locally convex algebras (Allan 1956) etc... and on the other hand, to non-associative case, like, topological jordan algebras with continuous inverse (C.Viola 1971), Banach-Jordan algebras (J.Martinez Moreno 1977) and non-commutative Banach-Jordan algebras (Kaïdi 1977). The purpose of this work is to develop a spectral theory of a special class of topological Jordan algebras.

C.Viola [5] has defined a topological Jordan algebra as a Jordan algebra J provided with a topology τ such that:

i) (J, τ) is a locally convex space.

ii) $(x, y) \rightarrow U_y(x) = 2R_y^2(x) - R_{y2}(x)$ is separately continuous, where $R_y(x) = xy$.

We think more natural to call this one a locally convex Jordan algebra, and to use the designation "topological Jordan algebra" when (J, τ) is a topological vector space, we adopt this last one.

When J is unital , ii) is equivalent to ii'):

ii') $(x, y) \rightarrow xy$ is separately continuous.

J will be said a Frechet-Jordan (resp. metric-Jordan) algebra, when (J, τ) is a Frechet (resp.metric) space.

It is clear that if A is an associative, non commutative locally convex (resp. Frechet, metric) algebra, then A^+ (which means A provided with the Jordan product $\circ$: $x \circ y = \frac{1}{2}(xy + yx)$) is a locally convex (resp.Frechet, metric) Jordan algebra.

1. . Topological Jordan algebras with continuous inverse

1.1. Definition

A topological Jordan algebra over the set of complex numbers **C**, with unit e is called with continuous inverse if:

i) There exists a neighborhood $V(e)$ of e such that:
every element x of $V(e)$ is invertible in the jacobson sens [2] (which means: $\exists y \in J / xy = e,\ x^2 y = x$)

ii) the application $x \rightarrow x^{-1}$ is continuous at e.

S. González (ed.), Non-Associative Algebra and Its Applications, 122–127.

In these algebras, the classical spectrum of an element x is defined, and we have the following properties:

1.2. SPECTRAL PROPERTIES

- The set Inv$J$ of the invertible elements of J is open.
- For every element x of J, the resolvent set $\rho(x) = C_{\mathbf{C}}$ Sp x is open.
- The resolvent $\lambda \to R(x,\lambda) = (x-\lambda e)^{-1}$ is holomorphic on $\rho(x)$ and $\lim_{\lambda\to\infty} R(x,\lambda) =$ 0

If, in addition , J has a total set of continuous linear forms, then $Sp\ x \neq \emptyset$, and the Gelfand-Mazur's theorem is true.

2. Locally convex Jordan algebras

In the associative case, Allan's work was motivated by the next remark: the spectrum of a closed operator T on a Banach space E is the set of complex numbers λ for wich $T - \lambda I$ has non bounded inverse [1].

Following his approach we developed an appropriated theory for the locally convex Jordan algebras.

In comparison with the Banach-Jordan case, the new is the use of the weak topology and the choice of a suitable notion of completeness.

From now on, J will be a complex locally convex Jordan algebra (l.c.J-a) with unit e.

2.1. BORNOLOGICAL SPECTRUM

2.1.1. . Bounded (or regular) element

An element a of J is said to be bounded iff for $\lambda \neq 0$, $\lambda \in \mathbf{C}$, the set $\{(\lambda a)^n, n = 1, 2, ...\}$ is bounded in J.

We will denote by J_0, the set of bounded element of J.

2.1.2. . Definition

For x$\in$ J, we call bornological spectrum of x, denoted by $\sigma(x)$, the subset of $\bar{\mathbf{C}}$ (compactification of $\mathbf{C}$) defined by:

* $\lambda \neq \infty$, $\lambda \in \sigma(x)$ iff $x - \lambda e$ is not invertible in the Jacobson way in J.
* $\infty \in \sigma(x)\ iff\ x \notin J_0$.

The set $\mathcal{B}$ of closed bounded idempotent and absolutely convex sets of J allows to give a Characterization of J_0 and a suitable notion of completeness.

2.1.3. Proposition

$$J_0 = \cup\{J(B)/\ B \in \mathcal{B}_0\}$$

Where J(B) is the subalgebra generated by B, and $\mathcal{B}_0$ a base of $\mathcal{B}$ (i.e $\mathcal{B}_0 \subset \mathcal{B}$ and $\forall B \in \mathcal{B}$, $\exists B_0 \in \mathcal{B}_0, B \subset B_0$)

2.1.4. Pseudo-completion

J is said pseudo-complete iff for every $B \in \mathcal{B}$, J(B) is a Jordan-Banach algebra. It suffices that $\mathcal{B}$ prossesses a basis $\mathcal{B}_0$ such that: $\forall B \in \mathcal{B}_0$, J(B) is a Jordan-Banach algebra.

The next result will be often used:

2.1.5. ***Proposition***

let be $\mathcal{B}_e = \{B \in \mathcal{B} / e \in B\}$

If J is pseudo-complete, then:

$$\forall B, C \in \mathcal{B}_e \quad \exists D \in \mathcal{B}_e / B \cup C \subset D.$$

In fact, let prove that BC is bounded; we consider $L_b : c \in J(C) \longrightarrow bc \in J$, for every b in B, L_b is continuous. $B' = \{L_b / b \in B\}$ is equicontinuous on J(C) since J(C), being a Jordan-Banach algebra, is barreled, and B is bounded for the point wise topology on J(C).

As in the associative case, the resolvent set of $x \in J$, denoted by ρ(x) is the complement of σ(x) in $\bar{C}$, and the resolvent is the mapping $R_\lambda : x \to R_\lambda(x) = (x - \lambda e)^{-1} = x_\lambda$
The use of the weak topology being more appropriated, we first proved.

2.1.6. Lemma

(J,$\sigma(J, J')$) is a locally convex Jordan algebra.

In the following we need the two lemmas:

2.1.7. Lemma

$x \in J$, if $x_\lambda = (x - \lambda e)^{-1}$ is weakly holomorphic at $\mu \neq \infty$, then x_λ has derivations for the weak topology $x_\lambda^{(n)} = n!\, x_\lambda^{n+1}$ n=1,2,...

2.1.8. Lemma

$x \in J$, if x_λ is weakly holomorphic at ∞ and if

$$y_\lambda = \begin{cases} x_{\frac{1}{\lambda}} & \text{if } \lambda \neq 0 \\ \lim\limits_{\lambda \to 0} x_{\frac{1}{\lambda}} & \text{if } \lambda = 0 \end{cases}$$

Then y_λ has weak derivations for all orders in a certain neighborhood of 0, given by:

$$y_\lambda^{(n)} = (-1)^n n! x^{n-1} (\lambda x - e)^{-(n+1)} \qquad n = 1, 2...$$

and $\quad y_\lambda = \lambda(\lambda x - e)^{-1}$

Finally we get the following theorem:

2.1.9. Theorem

$x \in J$, then:

1) If the resolvent of x is weakly holomorphic at μ, then $\mu \in \rho(x)$.

2) If $\mu \in \rho(x)$, then: $\exists V_\mu$ a neighborhood of μ, $\exists B \in \mathcal{B}_e$ such that: $x_\lambda \in J(B)$, $\forall \lambda \in V_\mu \cap \rho(x)$ for $\mu \neq \infty$. The resolvent is differentiable at μ (for the norm of J(B))

3) If J is pseudo-complete and $\mu \in \rho(x)$,then V_μ and B of 2) can be choosen such that:

$\forall \lambda \in V_\mu, \lambda \in \rho(x)$, $x_\lambda \in J(B)$ and the resolvent is holomorphic at μ.

2.1.10. Corollary

$\forall x \in J, \sigma(x) \neq \emptyset$; and if J is pseudo-complete then $\sigma(x)$ is closed.

2.1.11. Corollary

Suppose that J is a division algebra over **C** and for every element x of J, there exists λ in $\mathbf{C}^*$ such that: $(\lambda x)^n \longrightarrow 0$ *when* $n \to +\infty$, then J is isomorphic to **C**.

2.1.12. Remark

When J is a topological Jordan algebra over **R** with a unit e, separated, with continuous inverse (resp. l.c.J-a) then, the complexificated algebra $J_{\mathbf{C}}$ of J, provided with the product topology, is unital, separated, with continuous inverse (resp. l.c.J-a).

We recall that $J_{\mathbf{C}} = J \times J$ with the operations
$(a,b)+(c,d) = (a+c, b+d)$
$(\alpha + i\beta)(a,b) = (\alpha a - \beta b, \alpha b + \beta a)$
$(a,b)(c,d) = (ac - bd, ad + bc)$

One can find in Kaïdi's Doctorat Dissertation [3] some properties of $sp_J x = sp_{J_{\mathbf{C}}}(x,0)$ when J is a normed complete n.c Jordan algebra on **R**.

For the l.c.J-a J over R, an element x of J is said to be bounded iff (x,o) is bounded in J_C and we still have the characterization:

$$J_0 = \cup\{J(B) \;/ B \in \mathcal{B}_0\}$$

The bornological spectrum is defined by: $\sigma_J(x) = \sigma_{J_{\mathbf{C}}}(x,0)$

2.2. Spectral radius

By definition the spectral radius $r_J(x)$ is:

$$r_J(x) = sup\{|\lambda|; \lambda \in \sigma(x) \text{ with } |\infty| = +\infty\}$$

If J is pseudo-complete, for every $x \in J_0$, we have
$\sigma(J) = \cap\{\sigma_{J(B)}; B \in \mathcal{B}_e, x \in J(B)\}$
$r_J(x) = inf\{r_{J(B)}(x); B \in \mathcal{B}_e, x \in J(B)\}$

3. Locally convex Jordan algebras with continuous inverse

Let J be a l.c-J-a with continuous inverse, there are two notions of the spectrum, the classical Sp x defined by Viola and the bornological $\sigma(x)$ studied above. We have compared $\sigma(x)$ and Sp x:

3.1. PROPOSITION.

1) Sp x $\subset \sigma(x) \subset \bar{Sp}\ x$ (The closure in $\bar{C}$)
2) If J is pseudo-complete then: $\sigma(x) = \bar{Sp}\ x$

3.2. . CONSEQUENCE

Sp x is bounded iff x $\in J_0$

4. Functional calculus

A generalization of Martinez's theorem on functional calculus for Jordan-Banach algebras, is obtained for a pseudo-complete l.c.J-a.

4.1. NOTATIONS

$x \in J$, H_∞(x) denote the algebra of holomorphic complex functions in a neighborhood of $\sigma(x)$, i.e locally holomorphic every where in $\sigma(x) \cap C$ and holomorphic at ∞ if $\infty \in \sigma(x)$, which means that:

$$f_0(\lambda) = \begin{cases} f(\lambda^{-1}) & if\ \lambda \neq 0 \\ \lim\limits_{\lambda \to 0} f(\lambda^{-1}) & if\ \lambda = 0 \end{cases}$$

is holomorphic at 0

$\mathcal{O}_\infty(x)$ is the quotient algebra obtained by considering on $H_\infty(x)$ the germs equivalence relation $f(\infty)$ will denote the limit of f at infinity when $f \in \mathcal{O}_\infty(x)$

4.2. THEOREM

$x \in J$, then there exists an homomorphism $f \to f(x)$ from $\mathcal{O}_\infty(x)$ to J defined by:

1) $f(x) = \frac{1}{2i\pi} \int_{+\partial\Omega} f(\lambda) x_\lambda d\lambda$ $\qquad if\ \ x \in J_0$

2) $f(x) = f(\infty)e + \frac{1}{2i\pi} \int_{+\partial\Omega} f(\lambda) x_\lambda d\lambda$ $\qquad if\ \ x \notin J_0\ and\ \rho(x) \neq \emptyset$

3) $f(x) = Ke$ $\qquad if\ \ f(\lambda) = K\ for\ the\ case\ \rho(x) = \emptyset$

N.B : Ω is a Cauchy domain satisfying sepecial conditions.

References

1. **G.R. ALLAN** : "A spectral Theory of locally convex algebras" Proc. London. Math. Soc. (3) 15 (1965) 399-421.
2. **N.JACOBSON** : " Structure and Representations of Jordan algebras". American Mathematical Society (Colloquium Publications XXXIX). Providence, Rholde Island 1968.
3. [**A.KAIDI** : " Bases para una teoria de las algebras no-associativas normadas". Tesis doctoral Granada (1977).
4. **J.MARTINEZ MORENO** : "Sobre algebras de Jordan Normadas completas". Tesis doctoral . Universidad de Granada (1977).
5. **C. VIOLA DEVAPAKKIAM** : "Jordan algebras with continuous inverse." Math. Japonica 16 (1971) p. 115-125.

REPRESENTATIONS OF REDUCED ENVELOPING ALGEBRAS

ROLF FARNSTEINER
Department of Mathematics
University of Wisconsin
Milwaukee, WI 53201

Abstract.
In this paper we indicate how algebro-geometric techniques can be employed in order to obtain information concerning the representation finite reduced enveloping algebras of restricted Lie algebras.

In the representation theory of complex simple Lie algebras Weyl's Theorem affords a reduction to the consideration of irreducible modules. By contrast, Weyl's Theorem fails for a Lie algebra of positive characteristic and one is therefore forced to study indecomposable modules. An early result by Zassenhaus [15], however, indicates that the problem of classifying indecomposable modules is not tractable in this generality. On the other hand, irreducible representations of restricted Lie algebras give rise to a family of finite dimensional associative algebras that share important features with group algebras of finite groups. Accordingly, much of the recent work in modular representation theory has focused on these so-called reduced enveloping algebras.

This paper delineates some of the geometric and homological methods of modular representation theory and indicates how they can be exploited in the study of indecomposable modules of restricted Lie algebras.

Throughout, F will denote an algebraically closed field of characteristic $p > 2$. All algebras and modules are supposed to be finite dimensional. We are focusing on indecomposable modules of associative algebras and, accordingly, we shall consider those algebras that do not have too many indecomposables. In view of our applications the associative algebras under consideration will be assumed to be Frobenius algebras.

Definition. An associative algebra A has *finite representation type* if there are only finitely many nonisomorphic indecomposable A-modules.

In case $A = F[G]$, the group algebra of a finite group, a theorem due to Higman [8] asserts that $F[G]$ has finite representation type if and only if its p-Sylow subgroups are cyclic. Our main objective is to formulate an analogue of this result within the framework of restricted Lie algebras.

Let L be a Lie algebra over F. A mapping $[p] : L \longrightarrow L$ that satisfies the formal properties of a p-power operator is called a *p-mapping.* In that case the pair $(L, [p])$ is referred to as a *restricted Lie algebra.* The notions of a p-subalgebra and a p-ideal are defined in the obvious fashion (cf. [14] for the details).

Example: The algebra $Mat_n(F)$ together with the ordinary p-power is a re-

S. González (ed.), Non-Associative Algebra and Its Applications, 128–132.

stricted Lie algebra. Moreover, $sl(n,F) := \{a \in Mat_n(F)\ ;\ tr(a) = 0\}$ is a p-subalgebra of $Mat_n(F)$.

Let $(L,[p])$ be restricted with universal enveloping algebra $\mathcal{U}(L)$. For a linear form $\chi \in L^*$ we define the *χ-reduced universal enveloping algebra* via

$$u(L,\chi) := \mathcal{U}(L)/(\{x^p - x^{[p]} - \chi(x)^p 1\ ;\ x \in L\}).$$

These algebras play a prominent role in the representation theory of L. Any irreducible L-module V defines a linear form χ such that V is an irreducible $u(L,\chi)$-module.

Reduced enveloping algebras have the following basic properties :

- $dim_F u(L,\chi) = p^n$ if $n = dim_F L$
- $u(L,\chi)$ is a Frobenius algebra.
- The algebra $u(L) := u(L,0)$ is a Hopf algebra.

Work on the representation type of $u(L)$ was initiated by Pollack [11] in the late sixties who showed that $u(L)$ has infinite representation type, whenever L is classical. In 1988 Pfautsch and Voigt [10] announced the classification of those restricted Lie algebras L for which $u(L)$ has finite representation type. Recently, J. Feldvoss and H. Strade [4] gave another proof primarily by resorting to techniques from the classification theory of simple modular Lie algebras.

An element $x \in L$ is said to be *toral* if $x^{[p]} = x$, and *p-nilpotent* if $x^{[p]^n} = 0$ for some $n \geq 0$. We let $T(L)$ and $N(L)$ denote the largest toral and the largest nilpotent ideal of L, respectively.

Theorem 1 ([10, 4]): *The algebra $u(L)$ has finite representation type if and only if there exists a toral element $t \in L$ and a p-nilpotent element $x \in N(L)$ such that $L = Ft + N(L)$ and $N(L) = T(L) \oplus \sum_{i\geq 0} Fx^{[p]^i}$.*

The point is that the structure of the underlying restricted Lie algebras is fairly simple, i.e. the first derived algebra $[L,L]$ is abelian. While this result explicitly identifies restricted enveloping algebras of finite representation type, it also implies that for many interesting Lie algebras, such as the simple algebras, the restricted representation theory will be fairly complicated. The following example shows that it is the insistance on the condition $\chi = 0$, that eliminates many Lie algebras.

Example: The algebra $u(sl(2,F),\chi)$ has finite representation type if and only if $\chi \neq 0$.

We shall therefore change our point of view and ask the following questions:

1. Given a restricted Lie algebra $(L,[p])$, what can be said about the set $\mathcal{F}_L := \{\chi \in L^*\ ;\ u(L,\chi)$ has finite representation type $\}$?
2. What is the structure of $u(L,\chi)$ for $\chi \in \mathcal{F}_L$?

Before addressing these problems, we briefly recall some of the geometric and homological methods that have come to bear in modular representation theory. In the sequel A is assumed to be a Frobenius algebra.

Definition. Let M be an A-module. We say that M is *periodic* if there exists an exact sequence

$$(*)\qquad (0) \longrightarrow M \longrightarrow P_n \longrightarrow P_{n-1} \longrightarrow \cdots \longrightarrow P_1 \longrightarrow P_0 \longrightarrow M \longrightarrow (0),$$

with projective modules modules $P_0, \ldots, P_n$ If n is minimal, then $n+1$ is called the *period* of M.

Theorem 2 ([7]): *If A has finite representation type, then every A-module is periodic.*

The preceding result effectively links the representation theory to properties of *Ext*-functors. Information concerning their behaviour can be neatly stored in a power series.

Definition. Let M be an A-module. The complex power series $P_M(t) := \sum_{i=0}^{\infty} dim_F Ext_A^i(M,M)t^i \in \mathbb{C}[[t]]$ is called the *Poincaré-Series* of M.

In their paper [6] Friedlander and Parshall observed that the concept of a support variety, which plays an important role in the modular representation theory of finite groups, possesses an analogue in the setting of restricted Lie algebras. Let M be a $u(L,\chi)$-module. The symmetric algebra of L^* will be denoted $S(L^*)$. If $Ext^*_{u(L,\chi)}(M,M)$ is endowed with the Yoneda product, a theorem by Hochschild [9, p. 575] provides a homomorphism $S(L^*) \longrightarrow Ext^*_{u(L,\chi)}(M,M)^{(-1)}$ of F-algebras that is natural with respect to L and sends L^* into $Ext^2_{u(L,\chi)}(M,M)^{(-1)}$. (Here, the superscript indicates that the field acts on the pertinent groups via $\alpha\, x := \alpha^p\, x$). In particular, $Ext^*_{u(L,\chi)}(M,M)^{(-1)}$ obtains the structure of an $S(L^*)$-module whose annihilator we denote by $Ann_L(M)$. The corresponding zero set $\mathcal{V}_L(M) := Z(Ann_L(M)) \subset L$ is called the *support variety* of M.

Theorem 3 ([5, 6]): *The following statements hold :*
(1) $Ext^*_{u(L,\chi)}(M,M)^{(-1)}$ is a noetherian $S(L^*)$-module.
(2) $\mathcal{V}_L(M)$ is conical and $dim\mathcal{V}_L(M)$ is the order of the pole of $P_M(t)$ at $t=1$.
(3) If M is indecomposable, then $Proj(\mathcal{V}_L(M))$ is connected.

If M is periodic and indecomposable, then Theorem 2 in conjunction with (2) of Theorem 3 yields $dim\mathcal{V}_L(M) = 1$. Owing to (3) we conclude that $\mathcal{V}_L(M)$ is a line. This fact can be utilized to obtain information concerning periodic modules.

Theorem 4 ([2]): *Let M be a $u(L,\chi)$-module. If M is periodic, then its period is bounded by 2.*

With regard to periodicity, Lie algebras thus behave like abelian p-groups (cf. [1]). Theorem 4 is actually the key to the solution of problem 2. We first recall a basic definition.

Definition. The algebra A is called a *Nakayama algebra,* if every principal indecomposable A-module has exactly one composition series.

Theorem 5 ([2]): *Let V be an irreducible, periodic $u(L,\chi)$-module with block $\mathcal{B}(V)$. Then the following statements hold :*
(1) $\mathcal{B}(V)$ is a Nakayama algebra and $P_V(t) \in \{1, \frac{1}{1-t}, \frac{1}{1-t^2}\}$
(2) If $P_V(t) = 1$ or $\frac{1}{1-t}$, then $\mathcal{B}(V) \cong Mat_n(F[X]/(X^c))$, where c is the Cartan invariant of V and $n = dim_F V$.
(3) If $P_V(t) = \frac{1}{1-t^2}$, then $\mathcal{B}(V)$ possesses exactly p irreducible modules all of which have the same dimension.

The foregoing result, which incidentally affords a short proof of Theorem 1, actually provides very detailed information concerning the irreducibles belonging to the block $\mathcal{B}(V)$. These are obtained from V by twisting the action of V by some iterate of the Nakayama automorphism of $u(L,\chi)$. Moreover, the Gabriel quiver of $\mathcal{B}(V)$, i.e. the directed Graph with the isomorphism classes of irreducible $\mathcal{B}(V)$-modules as vertices and $dim_F Ext^1_{\mathcal{B}(V)}(M,N)$ arrows leading from the class of M to the class of N, has one of the following associated undirected graphs:

- If $P_V(t) = 1$, a single point without a bond.
- If $P_V(t) = \frac{1}{1-t}$, a single point with a bond.
- If $P_V(t) = \frac{1}{1-t^2}$, a Euclidian diagram of type $\tilde{A}_{p-1}$

We note that blocks with p vertices cannot occur if the algebra $u(L,\chi)$ is symmetric. In particular, (2) of Theorem 5 applies for nilpotent and simple Lie algebras.

Having answered question 2, we turn to properties of the set $\mathcal{F}_L$.

Theorem 6 ([2]): *The set $\mathcal{F}_L$ is Zariski-open in L^*.*

We may thus summarize our answers to questions 1 and 2 as follows. If $\mathcal{F}_L$ is not empty, then $u(L,\chi)$ has finite representation type for "most" characters. Moreover, for every $\chi \in \mathcal{F}_L$ the algebra $u(L,\chi)$ is a Nakayama algebra. Consequently, the indecomposable $u(L,\chi)$-modules are images of the principal indecomposables and can thus be obtained from the Loewy series of the algebra $u(L,\chi)$.

According to Theorem 5, the determination of the representation type of $u(L,\chi)$ necessitates a fairly detailed knowledge of its irreducible modules. This information is, for instance, available for solvable restricted Lie algebras [13], and for the simple restricted Lie algebra $W(1)$ (cf. [12]). By way of illustration we consider the Witt algebra $W(1) := \bigoplus_{i=-1}^{p-2} Fe_i$ with multiplication $[e_i, e_j] = (j-i)\, e_{i+j} \quad -1 \leq i+j \leq p-2$ and 0 otherwise. Its p-map is given by $e_i^{[p]} = \delta_{i,0}\, e_i \quad -1 \leq i \leq p-2$. The algebra $W(1)$ is filtered by means of $W(1)_{(i)} := \sum_{j \geq i} Fe_j$ and for a linear form $\chi \in W(1)^*$ we put, following Strade [12], $r(\chi) := min(\{i \geq -1 \; ; \; \chi(W(1)_{(i)}) = (0)\})$.

Theorem 7 ([3]): *The following statements hold :*

(1) Let V be an irreducible $u(W(1),\chi)$-module. If $0 \leq r(\chi) \leq p-2$ and $r(\chi) \equiv 0 \; mod(2)$, then $\mathcal{V}_{W(1)}(V) = W(1)_{(r(\chi)+1)}$.

(2) $\mathcal{F}_{W(1)} = \{\chi \in W(1)^* \; ; \; r(\chi) \geq p-3\}$.

References

1. Carlson, J. *The Dimensions of Periodic Modules over Modular Group Algebras.* Ill. J. Math. **23** (1979), 295-306
2. Farnsteiner, R. *Periodicity and Representation Type of Modular Lie Algebras.* Preprint 1994
3. Farnsteiner, R. *On the Representation Type of Semiprimary Reduced Enveloping Algebras.* (Manuscript in preparation)
4. Feldvoss, J., Strade, H. *Restricted Lie Algebras with Bounded Cohomology and Related Classes of Lie Algebras.* Manuscripta math. **74** (1992), 47-67
5. Friedlander, E., Parshall, B. *Support Varieties for Restricted Lie Algebras.* Invent. math. **86** (1986), 553-562
6. Friedlander, E., Parshall, B. *Geometry of p-Unipotent Algebras.* J. Algebra **109** (1987), 25-45

7. Heller, A. *Indecomposable Representations and the Loop Space Operation.* Proc. Amer. Math. Soc. **12** (1961), 640-643
8. Higman, D.G. *Indecomposable Representations at Characteristic p.* Duke J. Math. **21** (1954), 377-381
9. Hochschild, G.P. *Cohomology of Restricted Lie Algebras.* Amer. J. Math. **76** (1954), 555-580
10. Pfautsch, W., Voigt, D. *The Representation-Finite Algebraic Groups of Dimension Zero.* C. R. Acad. Sci. Paris **306** (1988), 685-689
11. Pollack, R. *Restricted Lie Algebras of Bounded Type.* Bull. Amer. Math. Soc. **74** (1968), 326-331
12. Strade, H. *Representations of the Witt Algebra.* J. Algebra **49** (1977), 595-605
13. Strade, H. *Darstellungen Auflösbarer Lie-Algebren.* Math. Ann. **232** (1978), 15-32
14. Strade, H., Farnsteiner, R. *Modular Lie Algebras and their Representations.* Dekker Monographs **116** New York 1988
15. Zassenhaus, H. *Representation Theory of Lie Algebras of Characteristic p.* Bull. Amer. Math. Soc. **60** (1954), 463-469

A COHOMOLOGICAL CHARACTERIZATION OF SOLVABLE MODULAR LIE ALGEBRAS

JÖRG FELDVOSS
Mathematisches Seminar der Universität
Bundesstraße 55, D-20146 Hamburg,
Federal Republic of Germany

Abstract. We give a unifying approach to characterizations of some classes of finite dimensional solvable modular Lie algebras by the vanishing of their cohomology and the structure of their principal block.

Key words: Modular Lie algebra, p-envelope, restricted cohomology, principal block, chief factor

1. Introduction

In this note we present a unifying approach to a cohomological characterization of some classes of finite dimensional modular Lie algebras. We are starting with a cohomological criterion for the solvability of a finite dimensional modular Lie algebra which implicitly is already contained in the work of Barnes (compare the proof of [2, Theorem 4]). From that we derive a refined version of [9, Proposition 5.12] that allows us to prove a cohomological characterization of finite dimensional supersolvable resp. nilpotent restricted Lie algebras without using their non-restricted analogues as in [9, §5] by applying [10, Proposition 6] instead of [2, Theorem 1]. If we use a recent formula of Farnsteiner relating the cohomology of an arbitrary (i.e., non-restricted) modular Lie algebra with the (ordinary) cohomology of its finite dimensional p-envelope, cohomological characterizations of finite dimensional supersolvable resp. nilpotent modular Lie algebras due to Barnes resp. Dzumadil'daev turn out to be easy consequences of our main results. This approach was motivated by a similar one due to Stammbach for finite modular group algebras (cf. [14, Theorem B and the remark on p. 296]). It is a pleasure to thank the organizers for their hospitality during the conference and for the opportunity to present the results of [10] at the meeting in Oviedo. In fact, this gave the author another possibility to think about the cohomological characterization of modular Lie algebras finally leading to the present paper.

2. Main Results

In the following F will always denote a commutative field of prime characteristic p. The proof of our first result is along the lines of the proof of [2, Theorem 4]. For the convenience of the reader we will repeat the arguments.

S. González (ed.), Non-Associative Algebra and Its Applications, 133–139.

Proposition 1 *A finite dimensional Lie algebra L is solvable if and only if $H^1(L/\mathrm{Ann}_L(S), S)$ vanishes for every simple L-module S.*

Proof. The "only if"-part is an immediate consequence of [1, Theorem 2]. In order to prove the "if"-part, we start to assume that L is simple. In this case the only non-faithful simple L-module is the one-dimensional trivial module and thus we obtain by assumption that $H^1(L, S) = 0$ for every *simple* L-module S. Then the long exact cohomology sequence yields $H^1(L, M) = 0$ for every *finite dimensional* L-module M in contradiction to [4, Corollary 2 in §1] or [7, Corollary 2.2]. Hence L is not simple and we (can) proceed by induction on $\dim L$.

Let I be a (non-zero) minimal ideal of L. Then every simple L/I-module S is via

$$x \cdot s := (x + I) \cdot s \qquad \forall\, x \in L, s \in S$$

a simple L-module with $\mathrm{Ann}_{L/I}(S) = \mathrm{Ann}_L(S)/I$. From our assumption we obtain

$$H^1((L/I)/\mathrm{Ann}_{L/I}(S), S) \cong H^1(L/\mathrm{Ann}_L(S), S) = 0$$

and by induction, L/I is solvable.

If J is another minimal ideal of L, then L can be embedded into $L/I \times L/J$. Since the latter Lie algebra is solvable, L is also solvable and we would be done. Therefore we may assume in the following that I is the *only* minimal ideal of L.

If $\mathrm{Ann}_L(I) = 0$, then $I^I = 0$ and the cohomological five-term exact sequence yields

$$H^1(I, I)^{L/I} \cong H^1(L, I) = 0.$$

By using again the vanishing of $\mathrm{Ann}_L(I)$, we can derive by a short computation that $[L, L] \subseteq I$. Then $H^1(L, I) = 0$ in conjunction with another easy computation yields that every inner derivation of L comes from an element of I. Finally, $C(L) = 0$ implies that $L = I$ is a (non-abelian) simple Lie algebra, a contradiction. Hence $\mathrm{Ann}_L(I) \neq 0$ and therefore the unique minimal ideal I of L is contained in $\mathrm{Ann}_L(I)$. Thus I is abelian and by the above we obtain that L is solvable. □

Remark. Using Whitehead's Theorem and the argument in [1, Theorem 2], it is easy to show that $H^n(L, S) = 0 \;\; \forall\, n \geq 0$ for every finite dimensional faithful simple module S over *any* finite dimensional Lie algebra L if the characteristic of the underlying ground field is zero.

Let L be a restricted Lie algebra. Then the *restricted universal enveloping algebra* $u(L)$ of L is a finite dimensional augmented associative algebra and the restricted (left) L-modules are in one-to-one correspondence with the (left, unitary) $u(L)$-modules (cf. [15, (V.2) and (V.3)]). Following Hochschild [12] we define the *restricted cohomology* of L with coefficients in a restricted L-module M by means of

$$h^n(L, M) := \mathrm{Ext}^n_{u(L)}(F, M) \qquad \forall\, n \geq 0,$$

where F becomes a $u(L)$-module via the augmentation mapping of $u(L)$. Since $u(L)$ is finite dimensional, one can decompose it into a direct sum of finitely many indecomposable two-sided ideals, the so-called *block ideals* of $u(L)$. In particular, this induces an equivalence relation *"belonging to a block"* on the finite set of isomorphism classes of simple restricted L-modules (see e.g. [10]). We refer to the block containing the one-dimensional trivial L-module as the *principal block* of L. It is well-known that every simple restricted L-module not belonging to the principal block of L has vanishing restricted cohomology. This is a special case of the first part of the following result.

Lemma 1 *Let L be a finite dimensional restricted Lie algebra and M, N be restricted L-modules. Then the following statements hold:*
(a) If M and N belong to different blocks, then $\mathrm{Ext}^n_{u(L)}(M,N)$ vanishes for every integer $n \geq 0$.
(b) Assume in addition that M and N are simple. Then M and N belong to the same block if and only if there exists a finite sequence $S_1, ..., S_n$ of simple restricted L-modules such that $M = S_1$, $S_n = N$ and

$$\mathrm{Ext}^1_{u(L)}(S_j, S_{j+1}) \neq 0 \text{ or } \mathrm{Ext}^1_{u(L)}(S_{j+1}, S_j) \neq 0 \text{ for every } 1 \leq j \leq n-1.$$

Proof. (a) is an immediate consequence of [5, Corollary 4.10] applied to the primitive central idempotents corresponding to M resp. N and (b) is folklore (see e.g. [14, Corollary 1]). □

In order to apply Proposition 1 we will need a description of restricted 1-cohomology due to Hochschild. It is well-known that $H^1(L,M) \cong \mathrm{Der}(L,M)/\mathrm{Ider(L,M)}$ for every L-module M, where $\mathrm{Der}(L,M)$ denotes the vector space of *derivations* from L into M (i.e., $d([x,y]) = x \cdot d(y) - y \cdot d(x) \ \ \forall\, x,y \in L$) resp. $\mathrm{Ider}(L,M)$ denotes the subspace of *inner derivations* from L into M (i.e., there exists an element $m \in M$ such that $d(x) = x \cdot m \ \forall\, x \in L$). In the case that L and M are restricted, a derivation d from L into M is called *restricted* if $d(x^{[p]}) = x^{p-1} \cdot d(x) \ \ \forall\, x \in L$ and we denote by $\mathrm{der}(L,M)$ the subspace of all *restricted derivations* from L into M. Note that every inner derivation is restricted! Finally, we put $M^L := \{m \in M \mid L \cdot m = 0\}$. The next (well-known) result will be used several times in proving the results below.

Lemma 2 *Let L be a restricted Lie algebra and M be a restricted L-module. Then the following statements hold:*
(a) $h^1(L,M) \cong \mathrm{der}(L,M)/\mathrm{Ider(L,M)}$. In particular, $h^1(L,M) \hookrightarrow H^1(L,M)$.
(b) If $M^L = 0$, then $h^1(L,M) \cong H^1(L,M)$.

Proof. (a) is just [12, Theorem 2.1] (see also [8, Korollar I.2.14]) and (b) is an immediate consequence of Hochschild's six-term exact sequence relating restricted cohomology with ordinary cohomology [12, p. 575]. □

The equivalence of (a) and (b) in the next result was already obtained in [9, Proposition 5.12].

Theorem 1 *For any finite dimensional restricted Lie algebra L the following statements are equivalent:*

(a) L is solvable.
(b) $h^1(L/\mathrm{Ann}_L(S), S) = 0$ for every simple restricted L-module S.
(c) $h^1(L/\mathrm{Ann}_L(S), S) = 0$ for every simple restricted L-module S in the principal block of L.

Proof. (a)$\Longrightarrow$(b) is an immediate consequence of Proposition 1 in conjunction with Lemma 2(a) and (b)$\Longrightarrow$(c) is trivial.

(c)$\Longrightarrow$(a): In order to apply the "only if"-part of Proposition 1, we have to show the vanishing of $H^1(L/\mathrm{Ann}_L(S), S)$ for every simple L-module S. If S is not restricted, the assertion follows from [3, Theorem 2]. Therefore we can assume that S is restricted. If S is in the principal block of L, then our assumption and Lemma 2(b) yield the assertion and if S is not in the principal block of L, then it is easy to see (cf. [10, Lemma 3]) that S is also not in the principal block of $L/\mathrm{Ann}_L(S)$ and thus Lemma 1(a) and Lemma 2(b) imply the assertion. □

It is very likely that the following stronger version of Theorem 1 is true:

Conjecture 1 *For any finite dimensional restricted Lie algebra L the following statements are equivalent:*
(a) L is solvable.
(b) There exists an odd integer n such that $h^n(L/\mathrm{Ann}_L(S), S) = 0$ for every simple restricted L-module S.
(c) There exists an odd integer n such that $h^n(L/\mathrm{Ann}_L(S), S) = 0$ for every simple restricted L-module S in the principal block of L.

A Lie algebra L is called *supersolvable* if there is a (descending) chain (from L to 0) of ideals (in L) such that all factors are one-dimensional. A *chief factor* of L is just a composition factor of the adjoint module of L. It is clear that L is supersolvable if and only if every chief factor of L is one-dimensional and one readily verifies that in this case every subalgebra of L is supersolvable.

Theorem 2 *For any finite dimensional restricted Lie algebra L the following statements are equivalent:*
(a) L is supersolvable.
(b) $h^1(L, S) = 0$ for every simple restricted L-module S with $\dim S \neq 1$.
(c) Every simple module in the principal block of L is one-dimensional.

Proof. (a)$\Longrightarrow$(b) is an immediate consequence of [1, Theorem 3] and Lemma 2(a).

(b)$\Longrightarrow$(c): Let S and S' be simple restricted L-modules such that $\dim S = 1$ and $\dim S' \neq 1$. Then $\mathrm{Hom}(S, S') \cong S^* \otimes S'$ and $\mathrm{Hom}(S', S) \cong S'^* \otimes S$ are also simple restricted L-modules such that $\dim \mathrm{Hom}(S, S') \neq 1 \neq \dim \mathrm{Hom}(S', S)$. From our assumption we derive

$$\mathrm{Ext}^1_{u(L)}(S, S') \cong h^1(L, \mathrm{Hom}(S, S')) = 0 = h^1(L, \mathrm{Hom}(S', S)) \cong \mathrm{Ext}^1_{u(L)}(S', S).$$

If we apply this sucessively to the finite sequence $F := S_1, ..., S_n =: M$ of simple $u(L)$-modules satisfying the condition in Lemma 1(b), we can conclude that M is one-dimensional.

(c)$\Longrightarrow$(a): In order to apply Theorem 1, we have to show the vanishing of $h^1(L/\mathrm{Ann}_L(S), S)$ for every restricted simple L-module S in the principal block of L. By assumption, S is one-dimensional and thus $L/\mathrm{Ann}_L(S)$ is abelian. Either $L = \mathrm{Ann}_L(S)$ or S is a non-trivial simple $L/\mathrm{Ann}_L(S)$-module and the assertion follows from [9, Proposition 5.5]. According to Theorem 1, L is solvable and [10, Proposition 6] in conjunction with our assumption shows that every chief factor of L is one-dimensional, i.e., L is supersolvable. □

Remark. The equivalence of (a) and (c) in Theorem 2 for algebraically closed ground fields was already obtained by Voigt in the more general context of infinitesimal algebraic groups [16, Satz 2.40]. Moreover, we should remark that the proof of [9, Theorem 5.10] is not correct, but that for *finite dimensional* modules the assertion follows from [5, Theorem 4.7] and the nilpotency of $[L, L]$. This can also be used to prove the implication (a)$\Longrightarrow$(b) of the preceding result.

Conjecture 2 *A finite dimensional restricted Lie algebra L is supersolvable if and only if there exists an odd integer n such that* $h^n(L, S) = 0$ *for every simple restricted L-module S with* $\dim S \neq 1$.

Corollary 1 (D.W. Barnes [2]) *A finite dimensional Lie algebra L is supersolvable if and only if* $H^1(L, S) = 0$ *for every simple L-module S with* $\dim S \neq 1$.

Proof. The "only if"-part is an immediate consequence of [1, Theorem 3]. In order to prove the "if"-part, we consider a *finite dimensional* p-envelope E of L (cf. [15, Proposition II.5.3(2)]). Let S be a simple restricted E-module S with $\dim S \neq 1$. According to [5, Corollary 5.1], all the elements c_i in [6, Corollary 2.4(2)] act trivially on S or $H^1(E, S) = 0$. By virtue of [15, Proposition II.1.3(1)], S is a (non-trivial) simple L-module and our assumption in conjunction with [6, Corollary 2.4(2)] implies also in the first case the vanishing of $H^1(E, S)$. Then Lemma 2(a) and Theorem 2 show that E and thus L is supersolvable. □

The equivalence of (a) and (c) in the next result was already obtained by Voigt in the more general context of infinitesimal algebraic groups (over an algebraically closed field) [16, Satz 2.41].

Theorem 3 *For any finite dimensional restricted Lie algebra L the following statements are equivalent:*
(a) L is nilpotent.
(b) $h^1(L, S) = 0$ *for every non-trivial simple restricted L-module S.*
(c) Every simple module in the principal block of L is trivial.

Proof. (a)$\Longrightarrow$(b) is a special case of [9, Proposition 5.5]) and (b)$\Longrightarrow$(c) can be obtained in a completely similar way to the proof of (b)$\Longrightarrow$(c) in Theorem 2.

(c)$\Longrightarrow$(a): In order to apply Theorem 1, we have to show the vanishing of $h^1(L/\mathrm{Ann}_L(S), S)$ for every restricted simple L-module S in the principal block of L. By assumption, S is trivial and thus $L/\mathrm{Ann}_L(S) = 0$. According to Theorem 1, L is solvable and [10, Proposition 6] in conjunction with our assumption shows that every chief factor of L is trivial, i.e., L is nilpotent. □

Conjecture 3 *A finite dimensional restricted Lie algebra L is nilpotent if and only if there exists an odd integer n such that $h^n(L,S) = 0$ for every non-trivial simple restricted L-module S.*

Remark. Note that a finite dimensional restricted Lie algebra L is a torus if and only if there exists an integer n such that $h^n(L,S) = 0$ for every simple restricted L-module S (cf. [13]).

The "only if"-part of our last result is an immediate consequence of [1, Theorem 1] and the "if"-part is completely analogous to the proof of Corollary 1.

Corollary 2 (A.S. Dzumadil'daev [4]) *A finite dimensional Lie algebra L is nilpotent if and only if $H^1(L,S) = 0$ for every non-trivial simple L-module S.* □

Remark. Since $H^3(L,S)$ vanishes for every simple L-module $S \not\cong F$ of the three-dimensional simple Lie algebra L over an algebraically closed field F of characteristic $\neq 2$ (cf. [3]), the analogues of Conjectures 1, 2 and 3 are not true for ordinary cohomology. Moreover, $h^n(L,S) = 0$ for every *even* integer n and every *restricted* simple L-module $S \not\cong F$ (see [11, Korollar zu Satz 3.I] or [8, Korollar III.1.8]), which shows that the odd integer in Conditions (b) resp. (c) of Conjecture 1 and in Conjectures 2 resp. 3 cannot replaced by any (or even *all*) even integers (simultaneously)!

Note added in proof

There is the following direct proof of Lemma 2(b):

Let $U(L)^+$ denote the augmention ideal of the (ordinary) *universal enveloping algebra* $U(L)$ of L. Then it is well-known (cf. p. 234 in P.J. Hilton, U. Stammbach: A Course in Homological Algebra. GTM 4, Springer-Verlag, New York-Heidelberg-Berlin, 1971) that every derivation d from L into an arbitrary L-module M can be extended in a unique way to a $U(L)$-module homomorphism $\tilde{d}$ from $U(L)^+$ into M. (In fact, this induces an isomorphism from $\mathrm{Der}(L,M)$ onto $\mathrm{Hom}_L(U(L)^+,M)$!) In particular, we obtain:

$$\tilde{d}(x^p) = (x)_M^{p-1}d(x) \quad \forall\, x \in L, d \in \mathrm{Der}(L,M).$$

If M is restricted, we have $x^{[p]} - x^p \in \mathrm{Ann}_{C(U(L))}(M)$, where $C(U(L))$ denotes the *center* of $U(L)$, and thus $\tilde{d}(\mathrm{Ann}_{C(U(L))}(M)) \subseteq M^{U(L)^+} = M^L$ for any $d \in \mathrm{Der}(L,M)$ finally yields:

$$d(x^{[p]}) - (x)_M^{p-1}d(x) = \tilde{d}(x^{[p]} - x^p) \in M^L \quad \forall\, x \in L, d \in \mathrm{Der}(L,M). \quad \Box$$

References

1. Barnes, D.W.: On the cohomology of soluble Lie algebras. *Math. Z.* **101**, 343-349 (1967).
2. Barnes, D.W.: First cohomology groups of soluble Lie algebras. *J. Algebra* **46**, 292-297 (1977).
3. Dzumadil'daev, A.S.: On the cohomology of modular Lie algebras. *Math. USSR Sbornik* **47** (1), 127-143 (1984).

4. Dzumadil'daev, A.S.: Cohomology of truncated coinduced representations of Lie algebras of positive characteristic. *Math. USSR Sbornik* **66** (2), 461-473 (1990).
5. Farnsteiner, R.: On the vanishing of homology and cohomology groups of associative algebras. *Trans. Amer. Math. Soc.* **306** (2), 651-665 (1988).
6. Farnsteiner, R.: Recent developments in the cohomology theory of modular Lie algebras. In: *International Symposium on Non-Associative Algebras and Related Topics, Hiroshima 1990* (eds. K. Yamaguti and N. Kawamoto), 19-48. World Scientific, Singapore, 1991.
7. Farnsteiner, R. and Strade, H.: Shapiro's lemma and its consequences in the cohomology theory of modular Lie algebras. *Math. Z.* **106**, 153-168 (1991).
8. Feldvoss, J.: *Homologische Aspekte der Darstellungstheorie modularer Lie-Algebren.* Dissertation, Universität Hamburg, 1989.
9. Feldvoss, J.: On the cohomology of restricted Lie algebras. *Comm. Algebra* **19** (10), 2865-2906 (1991).
10. Feldvoss, J.: On the block structure of solvable restricted Lie algebras. Submitted.
11. Fischer, G.: *Darstellungstheorie des ersten Frobeniuskerns der SL_2.* Dissertation, Universität Bielefeld, 1982.
12. Hochschild, G.P.: Cohomology of restricted Lie algebras. *Amer. J. Math.* **76**, 555-580 (1954).
13. Hochschild, G.P.: Representations of restricted Lie algebras of characteristic p. *Proc. Amer. Math. Soc.* **5**, 603-605 (1954).
14. Stammbach, U.: Cohomological characterisations of finite solvable and nilpotent groups. *J. Pure Appl. Algebra* **11**, 293-301 (1977).
15. Strade, H. and Farnsteiner, R.: *Modular Lie Algebras and Their Representations.* Monographs and Textbooks in Pure and Applied Mathematics **116**, Marcel Dekker, Inc., New York and Basel, 1988.
16. Voigt, D.: *Induzierte Darstellungen in der Theorie der endlichen, algebraischen Gruppen.* Lecture Notes in Mathematics **592**, Springer-Verlag, Berlin-Heidelberg-New York, 1977.

AN EXTENSION OF THE ZEL'MANOV-GOLDIE THEOREM

ANTONIO FERNÁNDEZ LÓPEZ and EULALIA GARCÍA RUS
Departamento de Algebra, Geometría y Topología.
Facultad de Ciencias, Universidad de Málaga, 29071 Málaga, Spain.

Abstract. A celebrated theorem for Jordan algebras due to Zel'manov [19, 20] states that a (linear) Jordan algebra J is an order in a semisimple artinian Jordan algebra Q if and only if J is semiprime, satisfies the annihilator chain condition, and does not contain infinite direct sums of inner ideals. Reciently [10], a notion of local order has been introduced in a (not necessarily unital) associative ring and proved a Goldie-like characterization of local orders in semiprime rings with dcc on principal one-sided ideals [11], equivalently, coinciding with their socles. Inspired by these ideas, we developed in [5, 6] a theory of local orders in Jordan algebras which need not have a unit and proved a natural extension of Zel'manov-Goldie theorem.

In this note we provide a quick approach to this result, giving the relevant definitions and the main tools used in its proof.

Key words: Jordan algebra, annihilator, inner ideal, uniform element, local order

1. Orders and local orders in associative algebras

Throughout this section A and Q will denote associative algebras. Recall that a subalgebra A of an associative algebra Q with unit element is called a *right order* if every regular element in A is invertible in Q, and each element $q \in Q$ can be written as $q = ab^{-1}$, where $a, b \in A$ with b regular. Left and two-sided orders are defined similarly.

Theorem 1 (Goldie theorem) *An associative algebra A is a right order in a semisimple artinian algebra Q if and only if A is semiprime, satisfies maximality condition for right annihilators, and contains no infinite direct sum of right ideals. Moreover, A is prime if and only if Q is simple.*

An element $a \in A$ is called *semiregular* if

$$ax^2 = 0 \Rightarrow ax = 0 \quad \text{and} \quad x^2a = 0 \Rightarrow xa = 0,$$

for $x \in R \cup \{1\}$. An element $a \in A$ is *locally invertible* if there exists an idempotent e such that a is invertible in the unital algebra eAe. In this case, the local inverse a' of a in eAe is precisely the group inverse of a, which is intrinsically characterized by the conditions:

$$aa' = a'a, \quad a = aa'a, \quad a' = a'aa'.$$

The idempotent e relative to the element a is also unique, $e = aa' = a'a$. Moreover, a is locally invertible if and only if $a \in a^2Aa^2$. A subalgebra A of a (not necessarily unital) associative algebra Q is said to be a *local order* in Q if

S. González (ed.), Non-Associative Algebra and Its Applications, 140–146.

(i) every semiregular element in A is locally invertible in Q, and

(ii) for each $q \in Q$, there exists a semiregular element $x \in A$ such that $q \in xQx$ and xRx is a two-sided order in the unital algebra $xQx = eQe$ ($e = xx'$).

Theorem 2 (Local Goldie theorem) ([11, 2, 8])

(i) An associative algebra A is a local order in a semiprime associative algebra Q satisfying dcc on principal one-sided ideals if and only if A satisfies the following conditions and their duals: (α) acc on left annihilators $lan(x)$, $x \in A$, and (β) each $x \in A$ has finite left uniform-dimension.

(ii) An associative algebra A is a local order in a simple algebra Q with minimal one-sided ideals if and only if A is prime, nonsingular and satisfies (β) and its dual.

In this last case, the symmetric ring of quotients $Q_s(A)$, is contained in $Q_s(Q)$, and the right ideal $I_r(Q_s(A))$ of $Q_s(A)$ of those elements having finite right uniform-dimension is contained in Q.

Another important tool in the proof of the main result of this paper is the notion of *generalized polynomial identity with involution*, in short, GPI*. By [3], if $(A, *)$ is a prime associative algebra with involution satisfying a GPI*, then $Q_s(A)$, with the unique involution extending that of A, has non-zero socle and satisfies the same GPI* as A. Moreover, by [8],

(1.1) $I_r(A)$ is a local order in the socle of $Q_s(A)$.

2. Orders in Jordan algebras

From now on all the algebras we consider here are over a field Φ of characteristic $\neq 2$. In particular, we will exclusively deal with (linear) Jordan algebras. Our standard references for Jordan algebras are [14, 21]. Given an associative algebra A with product ab, we denote as usual by A^+ the Jordan algebra defined on A by the new multiplication, $a \cdot b = \frac{1}{2}(ab + ba)$. The *annihilator* of a subset M of a Jordan algebra J is defined to be the set

$$ann_J(M) = \{x \in J : x \cdot M = 0 = (x, J, M)\}$$

where $(a, b, c) = (a \cdot b) \cdot c - a \cdot (b \cdot c)$ is the associator of $a, b, c \in J$. A subset S of J is called a *monad* if $U_x y$ and z^2 are in S, for all $x, y, z \in S$. Now J is an *order* in a Jordan algebra with unit element Q relative to a monad S if

(i) every element $x \in S$ is invertible in Q,

(ii) each element $q \in Q$ has a J – *denominator* $s \in S$, that is, $q \cdot s \in J$ and $(q, J, s) \subset J$,

(iii) for all $a, b \in S$, $U_a S \cap U_b S \neq \emptyset$.

Notice that for any Jordan algebra J the set

$$Reg(J) = \{x \in J : U_x : J \to J \quad \text{is injective}\}$$

is a monad in J. This notion of order in Jordan algebras agrees with the classical one in associative algebras. In fact, if A is an associative algebra which is a two-sided order in a unital associative algebra Q, then for any Jordan algebra J trapped between A^+ and Q^+, J is an order in the unital Jordan algebra Q^+. A similar result holds if A has an involution $*$, and J is between $H(A,*)$ and $H(Q,*)$.

A Jordan algebra Q is said to be *artinian* if it satisfies dcc on all inner ideals. By [16], a simple Jordan algebra with finite capacity is artinian if and only if it is not the Jordan algebra defined by a quadratic form containing an infinite dimensional totally isotropic vector space, in short, a *non-artinian quadratic factor*. A Jordan algebra J is said to be a *Goldie algebra* if it satisfies the annihilator chain condition (acc and dcc are equivalent for annihilators) and contains no infinite direct sum of inner ideals.

Theorem 3 (Zel'manov-Goldie theorem) ([19, 20]) *A Jordan algebra J is an order in a semisimple artinian Jordan algebra Q if and only if J is a semiprime Goldie algebra.*

3. Weak local orders in Jordan algebras

Recall that an element x in a Jordan algebra J is said to be if $x \in U_x^2 J$. An element $x \in J$ is strongly regular if and only if it is *locally invertible* ($x \in LocInv(J)$), that is there is a (unique) idempotent $e = P(x)$ such that x is invertible (in the usual sense) in the unital Jordan algebra $U_e J$. Then the inverse $x' \in U_e J$ of x is called the *generalized inverse* of x, and can be characterized (among others) by the following equivalent conditions (see [9]) :

(i) $U_x x' = x, \quad U_{x'} x = x' \quad \text{and} \quad U_{x'} U_x = U_x U_{x'}$,

(ii) $U_x x' = x \quad \text{and} \quad U_x U_{x'} x' = x'$,

(iii) $(x')^2 \cdot x = x' \quad \text{and} \quad U_x x' = x$.

Now J is said to be a *weak local order* in Q if for each $q \in Q$ there exists $x \in LocInv(Q) \cap J$ such that $q \in U_x Q$ with $U_x J$ being an order in the unital Jordan algebra $U_x Q = U_e Q \quad (e = P(x))$ relative to a monad S of $U_x J$.

As proved in [5], orders in unital Jordan algebras are actually weak local orders. Moreover, by [5], weak local orders J of Q are inner-tight covers compatible with annihilators. More precisely,

(3.1) Let J be a nondegenerate Jordan algebra. For $0 \neq q \in Q$, there exists $s \in J \cap LocInv(Q)$ such that $q \in U_e Q$ with $0 \neq U_q U_s J \subset J$ and $e = P(s)$. In particular, $I \cap J \neq 0$ for every non-zero inner ideal I of Q.

(3.2) $ann_J(M) = ann_Q(M) \cap J$ for every $M \subset J$.

If additionally Q agrees with its socle (definition below) then

(3.3) $ann_J(M) \subset ann_J(N)$ iff $ann_Q(M) \subset ann_Q(N)$, $\quad M, N \subset J$.

4. Local orders in Jordan algebras with dcc on principal inner ideals

Recall that the *socle* $Soc(J)$ of a nondegenerate Jordan algebra J is defined as the sum of all minimal inner ideals of J [17]. A nondegenerate Jordan algebra J satisfying dcc on principal inner ideals (equivalently, coinciding with its socle [15]) also satisfies acc on $ann(x)$, $x \in J$ [4]. Actually, this is a consequence of the following more general result [7]: For each element x in the socle of a nondegenerate Jordan algebra J, $ann(ann(x)) = U_xJ$.

An element x in a Jordan algebra J will be called *semiregular* if $ann(x) = ann(x^2)$. The set of all semiregular elements of J will be denoted by $SemiReg(J)$. In general, $LocInv(J) \subset SemiReg(J)$. The reverse inclusion holds in Jordan algebras satisfying dcc on principal inner ideals [5].

A subalgebra J of a Jordan algebra Q will be called a *local order* in Q if

(i) $SemiReg(J) \subset LocInv(Q)$ and

(ii) for every $q \in Q$ there exists $x \in SemiReg(J)$ such that $q \in U_xQ$ with U_xJ being an order in the unital Jordan algebra $U_xQ = U_eQ$, $(e = P(x))$ relative to $Reg(U_xJ)$.

By [5], local orders in unital Jordan algebras are actually orders.

An element x in a Jordan algebra J is said to have *finite Goldie dimension* if J contains no infinite direct sum of inner ideals of J inside U_xJ. It follows from the Litoff theorem for Jordan algebras [1], and from the structure of inner ideals in Jordan algebras of finite capacity [16], that if Q is a simple Jordan algebra with dcc on principal inner ideals and which is not a non-artinian quadratic factor, then every element $x \in Q$ has finite Goldie dimension.

Theorem 4 (Local Zel'manov-Goldie theorem) ([5, 6]) *Let J be a Jordan algebra which is a weak local order in a nondegenerate Jordan algebra Q with dcc on principal inner ideals. Then*

(i) J is a local order in Q,

(ii) J is a nondegenerate Jordan algebra satisfying acc on $ann_J(x)$.

If additionally Q contains no non-artinian quadratic factor, then

(iii) every element $x \in J$ has finite Goldie dimension.

Conversely, every nondegenerate Jordan algebra satisfying local Goldie conditions (ii) and (iii) is a local order in a nondegenerate Jordan algebra with dcc on principal inner ideals and without non-artinian quadratic factors.

Sketch of the proof: Suppose first that J is a weak local order in a nondegenerate Jordan algebra Q satisfying dcc on principal inner ideals. That J is nondegenerate can be shown by reducing the problem, via the Litoff theorem for Jordan algebras [1], to orders in Jordan algebras of finite capacity, where Zel'manov's result [20] can be used. Now (3.3) allows us to prove that J is actually and order in Q and that J satisfies acc on annihilators of elements. If additionally Q does not contain non-artinian quadratic factors, then it follows from (3.1), via the corresponding Zel'manov's result [20], that every element $x \in J$ has finite Goldie dimension.

Suppose conversely that J is a nondegenerate Jordan algebra satisfying local Goldie conditions (ii) and (iii) of Theorem.

(I) Reduction to the case when J is prime

An ideal I of J will be called *uniform* if for any nonzero ideals B, C of J inside I, the intersection $B \cap C$ is nonzero. By [6], for each uniform ideal I of a nondegenerate Jordan algebra J there exists a maximal uniform ideal M of J containing I, actually $M = ann(ann(I))$. Moreover, the sum of all maximal uniform ideals of J is direct.

Since annihilators are inner ideals, for any element u in a Jordan algebra J, $ann(u) \subset ann(x)$ for $x \in U_u J$. Now a nonzero element $u \in J$ is called *uniform* if

$$ann(u) = ann(x) \quad \text{for any} \quad 0 \neq x \in U_u J.$$

Notice that if $u \in J$ is uniform then every nonzero element in the inner ideal generated by u is uniform as well. Clearly every nonzero element $u \in J$ such that $ann(u)$ is maximal is a uniform element. In particular, if J satisfies acc on the annihilators of its elements, then every nonzero inner ideal of J contains a uniform element. By using socle theory it is not difficult to see that an element x in the socle of J is uniform if and only if x is minimal, that is, it generates a minimal inner ideals. The following results have been proved in [6].

(4.1) Let J be a nondegenerate Jordan algebra. Then every uniform element $u \in J$ generates a uniform ideal.

Let J be nondegenerate and $u \in J$ uniform. The maximal uniform ideal $M(u)$ of J containing u will be called the *uniform component* of u. As pointed out above, the sum $\sum M(u)$ of all uniform components is direct.

Following [12], an *essential direct product* J of a collection $\{J_\alpha\}$ of Jordan algebras is any subdirect product of the J_α which contains an essential ideal of the direct product of the J_α. If J is actually contained in the direct sum of the J_α, then J will be called an *essential direct sum*.

(4.2) Let J be a nondegenerate Jordan algebra such that every nonzero ideal of J contains a uniform element. Then J is an essential direct product of prime nondegenerate Jordan algebras J_α each of which contains a uniform element.

If additionally every element $x \in J$ has finite Goldie dimension, then J is actually an essential direct sum of the J_α and each element $x_\alpha \in J_\alpha$ has finite Goldie dimension.

(II) The case when J is prime

By Zel'manov's structure theorem for prime nondegenerate Jordan algebras [18], we can focus attention on the hermitian case, i.e., we may assume that J contains a hermitian ideal of the form $H(A,*)$ and that it is contained in $H(Q_s(A),*)$, where A is a *-prime associative algebra with involution which is a *-envelope of J, and $Q_s(A)$ is the symmetric ring of quotients of A with the unique involution * extending that of A. If A is not prime then

(4.3) J contains an ideal of the form B^+ and it is contained in $Q_s(B)^+$, where B is prime associative algebra.

We can prove in this case that if J contains a uniform element then B is nonsingular, and that if every element in J has finite Goldie dimension then every element in B has finite (left and right) uniform-dimension. Now it follows from Local Goldie theorem for prime associative algebras that B is a local order in a simple associative algebra Q with minimal one-sided ideals, and that J is contained in Q^+. Hence it is not difficult to show that J is actually a local order in Q^+. Suppose now that A is prime, i.e. that

(4.4) J contains an ideal $H(A,*)$ and it is contained in $H(Q_s(A),*)$, where A is a prime *-envelope of J.

If the involution $*$ is *diagonal*, i.e., $aH(A,*)a^* = 0$ implies $a = 0$, we can use Zel'manov's methods to prove that if J contains a uniform element then A is nonsingular, and that if every element in J has finite Goldie dimension then every element in A has finite (left and right) uniform-dimension. Again, by Local Goldie theorem, A is a local order in a simple associative algebra Q with minimal one-sided ideals and with diagonal involution extending that of A. Hence $(Q,*)$ is actually the ring $(F_V(V),*)$ of all finite rank continuous linear operators relative to an hermitian nondegenerate self-dual vector space $(V, <.,.>)$, with $*$ being the adjoint involution.

If the involution $*$ is no diagonal, the methods of [19] do not work. But in this case, A satisfies the *generalized polynomial identity with involution* $p(X,X^*) = a(X+X^*)a^*$ for some nonzero $a \in A$, and hence [3], $Q_s(A)$ has nonzero socle and satisfies the same generalized identity. Then, by the structure theorem of prime rings with involution containing minimal one-sided ideals, $(Q_s(A),*)$ is a self-adjoint subring of $(L_V(V),*)$ whose socle is $F_V(V)$, where $(V, <.,.>)$ is now an alternate nondegenerate self-dual vector space and $L_V(V)$ denotes the ring of all continuous linear operators. If each element in J has finite Goldie dimension, then $J \subset H(F_V(V),*)$ and hence every element $a \in A$ has finite uniform dimension, because A is generated by J. Now it follows from (1.1) that A is a local order in $F_V(V)$, which implies that J is a local order in $H(F_V(V),*)$. This completes the proof.

References

1. Anh P.N.: 1986 'Simple Jordan algebras with minimal inner ideals', *Commun. in Algebra* Vol. no. **14**, pp. 489–492.

2. Anh P.N. and Marki L.: 1991 'Left orders in regular rings with minimum condition for principal one-sided ideals', *Math. Proc. Camb. Phil. Soc.* Vol. no. **109**, pp. 323–333.
3. Beidar K.I., Mikhalev A.V., and Salavova K.: 1981 'Generalized identities and semiprime rings with involution', *Math. Z.* Vol. no. **178**, pp. 37–62.
4. Fernández López A.: 1992 'On annihilators in Jordan algebras', *Publicacions Matemàtiques* Vol. no. **36**, pp. 569–589.
5. Fernández López A. and García Rus E.: 1993 'Prime Jordan algebras satisfying local Goldie conditions', *J. Algebra* (to appear).
6. Fernández López A. and García Rus E.: 1993 'Nondegenerate Jordan algebras satisfying local Goldie conditions', Preprint.
7. Fernández López A., García Rus E. and Loos O.: 1993 'Annihilators of elements of the socle of a Jordan algebra', *Commun. in Algebra* (to appear).
8. Fernández López A., García Rus E. and Sánchez Campos E.: 1993 'Structure theorem for prime rings satisfying a generalized polynomial identity', *Commun. in Algebra* (to appear).
9. Fernández López A., García Rus E., Sánchez Campos E. and Siles Molina, M.: 1992 'Strong regularity and generalized inverses in Jordan systems', *Commun. in Algebra* Vol. no. **20**, pp. 1917–1936.
10. Fountain J., and Gould V.: 1990 'Orders in rings without identity', *Commun. in Algebra* Vol. no. **18**, pp. 3085–3110.
11. Fountain J., and Gould V.: 1991 'Orders in regular rings with minimal condition for principal right ideals', *Commun. in Algebra* Vol. no. **19**, pp. 1501–1527.
12. Goodearl K.R.: 1976 'Ring theory. Nonsingular rings and modules', Marcel Dekker, Inc., New York and Basel.
13. Herstein I.N.: 1976 'Rings with involutions', The University of Chicago Press, Chicago.
14. Jacobson N.: 1968 'Structure and representations of Jordan algebras', *Amer. Math. Soc. Colloq. Publ.* Vol.no. **39**, Amer. Math. Soc.
15. Loos O.: 1989 'On the socle of a Jordan pair', *Collect. Math.* Vol. no. **40**, pp. 109–125.
16. McCrimmon K.: 1971 'Inner ideals in quadratic Jordan algebras', *Trans. Amer. Math. Soc.* Vol. no. **159**, pp. 445–468.
17. Osborn J.M. and Racine M.L.: 1979 'Jordan rings with nonzero socle', *Trans. Amer. Math. Soc.* Vol. no. **251**, pp. 375–387.
18. Zel'manov E.: 1983 'On prime Jordan algebras II', *Siberian Math. J.* Vol. no. **24**, pp. 89–104.
19. Zel'manov E.: 1987 'Goldie's theorem for Jordan algebras', *Siberian Math. J.* Vol. no. **28**, pp. 44–52.
20. Zel'manov E.: 1988 'Goldie's theorem for Jordan algebras II', *Siberian Math. J.* Vol. no. **29**, pp. 68–74.
21. Zhevlakov K.A., Slin'ko A.M., Shestakov I.P. and Shirshov A.I.: 1982 'Rings that are nearly associative', Academic Press, New York-London.

ON THE UNITARIZATION OF HIGHEST WEIGHT REPRESENTATIONS FOR AFFINE KAC-MOODY ALGEBRAS

J. GARCIA ESCUDERO and M. LORENTE
Departamento de Física, Universidad de Oviedo
33007, Oviedo, Spain

Abstract. In a recent paper ([1],[2]) we have classified explicitely all the unitary highest weight representations of non compact real forms of semisimple Lie Algebras on Hermitian symmetric space. These results are necessary in order to construct all the unitary highest weight representations of affine Kac-Moody Algebras following some theorems proved by Jakobsen and Kac ([3],[4]).

1. The Affine Kac-Moody Algebra

Let $\dot{g}$ be a finite dimensional semisimple complex Lie algebra with Chevalley basis $\{H_\alpha, E_\alpha, F_\alpha\}$ with α belonging to the set of simple roots. The elements of the so-called Cartan matrix A are defined by

$$A_{jk} = \alpha_k(H_{\alpha j}) = \frac{2(\alpha_k, \alpha_j)}{(\alpha_j, \alpha_j)}, \qquad j, k = 1, 2, \ldots, \ell$$

where, $(\alpha_j, \alpha_k) = B(H_{\alpha_j}, H_{\alpha_k})$, $B(\ ,\ )$ being the Killing form of $\dot{g}$ and H_α an element of the Cartan subalgebra h which is assigned uniquely to each root $\alpha \in \Delta$ by the requirement that $B(H_\alpha, H) = \alpha(H)$ for all $H \in h$.

The elements in the Chevalley basis satisfy

$$[H_{\alpha_j}, E_{\alpha_k}] = A_{jk} E_{\alpha_k}, \quad [H_{\alpha_j}, F_{\alpha_k}] = -A_{jk} F_{\alpha_k}$$
$$[E_{\alpha_j}, F_{\alpha_k}] = \delta_{jk} H_{\alpha_k}, \quad \text{for all} \quad \alpha_j, \alpha_k \in \Delta$$

Every finite dimensional semisimple complex Lie algebra can be constructed from its Cartan matrix A which satisfies the following propierties.

a) $A_{ii} = 2 \quad \forall i, \quad i = 1, \ldots, \ell$
b) $A_{ij} = 0, -1, -2, \quad \text{or} \ -3$ if $i \neq j, \quad i, j = 1, \ldots, \ell$
c) $A_{ij} = 0$ if and only if $A_{ji} = 0$
d) det A and all proper principal minors of A are positive.

The starting point on the construction of infinite dimensional Kac-Moody algebras is the definition of a generalized Cartan-matrix with elements A_{ij} $(i, j \in I,\ I = 0, 1, \ldots, \ell)$ satisfying

a) $A_{ii} = 2 \qquad \forall i \in I$
b) A_{ij} is either zero or a negative integer for $i \neq j$.
c) $A_{ij} = 0$ if and ony if $A_{ji} = 0$.

A Lie algebra whose Cartan matrix is a generalized Cartan matrix is called a Kac-Moody algebra ([5],[6]). A Kac-Moody algebra is affine if its generalized Cartan

S. González (ed.), Non-Associative Algebra and Its Applications, 147–152.

matrix is such that det $A = 0$ and all the proper principal minors of A are positive. In the following we restrict ourselves to the affine case.

For a Kac-Moody algebra the Cartan subalgebra h is divided into two parts $h = h' \oplus h''$.

The basis elements of h' are H_{a_j} $(j \in I)$ and h'' is the one-dimensional complementary subspace of h' in h generated by one element that we call d.

The center C in the affine case is one-dimensional. Every element of C is a multiple of h_δ where δ is defined by the following conditions

$$\begin{aligned} \delta(h) &= 0 \quad \text{for } H \in h' \\ \delta(d) &= 1 \quad \text{for } d \in h'' \end{aligned}$$

The Cartan subalgebra h has dimension $\ell + 2$. In order to obtain a basis for h^* we need $\ell + 2$ linear functionals. We can take as basis α_k $(k = 0, 1, \ldots, \ell)$, and the linear functional Λ_0 defined as:

$$\begin{aligned} (\Lambda_0, \alpha_k) &= \begin{cases} \frac{1}{2}(\alpha_0, \alpha_0) & \text{if } k = 0 \\ 0 & \text{if } k = 1, 2, \ldots, \ell \end{cases} \\ (\Lambda_0, \delta) &= \tfrac{1}{2}(\alpha_0, \alpha_0) \end{aligned}$$

There exists two types of affine complex Kac-Moody algebras: Untwisted and Twisted. The untwisted ones $g^{(1)}$ may be constructed starting from any simple complex Lie algebra. The twisted affine Kac-Moody algebras $g^{(q)}$ $(q = 2, 3)$ can all be constructed as subalgebras of certain of these untwisted algebras.

i) Untwisted affine Kac-Moody algebras

Let $\dot{g}$ be a simple complex Lie algebra of rank ℓ. A realization of the complex untwisted affine Kac-Moody algebra $g^{(1)}$ is given by

$$g^{(1)} = \mathbb{C}c \oplus \mathbb{C}d \oplus \sum_{j \in Z} z^j \otimes \dot{g}$$

with the following conmutation relations

$$\begin{aligned} \left[z^j \otimes a, z^k \otimes b\right] &= z^{j+k} \otimes [a, b] + j\delta_{j,-k} B(a, b)\, c \quad \forall a, b \in \dot{g} \\ \left[z^j \otimes a, c\right] &= 0 \\ \left[d, z^j \otimes a\right] &= j z^j \otimes a \\ [d, c] &= 0 \end{aligned}$$

ii) Twisted Affine Kac-Moody algebras

Let $\dot{g}$ be a simple complex Lie algebra and τ a rotation of the set of roots of $\dot{g}$. If the rotation τ is not an element of the Weyl group of $\dot{g}$ then there exist an associated outer automorphism Ψ_τ such that $\Psi_\tau(h_\alpha) = h_{\tau(\alpha)}$. We have $\tau^q = -1$ and also $(\Psi_\tau)^q = 1$ with $q = 2, 3$. The eigenvalues of Ψ_τ are $e^{2\pi i p/q}$, $p = 0, 1, \ldots, q-1$. Let $\dot{g}_p^{(q)}$ be the eigenspace corresponding to the eigenvalue $e^{2\pi i p/q}$. The Twisted Affine Kac-Moody algebra is then:

$$g^{(q)} = (\mathbb{C}c) \oplus (\mathbb{C}d) \oplus \sum_{p=0}^{q-1} \sum_{\substack{j \\ j \bmod q = p}} \left(z^j \otimes \dot{g}_p^{(q)}\right)$$

2. Highest Weight Representations. The Contravariant Hermitian Form

A subset Δ_+ of Δ is called a set of positive roots if the following propierties are satisfied

i) If $\alpha, \beta \in \Delta_+$ and $\alpha+\beta \in \Delta$ then $\alpha+\beta \in \Delta_+$
ii) If $\alpha \in \Delta$ then either α or -α belongs to Δ_+
iii) If $\alpha \in \Delta_+$ then $-\alpha \notin \Delta_+$

For each set of positive roots we may construct a Borel subalgebra $b = \oplus_{\alpha \in \Delta_+ \cup \{0\}} g_\alpha$. A subalgebra $p \subset g$ such that $b \subset p$ is called a parabolic subalgebra.

Let $U(g)$ denote the universal enveloping algebra of g and let ω be an antilinear anti-involution of g $\left(\text{i.e. } \omega[x,y] = [\omega y, \omega x] \quad \text{and } \omega(\lambda x) = \overline{\lambda}\omega(x)\right)$ such that

$$g = p + \omega p$$

An antilinear anti-involution ω of g is called consistent if $\forall \alpha \in \Delta$, $\omega g_\alpha = g_{-\alpha}$. Now let e_i, f_i, h_i $\quad (i = 0, \ldots, l)$ be defined as ([5])

$$\begin{array}{lll} e_0 = z \otimes F_{\gamma_r} & f_0 = z^{-1} \otimes E_{\gamma_r} & h_0 = \frac{2}{(\gamma_r,\gamma_r)} c - 1 \otimes H_{\gamma_r} \\ e_i = 1 \otimes E_i & f_i = 1 \otimes F_i & h_i = 1 \otimes H_i \quad i = 1, \ldots, \ell \end{array}$$

where γ_r is the highest positive root.

When $\omega e_i = f_i$ and $\omega h_i = h_i \quad (i = 0, \ldots, l)$ then ω is called the compact antilinear anti-involution and is denoted by w_c.

Let now $\Lambda : p \longrightarrow \not{c}$ be a 1-dimensional representation of p. A representation $\Pi : g \longrightarrow g\ell(V)$ is called a highest weight representation with highest weight Λ if there exists a vector $\vartheta_\Lambda \in V$ satisfying

a) $\Pi(u(g))\vartheta_\Lambda = V$
b) $\Pi(x)\vartheta_\Lambda = \Lambda(x)\vartheta_\Lambda \forall x \in p$

An Hermitian form H on V such that

$$\begin{array}{l} H(\vartheta_\Lambda, \vartheta_\Lambda) = 1 \\ H(\Pi(x)u, v) = H(u, \Pi(\omega x)v) \quad \forall x \in g; \quad u, v \in V \end{array}$$

is called contravariant. When H is positive semi-definite, Π is said to be unitarizable.

In the following we construct the Hermitian form H ([3]). We choose a subspace $n \subset g$ such that $g = p \oplus n$. Then we have $U(g) = nU(g) \oplus U(p)$. Let β be the proyection on the second sumand. Let Λ be a 1-dimensional representation of a parabolic subalgebra p (in particular a Borel subalgebra b as in the integrable representations case) satisfying $\Lambda(\beta(u)) = \overline{\Lambda(\beta(\omega u))} \quad \forall u \in U(g)$.

Let $p^\Lambda = \{x \in p / \Lambda(x) = 0\}$. The space

$$M_{p,\omega}(\Lambda) = U(g) / U(g) p^\Lambda$$

defines a representation of g on $M_{p,\omega}(\Lambda)$ via left multiplication that is called a (generalized) Verma module and that is a highest weight representation. In addition it can be shown that there exists a unique contravariant hermitian form defined by

$$H(u, v) = \Lambda(\beta(\omega(v))u) \text{ for } u, v \in U(g)$$

which is independent of the choice of p.

Let $I(\Lambda)$ denote the Kernel of H on $M_{p,\omega}(\Lambda)$. Then H is nondegenerate on the highest weight module

$$L_{p,\omega}(\Lambda) = M(\Lambda)/I(\Lambda)$$

In the following we will give for each of the unitarizable representations (integrable, elementary and exceptional) the choice of p and ω for which the hermitian form is nondegenerate and positive definite.

3. Integrable representations

Let $\Pi^{st} = \{\alpha_0, \ldots \alpha_\ell\}$ be the standard set of simple roots. The standard set of positive roots ([4]) is $\Delta_+^{st} = \left\{\sum k_i\alpha_i / k_i = 0, 1, 2, \ldots \, ; \, \alpha_i \in \Pi^{st}\right\}$ and the corresponding Borel subalgebra is denoted by b^{st}:

$$\begin{aligned} b^{st} &= \mathbb{C}c \oplus \left(1 \otimes \dot{b}\right) \oplus (z \otimes \dot{g}) \oplus (z^2 \otimes \dot{g}) \oplus \ldots \\ &= \text{span } \{z^k \otimes h_i / k \geq 0, i = 0, \ldots \ell\} \oplus \text{span } \{z^k \otimes e_i / k \geq 0, i = 0, \ldots \ell\} \\ &\oplus \text{span } \{z^k \otimes f_i / k > 0, i = 0, \ldots \ell\} \end{aligned}$$

Let $\omega = \omega_c$ and let $\Lambda : b^{st} \to \mathbb{C}$ be a 1-dimensional representation of b^{st} defined by

$$\Lambda(e_i) = 0 \quad \Lambda(h_i) = m_i \in Z_+ \quad (i = 0, \ldots, \ell)$$

These representations are called the integrable highest weight representations. In particular if g is finite-dimensional, these are the finite dimensional representations. The fundamental weights $\Lambda_0, \Lambda_1, \ldots, \Lambda_l$ are such that

$$\left.\begin{aligned} \Lambda_j(H_k) &= \tfrac{2(\Lambda_j, \alpha_k)}{(\alpha_k, \alpha_k)} = \delta_{jk} \\ \Lambda_j(d) &= 0 \end{aligned}\right\} \quad j, k = 0, \ldots, \ell$$

In this way given the fundamental weights of a finite dimensional Lie algebra $\dot{g}$

$$\dot{\Lambda}_j(H_k) = \frac{2\left(\dot{\Lambda}_j, \alpha_k\right)}{(\alpha_k, \alpha_k)} = \delta_{jk} \quad j, k = 1, \ldots, \ell$$

we can construct the fundamental weights of the Kac-Moody algebra g as extensions of the fundamental weights $\dot{\Lambda}$ in the following way:

$$\begin{aligned} \Lambda_j &= \dot{\Lambda}_j + \mu_j \Lambda_0 \\ \mu_j &= -\sum_{k=1}^{\ell} A_{ok} \left((\dot{A})^{-1}\right)_{kj} \end{aligned}$$

$\dot{A}$ being the Cartan matrix of $\dot{g}$ and Λ_0 the linear functional defined in paragraph 1.

Every integrable highest weight representation is unitarizable.

4. Elementary representations

We know that for a finite dimensional simple Lie algebra $\dot{g}$ an infinite dimensional highest weight representation is unitarizable only if $\dot{\omega}$ is a consistent antilinear anti-involution corresponding to a hermitian symmetric space (see [7],[8]).

The remaining unitarizable representations can be constructed only for Kac-Moody algebras related to these type of finite dimensional Lie algebras.

Let $\dot{b} = \bigoplus_{\alpha \in \dot{\Delta}_+ \cup \{0\}} \dot{g}_\alpha$ be a Borel subalgebra of the finite dimensional Lie algebra $\dot{g}$.

Consider the parabolic subalgebra (called "natural")

$$p^{\text{nat}} = z^n \otimes \dot{b} = \text{span}\left\{z^n \otimes h_i, z^n \otimes e_i\right\}, \quad n \in Z, \quad i = 1, \ldots, \ell$$

Take a Cartan decomposition of the Lie algebra $\dot{g}$ corresponding to a hermitian symmetric space:

$$\dot{g} = \dot{p}^- \oplus \dot{k} \oplus \dot{p}^+, \quad \dot{k} = \dot{k}^- \oplus \dot{\eta} \oplus \dot{k}^+$$

where $\dot{p}$ and $\dot{k}$ are the subspace and the subalgebra corresponding to non-compact and compact roots respectively.

We define an antilinear anti-involution ω of g by

$$\left.\begin{array}{lll} \omega\left(z^n \otimes k_i^+\right) & = z^{-n} \otimes k_i^- & i = 2, \ldots, \ell \\ \omega\left(z^n \otimes p_1^+\right) & = -z^{-n} \otimes p_1^- & \\ \omega\left(z^n \otimes h_i\right) & = z^{-n} \otimes h_i & i = 0, 1, \ldots, \ell \end{array}\right\} \quad n \in Z$$

where $k_2^+, \ldots, k_l^+$ belong to $\dot{k}^+$ and where p_1^+ belongs to the root space corresponding to the unique simple non compact root. In the previous notation $e_1 = p_1^+$ and $e_i = k_i^+$ for $i = 2, \ldots, \ell$.

Consider now a set of highest weights $\Lambda_1, \ldots, \Lambda_N$ corresponding to unitarizable highest weight modules for the hermitian symmetric space $\dot{g}$ (for an explicit calculation see [1],[2]). Define a representation $p^{\text{nat}} \to \mathbb{C}$ by

$$\Lambda\left(z^k \otimes x\right) = \sum_{i=1}^{N} C_i^k \Lambda_i(x)$$

for $x \in \dot{b}$ and $C_k^i \in \mathbb{C}$ with $\left|C_k^i\right| = 1$.

Then the resulting representation is called "elementary" and it is unitarizable.

5. Exceptional representations

Another class of unitary representations (called "exceptional") are constructed in this paragraph for the Kac-Moody algebra $z^k \otimes su(n,1) \quad k \in Z, \quad n \geq 1$.

Let $\dot{g} = su(n,1)$ and let be a Cartan decomposition $\dot{g} = \dot{k} \oplus \dot{p}$. Then $\dot{h} = \dot{k}_1 \cap \dot{h} \oplus R\dot{h}_c$ where $\dot{k}_1 = \left[\dot{k}, \dot{k}\right]$ and $\dot{h}_c$ belongs to the center of $\dot{g}$.

We take a realization of $g = z^k \otimes su(n,1)$ in terms of matrices $(a_{ij}(z)) \quad i,j = 0, \ldots, n$. The matrix elements are of the form $a(z) = \sum_{n \in Z} a_n e^{in\theta}$ with $z = e^{i\theta}$. We will use the notation $\overline{a}(z) = \sum_{n \in Z} \overline{a}_n e^{in\theta}$. Let $p = \{(a_{ij}(z)) \in g / a_{ij} = 0 \quad \text{if } i > j\}$ be the parabolic subalgebra. The antilinear anti-involution acts in this case as

$$\begin{aligned} \omega\left(z^k \otimes h_c\right) &: \omega\left(a_{00}(z)\right) = \overline{a}_{00}\left(z^{-1}\right) \\ \omega\left(z^k \otimes p^+\right) &: \omega\left(a_{0j}(z)\right) = -\overline{a}_{j0}\left(z^{-1}\right) \quad j = 1, \dots, n \\ \omega\left(z^k \otimes k_1\right) &: \omega\left(a_{ij}(z)\right) = \overline{a}_{ji}\left(z^{-1}\right) \quad i, j = 1, \dots, n \end{aligned}$$

Define a representation $\Lambda : p \to \mathbb{C}$ by

$$\begin{aligned} \Lambda\left(a_{ij}(z)\right) &= 0 \qquad i, j = 1, \dots, n \\ \Lambda\left(a_{0j}(z)\right) &= 0 \qquad j = 1, \dots, n \\ \Lambda\left(a_{00}(z)\right) &= -\int_{s^1} a_{00}\left(e^{i\theta}\right) d\mu(\theta) = -\varphi\left(a_{00}(z)\right) \end{aligned}$$

where $\mu(\theta)$ is a positive Radon mesure defined in the unit circle s^1 and infinitely supported.

It can be shown (see [3]) that the hermitian form H (we remind that it is completely determined by giving w, p and a representation of p) is positive definite and then the corresponding representation in the space $L_{p,w}(L)$ is unitary.

6. Tensor products

The only remaining possibility in order to complete the set of all unitarizable representations for affine Kac-Moody algebras is the corresponding to the highest component of a tensor product of an elementary with an exceptional representation for $z^k \otimes su(n, 1)$ (see [4]).

Explicit results concerning the construction of these unitary representations will be given in a forthcoming paper.

References

1. García-Escudero, J. and Lorente, M.: 'Highest Weight Unitary Modules for Non-Compact Groups and Applications to Physical Problems', *Symmetries in Science V* (ed. B. Gruber et al.) Plenum Publishing. New York (1991)
2. García-Escudero, J. and Lorente, M.: 'Classification of Unitary Highest Weight Representations for Non-compact Real Forms'. *J. Math. Phys.*, 781 – 790 (1990).
3. Jakobsen, H.P. and Kac, V.: 'A new class of unitarizable highest weight representations of infinite dimensional Lie algebras', in *Non-Linear Equations in Classical and Quantum Field Theory* (ed. N. Sánchez). Lect. Notes in Phys. 226, Springer Verlag (1985).
4. Jakobsen, H.P. and Kac, V. 'A new class of unitarizable highest weight representations of infinite dimensional Lie algebras II'. *J. Funct. Anal.* 82 (1989).
5. Kac, V.G. Infinite dimensional Lie algebras; Cambridge University Press third edition, Cambridge 1990 .
6. Cornwell, J.F. Group Theory in Physics, Vol III. Academic Press (1989).
7. Jakobsen, H.P. 'Hermitian Symmetric Spaces and their Unitary Highest Weight Modules'. *J. Funct. Anal.* 52, 385 – 412 (1983) .
8. Enright, T; Howe, R.; Wallach, N. 'A classification of Unitary Highest Weight Modules, in Representation Theory of Reductive Groups' (ed. P. Trombi) *Progress in Math.* 40. Birkhaüser, Boston (1983).

COHN'S THEOREM FOR SUPERALGEBRAS*

CARLOS GÓMEZ-AMBROSI
Departamento de Matemáticas, Universidad de Zaragoza
50009 Zaragoza, Spain

It is known that Cohn's theorem [1] plays an essential role in the structure theory of Jordan algebras, specially in Zelmanov's theory (for a very readable account of the theory of Jordan algebras which emphasizes the contributions of the russian school see [2]).

It seems plausible that the generalization of Cohn's theorem to superalgebras may prove useful in understanding the structure of special Jordan superalgebras. In this paper we will provide such a generalization.

In the first section we discuss the general background of superalgebra theory (as it is done, for example, in [3] and [4]), and introduce the notions of a free superalgebra and a free special Jordan superalgebra (some ideas related to this may be found in [5]). In the second section we introduce the notion of a supertetrad and prove Cohn's theorem for superalgebras.

All algebras will be taken over the same field F of characteristic different from two.

1. Free special Jordan superalgebras

By a *superset* we mean a set $X = \{0\} \cup X_0 \cup X_1$ which is the disjoint union of a base point 0 and two subsets X_0, X_1. Elements from the set X_0 (respectively X_1) will be called even (respectively odd). A *supermap* will be a map $f: X \to Y$ between two supersets X, Y such that $f(0) = 0$, $f(X_0) \subseteq \{0\} \cup Y_0$, $f(X_1) \subseteq \{0\} \cup Y_1$. In this way we obtain the category Set_2 of supersets.

As usual, a superalgebra is a Z_2-graded algebra $A = A_0 \oplus A_1$, $A_i A_j \subseteq A_{i+j}$ $(i, j \in Z_2)$, and a (super)homomorphism $f: A \to B$ between two superalgebras is a homomorphism of algebras which preserves gradings. We thus obtain the category Alg_2 of superalgebras.

Now let $A = A_0 \oplus A_1$ be a superalgebra. The set $A_0 \cup A_1$ of homogeneous elements of A is a superset in a natural way: $A_0 \cup A_1 = \{0\} \cup (A_0 - \{0\}) \cup (A_1 - \{0\})$. We have thus defined a forgetful functor from Alg_2 to Set_2.

Given an arbitrary superset $X = \{0\} \cup X_0 \cup X_1$, we consider the free nonassociative algebra $Alg[X_0 \cup X_1]$ from set of generators $X_0 \cup X_1$. A nonassociative word from elements of the set $X_0 \cup X_1$ is called even (respectively odd) if the number of odd elements, counting repetitions, which appear in it is even (respec-

* This paper has been written under the direction of Professors Santos González and Fernando Montaner and it will be a part of the author's Doctoral Thesis. The author has been supported by the Diputación General de Aragón.

S. González (ed.), Non-Associative Algebra and Its Applications, 153–157.

tively odd). This is consistent with elements from $X_0 \cup X_1$ being even or odd. Let $Alg[X_0 \cup X_1]_0$ (respectively $Alg[X_0 \cup X_1]_1$) be the subspace of $Alg[X_0 \cup X_1]$ spanned by all nonassociative words which are even (respectively odd). It is easy to check that $Alg[X_0 \cup X_1] = Alg[X_0 \cup X_1]_0 \oplus Alg[X_0 \cup X_1]_1$ is a superalgebra. This superalgebra is called the *free (nonassociative) superalgebra* from superset of generators X, and it is denoted by $Alg_2[X] = Alg_2[X]_0 \oplus Alg_2[X]_1$. Note that X is included in $Alg_2[X]$ by identifying $0 \in X$ with $0 \in Alg_2[X]$. The free superalgebra $Alg_2[X]$ satisfies the following universal property: *for any superalgebra $A = A_0 \oplus A_1$ and any supermap $f: X \to \{0\} \cup (A_0 - \{0\}) \cup (A_1 - \{0\})$, there exists a unique homomorphism $\tilde{f}: Alg_2[X] \to A$ which extends f.* Therefore we have a free functor from Set_2 to Alg_2, which is left adjoint to the foregoing forgetful functor.

One of the most important examples of a superalgebra is the Grassman algebra $G = \langle e_0, e_1, e_2, \ldots \mid e_i e_j + e_j e_i = 0 \rangle$ endowed with the natural Z_2–grading $G = G_0 \oplus G_1$, where $e_{i_1} \ldots e_{i_n} \in G_0$ if n is even and $e_{i_1} \ldots e_{i_n} \in G_1$ if n is odd. Given a superalgebra $A = A_0 \oplus A_1$, we consider the algebra tensor product $A \otimes G$. Its subalgebra $G(A) = A_0 \otimes G_0 + A_1 \otimes G_1$ is called the *Grassman envelope* of the superalgebra A. Now let V be a variety of algebras. We say that a superalgebra $A = A_0 \oplus A_1$ is a *V–superalgebra* if $G(A) \in V$. In this sense, it is easy to see that an associative superalgebra is exactly the same as a Z_2–graded associative algebra.

Given a superalgebra $A = A_0 \oplus A_1$, we can define a new multiplication by

$$x_i \circ y_j = \frac{1}{2}(x_i y_j + (-1)^{ij} y_j x_i),$$

where $x_i \in A_i$, $y_j \in A_j$ $(i, j \in Z_2)$. In this way we obtain another superalgebra $A^+ = A_0 \oplus A_1$. Straightforward calculations show that $G(A^+) = G(A)^+$ (in the second member + denotes the usual symmetrization procedure). Hence if A is an associative superalgebra then A^+ is a Jordan superalgebra. A Jordan superalgebra is called *special* if it is embeddable in A^+ for some associative superalgebra A. Otherwise it is called *exceptional.*

Given an arbitrary superset X we can construct the *free associative superalgebra* from superset of generators X, denoted by $Ass_2[X]$. The procedure is similar to that for the free superalgebra and $Ass_2[X]$ satisfies the analogous universal property. We now consider the corresponding Jordan superalgebra $Ass_2[X]^+$ and we define the *free special Jordan superalgebra* from superset of generators X to be the sub–superalgebra $FSJ_2[X]$ of $Ass_2[X]^+$ generated by the set X. $FSJ_2[X]$ verifies the following universal property: *for any special Jordan superalgebra $J = J_0 \oplus J_1$ and any supermap $f: X \to \{0\} \cup (J_0 - \{0\}) \cup (J_1 - \{0\})$, there exists a unique homomorphism $\tilde{f}: FSJ_2[X] \to J$ which extends f.*

Other examples of special Jordan superalgebras are those given by superinvolutions. Let $A = A_0 \oplus A_1$ be a superalgebra. A Z_2–graded linear map $*: A \to A$ is called a *superinvolution* if $(x_i^*)^* = x_i$, $(x_i y_j)^* = (-1)^{ij} y_j^* x_i^*$ for any $x_i \in A_i$, $y_j \in A_j$ $(i, j \in Z_2)$. A homogeneous element $x_i \in A_i$ is called *supersymmetric* if $x_i^* = (-1)^i x_i$. The subspace $H(A, *) = H(A_0, *) \oplus H(A_1, *)$ of supersymmetric elements is a sub–superalgebra of A^+. Thus in case A be associative $H(A, *)$ is a special Jordan superalgebra. If $x_i \in A_i$ we define $\{x_i\} = \frac{1}{2}(x_i + (-1)^i x_i^*)$. Then the elements $\{x_i\}$ span the space of supersymmetric elements $H(A, *)$. The *opposite superalgebra* $A^{op} = A_0 \oplus A_1$ of the superalgebra $A = A_0 \oplus A_1$ is obtained from A by changing the product in the following way: $x_i \diamond y_j = (-1)^{ij} y_j x_i$. It is easy to check that a superinvolution $*$ on A is the same as a homomorphism $*: A \to A^{op}$ such that $(x_i^*)^* = x_i$.

2. Cohn's theorem for superalgebras

From now on our superset will always be $X = \{0\} \cup \{x_0, x_2, x_4, \ldots\} \cup \{x_1, x_3, x_5, \ldots\}$. The *degree* g of an associative word w from elements of the set $X - \{0\}$, denoted by $g = deg(w)$, will be the number of odd variables, counting repetitions, which appear in it. Thus such a word is even or odd, as defined before, accordingly to g being even or odd.

In our generalization of Cohn's theorem to superalgebras we will need the following result.

Proposition. *There exists a unique superinvolution $*$ on $Ass_2[X]$ such that $FSJ_2[X] \subseteq H(Ass_2[X], *)$. This superinvolution is given by*

$$(x_{i_1} \ldots x_{i_n})^* = (-1)^{\frac{1}{2}g(g+1)} x_{i_n} \ldots x_{i_1},$$

where $g = deg(x_{i_1} \ldots x_{i_n})$.

Proof. Suppose that $*$ is a superinvolution on $Ass_2[X]$ such that $FSJ_2[X] \subseteq H(Ass_2[X], *)$. As the elements from X are supersymmetric, we must have $x_n^* = (-1)^n x_n$, $\forall n \in N$.

Consider the assignation $x_n \mapsto (-1)^n x_n$ from X to the homogeneous elements of $Ass_2[X]^{op}$, which is clearly a supermap. By the universal property of $Ass_2[X]$, there exists a unique homomorphism $*: Ass_2[X] \to Ass_2[X]^{op}$ such that $x_n^* = (-1)^n x_n$. Moreover $(x_n^*)^* = x_n$, and so we have the desired superinvolution.

Finally, $(x_{i_1} \ldots x_{i_n})^* = (-1)^{\sum_{p<q} i_p i_q} x_{i_n}^* \ldots x_{i_1}^* = (-1)^{\sum_{p \leq q} i_p i_q} x_{i_n} \ldots x_{i_1} = $ $= (-1)^{\frac{1}{2}g(g+1)} x_{i_n} \ldots x_{i_1}$.

QED

The superinvolution in the proposition is called the *natural superinvolution* on $Ass_2[X]$.

If $x_{i_1} \ldots x_{i_n}$ is an associative word of degree g we have already defined $\{x_{i_1} \ldots x_{i_n}\}$ $= \frac{1}{2}(x_{i_1} \ldots x_{i_n} + (-1)^g (x_{i_1} \ldots x_{i_n})^*) = \frac{1}{2}(x_{i_1} \ldots x_{i_n} + (-1)^{g + \frac{1}{2}g(g+1)} x_{i_n} \ldots x_{i_1})$. Now we can introduce the notion of a supertetrad.

Definition. A *supertetrad* is an element from $Ass_2[X]$ of the form $\{x_{i_1} \ldots x_{i_4}\}$, where

(i) $i_1 (mod\, 2) \leq \ldots \leq i_4 (mod\, 2)$, i.e. the even variables, if any, are at the beginning and the odd variables, if any, are at the end.

(ii) if i_p and i_q are both even and $p < q$ then $i_p < i_q$, i.e. the even variables, if any, are increasingly ordered (in particular, there are no repetitions of even variables).

(iii) if i_p and i_q are both odd and $p < q$ then $i_p \leq i_q$, i.e. the odd variables, if any, are non decreasingly ordered (in particular, there can be repetitions of odd variables).

The reason for allowing repetitions of odd variables in supertetrads is that supertetrads such as $x_1^4 = \{x_1x_1x_1x_1\}$ do not belong to $FSJ_2[X]$. To prove this, suppose on the contrary that $x_1^4 \in FSJ_2[X]$ and consider the usual algebra $F[z]$ of polinomials in one indeterminate z. If we put $F[z]_0 = F(1,$ monomials of even degree), $F[z]_1 = F$(monomials of odd degree) then we obtain an associative superalgebra $F[z] = F[z]_0 \oplus F[z]_1$. It is easily checked that $J = F1 \oplus Fz$ is a sub–superalgebra of $F[z]^+$ (notice that $z \circ z = \frac{1}{2}(z^2 - z^2) = 0$), and hence J is a special Jordan superalgebra. By virtue of the universal property of $FSJ_2[X]$, there exists a unique homomorphism $f\colon FSJ_2[X] \to J$ which respects the asignation $x_0 \mapsto 1$, $x_1 \mapsto z$, $x_n \mapsto 0$ $(n \geq 2)$. But then $x_1^4 \mapsto z^4 \in J$, which is a contradiction.

Next we present the superalgebra version of Cohn's theorem.

Theorem. *If $*$ is the natural superinvolution on $Ass_2[X]$ then the Jordan superalgebra $H(Ass_2[X], *)$ of supersymmetric elements coincides with the sub–superalgebra H' of $Ass_2[X]^+$ generated by the set X and all the supertetrads.*

Proof. Clearly $H' \subseteq H$. To show the reverse containment it is enough to show that if $x_{i_1} \ldots x_{i_k}$ is an associative word then $\{x_{i_1} \ldots x_{i_k}\} \in H'$. This is clear for $k = 1, 2$ and we may assume it for $k < n$, where $n \geq 3$. Take $1 \leq m < n$ and let $g = deg(x_{i_1} \ldots x_{i_n})$, $p = deg(x_{i_1} \ldots x_{i_m})$, $q = deg(x_{i_{m+1}} \ldots x_{i_n})$. Lengthy calculations show that

$$4\{x_{i_1} \ldots x_{i_m}\}\circ\{x_{i_{m+1}} \ldots x_{i_n}\} = \{x_{i_1} \ldots x_{i_n}\} + (-1)^{q+\frac{1}{2}q(q+1)}\{x_{i_1} \ldots x_{i_m} x_{i_n} \ldots x_{i_{m+1}}\}$$

$$+(-1)^{p+\frac{1}{2}p(p+1)}\{x_{i_m} \ldots x_{i_1} x_{i_{m+1}} \ldots x_{i_n}\} + (-1)^{pq+g+\frac{1}{2}g(g+1)}\{x_{i_m} \ldots x_{i_1} x_{i_n} \ldots x_{i_{m+1}}\}.$$

Hence, by the induction assumption,

$$\{x_{i_1} \ldots x_{i_n}\} + (-1)^{q+\frac{1}{2}q(q+1)}\{x_{i_1} \ldots x_{i_m} x_{i_n} \ldots x_{i_{m+1}}\} + (-1)^{p+\frac{1}{2}p(p+1)}\{x_{i_m} \ldots x_{i_1} x_{i_{m+1}} \ldots x_{i_n}\} + \quad (2.1)$$

$$(-1)^{pq+g+\frac{1}{2}g(g+1)}\{x_{i_m} \ldots x_{i_1} x_{i_n} \ldots x_{i_{m+1}}\} \equiv 0 \ (mod\ H'). \ (1) \quad (2.2)$$

Having in mind that $\{x_{i_n} \ldots x_{i_1}\} = (-1)^{g+\frac{1}{2}g(g+1)}\{x_{i_1} \ldots x_{i_n}\}$ and taking $m = 1$ in (1) we obtain

$$\{x_{i_1} \ldots x_{i_n}\} \equiv (-1)^{1+(1+g)i_1}\{x_{i_2} \ldots x_{i_n} x_{i_1}\} \ (mod\ H'). \quad (2)$$

We may iterate this to obtain $\{x_{i_1} \ldots x_{i_n}\} \equiv (-1)^n\{x_{i_1} \ldots x_{i_n}\}$ $(mod\ H')$. If n is odd this implies $\{x_{i_1} \ldots x_{i_n}\} \in H'$. Hence we may assume n is even and ≥ 4.

The case $m = 2$ in (1) leads to the formula

$$\{x_{i_1} \ldots x_{i_n}\} \equiv (-1)^{1+i_1i_2}\{x_{i_2} x_{i_1} x_{i_3} \ldots x_{i_n}\} \ (mod\ H'). \quad (3)$$

From (2) and (3) it is easy to see what happens when we transpose two consecutive variables:

$$\{x_{i_1} \ldots x_{i_k} x_{i_{k+1}} \ldots x_{i_n}\} \equiv (-1)^{1+i_k i_{k+1}}\{x_{i_1} \ldots x_{i_{k+1}} x_{i_k} \ldots x_{i_n}\} \ (mod\ H'). \quad (4)$$

It follows that if $i_j = i_k$ for some $j \neq k$ with i_j even then $\{x_{i_1} \ldots x_{i_n}\} \in H'$. We may therefore assume that there are not repeated even variables in $\{x_{i_1} \ldots x_{i_n}\}$.

Consider now the case $n = 4$. By the assumptions made above it becomes apparent that there exists $\sigma \in S_4$ (the symmetric group on 4 letters) such that $\{x_{i_{\sigma(1)}} \dots x_{i_{\sigma(4)}}\}$ is a supertetrad, and so it is in H'. Now the cycle (1234) and the transposition (12) generate the group S_4. Hence equations (2) and (3) imply that $\{x_{i_1} \dots x_{i_4}\} \equiv \pm\{x_{i_{\sigma(1)}} \dots x_{i_{\sigma(4)}}\}$ $(mod\ H')$. From this it is clear that $\{x_{i_1} \dots x_{i_4}\} \in H'$.

Finally, let n be even ≥ 6 and consider (1) for $m = 4$. This and equations (2)–(4) imply that $\{x_{i_1} \dots x_{i_n}\} \in H'$.

QED

Notice that classical Cohn's theorem can be obtained as a consequence of the preceeding result.

Acknowledgements

The author wishes to thank the University of Wisconsin for its hospitality during his visit in Autumn 1992. The work on the present paper was completed during that time.

References

1. N. Jacobson, *Structure and Representations of Jordan Algebras*, Amer. Math. Soc. Colloq. Publ., Vol. 39, AMS, Providence, RI, 1968.
2. K. McCrimmon, The Russian Revolution in Jordan Algebras, Algebras, Groups and Geometries **1** (1984), 1–61.
3. E. I. Zelmanov, Superalgebras and Identities, Amer. Math. Soc. Transl. Ser. 2, **148**, AMS, Providence, RI, 1991.
4. Yu. Medvedev, E. I. Zelmanov, Some Counter–Examples in the Theory of Jordan Algebras, Nonassociative Algebraic Models, Nova Science Publishers, 1992, 1–16.
5. A. R. Kemer, Varieties and Z_2–graded Algebras, Izv. Akad. Nauk SSSR **48** (1984), 1042–1059; english transl. in Math. USSR–Izv. **25** (1985), 359–374.

ON BERNSTEIN ALGEBRAS OF N-TH ORDER

S. GONZALEZ, J.C. GUTIERREZ and C. MARTINEZ*
Departamento de Matemáticas, Universidad de Oviedo
33007 Oviedo, Spain

Abstract. Let (A,ω) be a finite dimensional n-th order Bernstein algebra over an infinite field K ($charK \neq 2$). If $e \in A$ is a nontrivial idempotent then $A = K_e \oplus U_e \oplus V_e$ where $U_e = \{x \in Ker\omega / ex = \frac{1}{2}x\}$ and $V_e = \{x \in Ker\omega / R_e^n x = 0\}$. In this paper we show the following results:(1) The dimension of U_e does not depend on idempotent e,(2) if $K = \mathbb{R}$ and $dimU_e = r$ then there is an r-parametric family of idempotents of the A, (3) if $K = \mathbb{R}$ then $dim_{\mathbb{R}} V_e \geq n$.

1. Introduction

Let A be a finite dimensional commutative algebra over an infinite fiel K ($charK \neq 2$). The plenary powers of an element $x \in A$, are defined as follows

$$x^{[1]} = x \quad \text{and} \quad x^{[r]} = x^{[r-1]}x^{[r-1]}. \tag{1}$$

An algebra A over K is baric if it admits a nontrivial algebra homomorphism $\omega : A \longrightarrow K$, called weight homomorphism. A baric K-algebra (A,ω) is nth order Bernstein, $n \geq 0$, if n is the minimum natural number, such that the identity

$$x^{[n+2]} = \omega(x)^{2^n} x^{[n+1]} \tag{2}$$

holds for every $x \in A$.

The definition of Bernstein algebras of order n was proposed by V.M. Abrahan [1] as a generalization of Bernstein algebras of order 1; Berstein algebras of order 1 (so-called Bernstein algebras) emerged in connection with a problem in mathematical heredity theory raised by I.N. Bernstein and the formal definition was given by P. Holgate [3] as an algebraic equivalent to the notion of Bernstein quadratic operator introduced by Yu. I. Lyubich in [4].

If A is an stochastic algebra describing a population and the element y of A represents a frequency distribution of genotypes in the initial generation (in this case $\omega(y) = 1$), then $y^{[r]}$ represents the frequency distribution in the rth generation. Now, if A satisfies the identity (2) we have $y^{[n+2]} = y^{[n+1]}$, and this shows that the population reaches the equilibrium after $n+1$ generations.

2. The Peirce Decomposition

We study the structure and properties of nth order Bernstein algebras by using the idempotent elements as a tool. In the following, (A,ω) will be a nth order Bernstein algebra over a fiel K ($charK \neq 2$).

* Partially supported by D.G.A. P. CB-6/91.

S. González (ed.), Non-Associative Algebra and Its Applications, 158–163.

It is known that a nth order Bernstein algebra A possesses exactly one nontrivial weight homomorphism and the set of idempotents of A is given by

$$I_p(A) = \{x^{[n+1]}/x \in A, \omega(x) = 1\}, \tag{3}$$

(see Hentzel-Peresi [2] and Mallol [6]). It is also known that a nth order Bernstein algebra has at least one idempotent element,and if e is an idempotent element, then A splits into the direct sum

$$A = Ke \oplus Ker(\omega) = Ke \oplus U_e \oplus V_e \tag{4}$$

where

$$U_e = \{x \in Ker\omega/ex = \frac{1}{2}x\} \quad \text{and} \quad V_e = \{x \in Ker\omega/R_e^n x = 0\}. \tag{5}$$

This decomposition is called the Peirce decomposition of A and it is known that

$$U^2 \subset V \tag{6}$$

$$eV \subset V. \tag{7}$$

In the present paper some general properties of a nth order Bernstein algebra are established.

It is known (see Lyubich [5]) that if A is a first order Bernstein algebra (so-called Bernstein algebras), the dimension of U and V does not depend on e, i.e., these number are invariants of the algebra. In the present paper we prove this result in the general case, (question posed by Mallol in [6]).

Theorem 2.1 *The dimension of the U-component of a nth order Bernstein algebra does not depend on e, i.e., this number is an invariant of the algebra.*

Proof. Let $\{u_1, u_2, ..., u_m\}$ be a basis of $Ker(\omega)$. Now we define the transformation $f : K^m \longrightarrow I_p(A)$ in the following way

$$f(\lambda_1, \lambda_2, ..., \lambda_m) := (e + \sum_{k=1}^{m} \lambda_k u_k)^{[n+1]}.$$

By (3) f is surjective and if we consider the endomorphism of $Ker(\omega)$

$$(2L)_f(x) := 2f(\lambda_1, \lambda_2, ..., \lambda_m)x$$

it is easy to show that the characteristic polynomial of $(2L)_f$ can be written in the form

$$P(\lambda_1, \lambda_2, ..., \lambda_m)(\mu) = \sum_{k=0}^{m} a_k(\lambda_1, \lambda_2, ..., \lambda_m)\mu^k$$

where $a_k(\lambda_1, \lambda_2, ..., \lambda_m) \in K[\lambda_1, \lambda_2, ..., \lambda_m]$. On the other hand, relations (4) and (5) imply that the minimum polynomial of the right multiplication by an idempotent e

restricted to $Ker(\omega)$ divides $x^n(2x-1)$, hence for every $\lambda_1, ..., \lambda_m \in K$ there exists $i \in \{0, 1, ..., m\}$ such that

$$P(\lambda_1, \lambda_2, ..., \lambda_m)(\mu) = \mu^{m-i}(\mu-1)^i = \sum_{k=m-i}^{m} \binom{i}{k+i-m} (-1)^{m-k}\mu^k.$$

Consequently, from the above considerations,we have that $Im(a_k)$ is finite. Therefore a_k is a constant function (it does not depend on λ's).

Since the minimum polinomial of $(2L)_f$ divides n to $\mu^n(\mu-1)$, its Jordan matrix is lower triangular and its diagonal is $\{1, ..., 1, 0, ..., 0\}$ where the number of 1's is equal to $dimU_f$ and $m - min\{k \in \mathbb{N}/a_k \neq 0\}$. Since the second number is independent of the considered idempotent, the result follows. □

Using the above Theorem, we define:

Definition 2.1 *For any nth order Bernstein algebra* (A, ω) *the type of A is the ordered pair of positive integers* $(1 + dimU_e, dimV_e)$ *for any idempotent element e.*

This definition is the natural generalization of the one, given by Lyubich, in Bernstein algebras. Clearly the sum of two integers of the type of A is equal to the dimension of A.

Mallol in [6] proved that if $dimU = 0$ then the algebra A possesses a unique idempotent element. Moreover it is known that if A is a Bernstein algebra, e an idempotent of A and dimension of U_e is equal to r then there is an r-parametric family of idempotent elements. In the following Theorem we establih the same relationship between $dimU$ and the number of idempotents in A.

Theorem 2.2 *If* $K = \mathbb{R}$ *and* $dimU = r$ *then there exists an r-parametric family of idempotent elements of A.*

Proof. Let $\{u_1, u_2, ..., u_r\}$ be a basis of U_e and $\lambda_1, \lambda_2, ..., \lambda_r \in K$, then

$$(e + \sum_{i=1}^{r} \lambda_i u_i)^{[k+1]} = e + \sum_{i_1+...+i_r \leq 2^k} \lambda_1^{i_1} ... \lambda_r^{i_r} \Lambda_{(i_1,...,i_r),k} \tag{8}$$

where the $\Lambda_{(i_1,...,i_r),m}$ are polynomial functions in $u_1, u_2, \ldots, u_r$ (These polynomial functions are a generalization of the ones introduced by Mallol [6] to analyse the structure of A). We have the equation

$$\Lambda_{\bar{s},k} = 2e\Lambda_{\bar{s},k-1} + \sum_{\bar{s}_1+\bar{s}_2=\bar{s}} \Lambda_{\bar{s}_1,k-1}\Lambda_{\bar{s}_2,k-1} \tag{9}$$

for every $k \geq 0$ and $\bar{s} \in (\mathbb{N}^r)^+ = \{(i_1, ..., i_r) \in \mathbb{N}^r / i_1, ..., i_r \geq 0\}$. We will show that

$$\Lambda_{h_i,k} = u_i \tag{10}$$

where $h_i = (0, \ldots, 0, \overset{i)}{1}, 0, \ldots, 0)$. We will prove (10) by induction on k, $k \geq 0$. If $k = 0$ then (10) is true and if $k > 0$, then by (9) we obtain

$$\bigwedge_{h_i,k} = 2e\bigwedge_{h_i,k-1} = 2eu_i = u_i.$$

On the other hand, we define as in the above Theorem the transformation

$$f : K^r \longrightarrow I_p(A)$$

by

$$f(\lambda_1, \lambda_2, ..., \lambda_r) := (e + \sum_{k=1}^{r} \lambda_k u_k)^{[n+1]}.$$

Now if $\{v_{r+1}, ..., v_m\}$ a basis of V_e, then by (6), (8) and (9) there exist $P_i, P_i', Qj \in \mathbf{R}[\lambda_1, \lambda_2, ..., \lambda_r]$ for $1 \leq i \leq r < j \leq m$ such that $P_i = \lambda_i + P_i'$, $Order P_i' \geq 2$, $Order Q_j \geq 2$ and

$$f(\lambda_1, ..., \lambda_r) = e + \sum_{i=1}^{r} P_i u_i + \sum_{j=r+1}^{m} Q_j v_j = e + \sum_{i=1}^{r} \lambda_i u_i + \sum_{i=1}^{r} P_i' u_i + \sum_{j=r+1}^{m} Q_j v_j.$$

hence

$$\frac{d}{d\lambda_i} f(0, ..., 0) = u_i$$

and therefore $rank D(f)(0, \ldots, 0) = r$, so f is injective in some neighborhood of $(0, \ldots, 0)$. □

Mallol proved in [6] that if A is a second order Bernstein algebra then the dimension of V_e is greater than or equal to 2 and he poses in [6] this question in general, i.e., is it true that for every nth order Bernstein algebra (A, ω) and for every idempotent element e of A the dimension of V_e is greater than or equal to n?. In the present paper we will answer this question if $K = \mathbf{R}$.

Theorem 2.3 *Let (A, ω) be a nth order Bernstein algebra and e an idempotent element of A, then $dim(V_e) \geq n$.*

The proof is based on the following Lemmas, definitions and propositions.
The first Lemma is a well-known result in dimension theory of topological spaces.

Lemma 2.1 *A subset X of $\mathbf{R}^m$ is m-dimensional (topological dimension) if and only if X contains a nonempty subset which is open in $\mathbf{R}^m$.*

Lemma 2.2 *Let $\Omega \subseteq \mathbf{R}^m$ be open and let $f : \Omega \longrightarrow \mathbf{R}^m$ be a function in $C^1(\Omega)$. Let $D_f(x)$ be of rank r for every $x \in \Omega$ and let $b = f(a) \in \mathbf{R}^m$. Then there exists an open neighborhood B of a, such that $f(B)$ is homeomorphic to $\mathbf{R}^m$.*

Definition 2.2 *Let Ω be open and let $f : \Omega \rightarrow \mathbf{R}^m$ be a function in $C^1(\Omega)$. Then the $rank(f) = max\{rank D(f)(\alpha)/\alpha \in \mathbf{R}^m\}$.*

Lemma 2.3 *Let Ω and Ω' be open in $\mathbf{R}^m$ and let $f : \Omega \longrightarrow \mathbf{R}^m$ and $g : \Omega' \longrightarrow \mathbf{R}^m$ be a function in $C^1(\Omega)$ and $C^1(\Omega')$ respectively and let $f(\Omega) \subseteq \Omega'$. Then:*
(i) *$rank(g \circ f) \leq min\{rank(f), rank(g)\}$, and*
(ii) *if $rank(f) = rankDf(a)$ then there exists an open neighborhood B of a, such that $rankDf(a) = rankDf(x)$ for every $x \in B$.*

In what follows let $\{w_1, \ldots, w_m\}$ be a basis of $Ker(A)$, $H = \{x \in A/\omega(x) = 1\}$ and we define

$$f : M \longrightarrow M, \pi : \mathbf{R}^m \longrightarrow H \text{ and } g : \mathbf{R}^m \longrightarrow \mathbf{R}^m$$

by

$$f(x) := x^{[2]}, \pi(\lambda_1, \ldots, \lambda_m) := (e + \textstyle\sum_{i=1}^m \lambda_i w_i) \text{ and } g := \pi^{-1} \circ f \circ \pi$$

respectively. We will denote r_i the $rank(g^i)$ for every positive integer i.

Proposition 2.1 *If g is the mapping defined above, then:*
(i) *g is polynomial function ($g \in C^\infty$),*
(ii) *$g^{n+1} \equiv g^n$ and*
(iii) *$g^{i+1} \not\equiv g^i$ for $0 \leq i < n$,*
(iv) *$Im(g^{i+1}) \subset Im(g^i)$, for $0 \leq i < n$*
(v) *No one open set in $Im(g^{i+1})$ is an open set in $Im(g^i)$ for $0 \leq i < n$.*

Proof. The (i), (ii), (iii) and (iv) are obtained from the definition. The (v) from (i) and (iii). □

Proposition 2.2 *Its have the following results:*
(i) *$dim(Ker\omega) = m = r_0 \geq r_1 \geq \ldots \geq r_n \geq r = dimU$.*
(ii) *There exists $\alpha_0 \in \mathbf{R}^m$ such that $rank(g^i) = rank(Dg^i(\alpha_0))$ for $i = 0, 1, \ldots, n$.*

Proof. It is easy to see that (i) follows from Proposition 2.1 (i) and Theorem 2.2.
(ii) We have that g is a polynomial function and hence $Dg^i(x)$ is a matrix whose entries are polynomial functions. For each $i, 0 \leq i \leq n$, there exists M_i a minor of order r_i of $Dg_i(x)$ with noncero determinant. The set of determinants of these M_i forms a family of nonzero polynomials, then there exists $\alpha_0 \in \mathbf{R}^m$ that is not a zero of any polynomial in this family. □

Proposition 2.3 *$r_i \neq r_{i+1}$ for $i = 0, 1, \ldots, n-1$.*

Proof. Let k be the smallest positive integer such that $r_k = r_{k+1}$ (we will denote this number, r_k, for $\tilde{r}$, in what follows).
Now from Proposition 2.2 (ii) there exists $\alpha_0 \in \mathbf{R}^m$ such that $rank(g^i) = rank(Dg^i(\alpha_0))$ for $i = 0, 1, \ldots, n$. Let $\alpha_1 = g(\alpha_0)$. Now by

$$r_{k+1} = rank(Dg^{k+1}(\alpha_0)) = rank((Dg^k(\alpha_1))(Dg(\alpha_0))) \leq rank(Dg^k(\alpha_1)) \leq r_k$$

and Proposition 2.2 (i) it follows that $rank(Dg^k(\alpha_1)) = \tilde{r}$.

Next let B' be an open neighborhood of α_1 such that $g^k(B')$ is homeomorphic to $\mathbf{R}^{\tilde{r}}$ (from Lemma 1 and 2 there exists B') and let B" be an open neighborhood of α_0 such that $B" \subset g^{-1}(B')$ and $g^{k+1}(B")$ is homeomorphic to $\mathbf{R}^{\tilde{r}}$. Hence by Lemma 1 $g^{k+1}(B")$ contains an open of $g^k(B')$ (we denote this open Z).

If B is $g^{-k}(Z) \cap B'$, then B is open and

$$g^k(\alpha) = g^{k+1}(\alpha)$$

for every $\alpha \in B$, therefore $g^k \equiv g^{k+1}$ and $k \geq n$. □

Now, Theorem 2.3 follows from Propositions 2.2 and 2.3.

References

1. V.M. Abraham, Liniarizing quadratic transformation in genetic algebras, Proc. London Math. Soc. (3) 40 (1980), 346-363.
2. R. Hentzel, L.A. Peresi and P. Holgate, *k*th Order Bernstein Algebras and Stability at the k+1 Generation in Polyploids, J. of Math. Applied in Medicine and Biology 7 (1990) 33-40.
3. P. Holgate, Genetic algebras satisfying Bernstein's stationary principle, J. London Math. Soc. (2) 9,613-23.
4. Yu. I. Lyubich, Two-Level Bernsteinian Populations, Math. USSR Sbornik, Tom 95 (137) (1974), No. 4.
5. Yu. I. Lyubich, A Classification of some types of Bernstein Algebras, Selecta Mathematica Sovietica Vol. 6 No. 1 1987.
6. C. Mallol, *A propos des algébres de Bernstein*, Thèse d'Etat, Université de Montpellier II, France, Décembre 1989.

ON BERNSTEIN ALGEBRAS

S. GONZALEZ and C. MARTINEZ*
Departamento de Matemáticas. Universidad de Oviedo
33.007 Oviedo, Spain

Abstract. In this paper we will make a small review of some known results on Bernstein algebras, we will consider free nuclear Bernstein algebras and finally we will give some new results obtained aplying the constuction of free Bernstein algebras

1. Introduction

Non associative algebras appear in Genetics when we want to express as a symbolic product the way in which the biological characteristics of the two parents are communicated to their offspring.

Algebras arising in this way are called "genetic algebras". The first papers on genetic algebras appeared in 1934 and 1936, but the subject was properly stablished by Etherington between 1939 and 1941.

Algebras arising in Genetics are in general baric algebras, that is, they have a nonzero algebras homomorphism $\omega : A \to K$ called weight homomorphism.

Usually these algebras have other properties, as being train algebras, Gonshor algebras or special train algebras.

Bernstein algebras are an important class of genetic algebras. They have their origin in some papers by Bernstein studying the stationarity principle in Genetics (see [3]). In this way, Bernstein algebras represent populations reaching the equilibrium after the first generation. They were introduced by Lyubich, see [10], [11] and their present algebraic formulation was given by P. Holgate in 1971 ([9]).

Definition 1.1 *A Bernstein algebra A is a commutative baric algebra over a field K ($charK \neq 2$) satisfying the identity $(x^2)^2 = \omega(x)^2 x^2$ for every $x \in A$*

From an algebraic point of view many unsolved problems about these algebras can be found.We will list now, without proofs, some well known results about Bernstein algebras.

1. The weight homomorphism is unique, what is true in a baric algebra when elements in $Ker\omega$ satisfy some nilpotence condition.

Consequently in a Bernstein algebra there is a well defined ideal of codimension 1, whose elements satisfy $(x^2)^2 = 0$.

2. A Bernstein algebra A has always nontrivial idempotent elements and the set of idempotent elements is given by $\{x^2/\omega(x) = 1\}$.

If $e \in A$ is one of these idempotent elements then $A = Ke \oplus U_e \oplus Z_e$ where $U_e = \{x \in Ker\omega : ex = \frac{1}{2}x\}$ and $Z_e = \{x \in Ker\omega : ex = 0\}$.

* Partially sipported by D.G.A. CB-6/91.

S. González (ed.), Non-Associative Algebra and Its Applications, 164–170.

Products between elements of U_e and Z_e satisfy the following relations:
a) $U_e.U_e \subseteq Z_e$, $U_e.Z_e \subseteq U_e$ and $Z_e.Z_e \subseteq U_e$.
b) $u^3 = 0 = u(uz) = (uz)^2 = uz^2$ for all $u \in U_e$, $z \in Z_e$.

3. If e is one idempotent element, the set of idempotent elements is $E(A) = \{f = e + u + u^2 : u \in U_e\}$. If $f = e + u + u^2$ is other idempotent element, then relations between U_e and U_f and Z_e and Z_f are given by:

$$U_f = \{x + 2ux : x \in U_e\}, \quad Z_f = \{x - 2ux - 2u^2x : x \in Z_e\}. \tag{1}$$

Consequently, type$A = (r+1, s)$, where $r = dimU_e$ and $s = dimZ_e$ is well defined, that is, it doesn't depend on the particular idempotent element used in its definition.

4. A Bernstein algebra satisfying

$$x^2y = \omega(x)xy \tag{2}$$

for every $x, y \in A$ is called normal or conservative. If e is an idempotent element of A and $A = Ke \oplus U_e \oplus Z_e$, this condition is equivalent to $(Z_e)^2 + U_eZ_e = 0$.

For a Bernstein algebra A the following conditions are equivalent:
i) A is power-associative,
ii) A is a Jordan algebra
iii) $x^3 = \omega(x)x^2$ for all $x \in A$,
iv) If $A = Ke \oplus U_e \oplus Z_e$, then $z^2 = 0 = (uz)z$ for every $u \in U_e$, $z \in Z_e$,
v) $(Z_e)^2 = 0$ for every idempotent element e in the Bernstein algebra A.

Notice that $x^3 = 0$ for all $x \in Ker\omega$ in a Jordan-Bernstein algebra, but the converse is not true.

5. A Bernstein algebra A is called exclusive or exceptional if $U_e^2 = 0$ for some idempotent element. Then, the same condition holds for every idempotent element.

A Bernstein algebra is called orthogonal if $U_e^3 = 0$ holds for some idempotent element. Contrary to the exclusive case, this property is not independent of the considered idempotent element. So in [1] "totally orthogonal" Bernstein algebras are considered, that is, Bernstein algebras in which $U_e^3 = 0$ holds for every idempotent element e.

It is proved that A is totally orthogonal if and only if there is one idempotent element e such that $(U_e)^3 = 0$ and $(U_e)^2(U_e)^2 = 0$. Consequently, in Jordan-Bernstein algebras both notions (orthogonal and totally orthogonal) are equivalent.

Classifications for Bernstein algebras of dimensions 3 and 4 have been made in [4]. Also nonexceptional 5-dimensional Bernstein algebras have been classified, but there is not a classification in the general case. With the aim of classify this structure and get new examples of Bernstein algebras, new notions have been introduced. So in [5] the homotope algebra of a Bernstein algebra is defined and studied. S. González defines the homotope algebra of a Bernstein algebra in the same way that Teddy did for non-associative algebras and proves that the homotope algebra $A^{(a)}$, with $\omega(a) \neq 0$, is again a Bernstein algebra. However, a natural notion of isotopy doesn't exist. Considering some of the first properties of the homotope algebra by one idempotent element, S. González defines in [6] the notion of quasiisomorphism. In a similar way, in [13] the following characterization of isomorphisms of Bernstein algebras is given:

Theorem 1.1 *Let A and A' be Bernstein algebras. Then a bijective linear application $\Theta : A \to A'$ is an isomorphism if and only if it satisfies the following two conditions:*

i) An element e is an idempotent of A if and only if $\Theta(e)$ is an idempotent element of A'.

ii) $xy = 0$ in A if and only if $\Theta(x).\Theta(y) = 0$ in A'.

In [7] it was proved that if $A = Ke \oplus U_e \oplus Z_e$ is a Bernstein algebra and $U_0 = \{u \in U_e | uU_e = 0\}$, then U_0 is an ideal of A, it doesn't depend on the idempotent element and it is invariant by derivations. It is also proved that A/U_0 is a Jordan Bernstein algebra. If A is nuclear, that is $A^2 = A$, the above ideal U_0 is contained in the annhilator of $Ker\omega$.

Nuclear-Bernstein algebras have a good behaviour (they are genetic in the sense of Gonshor, that is, $Ker\omega$ is nilpotent) and they seem to reflect a genetical meaning. So we have defined and studied free nuclear Jordan-Bernstein algebras([14]).

It is known that the powers of $Ker\omega$, $((Ker\omega)^{i+1} = (Ker\omega)^i(Ker\omega))$ are ideals of A. If A is a Jordan-Bernstein algebra it may be proved that:

I) Every product of l elements of $Ker\omega$, with any arrangement of brackets belongs to $(Ker\omega)^l$.

II) If A is a Jordan-Bernstein algebra and e,f are two idempotent elements of A, then $dim(U_e^m) = dim(U_f^m)$ for all $m \geq 1$. Furthermore, if $f = e + u^* + (u^*)^2, u^* \in U_e$, then $U_f^m = \{x - 2u^*x \mid x \in (U_e)^m\}$ if m is odd and $(U_f)^m = \{x - 2u^*x \mid x \in (U_e)^m\}$ if m is even.

III) For all $e, f \in E(A)$ and $m \geq 1$, it is true that $(U_e)^m + (U_e)^{m+1} = (U_f)^m + (U_f)^{m+1}$ and it is an ideal of A.

IV) If A is a Jordan-Bernstein algebra of dimension n, then every product of n elements in A is equal to zero. It suffices consider the strictly decreasing chain: $Ker\omega \geq (Ker\omega)^2 \geq (Ker\omega)^3 \geq ...$, what assures that $(Ker\omega)^n = 0$. Now it suffices to apply (I). It may also been proved that $(Ker\omega)^{n-1} = 0$ if $dimA = n \geq 4$ and the bound may be reached for n = 4. If $dimA = n \geq 5$, then $(Ker\omega)^{n-2} = 0$.

If A is a nuclear-Bernstein algebra, then $x^4 = 0$ for all $x \in Ker\omega$. If $dimA = n$, then $(Ker\omega)^n = 0$ and every product of 2n-1 elements of $Ker\omega$ is zero.

2. Free Nuclear Jordan-Bernstein Algebra over a set of r generators $\{u_1, ..., u_r\}$

Let $L = K < X_1, ..., X_r >$ be the commutative nonassociative algebra of polynomials of degree ≥ 1 in the indeterminates $X_1, ..., X_r$. Let us consider the ideal I of L generated by all elements in one of the following forms:

(i) $(m_{2l+1}(X))^3$, where $m_{2l+1}(X)$ denotes a monomial of odd degree 2l+1.

(ii) $(n_{2h}(X))^2$, where $n_{2h}(X)$ is a monomial of even degree $2h > 0$.

(iii) $(n_{2h}(X)m_{2l+1}(X))n_{2h}(X)$

(iv) $(m_{2l+1}(X)n_{2h}(X))n_{2h}(X)$

(v) $(m_{2l+1}(X)n_{2h}(X))m_{2l+1}(X)$.

With the above relations $B = L/I$, the quotient algebra, is a 2-graded algebra that can be considered the nucleus of a nuclear Jordan-Bernstein algebra. In fact,

$B = B_0 + B_1$, where B_0 is generated by all monomials of even degree and B_1 by those of odd degree. We will denote $u_i^* = X_i + L$.

So the algebra $A_r^* = Ke^* + B$, with $e^*e^* = e^*$, $e^*x = \frac{1}{2}x$ if $x \in B_1$ and $e^*y = 0$ if $y \in B_0$ will be called free nuclear Jordan-Bernstein algebra of rank r and will be denoted $A_r^* = A\{u_1^*, ..., u_r^*\}$.

Theorem 2.1 *Let $A = Ke \oplus U_e \oplus Z_e$ a Jordan-Bernstein algebra and a map $\varphi : \{X_1, ..., X_r\} \in A$ such that $Im\varphi \subseteq U_e$. Then there is a unique homomorphism of Bernstein algebras $\Phi : A_r^* \in A$ such that $\Phi(e^*) = e$ and Φ extends φ.*

Theorem 2.2 *If $A = Ke + U_e + U_e^2$ is a nuclear Jordan-Bernstein algebra, $l = dimU_e - dimU_e^3$, then A is a quotient of the "free nuclear Jordan-Bernstein algebra" of rank l, A_l^*.*

We can construct a "canonical basis" of A, $B = B_0 \cup ... \cup B_i \cup ...$ a basis of U_e and $B^* = B_0^* \cup ... \cup B_1^* \cup ...$ a basis of U_e^2 such that elements in B_i are in U^{2i+1}, but not in U^{2i+3}, B_i is formed by products of 2i+1 elements in B_0 and $B_i \bigcup ...$ is a basis of U^{2i+1}. Similarly, B_i^* is formed by products of 2i elements of B_0 and $B_j^* \cup ...$ is a basis of U^{2j}.

Theorem 2.3 *If $A_l^* = Ke^* \oplus U_e \oplus Z_e$ is the free nuclear Jordan-Bernstein algebra of rank l, then $U^{2l+1} = 0$. Consequently every product of 2l+1 elements in $Ker\omega$ is equal to zero.*

3. Free Jordan-Bernstein algebra

In a similar way, the free Jordan-Bernstein algebra over a set $\{u_1, ...u_l, z_1, ..., z_t\}$ of generators may be constructed.

We start considering $J = K < X_1, ..., X_l, Y_1, ..., Y_t >$ the commutative nonassociative algebra of polynomials of degree ≥ 1 in the above unknowns. We will consider $A_0 = \{$ monomials of even degree in $X_1, ..., X_l\}$ (Notice that $\{Y_1, ..., Y_t\} \subseteq A_0$) and $A_1 = \{$ monomials of odd degree in $X_1, ..., X_l\}$. Let us construct the ideal N generated by the elements of the following forms:

(i) $\alpha\beta$, with α, $\beta \in A_0$
(ii) γ^3, with $\gamma \in A_1$
(iii) $(\alpha\gamma)^2$, with $\alpha \in A_0$, $\gamma \in A_1$
(iv) $(\alpha\gamma)\gamma$ and $(\alpha\gamma)\alpha$, with $\alpha \in A_0$, $\gamma \in A_1$.

The quotient algebra $K < X_1, ..., X_l, Y_1, ..., Y_t > /N = A_0^{(l,t)} + A_1^{(l,t)}$ is a 2-graded algebra that can be considered, as before, the nucleus of a Jordan-Bernstein algebra, that we will denote $\bar{A}_{l,t}$ and will be called the "free Jordan-Bernstein algebra" of rank (l,t).

Theorem 3.1 *If $A = Ke \oplus U_e \oplus Z_e$ is a Jordan-Bernstein algebra, $l = dimU_e - dimU_e^3$ and $t = dimZ_e - dimU_e^2$, then A is a quotient of $\bar{A}_{l,t}$.*

Theorem 3.2 *There is a monomorphism between $\bar{A}_{l,t}$ and A_{l+t}^*.*

4. Applications

Theorem 4.1 *If A is a nuclear Bernstein algebra generated by r elements, then* $(Ker\omega)^{2r+2} = 0$

Proof. Let us consider $U_0 = \{u \in U : uU = 0\}$. It is known that U_0 is an ideal of A and the quotient algebra $\bar{A} = A/U_0$ is a nuclear Jordan-Bernstein algebra generated by r elements. If $\bar{A} = Ke \oplus Ker\bar{\omega}$, since $\bar{A}$ is a quotient of A_s^*, for some $s \leq r$, the above theorem assures that $(Ker\bar{\omega})^{2s+1} = 0$ and consequently $(Ker\bar{\omega})^{2r+1} = 0$, that is $(Ker\omega)^{2r+1} \subseteq U_0$, what implies that $(Ker\omega)^{2r+2} = 0$.

In the same way, we can prove that every product of 4r+1 elements (with any arrangement of brackets) is equal to zero.

In the general case, we have:

If A is a Jordan Bernstein algebra, $r = dimU - dimU^3$, $s = dimZ - dimU^2$, then $(Ker\omega)^{2r+2s+1} = 0$. Furthemore every product of 2r+2s+1 elements in $Ker\omega$ is equal to zero.

If A is a Bernstein algebra, $r = dimU - dim(U^3 + U_0)$ and $s = dimZ - dimU^2$, then $(Ker\omega)^{2r+2s+1} \subseteq U_0$.

In [12] inner derivations of Jordan-Bernstein algebras were considered and it was conjectured that every Jordan-Bernstein algebras have non-inner derivations. Now it may be proved:

Theorem 4.2 *If A_m^* is the free nuclear Jordan-Bernstein algebra over a set $\{u_1, u_2, ..., u_m\}$, then A has derivations that are non-inner.*

Finally we will solve a problem posed by Grishkov, in which we will use some of the above ideas. We want to find all nuclear Bernstein algebras satisfying the three following conditions:

1. $(Ker\omega)^3 \neq 0$,
2. If B is a proper nuclear Bernstein subalgebra of A, then $(Ker\omega')^3 = 0$, where ω' denotes the restriction of the weight homomorphism ω of A to B,
3. If I is an ideal of A, $I \subset Ker\omega$, then the quotient Bernstein algebra $\widetilde{A} = A/I$ satisfies that $(Ker\widetilde{\omega})^3 = 0$. This is equivalent to say that $(Ker\omega)^3 \subseteq I$ for every ideal I of A, with $0 \neq I \subseteq Ker\omega$.

Theorem 4.3 *There are exactly two nuclear Bernstein algebras stisfying the above conditions, one of them is 6-dimensional and Jordan and the other one is 7-dimensional and is not Jordan.*

Proof. We will distinguish two cases:

i) If A is not Jordan, then we know that $U_0 \neq 0$, where $U_0 = \{u \in U : uU = 0\}$ and A/U_0 is a nuclear Jordan-Bernstein algebra. Consequently $(Ker\omega)^3 \subseteq U_0$ and so $(Ker\omega)^4 = (Ker\omega)^3 Ker\omega = 0$ (we are using that $U_0 Ker\omega = 0$ in a nuclear Bernstein algebra). But for every $u \in U_0$ the one dimensional vector space Ku is and ideal of A. So $U_0 = (Ker\omega)^3 = Ku$.

Let us consider a Peirce decomposition of A, $A = Ke \oplus U_e \oplus Z_e$. If $(U_e)^3 = 0$, then $U_e.Z_e = 0$ and since A is not Jordan we know that $(Z_e)^2 \neq 0$, that is,

$(U_e)^2.(U_e)^2 \neq 0$. But this assures that A is orthogonal, but not totally orthogonal and then there is one idempotent element f with $(U_f)^3 \neq 0$. Let us consider the Peirce decomposition of A related to f, $A = Kf \oplus U_f \oplus Z_f$. We know the existence of two elements $u_1, u_2 \in U_f$ with $u_1^2.u_2 \neq 0$ and we may assume that $u = u_1^2 u_2 = -2u_1(u_1u_2)$. Let us consider $z_1 = u_1^2$ and $z_2 = u_1u_2$, two nonzero elements of Z_f. They are linearly independent, because $2u_1(u_1u_2) = -u \neq 0$, but $u_1.u_1^2 = 0$. So $B = Ke + K(u_1, u_2, u) + K(u_1^2, u_1u_2, u_2^2)$ is a nuclear subalgebra of A of dimension 6 or 7 and $(Ker\omega')^3 \neq 0$. So, conditions on A assure that $A = B$.

Let us suppose that $dimB = 6$, what means that $u_2^2 = \alpha u_1^2 + \beta u_1u_2$. Then $0 = u_1^2.u_1^2 = u_1^2(u_1u_2)$ and $2(u_1u_2)^2 = (u_1^2)(u_2^2) = (u_1^2)(\alpha u_1^2 + \beta(u_1u_2)) = 0$, that is, $B = A$ would be Jordan, against the assumption. Consequently, $dimB = 7$ and u_2^2, u_1^2 and u_1u_2 are linearly independent, that is, $A = Kf + K(u_1, u_2, u) + K(z_1 = u_1^2, z_2 = u_1u_2, z_3 = u_3^2)$. Since A is not Jordan, $z_1^2 = z_1z_2 = z_2z_3 = z_3^2 = 0$ and $2z_2^2 = -z_1z_3$, we conclude that $z_2^2 = \lambda u$, with $\lambda \neq 0$.

By changing u by $\mu^3 u$, u_1 by μu_1, u_2 by μu_2 and z_1 by $\mu^2 z_1$, z_2 by $\mu^2 z_2$ and z_3 by $\mu^2 z_3$, where $\lambda\mu = 1$, we get that $A = Kf + K(u_1, u_2, u) + K(z_1, z_2, z_3)$ and products given by: $u_1^2 = z_1$, $u_1u_2 = z_2$, $u_2^2 = z_3$, $u^2 = uu_1 = uu_2 = 0$, $uZ = 0$, $u_1z_1 = 0 = u_2z_3$, $2u_1z_2 = 2u_2z_2 = -u$, $u_2z_1 = u_1z_3 = u$, $z_1^2 = z_3^2 = z_1z_2 = z_2z_3 = 0$, $2z_2^2 = -z_1z_3 = 2u$.

ii) If A is Jordan, so $(U_e)^3 \neq 0$. In fact, $(U_e)^3 = 0$ would imply that $(Ker\omega)^2 = (U_e)^2$ and $(Ker\omega)^3 = 0$. So, again, there are elements u_1, u_2 in U_e with $(u_1)^2u_2 \neq 0$. In the same way as before, considering the Bernstein subalgebra of A, $B = Ke + K(u_1, u_2, u_1^2u_2, u_2^2u_1) + K(u_1^2, u_2^2, u_1u_2)$, we conclude that $A = B \cong A_2^*/I$, where A_2^* is the free nuclear Jordan-Bernstein algebra of rank 2 generated by $\{x_1, x_2\}$ and I is an ideal of A_2^* with $(Ker\omega^*)^3 = K(x_1^2x_2, x_2^2x_1) \not\subseteq I$, but $(Ker\omega^*)^3 \subseteq J$ for every ideal J of A in which I is strictly contained.

So it can be proved that $I = K(u_2^2 + \gamma^2u_1^2 + 2\gamma u_1u_2) + K(u_2^2u_1 - \gamma u_1^2u_2)$. And changing basis in A, we get that $A = K(e) + K(u_1, u_2, u_3) + K(z_1, z_2)$ with $u_1^2 = z_1$, $u_1u_2 = z_2$, $u_2^2 = 0$ and $u_2z_1 = u_3$.

References

1. M. T. Alcalde, R. Baeza and C. Burgueño,Autour des algèbres de Bernstein, Arch. Math.53, (1989), 134-140.
2. R. Baeza, Bernstein Algebras: Linear Algebra and its Applications, 142, (1990), 19-23.
3. S. Bernstein, Principe de stationarité et généralization de la loi de Mendel, Comptes rendus de l'Acad. des Sci. Paris, 177 (1923), 581-584.
4. T. Cortés, Classification of 4-dimensional Bernstein algebras, Comm. in Algebra 19 (5),(1991),1429-1443.
5. S. González, Homotope algebra of a Bernstein algebra, Hadronic Mechanics and Nonpotential Interaction, H. C. Myung ed., Nova Science Publ. New York, (1992),186-200.
6. S. González, Quasiisomorphisms of Bernstein algebras, Comm. in Algebra 21 (11), (1992),4153-4166.
7. S. González and C. Martínez, Idempotent elements in a Bernstein algebra, J. London Math. Soc. 42, (1991), 430-436.
8. A. N. Griskhov, The geneticism of Bernstein algebras, Dokl. Akad. Nauk. SSSR, 294, (1987),27-30.
9. P. Holgate, Genetic algebras satisfying Bernstein's stationarity principle, J. London Math. Soc. 9 (2), (1975), 613-623.
10. Y. Lyubich, Basic concepts and theorems of the evolutionary genetics of free populations,

Russian Math. Surveys 26 (5), (1971), 51-123.

11. Y. Lyubich, Algebraic methods in evolutionary genetics, Biomat. J. 20, (1978), 511-529.
12. C. Martínez, Inner derivations in Jordan-Bernstein algebras, Hadronic Mechanics and Non-potential Interaction, H. C. Myung ed., Nova Science Publ. New York, (1992), 217-228.
13. C. Martínez, Isomorphisms of Bernstein algebras, Journal of Algebra 160, (1993), 419-423
14. C, Martínez, Free nuclear Bernstein Algebras, To appear in J. of Algebra.
15. M. Osborn, Varieties of algebras, Advances in Math. 8, (1972),1163-369.
16. S. Walcher, Bernstein's algebras which are Jordan algebras, Arch. Math. 50, (1988), 218-222.
17. A. Worz-Busekros, "Algebras in Genetics", Lecture Notes in Biomathematics, vol.36, Springer Verlag, Berlin-Heilderberg, (1980)
18. A. Worz-Busekros, Bernstein algebras, Arch. Math. 48, (1987), 388-398.

REALIZATION OF LIE ALGEBRAS WITH POLYNOMIAL VECTOR FIELDS

HANS GRADL
Mathematisches Institut, Technische Universität München, 80290 München, Germany.

Abstract. We show how a certain class of Lie algebras (including the semi-simple ones) can be realized with polynomial vector fields, thus giving systems of ordinary differential equations with the so-called fundamental solutions property.

Key words: semisimple Lie algebra, Lie algebra of vector fields, differential equations with fundamental solutions

1. Introduction

Througout this note, V denotes a finite-dimensional vector space over the field of complex numbers $\mathbb{C}$. By *Rat* V resp. *Pol* V we mean the Lie algebra of all rational resp. polynomial maps from V to V. The Lie bracket is given by $[f,g](x) = \mathrm{D}g(x)f(x) - \mathrm{D}f(x)g(x)$, where $\mathrm{D}f(x)$ is the derivative of f at x. On *Pol* V we have a natural $\mathbb{Z}$-grading: Let $\mathcal{P}_i$ consist of all polynomials of degree $i+1$. Then $Pol\, V = \oplus_{i\in\mathbb{Z}}\mathcal{P}_i$ and $[\mathcal{P}_i, \mathcal{P}_j] \subset \mathcal{P}_{i+j}$ for all i and j.

In general, a Lie algebra $\mathcal{L}$ is called *$\mathbb{Z}$-graded* if there exists a decomposition $\mathcal{L} = \oplus_{i\in\mathbb{Z}}\mathcal{L}_i$ such that $[\mathcal{L}_i, \mathcal{L}_j] \subset \mathcal{L}_{i+j}$ for all i and j. $\mathcal{L}$ is called $(2k+1)$-*graded* if it is $\mathbb{Z}$-graded and $\mathcal{L}_i$ is non-trivial if and only if $|i| \le k$.

For $k = 1$ one obtains the class of *3-graded* Lie algebras which have been (and still are) studied intensively. Since Kantor [3] and Koecher [4], [5] one knows how to realize 3-graded Lie algebras with polynomial vector fields (thus obtaining Jordan triple systems [8] and Jordan pairs [7]).

This has an interesting application given by a Theorem of Lie [6] (which dates back precisely 100 years!): The differential equation $\dot{x} = F(t,x)$ in V, the right hand side being a polynomial in x, has *fundamental solutions* (i.e. the general solution is – at least – locally determined by finitely many particular solutions "in general position") if and only if $F(t,x)$ can be written as $F(t,x) = \sum_i \alpha_i(t) f_i(x)$ such that the f_i generate a finite-dimensional subalgebra of *Pol* V. Thus the Kantor-Koecher construction gives rise to a class of quadratic differential equations with fundamental solutions.

As a matter of fact, there are complex (semi)simple Lie algebras that do not have a 3-grading: among the simple ones precisely those of types G_2, F_4 and E_8. However, they do have a 5-grading.

We hasten to add that 5-graded Lie algebras can be constructed from Jordan triple systems of second order introduced by Kantor [3] and vice versa.

In the following we will show how to realize all simple types with polynomial

S. González (ed.), Non-Associative Algebra and Its Applications, 171–175.

vector fields.

2. The "tunnel formula"

Throughout this section, let $\mathcal{L}$ be a simple $(2k+1)$-graded complex Lie algebra. We define

$$\mathrm{p} = \bigoplus_{i\leq 0} L_i, \quad \mathrm{q} = \bigoplus_{i>0} L_i, \tag{1}$$

the notation referring to the fact that p is a parabolic subalgebra of $\mathcal{L}$, q its natural complement: $\mathcal{L} = \mathrm{p} \oplus \mathrm{q}$. For $x \in \mathcal{L}$ we denote the component in p resp. q by x_- resp. x_+. Without loss of generality we may assume that $\mathcal{L} \subseteq \mathrm{sl}_m(\mathbb{C})$ for some $m \in \mathbb{N}$. Let G be the Lie subgroup of $\mathrm{SL}_m(\mathbb{C})$ generated by $\{\exp x : x \in \mathcal{L}\}$. From Lie group theory one knows that there are uniquely determined Lie subgroups P and Q of G such that $\mathrm{p} = Lie(P)$, $\mathrm{q} = Lie(Q)$. In fact, $Q = \{\exp x : x \in \mathrm{q}\}$, since q is nilpotent whence the Baker-Campbell-Hausdorff series is a finite sum.

The following properties are proved in [1]:

Proposition 1 *Let $\mathcal{L}, \mathrm{p}, \mathrm{q}, G, P$ and Q be as above.*

a) The map $\begin{cases} P \times Q & \to P \cdot Q \\ (g_-, g_+) & \mapsto g_- \cdot g_+ \end{cases}$ *is a local diffeomorphism.*

b) $P \cdot Q$ is open in G.

c) G acts transitively on the homogeneous space G/P. □

Thus G can be represented as a local transformation group, its Lie algebra being isomorphic to $\mathcal{L}$. The computation comes down to the following problem: For $x \in \mathcal{L}$ and $u \in \mathrm{q}$ find $p_x(u)$ such that

$$\exp(t \cdot x) \cdot \exp(u) = \underbrace{\exp\left(u + t \cdot p_x(u) + O(t^2)\right)}_{\in Q} \cdot \underbrace{g_-(t \cdot x, u)}_{\in P} \tag{2}$$

for t sufficiently close to 0.

In general, the map

$$\pi : \begin{cases} \mathcal{L} & \to \mathcal{A}(\mathrm{q}) \\ x & \mapsto p_x \end{cases} \tag{3}$$

where $\mathcal{A}(\mathrm{q})$ denotes the Lie algebra of analytic vector fields on q, will be a homomorphism of Lie algebras.

We obtain the following explicit formula:

Theorem 2 *("Tunnel formula")*

$$p_x(u) = \left(Id + \frac{1}{2}(\mathrm{ad}\, u) + \sum_{n\geq 1} (-1)^{n+1} \frac{B_n}{(2n)!} \cdot (\mathrm{ad}\, u)^{2n} \right) (\exp(-\mathrm{ad}\, u)x)_+$$

where B_n are the Bernoulli numbers.

Proof: We perform the operation $\frac{\partial}{\partial t}(\ \ )\cdot(\ \ )^{-1}\big|_{t=0}$ on both sides of (2):

$$x = \frac{\partial}{\partial t}\exp\left(u+t\cdot p_x(u)+O(t^2)\right)\cdot\exp\left(u+t\cdot p_x(u)+O(t^2)\right)^{-1}\Big|_{t=0} \qquad (4)$$

$$+\exp(u)\cdot\left(\frac{\partial}{\partial t}g_-\cdot g_-{}^{-1}\right)\Big|_{t=0}\cdot\exp(-u).$$

From Helgason [2] we take the formula

$$\frac{\partial}{\partial t}\exp(y)\exp(-y) = \sum_{n\geq 1}\frac{(\operatorname{ad} y)^{n-1}}{n!}\cdot\frac{\partial}{\partial t}y. \qquad (5)$$

Apply this and $\exp(-u)x\exp(u) = \exp(-\operatorname{ad} u)x$ to get

$$\exp(-\operatorname{ad} u)x = \exp(-\operatorname{ad} u)\left(\sum_{n\geq 1}\frac{(\operatorname{ad} u)^{n-1}}{n!}p_x(u)\right)+\left(\frac{\partial}{\partial t}g_-\cdot g_-{}^{-1}\right)\Big|_{t=0}. \qquad (6)$$

Now the first term on the right hand side is in $\mathbf{q}$, where as the second one is in $\mathbf{p}$. Thus comparing components yields:

$$\exp(-\operatorname{ad} u)\sum_{n\geq 1}\frac{(\operatorname{ad} u)^{n-1}}{n!}p_x(u) = (\exp(-\operatorname{ad} u)x)_+ \qquad (7)$$

Next, we use the power series expansion of the exponential map and compute:

$$\sum_{k\geq 1}\frac{(\operatorname{ad} u)^{k-1}}{k!}\sum_{n\geq 1}\frac{(\operatorname{ad} u)^{n-1}}{n!} = \sum_{\ell\geq 1}\left(\sum_{k=0}^{\ell-1}\frac{(-1)^k}{k!(\ell-k)!}\right)(\operatorname{ad} u)^{\ell-1}. \qquad (8)$$

The inner sum simplifies to

$$\frac{1}{\ell!}\sum_{k=0}^{\ell-1}(-1)^k\binom{\ell}{k} = \frac{1}{\ell!}\left((1-1)^\ell-(-1)^\ell\right) = -\frac{(-1)^\ell}{\ell!} \qquad (9)$$

Applying these simplifications to (7) gives

$$\sum_{n\geq 1}\frac{(\operatorname{ad} u)^{n-1}}{n!}p_x(u) = (\exp(-\operatorname{ad} u)x)_+ \ . \qquad (10)$$

Finally we note that $\sum_{n\geq 1}\frac{z^{n-1}}{n!}$ is the Taylor series expansion of the function $f(z) = \frac{e^z-1}{z}$ about 0, whereas the expansion of its reciprocal $g(z) = \frac{z}{e^z-1}$ is $1-\frac{z}{2}+\sum_{n\geq 1}(-1)^{n+1}\frac{B_n}{(2n)!}z^{2n}$, B_n the Bernoulli numbers. Since we are dealing with a finitely graded Lie algebra, the map $(\operatorname{ad} u)$ is nilpotent for any $u\in\mathbf{q}$, thus we may replace z by $(-\operatorname{ad} u)$ in the series and need not worry about convergence. This completes the proof. □

When deriving this formula the author thought of drilling a tunnel on the "flat" Lie algebra level and thus avoiding to "climb" on the group level.

The obvious application of the "tunnel formula" is

Corollary 3 *Every complex simple Lie algebra can be realized with polynomial vector fields of degree not greater than 4.*

Proof: Choose a 5-grading $\mathcal{L} = \mathcal{L}_{-2} \oplus \mathcal{L}_{-1} \oplus \mathcal{L}_0 \oplus \mathcal{L}_1 \oplus \mathcal{L}_2$ and define $\mathbf{p}$ and $\mathbf{q}$ as above. An immediate consequence of the grading is

$$(\operatorname{ad} u)^5 = 0 \tag{11}$$

for all $u \in \mathbf{q}$. Thus from the tunnel formula we see that p_x is a polynomial of degree ≤ 4 in u.

We need to check that the p_x are non-trivial. This can immediately be seen by computing the easiest ones: For $x \in \mathcal{L}_1 + \mathcal{L}_2$ one gets

$$p_x(u) = x + \frac{1}{2}[x, u]. \tag{12}$$

Since $\mathcal{L}$ was assumed to be simple the kernel of the homomorphism $\pi : x \mapsto p_x$ is zero, hence $\mathcal{L}$ is isomorphic to the Lie algebra of vector fields $\mathcal{P} := \{p_x : x \in \mathbf{q}\}$. □

In [1] it is shown that from the polynomial realization $\mathcal{P}$ one can obtain a relization $\mathcal{R}$ of $\mathcal{L}$ with homogeneous rational vector fields of degrees -2, -1, 0, 1 and 2.

3. Generalizations

For arbitrary Lie algebras, the map sending x to p_x will not be injective. We ask for necessary and sufficient conditions:

Lemma 4 *Let $x \in \mathcal{L}$.*

a) $p_x = 0$ if and only if $(\exp(\operatorname{ad} u)x)_+ = 0$ for all $u \in \mathbf{q}$.

b) If $p_x = 0$ then $x \in \mathbf{p}$.

c) The map $\pi : x \mapsto p_x$ is injective on $\mathcal{L}$ if and only if $\mathbf{p}$ contains no non-trivial ideal of $\mathcal{L}$.

Proof: a) The "if"-part is an immediate consequence of Thm. 2, the "only-if"-part follows directly from equation (7) in the proof of Thm. 2.

b) Set $u = 0$ in a).

c) Since π is a homomorphism, the kernel of π is an ideal which by b) is contained in $\mathbf{p}$. On the other hand, if $I \subset \mathbf{p}$ is an ideal of $\mathcal{L}$, then by the tunnel formula $p_x = 0$ for all $x \in I$. □

Thus we obtain the following generalization:

Theorem 5 *Let $\mathcal{L}$ be a Lie algebra satisfying the following conditions:*

- *There is a decomposition $\mathcal{L} = \mathbf{p} \oplus \mathbf{q}$, where $\mathbf{p}$ and $\mathbf{q}$ are subalgebras.*

- $\operatorname{ad} u$ *is nilpotent on* $\mathcal{L}$ *for every* $x \in \mathfrak{q}$.
- *There is no non-trivial ideal of* $\mathcal{L}$ *contained in* $\mathfrak{p}$.

Then $\mathcal{L}$ *is isomorphic to a subalgebra of* $\mathit{Pol}\,\mathfrak{q}$. □

The following class of Lie algebras attracts quite some interest: Call a Lie algebra $\mathcal{L}$ *root-graded*, if there is a decomposition of vector spaces

$$\mathcal{L} = \bigoplus_{\alpha \in \Delta \cup \{0\}} \mathcal{L}_\alpha,$$

where Δ is the root system of a simple finite-dimensional Lie algebra such that $[\mathcal{L}_\alpha, \mathcal{L}_\beta] = 0$ if $\alpha + \beta \notin \Delta$.

It is easy to see that root-graded Lie algebras satisfy the first two requirements of Theorem 5. Thus, Theorem 5 shows a way to realize also Lie algebras of possibly infinite dimension with polynomials.

References

1. H. Gradl. Realization of semisimple Lie algebras with polynomial and rational vector fields. *Comm. Algebra.*
2. S. Helgason. *Differential Geometry and Symmetric Spaces.* Academic Press, 1962.
3. I. L. Kantor. Models of exceptional Lie algebras. *Soviet Math. Dokl.*, 14:254–258, 1973.
4. M. Koecher. Imbedding of Jordan algebras into Lie algebras I. *Amer. J. Math.*, 89:787–816, 1967.
5. M. Koecher. Gruppen und Lie-Algebren von rationalen Funktionen. *Math. Z.*, 109:349–392, 1969.
6. S. Lie and G. Scheffers. *Vorlesungen über continuirliche Gruppen.* Teubner, Leipzig, 1893.
7. O. Loos. Jordan pairs. *Springer Lecture Notes in Mathematics*, 460, 1975.
8. K. Meyberg. Jordan-Tripelsysteme und die Koecher-Konstruktion von Lie-Algebren. *Math. Z.*, 115:58–78, 1970.

LES ALGÈBRES DE KAC-MOODY ET L'HOMOLOGIE DIÈDRALE

AZIZ HADDI
Faculté des Sciences de Tétouan
Département de Mathématiques
B.P. 2121
Tétouan- Maroc.

Abstract. Nous démontrons que l'espace des formes trilinéaires symétriques invariantes d'une algèbre de Kac-Moody est un invariant. En application, nous calculons le groupe d'homologie d'ordre trois des algèbres de Lie de la forme $A \underset{K}{\otimes} g$, où K est un corps de caractéristique nulle, A une K-algèbre commutative et g est une algèbre de Kac-Moody.

1- Dans tout le texte, K désigne un corps commutatif de caractéristique nulle. Les produits tensoriels et puissances extérieures sont relatifs à K et on note V_g l'espace des coinvariants d'un g-module V. Soit M une matrice de Cartan, indécomposable et symétrisable. On note $g(M)$ l'algèbre de Kac-Moody associée à M [2], $g'(M)$ la sous-algèbre dérivée.

Le résultat principal de cet article est le théorème suivant :

Théorème 1. Pour toute matrice de Cartan symétrisable et indécomposable M, la dimension de $\left(S^3 g'(M)\right)_{g(M)}$ est au plus égale à un et $\left(S^3 g'(M)\right)_{g(M)} = K$ si et seulement si M est de type A_l ou $A_l^{(1)}$, $l \geq 2$.

2- Rappel sur l'homologie cyclique et une application du théorème 1.

Si A est une K-algèbre associative, nous considérons $A^{\otimes n+1}$ comme un $Z/(n+1)$-module, où le générateur t de $Z/(n+1)$ agit par la formule :

$$t(a_0 \otimes a_1 \otimes \dots \otimes a_n) = (-1)^n a_n \otimes a_0 \otimes \dots \otimes a_{n-1}$$

On définit l'homologie cyclique de A, notée $HC_*(A)$ (voir [4] et [5]) comme celle du complexe $(C_*(A), b_*)$.
où

$$C_n(A) = A^{\otimes n+1}/Im(1-t)$$

et

$$b_n : C_n(A) \longrightarrow C_{n-1}(A)$$

S. González (ed.), Non-Associative Algebra and Its Applications, 176–178.

$$a_0 \otimes a_1 \otimes \ldots\ldots \otimes a_n \longrightarrow \sum_{i=0}^{n-1}(-1)^i a_0 \otimes \ldots \otimes a_i.a_{i+1} \otimes \ldots \otimes a_n + (-1)^n a_n.a_0 \otimes \ldots \otimes a_{n-1}$$

Le complexe $(C_*(A), b_*)$ est muni d'une involution u définie par :

$$u(a_0 \otimes a_1 \otimes \ldots\ldots \otimes a_n) = (-1)^{\frac{n(n+1)}{2}} a_0 \otimes a_n \otimes \ldots \otimes a_2 \otimes a_1$$

Considérons les deux sous complexes $(C_*(A)^+, b_*)$ et $(C_*(A)^-, b_*)$ définis par :

$$C_n(A)^+ = \{x \in C_n(A) \ : \ u(x) = x\}$$

et

$$C_n(A)^- = \{x \in C_n(A) \ : \ u(x) = -x\}$$

Les homologies correspondantes sont notées $HC_*(A)^+$ et $HC_*(A)^-$. On a la décomposition:

$$HC_*(A) = HC_*(A)^- \oplus HC_*(A)^+$$

$HC_*(A)^-$ est appelée homologie diédrale de l'algèbre A [5]. Si g est une algèbre de Lie arbitraire et A une K-algèbre commutative, associative et unitaire, on lui associe l'algèbre de Lie $A \otimes g$, où le crochet est défini par :

$$[a \otimes x, b \otimes y] = ab \otimes [x, y]$$

Le complexe de Koszul $(\Lambda^* (A \otimes g), d_*)$ (voir [3]) possède une structure naturelle de g-module. On note $\left((\Lambda^* (A \otimes g))_g, d_*\right)$ le complexe des coinvariants sous l'action de g.

Théorème 2. Pour toute matrice de Cartan symétrisable et indécomposable, on a:

$$\begin{aligned} H_3\left((\Lambda^* (A \otimes g(M)))_{g(M)}\right) &= HC_2(A)^- \oplus HC_2(A)^+ \otimes \left(S^3 g'(M)\right)_{g(M)} \\ &\oplus HC_1(A) \otimes (A \otimes g(M)/g'(M)) \oplus \Lambda^3 (A \otimes g(M)/g'(M)) \end{aligned}$$

Remarquons que si M est de type fini, on retrouve un résultat de J.L. Cathelineau [1]

En effet, dans ce cas, on a :

$H_3(A \otimes g(M)) =$

$$= H_3\left((\Lambda^* (A \otimes g(M)))_{g(M)}\right) = \begin{cases} HC_2(A) & \text{si } M \text{ est de type } A_l, l \geq 2 \\ HC_2(A)^- & \text{sinon} \end{cases}$$

(car $HC_2(A) = HC_2(A)^- \oplus HC_2(A)^+$)

Les théorèmes 1 et 2 sont démontrées dans [9].

References

1. J.-L.Cathelineau, Homologie de dgré trois d'algèbres de Lie simple déployées étendues à une algèbre commutative, l'enseignement Mathématiques. 33 (1987), 159-173.
2. V.Kac, Fourteen lectures on infinite dimensional Lie algebras Birkhauser, Boston 1983.
3. J.-L. Koszul, Homologie et cohomologie des algèbres de Lie, Bull. soc. math. France 78 (1950), 65-127.
4. J.-L. Loday, D.Quillen, cyclic homology and the lie algebra homology of matrices, comment. Math. Helvetici 59 (1984), 565-591.
5. J.-L. Loday, Homologies diédrale quaternionique adv, in Math 66 (1987), 119-148.
6. R.V. Moody, A new class of lie algebras, J. algebra 10 (1968), 211-230.
7. A Haddi, Détermination des extensions centrales des algèbres de Kac-Moody. C.R.A.S. 306 (1988), 691-694.
8. A Haddi, Homologie des algèbres de Lie étendues à une algèbre commutative, communications in algebra, 20 (4), (1992) 1145-1166.
9. A Haddi, Les algèbres de Kac-Moody et l'homologie diédrale, à paraître dans "Communications in algebra".

QUADRATIC DIFFERENTIAL EQUATIONS IN GRADED ALGEBRAS

NORA C. HOPKINS
Department of Mathematics and Computer Science
Indiana State University
Terre Haute, Indiana 47809, USA

Abstract. Gradings induced by algebra automorphisms are used to obtain qualitative results on solutions to quadratic differential equations occurring in the algebras.

Once upon a time, an algebraist attended a conference at the University of Northern Iowa where she met two mathematicians working on quadratic differential equations. Up to this point she had been very pure, but with sly smiles these two mathematicians enticed the algebraist into using her algebraic skills in the study of these differential equations, which include the Lorenz model of thermal convection, linear control systems with quadratic cost function, the Euler equation for the motion of a rotating rigid body with no external forces, and Lotka-Volterra predator-prey models [3], [4]. *Being somewhat simpleminded, she believed their promises of easily proven results and decided to use what she was most familiar with on the problem, namely gradings on algebras ....*

1. Preliminaries

Throughout $\mathfrak{a}$ will be a finite dimensional commutative nonassociative algebra over $\mathbf{R}$. $Z : \mathbf{R} \times \mathfrak{a} \longrightarrow \mathfrak{a}$ will be the solution to the quadratic differential equation

$$(*) \qquad \frac{dZ}{dt} = Z^2 \quad \text{with initial condition } Z(0,P) = P$$

where Z^2 is the square of Z in the algebra $\mathfrak{a}$. Normally the algebra multiplication is defined from the differential equation, rather than conversely, as can be seen in the following example.

Example 1.1: For the quadratic system

$$\begin{aligned} \frac{dx_1}{dt} &= x_1^2 + 2x_1x_2 \\ \frac{dx_2}{dt} &= x_2^2 - 4x_1x_2 \end{aligned}$$

$$\mathfrak{a} = \mathbb{R}^2 \text{ and for } Z = \begin{bmatrix} z_1 \\ z_2 \end{bmatrix} \ Z^2 = \begin{bmatrix} z_1^2 + 2z_1z_2 \\ z_2^2 - 4z_1z_2 \end{bmatrix}.$$

S. González (ed.), Non-Associative Algebra and Its Applications, 179–182.

The multiplication on $\mathfrak{a}$ is defined by

$$XY := \frac{1}{2}[(X+Y)^2 - X^2 - Y^2], \text{ i.e.}$$

$$\begin{bmatrix} x_1 \\ x_2 \end{bmatrix} \begin{bmatrix} y_1 \\ y_2 \end{bmatrix} = \begin{bmatrix} x_1y_1 + x_1y_2 + x_2y_1 \\ x_2y_2 - 2x_1y_2 - 2x_2y_1 \end{bmatrix}.$$

Such quadratic differential equations have been studied in [[1], 3-8]. If P is fixed we will denote $Z(t,P)$ by $Z(t)$. For $P_i \in \mathfrak{a}$, $i = 0,\dots,n, \mathfrak{a}(P_0,\dots,P_n)$ will be the subalgebra of $\mathfrak{a}$ generated by $P_0,\dots,P_n$. $\operatorname{Aut}\mathfrak{a} = \{A \in GL(\mathfrak{a}) | A(PQ) = A(P)\cdot A(Q)\ \forall P,Q \in \mathfrak{a}\}$ is the *automorphism group* of $\mathfrak{a}$. For $A \in \operatorname{Aut}\mathfrak{a}$ we will write $AZ(t,P)$ instead of $A(Z(t,P))$. The proofs of the following well known results can be found in [8].

Theorem 1.2: *Suppose $A \in Aut\ \mathfrak{a}$ and $s \in \mathbb{R}$.*

(i) $AZ(t,P) = Z(t, A(P))$.
(ii) $Z(t,sP) = sZ(st,P)$.
(iii) $Z(t,P) \in \mathfrak{a}(P) \quad \forall t$.
(iv) *If $\mathfrak{a}(P)$ is nilpotent, then $Z(t,P)$ is polynomial in t and exists for all $t \in \mathbb{R}$.*

We will be interested in both $\mathbb{Z}$-gradings and $\mathbb{Z}_2$-gradings on $\mathfrak{a}$, where $\mathfrak{a}$ is *$\mathbb{Z}$-graded* if there are subspaces $\mathfrak{a}_i$ for $i \in \mathbb{Z}$ with $\mathfrak{a} = \oplus_{i\in\mathbb{Z}} \mathfrak{a}_i$ such that $\mathfrak{a}_i\mathfrak{a}_j \subseteq \mathfrak{a}_{i+j}\ \forall i,j \in \mathbb{Z}$, and $\mathfrak{a}$ is *$\mathbb{Z}_2$-graded* if there are subspaces $\mathfrak{a}_0, \mathfrak{a}_1$ with $\mathfrak{a} = \mathfrak{a}_0 \oplus \mathfrak{a}_1$ such that $\mathfrak{a}_j\mathfrak{a}_j \subseteq \mathfrak{a}_{(i+j)(\text{mod } 2)}$ for $i,j = 0,1$. We will be particularly interested in gradings on $\mathfrak{a}(P)$ where P is an eigenvector for some $A \in \operatorname{Aut}\mathfrak{a}$.

Finally, $\operatorname{Der}\mathfrak{a} = \{D \in \operatorname{End}\mathfrak{a} | D(PQ) = D(P)\cdot Q + P\cdot D(Q) \quad \forall P,Q \in \mathfrak{a}\}$ is the *derivation algebra* of $\mathfrak{a}$.

The results in this paper are joint work with Michael Kinyon.

2. Eigenvalues Other Than ± 1

Theorem 2.1: *Suppose there is an $A \in Aut\,\mathfrak{a}$ with $A(P) = \alpha P$ where $\alpha \neq \pm 1$. Then $Z(t,P)$ is polynomial in t and hence exists for all time t.*

Proof: $\mathfrak{a}(P) = \oplus_{i>0}\mathfrak{a}_i(P)$ is a $\mathbb{Z}$-grading of $\mathfrak{a}(P)$ where $\mathfrak{a}_i(P) = \{Q \in \mathfrak{a}(P) | A(Q) = \alpha^i Q\}$. Since $\mathfrak{a}(P)$ is finite dimensional, this implies that $\mathfrak{a}(P)$ is nilpotent and the result follows from Theorem 1.2 (iv). □

Corollary 2.2: If the origin is stable and there is a $D \in \operatorname{Der}\mathfrak{a}$ with $D(P) = \alpha P$ where $\alpha \neq 0$, then $P^2 = 0$ so P is an equilibrium point.

Proof: $A \in \operatorname{Aut}\mathfrak{a}$ where $A = \exp(D)$ and $A(P) = e^\alpha P$ so $Z(t,P)$ is polynomial by Theorem 2.1. By Theorem 1.2 (ii) $Z(t,sP)$ is polynomial in t. Since the origin is stable by hypothesis, for s sufficiently small the coefficient of t in a MacClauren series fo $Z(t,sP)$ must be 0, i.e. $(sP)^2 = 0$ so $P^2 = 0$. □

3. Eigenvalue −1

Note that if $A \in \operatorname{Aut} \mathfrak{a}$ with $A(P) = -1$, then $\mathfrak{a}(P) = \mathfrak{a}_0(P) \oplus \mathfrak{a}_1(P)$ is a $\mathbb{Z}_2$-grading of $\mathfrak{a}(P)$ where $\mathfrak{a}_i(P) = \{Q \in \mathfrak{a}(P) | A(Q) = (-1)^i Q\}$ and $A^2 = id$ on $\mathfrak{a}(P)$. Conversely, if $\mathfrak{a} = \mathfrak{a}_0 \oplus \mathfrak{a}_1$ is a $\mathbb{Z}$-grading of $\mathfrak{a}$, then $A \in \operatorname{Aut} \mathfrak{a}$ with $A^2 = id$ where A is defined by $A(Z_0 + Z_1) := Z_0 - Z_1$ for $Z_i \in \mathfrak{a}_i$. **Lemma 3.1**: Suppose the maximum interval

of existence of $Z(t,P)$ is $(-\alpha,\omega)$ and there is an $A \in \operatorname{Aut} \mathfrak{a}$ with $A(P) = -P$.

(i) $AZ(t,P) = -Z(-t,P)$ for all $t \in (\alpha,\omega)$ for which $-t \in (\alpha,\omega)$.
(ii) $\alpha = \omega$.
(iii) If there is a $B \in \operatorname{Aut} \mathfrak{a}$ with $BZ(t_0,P) = -Z(t_0,P)$ for some $0 \neq t_0 \in (-\alpha,\omega)$, then $\alpha = \omega = \infty$, i.e. $Z(t,P)$ exists for all time t.

Proof: (i) If $t, -t \in (\alpha,\omega)$, then by Theorem 1.2 (i),(ii) $AZ(t,P) = Z(t, A(P)) = Z(t,-P) = -Z(-t,P)$.

(ii) Suppose not. Without loss of generality we can assume $\alpha < \omega$. Define $Y(t)$ on $(-\omega,\omega)$ by $Y(t) := \begin{cases} Z(t,P) & \text{for } t \in [0,\omega) \\ -AZ(-t,P) & \text{for } t \in (-\omega,0] \end{cases}$. It is easy to check that $\frac{dY}{dt} = Y^2 \quad \forall t \in (-\omega,\omega)$ and $Y(0) = P$. Hence $Y(t) \equiv Z(t,P)$ and $\alpha = \omega$.

(iii) Since the differential equation $(*)$ is autonomous, $BZ(t_0,P) = -Z(t_0,P)$ for $0 \neq t_0 \in (-\alpha,\omega)$ implies $(-\alpha,\omega)$ is symmetric about t_0 by (ii). Since $(-\alpha,\omega)$ is symmetric about 0 as well as t_0, $\alpha = \omega = \infty$.

□

Theorem 3.2: *Suppose there are $A, B \in \operatorname{Aut} \mathfrak{a}$ with $A(P) = -P$ and $BZ(t_0,P) = -Z(t_0,P)$ for some $t_0 \neq 0$. If $(AB)|_{\mathfrak{a}(P)}$ has finite order, then $Z(t,P)$ is periodic.*

Proof: By Lemma 3.1(i) $AZ(t,P) = -Z(-t,P)$ and $BZ(t,P) = -Z(-t + 2t_0,P)$ $\forall t$ so $ABZ(t,P) = Z(t - 2t_0,P)$ and by induction $(AB)^k Z(t,P) = Z(t - 2kt_0,P)$. If the order of $(AB)|_{\mathfrak{a}(P)}$ is n, this gives $Z(t,P) = Z(t - 2nt_0,P)$ so $Z(t,P)$ is periodic. □

Corollary 3.3: A trajectory of $(*)$ intersects the -1-eigenspace of $A \in \operatorname{Aut} \mathfrak{a}$ in at most two points.

Theorem 3.2 explains why the trajectories of solutions to the system

$$\frac{dx}{dt} = ax^2 + by^2$$
$$\frac{dy}{dt} = cxy$$

never cross the y-axis twice since it is the -1-eigenspace of $A \in \operatorname{Aut} \mathfrak{a}$ defined by $A\begin{bmatrix} x \\ y \end{bmatrix} = \begin{bmatrix} x \\ -y \end{bmatrix}$.

4. Some Other Results

If $\mathfrak{a} = \mathfrak{a}_0 \oplus \mathfrak{a}_1$ is a $\mathbf{Z}_2$-grading of $\mathfrak{a}$, then writing $Z = X + Y$ and $P = P_0 + P_1$ for $X, P_0 \in \mathfrak{a}_0, Y, P_1 \in \mathfrak{a}_1$ allows (*) to be rewritten as the system

$$(**) \qquad \frac{dX}{dt} = X^2 + Y^2 \quad \text{where } X(0,P) = P_0$$

$$\frac{dY}{dt} = 2XY \quad \text{where } Y(0,P) = P_1.$$

We mention without proof two more results. Let $B_i := \mathfrak{a}(P_0, P_1) \cap \mathfrak{a}_i$ for $i = 1, 2$.

Proposition 4.1: Suppose $UV = 0$ for all $U, V \in B_1$. Then (**) can be solved by first solving $\frac{dX}{dt} = X^2, X(0) = P_0$ on $\mathfrak{a}_0$ and then solving the linear differential equation $\frac{dY}{dt} = 2XY,\ Y(0) = P_1$ on $\mathfrak{a}_1$.

Proposition 4.2: (i) Suppose $UV = 0$ implies $U = 0$ or $V = 0 \quad \forall U, V \in B_1$ and $X(t,P)$ is periodic of period T. Then $Y(t,P)$ is periodic of period T or $2T$. Hence $Z(t,P)$ is periodic.

(ii) Suppose $UV = 0$ implies $U = 0$ or $Z = 0\ \forall\ U \in B_0,\ V \in B_1$ and $Y(t,P)$ is periodic of period S. Then $X(t,P)$ is periodic of period S. Hence $Z(t,P)$ is periodic.

... Thus the algebraist found her innocent faith in the power of gradings on algebras to be justified, and she continues to work on their effect on quadratic differential equations. Her co-worker, Michael Kinyon, found some interesting examples which he considers in his contribution to these proceedings [2].

MORAL: Gradings are good!

References

1. H. Gradl, K. Meyberg, and S. Walcher: 'R-algebras with an associative trace form', *Nova J. of Algebra and Geometry*, to appear.
2. M. Kinyon: 'Quadratic differential equations on graded structures', these proceedings.
3. M. Kinyon and A. Sagle: 'Quadratic dynamical systems', *International Symposium on Nonassociative Algebras and Related Topics, Hiroshima*, World Scientific, 1991.
4. M. Kinyon and A. Sagle: 'Quadratic dynamical systems and algebras', *J. Diff. Eq.*, to appear.
5. M. Koecher: 'Die Riccatishe Differentialgleichung und nicht-assoziative Algebren', 1977, *Abh. Math. Sem. Univ.* Hamburg **46**, pp. 129 - 141.
6. L. Markus: 'Quadratic differential equations and nonassociative algebras', 1960, *Ann. Math. Studies*, Princeton Univ. Press, **45**, pp. 185 - 213.
7. H. Röhrl, Algebras and differential equations, 1977, *Nagoya Math. J.*, **68**, pp. 59 - 122.
8. S. Walcher, 1991, *Algebras and Differential Equations*, Hadronic Press, Palm Harbor, FL.

PROJECTIVE DOUBLE LIE ALGEBRAS ON A LIE ALGEBRA

TOSHIHARU IKEDA
Department of Mathematics,
Kyushu Institute of Technology,
Tobata, Kitakyushu 804, Japan

Abstract. The notion of projective double Lie algebras on a Lie algebra $\mathfrak{g}$ has arisen in connection with certain local Lie loop structures on a Lie group with Lie algebra $\mathfrak{g}$. In this article, including semisimple Lie algebras, we are mainly concerned with certain group-graded Lie algebras which are not necessarily finite-dimensional, and we determine their projective double Lie algebras.

In a recent work [8] Kikkawa presented a way to determine the geodesic homogeneous local Lie loops which have projective relation with a given Lie group G, that is, their classification is reduced to the determination of a class of double Lie algebras on the Lie algebra of G (see [8, Theorem 3.3], [9, Theorem 7.3]). Succeedingly Sanami and Kikkawa [10] introduced the notion of projective double Lie algebras on a Lie algebra, and classified those on odd-dimensional real simple Lie algebras. However, there was a gap in their proof, and except this fact very little is known about this notion. Therefore it is desirable for us to know about projective double Lie algebras on some significant Lie algebras. In this article, including semisimple Lie algebras, we are mainly concerned with certain group-graded Lie algebras which are not necessarily finite-dimensional, and we determine their projective double Lie algebras.

Let $\mathfrak{g}$ be a Lie algebra over a field F with Lie bracket $[\ ,\ ]$ and $\mathfrak{h}$ another Lie algebra with Lie bracket $[\ ,\ ]_{\mathfrak{h}}$ on the same underlying vector space as $\mathfrak{g}$. If the relation

$$\mathrm{ad}_{\mathfrak{h}}(\mathfrak{h}) \subset \mathrm{Der}(\mathfrak{g})$$

holds, then $\mathfrak{h}$ is called a projective double Lie algebra on $\mathfrak{g}$, where $\mathrm{ad}_{\mathfrak{h}}$ denotes the adjoint representation of $\mathfrak{h}$ and $\mathrm{Der}(\mathfrak{g})$ the derivation algebra of $\mathfrak{g}$. The source of this notion is given in [8].

The centroid $\Gamma(\mathfrak{g})$ of $\mathfrak{g}$ is defined to be the set of all linear endomorphisms f such that f and $\mathrm{ad}_{\mathfrak{g}}(X)$ are commutative for all $X \in \mathfrak{g}$. For a linear endomorphism f of $\mathfrak{g}$ such that $f \in \Gamma(\mathfrak{g})$, we denote by $\mathfrak{g}_f$ the Lie algebra $\mathfrak{h}$ with the Lie multiplication

$$[X, Y]_{\mathfrak{h}} = f([X, Y]).$$

It is easy to see that $\mathfrak{g}_f$ is a projective double Lie algebra on $\mathfrak{g}$ for any $f \in \Gamma(\mathfrak{g})$. In particular, $\mathfrak{g}_p = \mathfrak{g}_{pI}$ is the most elementary example of a projective double Lie algebra on $\mathfrak{g}$, where I is the identity mapping of $\mathfrak{g}$ and $p \in F$. We shall call this elementary one scalar. A linear endomorphism J of a real Lie algebra $\mathfrak{g}$ is said to be

S. González (ed.), Non-Associative Algebra and Its Applications, 183–187.

a complex Lie structure if $J^2 = -I$ and $J \in \Gamma(\mathfrak{g})$. In this case, $\mathfrak{g}$ is even-dimensional if $\mathfrak{g}$ is finite-dimensional. It is well known that there are two distinct types of finite-dimensional real simple Lie algebras (cf. [6, Theorem 10.1 and 10.2], [2, §4.2]). One is a real form of a complex simple Lie algebra and the centroid $\Gamma = \boldsymbol{R}I$. The other has a complex Lie structure J and $\Gamma = \boldsymbol{R}I + \boldsymbol{R}J$.

A locally finite Lie algebra $\mathfrak{g}$ is called neoclassical semisimple if $\mathfrak{g}$ is decomposed as a direct sum of finite-dimensional simple ideals (see [1, Theorem13.4.2]). The following is one of the main results we would like to present.

Theorem 1. *Let $\mathfrak{g}$ be a neoclassical real semisimple Lie algebra with the simple components $\mathfrak{s}^\lambda$ $(\lambda \in \Lambda)$. Then any projective double Lie algebra on $\mathfrak{g}$ has the form $\bigoplus_{\lambda \in \Lambda} (\mathfrak{s}^\lambda)_{p_\lambda + q_\lambda J_\lambda}$ for some real numbers p_λ, q_λ and complex Lie structure J_λ of $\mathfrak{s}^\lambda$, where J_λ is null if $\mathfrak{s}^\lambda$ has no complex Lie structures.*

In this article we shall consider not necessarily finite-dimensional group-graded Lie algebras over an arbitrary field F. Let A be an additive torsion-free group and $\mathfrak{g} = \bigoplus_{a \in A} \mathfrak{g}^a$ an A-graded Lie algebra with a finite-dimensional abelian Cartan subalgebra $\mathfrak{c} = \mathfrak{g}^0$. We suppose that $\mathfrak{g}$ has finite-dimensional irreducible $\mathfrak{c}$-submodules $\mathfrak{a}_1, \cdots, \mathfrak{a}_n$ such that

(a) $\mathfrak{a}_i$ is contained in some $\mathfrak{g}^{e_i}$ $\quad (e_i \neq 0;\ \ i = 1, \cdots, n)$,

(b) $\mathfrak{g}$ is generated by $\mathfrak{c} + \sum_{i=1}^{n} \mathfrak{a}_i$,

(c) $\langle \mathfrak{a}_i^{\mathfrak{g}} \rangle$ is a faithful $\mathfrak{c}$-submodule of $\mathfrak{g}$ $\quad (i = 1, \cdots, n)$,

where $\langle \mathfrak{a}_i^{\mathfrak{g}} \rangle$ is the smallest ideal of $\mathfrak{g}$ containing $\mathfrak{a}_i$. We shall call these irreducible $\mathfrak{c}$-submodules $\mathfrak{a}_1, \cdots, \mathfrak{a}_n$ connected generators relative to $\mathfrak{c}$. We need the following lemma to see the centroid of $\mathfrak{g}$.

Lemma 1. *Let $\mathfrak{g} = \bigoplus_{a \in A} \mathfrak{g}^a$ be an A-graded Lie algebra with finite-dimensional abelian Cartan subalgebra $\mathfrak{c} = \mathfrak{g}^0$ and connected generators relative to $\mathfrak{c}$. Suppose that σ is a linear endomorphism of $\mathfrak{g}$ satisfying*

$$[\sigma(X), Y] = [X, \sigma(Y)]$$

for all $X, Y \in \mathfrak{g}$. Then σ is either a linear isomorphism or null.

We outline the proof. Let $\mathfrak{a}_1, \cdots, \mathfrak{a}_n$ be connected generators relative to $\mathfrak{c}$. Making an extension of the base field if need be, we can see that $\sigma(\mathfrak{c}) \subset \mathfrak{c}$ and $\sigma([\mathfrak{a}_{j_1}, \cdots, \mathfrak{a}_{j_r}]) \subset [\mathfrak{a}_{j_1}, \cdots, \mathfrak{a}_{j_r}] + \mathfrak{c}$ for any $r \in \boldsymbol{N}$ and $1 \leq j_1, \cdots, j_r \leq n$ by a standard root space calculation. If $\sigma(\mathfrak{c}) = \mathfrak{c}$, then it is easy to see that σ is bijective. So we assume $\sigma(\mathfrak{c}) \not\subset \mathfrak{c}$. Then $\sigma(T) = 0$ for some $T \in \mathfrak{c} \setminus \{0\}$. By the faithfulness of $\langle \mathfrak{a}_i^{\mathfrak{g}} \rangle$ and the irreducibility of $\mathfrak{a}_i$, it can be proved $[\mathfrak{a}_i, T] \neq 0$ and $\sigma(\mathfrak{a}_i) = 0$ for some i. Next it may be shown by induction that $\sigma([\mathfrak{a}_i, \mathfrak{a}_{j_1}, \cdots, \mathfrak{a}_{j_r}]) = 0$ for any $r \in \boldsymbol{N}$ and $1 \leq j_1, \cdots, j_r \leq n$. It follows that $[\langle \mathfrak{a}_i^{\mathfrak{g}} \rangle, \sigma(\mathfrak{c})] = [\sigma(\langle \mathfrak{a}_i^{\mathfrak{g}} \rangle), \mathfrak{c}] = 0$, and so $\sigma(\mathfrak{c}) = 0$ because of the faithfulness of $\langle \mathfrak{a}_i^{\mathfrak{g}} \rangle$. Now for any j we have $[\sigma(\mathfrak{a}_j), \mathfrak{c}] = [\mathfrak{a}_j, \sigma(\mathfrak{c})] = 0$, whence $\sigma(\mathfrak{a}_j) \subset \mathfrak{c}$. Then we can obtain $\sigma(\langle \mathfrak{a}_j^{\mathfrak{g}} \rangle) = 0$ in the same way as $\mathfrak{a}_i$. Consequently we have $\sigma = 0$ since $\mathfrak{g} = \mathfrak{c} + \sum_{j=1}^{n} \langle \mathfrak{a}_j^{\mathfrak{g}} \rangle$.

We note that if the linear endomorphism σ in Lemma 1 has degree zero, that is, $\sigma(\mathfrak{g}^a) \subset \mathfrak{g}^a$ for any $a \in A$, then $\sigma \in \Gamma(\mathfrak{g})$. We can also see the following lemma by using Lemma 1.

Lemma 2. *Let $\mathfrak{g}$ be the same A-graded Lie algebra as in Lemma 1. Then the centroid $\Gamma(\mathfrak{g})$ of $\mathfrak{g}$ is a finite extension field of $F = FI$. In particular, if $\mathfrak{c}$ is a split Cartan subalgebra of $\mathfrak{g}$, then $\Gamma(\mathfrak{g}) = FI$.*

A centerless Lie algebra $\mathfrak{g}$ is called complete if any derivation of $\mathfrak{g}$ is inner. The following is the main result of this article.

Theorem 2. *Let $\mathfrak{g} = \bigoplus_{a \in A} \mathfrak{g}^a$ be a complete A-graded Lie algebra with a finite-dimensional abelian Cartan subalgebra $\mathfrak{c} = \mathfrak{g}^0$ and connected generators relative to $\mathfrak{c}$. Then any projective double Lie algebra on $\mathfrak{g}$ has the form $\mathfrak{g}_\sigma$ for some $\sigma \in \Gamma(\mathfrak{g})$.*
In particular, if $\mathfrak{c}$ is a split Cartan subalgebra of $\mathfrak{g}$, then σ is scalar.

Proof. Let $\mathfrak{h}$ be a non-abelian projective double Lie algebra on $\mathfrak{g}$. By the completeness of $\mathfrak{g}$, for any $X \in \mathfrak{h}$ there exists a unique element $\sigma(X) \in \mathfrak{g}$ such that $\mathrm{ad}_{\mathfrak{h}}(X) = \mathrm{ad}_{\mathfrak{g}}(\sigma(X))$. Then σ gives a Lie homomorphism from $\mathfrak{h}$ to $\mathfrak{g}$. In fact, we have

$$\begin{aligned}
\mathrm{ad}_{\mathfrak{g}}(\sigma([X,Y]_{\mathfrak{h}})) &= \mathrm{ad}_{\mathfrak{h}}([X,Y]_{\mathfrak{h}}) \\
&= \mathrm{ad}_{\mathfrak{h}}(X)\mathrm{ad}_{\mathfrak{h}}(Y) - \mathrm{ad}_{\mathfrak{h}}(Y)\mathrm{ad}_{\mathfrak{h}}(X) \\
&= \mathrm{ad}_{\mathfrak{g}}(\sigma(X))\mathrm{ad}_{\mathfrak{g}}(\sigma(Y)) - \mathrm{ad}_{\mathfrak{g}}(\sigma(Y))\mathrm{ad}_{\mathfrak{g}}(\sigma(X)) \\
&= \mathrm{ad}_{\mathfrak{g}}([\sigma(X), \sigma(Y)])
\end{aligned}$$

for all $X, Y \in \mathfrak{h}$. Since $\mathfrak{g}$ is complete, we have $\sigma([X,Y]_{\mathfrak{h}}) = [\sigma(X), \sigma(Y)]$ for all $X, Y \in \mathfrak{h}$.

Now for any $X, Y \in \mathfrak{g}$, we have $[\sigma(X), Y] = [X, \sigma(Y)]$. From Lemma 1, it follows that the algebras $\mathfrak{h}$ and $\mathfrak{g}$ are isomorphic. On the other hand, since $\sigma([X,Y]_{\mathfrak{h}}) = [\sigma(X), \sigma(Y)] = [\sigma(X), Y]_{\mathfrak{h}}$ for any $X, Y \in \mathfrak{h}$, we have $\sigma \in \Gamma(\mathfrak{h})$. It is easily seen that $\Gamma(\mathfrak{h}) \subset \Gamma(\mathfrak{g})$. Thus we have $\mathfrak{h} = \mathfrak{g}_\sigma$. Moreover if $\mathfrak{c}$ is splitting, then σ is scalar from Lemma 2.

Corollary. *Let $\mathfrak{h}$ be a projective double Lie algebra on a finite-dimensional real simple Lie algebra $\mathfrak{g}$.*

(1) *If $\mathfrak{g}$ has no complex Lie structures, then $\mathfrak{h} = \mathfrak{g}_p$ for some real number p.*

(2) *If $\mathfrak{g}$ has a complex Lie structure J, then $\mathfrak{h} = \mathfrak{g}_{p+qJ}$ for some real numbers p and q.*

Evidently an odd-dimensional real simple Lie algebra $\mathfrak{g}$ has no complex Lie structures. Therefore any projective double Lie algebra on $\mathfrak{g}$ is of the form $\mathfrak{g}_p$ for some $p \in \boldsymbol{R}$, which was shown in [10].

Proof of Theorem 1. Let $\mathfrak{h}$ be a projective double Lie algebra on $\mathfrak{g}$. As in the proof of Theorem 2, there exists a Lie homomorphism $\sigma : \mathfrak{h} \longrightarrow \mathfrak{g}$ such that $\mathrm{ad}_{\mathfrak{h}}(X) = \mathrm{ad}_{\mathfrak{g}}(\sigma(X))$ for all $X \in \mathfrak{h}$. It is known that $\mathrm{Der}(\mathfrak{g})$ is isomorphic to the Cartesian sum $\prod_{\lambda \in \Lambda} \mathrm{ad}_{\mathfrak{g}}(\mathfrak{s}^\lambda)$ (see [1, Proposition 13.4.5]). Each component $\mathfrak{s}^\lambda$ is a characteristic ideal of $\mathfrak{g}$, and therefore $\mathfrak{s}^\lambda \lhd \mathfrak{h}$ since $\mathrm{ad}_{\mathfrak{h}}(\mathfrak{h}) \subset \mathrm{Der}(\mathfrak{g})$. Then for any $X_\lambda \in \mathfrak{s}^\lambda$ and $X_\mu \in \mathfrak{s}^\mu$ such that $\lambda \neq \mu$ $(\lambda, \mu \in \Lambda)$, we have $[\sigma(X_\lambda), X_\mu] = [X_\lambda, X_\mu]_{\mathfrak{h}} = 0$. It follows that $\sigma(\mathfrak{s}^\lambda) \subset \mathfrak{s}^\lambda$ since $\bigcap_{\mu \neq \lambda} C_{\mathfrak{g}}(\mathfrak{s}^\mu) = \mathfrak{s}^\lambda$, where $C_{\mathfrak{g}}(\mathfrak{a})$ is the centralizer of $\mathfrak{a}$ in $\mathfrak{g}$ for $\mathfrak{a} \leq \mathfrak{g}$. Then because the algebra $\mathfrak{s}^\lambda$ with Lie bracket $[\ ,\ ]_{\mathfrak{h}}$ is a projective double Lie algebra of the simple ideal $\mathfrak{s}^\lambda$ of $\mathfrak{g}$, using Theorem 2 we can deduce that $\sigma|_{\mathfrak{s}^\lambda} \in \Gamma(\mathfrak{s}^\lambda) = \boldsymbol{R} + \boldsymbol{R} J_\lambda$, where J_λ is a complex Lie structure of $\mathfrak{s}^\lambda$ and is null if $\mathfrak{s}^\lambda$ has no complex Lie structures. This completes the proof.

We finally recall two kinds of semisimple Lie algebras. The first one is a Kac-Moody Lie algebra with non-singular generalized Cartan matrix. Its standard Borel subalgebra is similar to the classical one. Theorem 1 holds for their subalgebras containing a standard Borel subalgebra. The second one is a generalized Witt algebra $W = W(A, I_n)$ with null center. If $A \subset F^n$ then W has a basis $\{w(a,i) \mid a = (a_i) \in A,\ 1 \leq i \leq n\}$ and Lie multiplication $[w(a,i), w(b,j)] = a_j w(a+b,i) - b_i w(a+b,j)$. If $e_1, \cdots, e_m$ are free generators of A then $\mathfrak{b} = \langle w(a,i) \mid a \in \sum_{j=1}^{m} \boldsymbol{Z}_+ e_j,\ 1 \leq i \leq n \rangle$ is a standard Borel subalgebra of W, where $\boldsymbol{Z}_+$ is the set of non-negative integers. If $\mathfrak{g}$ is a subalgebra of W containing $\mathfrak{b}$, then $\mathfrak{g}$ is not necessarily complete (cf. [4], [5], [7]). But by a similar argument to the proof of Theorem 2, we can see that any projective double Lie algebra on $\mathfrak{g}$ is scalar.

Acknowledgements

I would like to express my thanks to Professor Santos González and organizers of the Conference for their hospitality. I would also like to thank Dr. Sanami for his valuable comments.

References

1. R.K. Amayo and I. Stewart, *Infinite-dimensional Lie Algebras*, Noordhoff, Leyden, 1974.
2. M. Goto and F. D. Grosshans, *Semisimple Lie Algebras*, Marcel Dekker, New York, 1978.
3. J. E. Humphreys, *Introduction to Lie Algebra and Representation Theory*, Springer-Verlag, New York, 1972.
4. T. Ikeda, *Derivations and central extensions of a generalized Witt algebra*, Nonassociative algebras and related topics (Hiroshima, 1990), 47 – 57, World Sci. Publishing, River Edge, NJ, 1991.

5. T. Ikeda and N. Kawamoto, *On the derivations of generalized Witt algebras over a field of characteristic zero*, Hiroshima Math. J. **20** (1990), 47 - 55.
6. N. Jacobson, *Lie Algebras*, Interscience, New York, 1962.
7. N. Kawamoto, *Generalizations of Witt algebras over a field of characteristic zero*, Hiroshima Math. J. **16** (1986), 417 - 426.
8. M. Kikkawa, *Projectivity of homogeneous left loops on Lie groups II*, Mem. Fac. Sci. Shimane Univ. **24** (1990), 1 - 16.
9. M. Kikkawa, *Projectivity of homogeneous left loops*, Nonassociative algebras and related topics (Hiroshima, 1990), 77 - 99, World Sci. Publishing, River Edge, NJ, 1991.
10. M. Sanami and M. Kikkawa, *A class of double Lie algebras on simple Lie algebras and projectivity of simple Lie groups*, Mem. Fac. Sci. Shimane Univ. **25** (1991), 39 - 44.

ON DERIVATION ALGEBRAS OF GROUP ALGEBRAS

T. IKEDA
Department of Mathematics, Kyushu Institute of Technology
Kitakyushu 804, Japan

and

N. KAWAMOTO
Department of Mathematics, Maritime Safety Academy
Kure 737, Japan

Abstract. It is shown that if G is a locally finite group and k is a field of characteristic 0, then the derivations of kG arelocally inner but not necessarily inner.

Key words: derivation, locally inner, group algebra

1. Introduction

In this paper we consider derivations of group algebras over a field of characteristic 0. Their Lie algebraic structures are also considered for several groups.

We considered generalized Witt algebras in [3]. Recently they are recognized in a wide perspective of simple Lie algebras by Osborn [5].There is another generalization. Generalized Witt algebras are considered as derivation algebras of finitely generated torsion-free abelian groups [3, p. 424]. It seems natural for us to observe the other groups, and in this paper we consider finite groups and locally finite groups (cf. [4]). We note here that we used computer algebra system CAYLEY in preparation of this paper.

Assume that the ground field k is of characteristic 0 throughout this paper. Let kG be the group algebra of the group G, and $[kG]$ be the Lie algebra obtained from kG by defining the Lie multiplication $[x, y] = xy - yx$ for $x, y \in kG$.

Let ad$x$ be the inner derivation of an associative algebra A defined by $(\text{ad}x)(y) = xy - yx$ for $x, y \in A$. A derivation δ of A is called a locally inner derivation (cf. [2]) if for any finite subset F of A there exists an inner derivation ad$x$ (depending on δ and F) such that

$$\delta|_F = \text{ad}x|_F.$$

We denote by Inn(A) the Lie algebra of the inner derivations of A, and by Lin(A) the Lie algebra of the locally inner derivations of A.

2. Locally Finite Groups

We recall that a group G is locally finite if for any finite subset F of G there exists a finite subgroup H of G such that $F \subseteq H$.

S. González (ed.), Non-Associative Algebra and Its Applications, 188–192.

Theorem 2.1. *Let G be a locally finite group. Then any derivation of kG is a locally inner derivation.*

Proof. Let δ be a derivation of kG and F be a finite subset of kG. Then there exists a finite subgroup H of G such that $F \subseteq kH$. Let

$$x = \frac{1}{|H|}\sum_{a\in H} a^{-1}\delta(a).$$

Then for any $y \in F$ we have

$$\begin{aligned} xy - yx &= \frac{1}{|H|}\left\{\sum_{a\in H} a^{-1}\delta(a)y - \sum_{a\in H} ya^{-1}\delta(a)\right\} \\ &= \frac{1}{|H|}\left\{\sum_{a\in H} a^{-1}\delta(ay) - \sum_{a\in H} a^{-1}a\delta(y) - \sum_{a\in H} ya^{-1}a\delta(a)\right\} \\ &= \frac{1}{|H|}\left\{\sum_{b\in H} yb^{-1}\delta(b) - |H|\delta(y) - \sum_{a\in H} ya^{-1}a\delta(a)\right\} \\ &= -\delta(y). \end{aligned}$$

Therefore

$$\delta|_F = -\mathrm{ad}x|_F.$$

As a corollary we have a well-known result (e.g. [1, p. 490]).

Corollary 2.2. *Let G be a finite group. Then the derivations of kG are inner.*

For a locally finite group G, derivations of kG are not necessarily inner. Moreover we can show the following

Theorem 2.3. *Let G be a countable locally finite group with ascending series $\{G_n\}_{n\in \mathbf{N}}$ of finite groups. Suppose that the center $\zeta(G_n)$ is trivial and the centralizer $C_G(G_n)$ is non-trivial for any $n \in \mathbf{N}$. Then* $\dim \mathrm{Der}(kG) = 2^{\aleph_0}$ *and, in particular,* $\mathrm{Inn}(kG) < \mathrm{Der}(kG)$.

Proof. Choose any $z_1 \in C_G(G_1)$ such that $z_1 \neq e$ and let $n(1) = \min\{n | z_1 \in G_n\}$. Next let $e \neq z_2 \in C_G(G_{n(1)})$ and $n(2) = \min\{n|z_2 \in G_n\}$. In this way we can choose a sequence $\{z_i\}_{i\in\mathbf{N}}$ of elements of G and a increasing sequence $\{n(i)\}_{i\in\mathbf{N}}$ of integers such that $e \neq z_i \in G_{n(i)}$.

Now for any $\alpha = (\alpha_n) \in k^{\mathbf{N}}$ we define a derivation δ_α as follows. If $x \in kG_n$ then $n \leq n(m)$ for some $m \in \mathbf{N}$. We define $\delta_\alpha(x) = \sum_{i=1}^{m} \alpha_i[x, z_i]$. It is easy to see that δ_α is well-defined and is a derivation of kG.

The mapping $\alpha \longmapsto \delta_\alpha$ is injective. In fact, suppose that $\delta_\alpha = \delta_\beta$ for $\alpha = (\alpha_n)$, $\beta = (\beta_n) \in k^{\mathbf{N}}$. For any $x \in G_{n(1)}$ we have $\alpha_1[x, z_1] = \beta_1[x, z_1]$. If $\alpha_1 \neq \beta_1$ then $z_1 \in \zeta(G_{n(1)})$. This contradicts the fact $\zeta(G_{n(1)}) = \{e\}$, whence $\alpha_1 = \beta_1$. In the same way we can see that $\alpha_i = \beta_i$ for any $i \in \mathbf{N}$. Thus we have $\alpha = \beta$ and therefore

$$2^{\aleph_0} = \dim_k k^{\mathbf{N}} \leq \dim \mathrm{Der}(kG) \leq \dim \mathrm{End}_k(kG) = 2^{\aleph_0}.$$

On the other hand it is clear that $\dim \mathrm{Inn}(kG) \leq \aleph_0$ since $\dim(kG) = \aleph_0$.

For example, let $G = S_\infty = \bigcup_{n\geq 1} S_n$, where S_n are symmetric groups of degree n with natural inclusions $S_1 \subset S_2 \subset S_3 \subset \cdots$. Then S_∞ is a locally finite group and $kS_\infty = \bigcup_{n\geq 1} kS_n$. We define a linear map $\delta : kS_\infty \longrightarrow kS_\infty$ by $\delta(x) = xy_n - y_n x$ for any $x \in kS_n$, where $y_n = (1\ 2) + (2\ 3) + \cdots + (n\ n+1)$. Then it is easy to see that δ is well-defined and is a derivation of kS_∞. We claim that δ is not an inner derivation of kG. Assume for the contrary that δ is inner and $\delta = \mathrm{ad}\, y$ for some $y \in S_\infty$. Then $y \in S_n$ for some n. Let $x = (n+1\ n+2)$. Then clearly $xy - yx = 0$. On the other hand we have

$$\begin{aligned} \delta(x) &= xy_{n+2} - y_{n+2}x \\ &= (n+1\ n+2)\{(n\ n+1) + (n+1\ n+2) + (n+2\ n+3)\} \\ &\quad - \{(n\ n+1) + (n+1\ n+2) + (n+2\ n+3)\}(n+1\ n+2) \\ &= (n\ n+1\ n+2) - (n\ n+2\ n+1) \\ &\quad + (n+1\ n+3\ n+2) - (n+1\ n+2\ n+3) \\ &\neq 0, \end{aligned}$$

which is a contradiction.

Remark 2.4. In general it holds that $\mathrm{Inn}(A)$ and $\mathrm{Lin}(A)$ are ideals of the Lie algebra $\mathrm{Der}(A)$ for an associative algebra A (cf. [2, p. 495]). If G is a locally finite group, then clearly kG is a locally finite algebra and $\mathrm{Inn}(kG)$ is a locally finite Lie algebra.

3. Finite Groups

Let G be a finite group and $k = \mathbf{C}$ in this section. Then the Lie algebraic structure of $\mathrm{Der}(\mathbf{C}G)$ is determined by the character table of G. This seems not mentioned in literature.

Since G is finite, the derivations of $\mathbf{C}G$ are inner. Hence we have

$$\begin{aligned} \dim \mathrm{Der}(\mathbf{C}G) &= |G| - \dim(\text{center of } \mathbf{C}G) \\ &= |G| - \#(\text{conjugacy classes of } G). \end{aligned}$$

Since $\mathbf{C}G$ is semisimple, we have

$$\mathbf{C}G = I_1 \oplus \cdots \oplus I_r \qquad (r \geq 1),$$

where I_i are simple ideals of $\mathbf{C}G$ and $I_i \simeq M_{n_i}(\mathbf{C})$ for some n_i $(i = 1, \cdots, r)$. G has r irreducible representations and their dimensions are $n_1, \cdots, n_r$.

Since a derivation δ of $\mathbf{C}G$ is inner, we have $\delta = \mathrm{ad}x$ for some $x \in \mathbf{C}G$, and hence

$$\begin{aligned}\mathrm{Der}\mathbf{C}G &= \mathrm{Der}(I_1) \oplus \cdots \oplus \mathrm{Der}(I_r) \\ &= \mathrm{ad}M_{n_1}(\mathbf{C}) \oplus \cdots \oplus \mathrm{ad}M_{n_r}(\mathbf{C}) \\ &\simeq sl_{n_1}(\mathbf{C}) \oplus \cdots \oplus sl_{n_r}(\mathbf{C}).\end{aligned}$$

Let χ be the character of a representation $\rho : G \longrightarrow GL(V)$. Then $\dim V = \chi(1)$ and hence $n_1, \cdots, n_r$ can be obtained from the character table of G.

We give some examples for groups of small orders. It is clear if G is a finite abelian group then $\mathrm{Der}(\mathbf{C}G) = 0$. In the following S_n denotes the symmetric group of degree n, and A_n is the alternating group of degree n. Q is the quaternion group, and D_n is a dihedral group of order $2n$.

Example 3.1

$$\begin{aligned}\mathrm{Der}\mathbf{C}S_3 &= sl_2(\mathbf{C}). \\ \mathrm{Der}\mathbf{C}D_4 &= sl_2(\mathbf{C}). \\ \mathrm{Der}\mathbf{C}Q &= sl_2(\mathbf{C}). \\ \mathrm{Der}\mathbf{C}D_5 &= sl_2(\mathbf{C}) \oplus sl_2(\mathbf{C}). \\ \mathrm{Der}\mathbf{C}D_6 &= sl_2(\mathbf{C}) \oplus sl_2(\mathbf{C}). \\ \mathrm{Der}\mathbf{C}S_4 &= sl_3(\mathbf{C}) \oplus sl_2(\mathbf{C}) \oplus sl_3(\mathbf{C}). \\ \mathrm{Der}\mathbf{C}A_4 &= sl_3(\mathbf{C}). \\ \mathrm{Der}\mathbf{C}S_5 &= sl_4(\mathbf{C}) \oplus sl_5(\mathbf{C}) \oplus sl_6(\mathbf{C}) \oplus sl_5(\mathbf{C}) \oplus sl_4(\mathbf{C}). \\ \mathrm{Der}\mathbf{C}A_5 &= sl_4(\mathbf{C}) \oplus sl_5(\mathbf{C}) \oplus sl_3(\mathbf{C}) \oplus sl_3(\mathbf{C}). \\ &\vdots\end{aligned}$$

Example 3.2 Let $G = \prod_{j=1}^{N} G_j$ be the direct product of finite groups $G_1, \cdots, G_N$, and $n_1^{(j)}, n_2^{(j)}, \cdots, n_{m_j}^{(j)}$ the degrees of the irreducible representations of G_j ($j = 1, 2, \cdots, N$). We consider sequences $(n_{i_j}^{(j)})_{j=1}^{N}$ such that $1 \leq i_j \leq m_j$, and for any $r \geq 2$ we put

$$d_r = \#\{\ (n_{i_j}^{(j)})_{j=1}^{N} \mid \prod_{j=1}^{N} n_{i_j}^{(j)} = r\}.$$

Then $\mathrm{Der}(\mathbf{C}G)$ can be expressed by the following direct sum.

$$\mathrm{Der}(\mathbf{C}G) \simeq \bigoplus_{r \geq 2} \underbrace{(sl_r(\mathbf{C}) \oplus \cdots \oplus sl_r(\mathbf{C}))}_{d_r}.$$

Acknowledgements

The authors would like to express their thanks to Professor Santos González and organizers of the Conference for their hospitality.

References

1. Curtis, C.W. and Reiner, I.: 'Representation Theory of Finite Groups and Associative Algebras', Interscience, New York, 1962.
2. Ikeda, T.: 'Locally inner derivations of ideally finite Lie algebras', *Hiroshima Math. J.* **17**(1987), 495-503.
3. Kawamoto, N.: 'Generalizations of Witt algebras over a field of characteristic zero', *Hiroshima Math. J.* **16** (1986), 417-426.
4. Kegel, O.H. and Wehrfritz, B.A.F.: 'Locally Finite Groups', North-Holland, Amsterdam, 1973.
5. Osborn, J.M.: 'Examples and conjectures relating to Lie algebras', Preprint, 1993.

J-DIVISEURS TOPOLOGIQUES DE ZÉRO DANS UNE ALGÈBRE DE JORDAN N.C. NORMÉE *

EL-AMIN KAIDI[†] and ANTONIO SÁNCHEZ SÁNCHEZ.
Departamento de Algebra Geometría y Topología
Universidad de Málaga.
Apartado 59. 29080 Málaga. Spain.

Abstract. Let A be a normed non commutative Jordan K-algebra ($K = \mathcal{R}$ or $\mathcal{C}$). An element $a \in A$ is said to be a J-topological divisor of zero in A, is there is a sequence $\{x_n\}$ in A, $\|x_n\| = 1$ such that $U_a(x_n) \to 0$ ($U_a(x) = axa + (xa)a - xa^2$). In this work we prove: if D is a normed, Jordan non commutative K-algebra $\neq 0$, with no nonzero J-topological divisor of zero, then D is a division algebra, which is isomorphic to $\mathcal{C}$ when D is complex, and a flexible quadratic algebra when D is real.

1. Introduction

Dans toute la suite K designera le corps $\mathcal{R}$, des nombres réels ou le corps $\mathcal{C}$, des nombres complexes. Une K-algèbre A est dite normée si le K-espace vectoriel A est muni d'une norme $\|\cdot\|$ verifiant $\|xy\| \leq \|x\| \, \|y\|$ pour tout $x, y \in A$. Si de plus, cette norme est complète on dira que A est de Banach. Soit A une K-algèbre normée et $a \in A$, a est dit un diviseur topologique de zéro (en abrégé d.t.z.) dans A, s'il existe une suite $\{x_n\} \subset A$ telle que $\|x_n\| = 1$ et $\{x_n a\} \to 0$ ou $\{a x_n\} \to 0$. (Evidemment tout diviseur de zéro est un d.t.z.). Cette notion a été introduite par Shilov, en 1940, dans le cas associatif unitaire [16]; Kaplanski a etudié le cas associatif non unitaire [9]. En [7] le premier auteur a traité le cas normé complèt unitaire non associatif général. Ultérieurment et indépendamment El-Mallah et Micali ont considéré le cas des algèbres réelles normées sans diviseurs topologiques de zéro [4]. Récemment [2] Cabrera et Rodríguez ont donné une nouvelle demonstration du resultat suivant, dû a Kaplansky:

Théorème 1 ([9]) *Une $\mathcal{R}$-algèbre normeé A, associative non nulle sans diviseurs topologiques de zéro non nules est isomorphe á $\mathcal{R}, \mathcal{C}$ ou $\mathcal{H}$ (le corps des quaternions réels).*

Si dans le théoreme de Kaplansky l'on suppose A alternative, il faut ajouter a la liste $\mathcal{O}$, l'algèbre de division des octonions réels, et s'i l'on suppose A de Jordan on obtiént $\mathcal{R}$ et $\mathcal{C}$ [2,4,7,8]. Le cas Jordan non commutatif est plus delicat, on a seulement que l'algèbre A est quadratique ($\dim \mathcal{R}[a] \leq 2, \forall a \in A$ où $\mathcal{R}[a]$ note la sous algèbre engendreé par a et l'unité), de division *linéaire* (L_a et R_a inversibles dans

* Cette recherche a ètè partiellement subventionné par 'Plan andaluz para la consolidación de grupos de investigación y desarrollo tecnológico'.

† Adresse habituelle: Dèpartement de Mathématiques. Université Mohammed V. Faculté des Sciences B.P. 1014 Rabat-Morocco

S. González (ed.), Non-Associative Algebra and Its Applications, 193–197.

$L(A)$, l'algèbre des opérateurs linéaires de A $\forall a \in A$, $a \neq 0$) [6]. C'est un probléme ouvert, apparemment difficile, de savoir si les algèbres de Jordan non commutatives réelles, quadratiques, de division linéaire sont de dimension finie. En [3] Cuenca a donné des exemples des algèbres normées complètes de division *linéaire á gauche* (L_a inversible pour tout $a \in A$, $a \neq 0$) et Rodríguez les a décrit complétement en [14]. L'etude de la notion de d.t.z. précédente est liée, d'une maniére essentielle, á la notion d'inverse *linéaire* dans A et les propriétés du spectre qui á elle sont associées. On rappelle que si A est une C-algèbre associative unitaire

$$Sp(a, A) = \{\lambda \in C : a - \lambda \notin inv(A)\}$$

oú $inv(A)$ est l'ensemble des eléments inversibles de A. si A est réelle le spectre est defini par

$$Sp(a, A) := Sp(a, A_C)$$

oú A_C est la complexifiée de A. L'on sait, si A est de Banach associative unitaire, que $Sp(a, A) \neq \emptyset$ et compact, et l'ensemble des éléments inversibles, $inv(A)$, est ouvert [1]. En ce basant dans ces résultats on montre:

– Un d.t.z. n'est pas inversible [1].
– La frontiére de l'ensemble des éléments inversibles de A, $Fr(inv(A))$, est constituée par des d.t.z., $\forall z \in Fr(Sp(a, A)), a - z$ est d.t.z. etc...[1,13].

Dans les algèbres de Jordan, Jacboson, en 1956, a introduit une notion, plus interessante et adéquate, d'inversibilité [5]. Un peu plus tard (1965) Mc Crimmon a éttendue cette notion au cas non commutatif [12] :

Définition 1 *Un élément x d'une K-algébre de Jordan n.c. unitaire A est inversible, s'il existe $y \in A$ tel que $xy = yx = 1$, $x^2y = yx^2 = x$.*

En 1976 Martínez établit les propriétés du spectre associé á cette notion dans le cas Jordan-Banach [10]. Une année plus tard, cette étude est effectuée en [7] pour le cas non commutatif.

Pour cette nouvelle inversibilité la notion de d.t.z. anterioure est inadaptée, les resultats associatifs ne subsistent plus. Par exemple dans $\mathcal{H}^+$ (l'espace vectoriel sous-jacent á $\mathcal{H}$ muni du produit $x \cdot y = \frac{1}{2}(xy + yx)$) on a : $i \cdot j = 0$, $i^{-1} = -i$ c'est á dire i est un d.t.z. inversible, ce qui impose la recherche d'une notion appropriée. En suivant la même démarche que Jacobson [6], la remarque suivante permet de jordaniser la notion associative de d.t.z.

Proposition 1 *Soit A une K-algèbre associative normée et $a \in A$, alors les affirmations suivantes sont equivalentes :*

– *a est d.t.z. dans A.*
– *Il existe une suite $\{x_n\} \subset A$ telle que $\|x_n\| = 1$ et $\{ax_na\} \to 0$.*

L'on voit clairemant que la deuxiéme condition est jordanisable, ce qui nous a conduit á poser la definition suivante [7]:

Définition 2 *Soit A une K-algèbre de Jordan n.c. (pour tout $x, y \in A$, on a : $(xy)x = x(yx)$, $(x^2y)x = x^2(yx)$) normée, un element $a \in A$ est dit diviseur topologique de zéro au sens de Jacobson ou J-diviseur topologique de zéro (en abrégé J-d.t.z.) s'il existe $\{x_n\} \subset A$ telle que $\|x_n\| = 1$ et $\{U_a(x_n)\} \to 0$ avec $U_a(x_n) = a(ax_n) + ax_na - a^2x_n$.*

Remarque 1 *– $a \in A$ est J-d.t.z $\Longleftrightarrow$ U_a est un d.t.z á gauche dans $BL(A)$ [1,7] ($BL(A)$ est l'algèbre des opérateurs linéaires continus de A).*

– Soit A une K-algèbre alternative, $a \in A$ est un d.t.z. si et seulemant si a es un J-d.t.z. ($U_a(x) = axa$).

– Si l'on appelle un élément $a \in A$ tel que U_a n'est pas injectif J-diviseur de zéro [6] on a, tout J-d.z. est un J-d.t.z. Si $\dim(A) < \infty$ *il y a equivalence entre les deux notions. En général tout J-d.z est un J-d.t.z.*

2. Algèbres de Jordan non commutatives sans J-diviseurs topologiques de zéro

Dans ce qui suit nous allons développer la notion introduite anterieurment, en etablissant les analogues des résultats classiques sur les d.t.z. et *l'extension* du théoreme de Kaplanky.

Proposition 2 *Dans une K-algèbre de Jordan n.c. unitaire normée A, un J-d.t.z. est non inversible.*

Preuve. Soit $a \in A$ inversible et supposons qu'il existe $\{x_n\} \subset A$, $\|x_n\| = 1$ et $U_a(x_n) \to 0$. Comme a est inversible U_a est inversible et $U_a^{-1} = U_{a^{-1}}$ [6,12]. La continuité de $U_{a^{-1}}$ implique $\lim U_{a^{-1}}(U_a(x_n)) = 0 = \lim x_n$, ce qui est absurde.

Proposition 3 *Soit A une $\mathcal{R}$-algèbre de Jordan n.c. normée unitaire, si $a \in A$ est J-d.t.z. dans i) la complexifieé normée A_C de A [7,9,13], ou ii) dans la complétion $\hat{A}$, de A [1,7], alors a est un J-d.t.z. dans A.*

Preuve. i) Soient $a \in A$ et $\{x_n + iy_n\} \subset A_C$ verifiant $\|x_n + iy_n\| = 1$ et $U_a(x_n + iy_n) \to 0$. En utilisant les propriétés de la complexifiée normée de A [7,13], on déduit de $\|x_n + iy_n\| = 1$, $\forall n \geq 1$ que l'une des suites $\{x_n\}$, $\{y_n\}$ ne converge pas vers zéro. Si en plus on tient compte de la linéareté de U_a, on obtient de $U_a(x_n + iy_n) \to 0$ que $U_a(x_n)$ et $U_a(y_n)$ convergent vers zéro. D'oú on peut conclure fácilment que a est un J-d.t.z. dans A.

ii) Soit $a \in A$ J-d.t.z. dans $\hat{A}$, et soit $\{x_n\} \subset \hat{A}$ telle que $\|x_n\| = 1$ et $U_a(x_n) \to 0$. Etant donné que A est dense dans sa complétée $\hat{A}$, on déduit que pour tout $n \geq 1$, il existe $x'_n \in A$ tel que $\|x_n - x'_n\| \leq 1/n$, d'oú:

$$1 - \frac{1}{n} \leq \|x'_n\| \leq 1 + \frac{1}{n}.$$

La continuité de U_a implique $\|U_a(x_n - x'_n)\| \leq \|U_a\|\|x_n - x'_n\| \leq \frac{1}{n}\|U_a\|$, oú $\|U_a\|$ note le norme habituelle de U_a dans $BL(\hat{A})$, ce qui donne $\lim U_a(x'_n) = 0$.

Proposition 4 *Soit A unc K-algèbre de Jordan-Banach n.c. unitaire et soit $a \in Fr(inv(A))$, alors a est un J-d.t.z. dans A.*

Preuve. Soit $a \in Fr(inv(A))$, comme $inv(A)$ est ouvert [7,10], $a \notin inv(A)$ et il existe une suite $\{a_n\} \subset inv(A)$ telle que $\lim a_n = a$. Etant donné que $BL(A)$ est de Banach associative et que l'application $a \mapsto U_a$ de A dans $BL(A)$ est continue, on déduit que $U_a \in Fr(inv(BL(A)))$, l'analogue associatif [1, lemma 56.3] et la remarque 1 terminent la preuve.

Corollaire 1 *Soit A une $\mathcal{C}$-algèbre (resp. $\mathcal{R}$-algèbre) de Jordan-Banach n.c. unitaire; $a \in A$ et $z \in Fr(Sp(a, A))$, alors $z - a$ (resp. $|z|^2 - (z + \bar{z})a + a^2$ est un J-d.t.z. dans A.*

Preuve. Soit $z \in Fr(Sp(a, A))$, du fait que $Sp(a, A)$ est fermé [7,10] on déduit fácilement que $z - a \in Fr(inv(A))$ dans le cas complexe ou bien $z - a \in Fr(inv(A_{\mathcal{C}}))$ dans le cas réel. La proposition précedente complète la preuve dans le premiér cas. Pour le derniér, en utilisant [6, Lemma 8, p. 38] á $\mathcal{C}[a]$ on obtient:

$$U_{z-a}U_{\bar{z}-a} = U_{(z-a)(\bar{z}-a)} = U_{|z|^2-(z+\bar{z})a+a^2}$$

d'oú on a que $|z|^2 - (z + \bar{z})a + a^2$ est un J-d.t.z. dans $A_{\mathcal{C}}$ et le résultat anterieur termine la démonstration.

Corollaire 2 *Soit A une K-algèbre de Jordan n.c. normée unitaire et soit $a \in A$, $a \neq 0$ tel que $Sp(a, A) = \{0\}$. Alors a est un J-d.t.z. dans A.*

Preuve. En remarquant, qui l'on peut supposer A de Jordan Banach complexe et que $a_n = a - 1/n \in inv(A)$ pour tout $n \geq 1$ on a que $a \in Fr(inv(A))$, la proposition 4 permet de conclure.

Lemme 1 *Soit A une K-algèbre de Jordan n.c. sans J-d.z.$\neq 0$, si $e \in A$ est un idempotent non nul, alors A est unitaire d'unité e.*

Preuve. Soit $A = A_0 + A_{1/2} + A_1$ la décomposition de Peirce de A associée á e [15] oú les A_i sont les sous-espaces de Peirce

$$A_i := \{x_i \in A : x_i e = ex_i = ix\}$$

avec $i = 0, 1$ et

$$A_{1/2} := \{x_{1/2} \in A : x_{1/2}e + ex_{1/2} = x_{1/2}\}.$$

On rappelle que $A_{1/2}A_{1/2} \subset A_1 + A_0$ [15]. Soit $x_0 \in A_0$, du fait que $U_e(x_0) = e(ex_0) + ex_0e - e^2x_0 = 0$ on a $x_0 = 0$. Comme A est a puissances associatives, par linéarisation de $(x, x, x^2) = 0$ on a

$$(x, x, xy + yx) + (y, x, x^2) = 0$$

en posant $y = e$ et prenant $x \in A_{1/2}$ on a $(ex)x^2 - ex^3 = 0$, comme $ex = x - xe$ on obtient

$$0 = (x - xe)x^2 - ex^3 = x^3 - (xe)x^2 - ex^3 = x^3 - x(ex^2) - ex^3.$$

Comme $x^2 \in (A_{1/2})^2 \subset A_1 + A_0 = A_1$ on a $ex^2 = x^2$ d'oú $-ex^3 = 0 = U_e(x^3) \Rightarrow x^3 = 0 = U_x(x^2) \Rightarrow x^2 = 0$ finalement $U_x(e) = x(xe) + xex - x^2e = x(xe + ex) = x^2 = 0 \Rightarrow x = 0$ puisque $e \neq 0$.

Théorème 2 *(Théorème principal) Soit $A \neq 0$ une K-algèbre de Jordan n.c. normée, sans J-d.t.z. non nuls. Alors A est isomorphe á $\mathcal{C}$ si $K = \mathcal{C}$, ou A est de division quadratique si $K = \mathcal{R}$.*

Preuve. Soit $0 \neq a \in A$, et $[a]$ la sous-algèbre engendrée par a, l'on sait que $[a]$ est associative et commutative; en plus $[a]$ est sans diviseurs topologiques de zéro, d'aprés Kaplansky $[a]$ est isomorphe á $\mathcal{R}$ ou $\mathcal{C}$, l'unité de $[a]$ est un idempotent non nul de A. Si $K = \mathcal{C}$, on a $\emptyset \neq Sp(a, \hat{A}) \neq \{0\}$, soit $0 \neq z \in Fr(Sp(a, \hat{A}))$ alors $z \cdot 1 - a$ est un J-d.t.z. et par conséquent $a = z \cdot 1$ d'oú $A \cong \mathcal{C}$.

Dans le cas reél, soit $\hat{A}_{\mathcal{C}}$ la complexifiée normée de la complétion $\hat{A}$ de A et soit $0 \neq a \in A$, on a $\emptyset \neq Sp(a, \hat{A}_{\mathcal{C}}) \neq \{0\}$. Soit $z \in Fr(Sp(a, \hat{A}_{\mathcal{C}})) \neq \{0\}$ et $z \neq 0$, d'aprés le corollaire on a $|z|^2 \cdot 1 - (z + \bar{z})a + a^2$ est un J-d.t.z. dans $\hat{A}_{\mathcal{C}}$, comme $|z|^2 \cdot 1 - (z + \bar{z})a + a^2 \in A$, alors $|z|^2 \cdot 1 - (z + \bar{z})a + a^2 = 0$ d'oú A est quadratique de division.

Remarque 2 [8] *Les algèbres de Jordan réelles normées de division D sont de la forme $D = \mathcal{R} \oplus V$ oú V est un espace vectoriel réel normé muni d'une forme bilinéaire symetrique $< , >$ definie negative continue, et le produit donné par :*

$$(\alpha + x)(\beta + y) = \alpha\beta + < x, y > + \alpha y + \beta x$$

et la norme $\|\alpha + x\| = |\alpha| + \|x\|$ oú $\| \cdot \|$ est la norme de V.

Remerciement Les auteurs remercient vivement le Professeur Rodríguez Palacios pour ces suggestions et remarques.

References

1. Berberian S. K.:1973, Lectures in Funtional Analysis and Operator Theory *Springer-Verlag.* G.T.M.
2. Cabrera García M. and Rodríguez Palacios A. A new proof of the Gelfand-Mazur-Kaplansky Theorem. á paraitre dans *Proc. Amer. Math. Soc.*
3. Cuenca Mira J. A.:1992, On one-sided division infinite-dimensional normed real algebras. *Publications matemátiques*, vol.**36**, pp.485–488.
4. El-Mallah M. L. et Micali A.:1980, Sur les algébres normées sans diviseurs topologiques de zéro. *Boletín de la Sociedad Matemática Mexicana*, vol **25**, pp.23–28.
5. Jacobson N.:1956, A theorem on the structure of Jordan algebras, *Proc. Nat. Acad. Sci. U.S.A.*, vol **42**, pp.140–147.
6. Jacobson N.:1968, Structure and Representations of Jordan Algebras, *Amer. Math. Soc. Coll. Publ.* vol **39**
7. Kaidi A. M.:1977, Bases para una teoría de las álgebras no asociativas normadas. Tesis doctoral, Universidad de Granada.
8. Kaidi A. M.:1991, Structure des algèbres de Jordan-Banach non commutatives réelles de division. *Ann. Sci. Univ. "Blaise Pascal", Clérmont II*, Sér. Math.,Fasc. **27** éme, pp.119–124
9. Kaplansky I.:1949, Normed Algebra, *Duke Math. J.* vol **16**, pp.399-418.
10. Martínez Moreno J.:1977, Sobre Algebras de Jordan Normadas Completas. Tesis doctorales de la Universidad de Granada. No. 149.
11. Martínez Moreno J.:1991, Holomorphic Functional Calculus in Jordan-Banach Algebras, *Ann. Sci. Univ. "Blaise Pascal", Clérmont II, Sér. Math.* Fasc **27** éme, pp.125-134.
12. McCrimmon K.:1965, Norms and non commutative Jordan algebras, *Pacific J. Math.* vol **15**, pp.925–956.
13. Rickart C. E.:1960, General Theory of Banach Algebras, Princeton, N.Y.: Van Nostrand.
14. Rodríguez Palacios A.:1992, One-sided division absolute valued algebras, *Publications Matemátiques*, vol **36**, pp.925–954.
15. Schafer R.D.:1966, An introduction to nonassociative algebras, *Pure and Applied Mathematics* 22, Academic Press, N-Y.
16. Shilov G.:1940, On the extension of maximal ideals, *Comptes Rendus (Doklay) de l'Académie des Sciences de l'URSS* (N.S.), vol **29**, pp.83–84

ON FREUDENTHAL–KANTOR TRIPLE SYSTEMS AND GENERALIZED STRUCTURABLE ALGEBRAS

NORIAKI KAMIYA
Department of Mathematics
Shimane University, 690 Japan

Abstract. In this paper, we shall investigate to the relationship of a Freudenthal–Kantor triple system with a generalized structurable algebra, and to the standard embedding Lie algebra associated with the generalized structurable algebra.

Key words: Freudenthal–Kantor triple systems, generalized structurable algebras

1. Introduction

Our work was motivated form the construction of Lie algebras due to N.Jacobson [3] and H.Freudenthal [2]. That is, we have been investigated the construction of Lie algebras or Lie superalgebras from triple systems [6,7,8,9,10 etc.].

On the other hand, we have been considered to a generalized structurable algebra as a generalized class of Lie algebras, commutative algebras and structurable algebras [12,13].

Thus this article is a continuation of these study. We shall be concerned with algebras and triple systems which are finite dimensional over a field Φ of characteristic $\neq 2$ *or* 3, unless otherwise specifid.

2. Lie Algebras Constructed by Generalized Structurable Algebras

In this section, we shall recall some of the definition and the basic results about triple systems or algebras which are need in the following.

For $\varepsilon = \pm 1$, $\delta = \pm 1$ a vecter space $U(\varepsilon, \delta)$ over a field Φ with triple product $< -,-,- >$ is called a (ε, δ)–*Freudenthal-Kantor triple system* if

$$[L(a,b), L(c,d)] = L(< abc >, d) + \varepsilon L(c, < bad >)$$

$$K(< abc >, d) + K(c, < abd >) + K(a, K(c,d)b) = 0$$

where $L(a,b)c =< abc >$ and $K(a,b)c =< acb > -\delta < bca >$.

Definition A Freudenthal-Kantor triple system with $\delta = 1$ is *balanced* if there exists an anti–symmetric bilinear form $< \, , \, >$ such that $K(x,y) =< x,y > Id$, $< x,y > \in \Phi^*$

Example 1. Let $U(\varepsilon, \delta)$ be a vector space over Φ equipped with a bilinear form $< x,y >= -\varepsilon < y,x > \in \Phi$. Then$(U(\varepsilon,\delta), < -,-,- >)$ is a (ε, δ) – Freudenthal–Kantor triple system with respect to

$$< xyz > := < y,z > x.$$

S. González (ed.), Non-Associative Algebra and Its Applications, 198–203.

Example 2. Let U be a vector space over Φ equipped with an anti–symmetric bilinear form $< x, y > \in \Phi$. Then $(U, < -, -, - >)$ is a balanced Freudenthal–Kantor triple systems with respect to

$$< xyz > := \frac{1}{2}(< y, z > x + < y, x > z + < x, z > y.)$$

Next we shall define a *generalized structurable superalgebra* over a field Φ to be a nonassociative algebra A equipped with a bilinear superderivation $D(x, y)$ in which the following conditions are satisfied

$$D(x, y) = -(-1)^{degx degy} D(y, x)$$

$$(-1)^{degx degz} D(xy, z) + (-1)^{degy degx} D(yz, x) + (-1)^{degz degy} D(zx, y) = 0.$$

For the case of degree 0, we call it to a *generalized structurable algebra* (for example, to see [12],[13])

Example 3. Let $(A, [\ ,\ ])$ be a Lie superalgebra over Φ. Then A is a generalized structurable superalgebra equipped with a superderivation

$$D(x, y) := ad\ [x, y].$$

Example 4.[12] A structurable algebra due to B.N.Allison is a generalized structurable algebra equipped with a derivation

$$D(x, y) := \frac{1}{3}[[x, y] + [\overline{x}, \overline{y}], z] + [z, y, x] - [z, \overline{x}, \overline{y}].$$

Example 5. Let A be a Poisson superalgebra. Then A is a generalized structurable superalgebra equipped with a superderivation $D(x, y)z = \{\{x, y\}, z\}$.

Example 6. Let A be a Jordan superalgebra. Then A be a generalized structurable superalgebra equipped with $D(x, y)z = (L(x)L(y) - (-1)^{degy degx} L(y)L(x))z$.

Proposition 7.[7,10] *Let $U(\varepsilon, \delta)$ be a (ε, δ)-Freudenthal–Kantor triple system. Then the vector space $U(\varepsilon, \delta) \oplus U(\varepsilon, \delta)$ becomes a Lie triple system (the case of $\delta = 1$) or an anti–Lie triple system (the case of $\delta = -1$)with respect to the triple product defined by*

$$[\begin{pmatrix} a \\ b \end{pmatrix} \begin{pmatrix} c \\ d \end{pmatrix} \begin{pmatrix} e \\ f \end{pmatrix}] = \begin{pmatrix} L(a, d) - \delta L(c, b) & \delta K(a, c) \\ -\varepsilon K(b, d) & \varepsilon(L(d, a) - \delta L(b, c)) \end{pmatrix} \begin{pmatrix} e \\ f \end{pmatrix}.$$

From this proposition, we may obtain the Lie or anti–Lie triple system $U(\varepsilon, \delta) \oplus U(\varepsilon, \delta)$ associated with $U(\varepsilon, \delta)$ and denote it by $T(\varepsilon, \delta)$. Thus we have the standard embedding Lie algebra $(\delta = 1)$ or superalgebra $(\delta = -1)$ as follows;

$$L(\varepsilon, \delta) = D(T(\varepsilon, \delta), T(\varepsilon, \delta)) \oplus T(\varepsilon, \delta)$$

where $D(T(\varepsilon, \delta), T(\varepsilon, \delta))$ is the Lie algebra of the inner derivation of Lie or anti–Lie triple system $T(\varepsilon, \delta)$. Furtheremore we get the decomposition

$$L(\varepsilon, \delta) = L_{-2}(\varepsilon, \delta) \oplus L_{-1}(\varepsilon, \delta) \oplus L_0(\varepsilon, \delta) \oplus L_1(\varepsilon, \delta) \oplus L_2(\varepsilon, \delta)$$

where

$L_{-2}(\varepsilon,\delta)$ =the linear span of all $\begin{pmatrix} 0 & \delta K(c,d) \\ 0 & 0 \end{pmatrix}$

$L_{-1}(\varepsilon,\delta) = U(\varepsilon,\delta) \oplus (0)$

$L_0(\varepsilon,\delta)$ = the linear span of all $\begin{pmatrix} L(a,b) & 0 \\ 0 & \varepsilon L(b,a) \end{pmatrix}$

$L_1(\varepsilon,\delta) = (0) \oplus U(\varepsilon,\delta)$

$L_2(\varepsilon,\delta)$= the linear span of all $\begin{pmatrix} 0 & 0 \\ -\varepsilon K(e,f) & 0 \end{pmatrix}$.

Remark For the case of $\delta = -1$, we have a decomposition of $L(\varepsilon,\delta) := V_0 \oplus V_1$ such that $degV_0 = 0$ and $degV_1 = -1$, where $V_0 = L_0(\varepsilon,\delta)$, $V_1 = L_{-2}(\varepsilon,\delta) \oplus L_{-1}(\varepsilon,\delta) \oplus L_1(\varepsilon,\delta) \oplus L_2(\varepsilon,\delta)$.

Theorem 8. *Let $U(\varepsilon,\delta)$ be a (ε,δ)-Freudenthal-Kantor triple system and $L_i(\varepsilon,\delta)$ be as in above. Then the vector space $B := L_{-2}(\varepsilon,\delta)\oplus L_{-1}(\varepsilon,\delta)\oplus L_1(\varepsilon,\delta)\oplus L_2(\varepsilon,\delta)$ is a generalized structurable algebra ($\delta = 1$) or superalgebra ($\delta = -1$) with respect to a product*

$$X \circ Y := [x_{-2}, y_1] + [x_{-1}, y_{-1}] + [x_{-1}, y_2] + [x_1, y_{-2}] + [x_1, y_1] + [x_2, y_{-1}]$$

and a derivation

$$D(X,Y) := ad\ ([x_{-2}, y_2] + [x_{-1}, y_1] + [x_1, y_{-1}] + [x_2, y_{-2}])$$

where

$$x_i, y_i \in L_i(\varepsilon,\delta), X = \sum_i x_i, Y = \sum_i y_i . (i = -2,-1,1,2)$$

The generalized structurable algebra or superalgera B defined by the above product is called the *generalized structurable algera* or *superalgebra associated with* a (ε,δ) –Freudenthal–Kantor triple system $U(\varepsilon,\delta)$.

From now on, we denote the sets of above derivations $D(X,Y)$ by $Inn\ Der\ B$.

Theorem 9. *Let B be as in above theorem. Then the vector space $L(B) := Inn\ Der\ B \oplus B$ is a Lie algebra ($\delta = 1$) or Lie superalgebra ($\delta = -1$) with respect to*

$$[D+X, E+Y]^* = [D,E] + D(X,Y) + DY - EX + X \circ Y$$

where $D, E \in Inn\ Der\ B, \quad X, Y \in B$.

3. Generalized Structurable Algebras Associated with Balanced Freudenthal–Kantor Triple Systems

In this section, we shall discuss to a finite dimensional balanced Freudenthal–Kantor triple system U over the complex number C and the generalized structurable algebra B associated with U.

Before going ahead, from [8,11], we recall the examples of simple balanced Freudenthal–Kantor triple systems as follows.

A_n – type :

Let $M_A(n)$ be the set of

$$\{\begin{pmatrix} 0 & a \\ b & 0 \end{pmatrix} | a, b \in Mat(1, n-1 : C)\}$$

$$< xyz > := x \circ (Py \circ z) + z \circ (Py \circ x) - Py \circ (x \circ z)$$

where

$$x \circ y = \begin{pmatrix} 0 & x_1 \\ x_2 & 0 \end{pmatrix} \circ \begin{pmatrix} 0 & y_1 \\ y_2 & 0 \end{pmatrix}$$

$$= \begin{pmatrix} B(x_1, y_2) & 0 \\ 0 & B(y_1, x_2) \end{pmatrix}, P : \begin{pmatrix} 0 & x \\ y & 0 \end{pmatrix} \to \begin{pmatrix} 0 & x \\ -y & 0 \end{pmatrix}$$

and $B(x, y) = xy^T$. (y^T is the transpose matrix of y)

The standard embedding Lie algebra which is obtained from this triple system $(M_A(n), < -, -, - >)$ is of type A_n.

$C_n - type$:

Let $M_C(n)$ be the set of

$$\{\begin{pmatrix} 0 & x \\ x & 0 \end{pmatrix} |x \in Mat(1, 2(n-1); C)\}$$

$$< xyz > := x \circ (Py \circ z) + z \circ (Py \circ x) - Py \circ (x \circ Z)$$

where

$$x \circ y = \begin{pmatrix} 0 & x \\ x & 0 \end{pmatrix} \circ \begin{pmatrix} 0 & y \\ y & 0 \end{pmatrix} = \begin{pmatrix} B(x, y) & 0 \\ 0 & B(y, x) \end{pmatrix}$$

and $B(x, y) = \frac{1}{2} < x, y >$ ($<\ ,\ >$is an anti–symmetric bilinear form).

The standard embedding Lie algebra which is obatined from this triple system $(M_C(n), < -, -, - >)$ is of type C_n.

$B_n, D_n - type$:

Let $M_{B,D}(p)$ be the set of

$$\{\begin{pmatrix} 0 & x \\ x & 0 \end{pmatrix} |x \in Mat(1, p : H)\}$$

$$< xyz > := x \circ (Py \circ z) + Z \circ (Py \circ x) - Py \circ (x \circ z)$$

where

$$x \circ y = \begin{pmatrix} 0 & x \\ x & 0 \end{pmatrix} \circ \begin{pmatrix} 0 & y \\ y & 0 \end{pmatrix} = \begin{pmatrix} \overline{jB(x, y)} & 0 \\ 0 & B(y, x)j \end{pmatrix}$$

and $B(x, y) = x\bar{y}^T$ ($-$ is the extension to $Mat(1, p : H)$ of an involution of the quaternion algebra H over the real number field R such that $\overline{\alpha\beta} = \overline{\beta}\overline{\alpha}, \bar{j} = -j$, and $j^2 = -1$).

The standard embedding Lie algebra which is obtained from this triple system $(M_{B,D}(p), < -, -, - >)$ is of type B_n (resp. D_n) if $p = 2n - 3$ (resp. if $p = 2n - 4$).

$G_2 - type$:

Let $M_2(C)$be the set of

$$\{\begin{pmatrix} \alpha & \gamma \\ \delta & \beta \end{pmatrix} \mid \alpha, \beta, \gamma, \delta \in C\}$$

$$< xyz > := x \circ (\overline{Py} \circ z) + z \circ (\overline{Py} \circ x) - Py \circ (\overline{x} \circ z)$$

where

$$x \circ y = \begin{pmatrix} \alpha & \gamma \\ \delta & \beta \end{pmatrix} \circ \begin{pmatrix} \alpha_1 & \gamma_1 \\ \delta_1 & \beta_1 \end{pmatrix} = \begin{pmatrix} \alpha\alpha_1 + \gamma\delta_1 & \alpha\gamma_1 + \gamma\beta_1 \\ \delta\alpha_1 + \beta\delta_1 & \delta\gamma_1 + \beta\beta_1 \end{pmatrix}$$

$$- : \begin{pmatrix} \alpha & \gamma \\ \delta & \beta \end{pmatrix} \longrightarrow \begin{pmatrix} \beta & \gamma \\ \delta & \alpha \end{pmatrix}, P : \begin{pmatrix} \alpha & \gamma \\ \delta & \beta \end{pmatrix} \longrightarrow \begin{pmatrix} -\alpha & \gamma \\ -\delta & \beta \end{pmatrix}.$$

The standard embedding Lie algebra which is obtained from this triple system $(M_2(C), < -, -, - >)$ is of type G_2.

$F_4, E_6, E_7, E_8 - type$:

We consider a composition algebra A over the complex number field C. We denote the reduced simple Jordan algebra by $H_3(A, \Gamma_G)$. (cf.[4],[8])

Let $M(H_3(A, \Gamma_G))$ be the set of

$$\{\begin{pmatrix} \alpha & a \\ b & \beta \end{pmatrix} \mid \alpha, \beta \in C,\ a, b \in H_3(A, \Gamma_G)\}$$

$$< x_1 x_2 x_3 > := x_1 \circ (\overline{Px_2} \circ x_3) + x_3 \circ (\overline{Px_2} \circ x_1) - Px_2 \circ (\overline{x_1} \circ x_3)$$

where for

$$x_i = \begin{pmatrix} \alpha_i & a_i \\ b_i & \beta_i \end{pmatrix} \in M(H_3(A, \Gamma_G)),$$

$$\begin{pmatrix} \alpha_1 & a_1 \\ b_1 & \beta_1 \end{pmatrix} \circ \begin{pmatrix} \alpha_2 & a_2 \\ b_2 & \beta_2 \end{pmatrix} = \begin{pmatrix} \alpha_1\alpha_2 + Tr(a_1 \cdot b_2) & \alpha_1 a_2 + \beta_2 a_1 + b_1 \times b_2 \\ \alpha_2 b_1 + \beta_1 b_2 + a_1 \times a_2 & \beta_2\beta_1 + Tr(a_2 \cdot b_1) \end{pmatrix}$$

Then we have the following table.

dim A	1	2	4	8
dim $H_3(A, \Gamma_G)$	6	9	15	27
dim $M(H_3(A, \Gamma_G))$	14	20	32	56

The standard embedding Lie algebras which are obtained from these triple systems $(M(H_3(A, \Gamma_G)), < -, -, - >)$ are of type F_4, E_6, E_7, E_8 with regard to dim $A = 1, 2, 4, 8$, respectively.

For the generalized structurable algebra B associated with the above triple systems U, we have the following table.

	dim L(U)	dim U	dim B	dim InnDer B
A_n	n(n+2)	2(n-1)	4n-2	n(n-2)+2
B_n	n(2n+1)	4n-6 (p=2n-3)	8n-10	(n-2)(2(n-2)+1)+3+1
C_n	n(2n+1)	2(n-1)	4n-2	(n-1)(2(n-1)+1)+1
D_n	n(2n-1)	4n-8 (p=2n-4)	8n-14	(n-2)(2(n-2)-1)+3+1
G_2	14	4	10	3+1
F_4	52	14	30	21+1
E_6	78	20	42	35+1
E_7	133	32	66	66+1
E_8	248	56	114	133+1

Theorem 10. *For the simplicity of the generalized structurable algebra B associated with a balanced Freudenthal-Kantor triple system U over a field characteristic 0, the following are equivalent;*

(i) *U is simple*

(ii) *B is simple.*

4. Root Systems of Inner Der B

In this section, we shall investigate to the diagrams of root systems of the Inn Der B of the generalized structurable algebra B associated with the balanced Freudenthal–Kantor triple systems U over C of type $A_n, B_n, \ldots, E_8$.

We note that the big circle of below diagrams should be omitted.

A_n : ○ — o ··· o — ○ (A_{n-2})

B_n : o — ○ — o ··· o ⇒ o (A_1, B_{n-2})

C_n : o ⇒ o — o ··· o — ○ (C_{n-1})

D_n : o — ○ — o ··· o (branching to o / o) (A_1, D_{n-2})

G_2 : ○ — o (A_1)

F_4 : ○ — o ⇒ o — o (C_3)

E_6 : o — o — o — o — o, with ○ attached to the middle node (A_5)

E_7 : o — o — o — o — o — ○, with o attached to the fourth node (D_6)

E_8 : ○ — o — o — o — o — o — o, with o attached to the fifth node (E_7)

References

1. B.N.Allison,; a class of nonassociative algebras with involution containing the class of Jordan algebras,Math.Ann.237(1978)133–156.
2. H.Freudenthal,; Beziehungen der E_7 und E_8 zur Okavenbene I, II. I: Indag. Math. 16 (1954) 218 – 230, II: Indag. Math. 16 (1954) 363–386.
3. N.Jacobson,; Lie and Jordan triple systems, Amer.J.Math.71 (1949)149– 170.
4. N.Jacobson,; Structure and representation sof Jordan algebras, Amer. Math. Soc. Colloq. Publ. vol.39,1968.
5. V.G.Kac,; Infinite dimensional Lie algebras, Birkhaüser 1983.
6. N.Kamiya,; A structure theory of Freudenthal–Kantor triple systems, J. Alg. 110 (1987) 108 – 123.
7. N.Kamiya,; A structure theory of Freudenthal–Kantor triple systems II. Comm. Math. Univ. Sancti Pauli 38 (1989) 41–60.
8. N.Kamiya,; A structure theory of Freudenthal–Kantor triple systems III. Mem. Fac. Sci. Shimane Univ. 23 (1989) 33–51.
9. N.Kamiya,; A structure theory of Freudenthal–Kantor triple systems IV. Mem. Fac. Sci. Shimane Univ. 25 (1991) 45–51.
10. N.Kamiya,; On (ε, δ)– Freudenthal-Kantor triple systems. International Symposium on a nonassociative algebras and related topics, Hiroshima,Japan(1991) 65–75,World Scientific.
11. N.Kamiya,; The construction of all simple Lie algebras over C from balanced Freudenthal–Kantor triple systems, Contributions to General Algebra 7,(1991) 205–213. Verlag Hölder–Pichler–Tempsky, Wien,Verlag G.Teubner,Stutugart.
12. N.Kamiya,; On a generalization of structurable algebras, Algebras,Groups and Geometries 9(1992) 35–47.
13. N.Kamiya,and R.M.Santilli,; Embedding of Lie algebras in isogeneralized structurable algebras, Contributed paper to the International conference in honor of Lie and Lobachwski, Tartu, Estonia, 1992. to appear.
14. N.Kamiya,; On isofields and their isoduals, Abstract, Symmetric Methods in Physics, In memory of Professor YA.A. Smorodinsky, Dubna, Russia, July 1993.
15. N.Kamiya,; A characterization of Kac–Moody algebras from generalized structurable algebras, preprint.

HERMITIAN JORDAN TRIPLE SYSTEMS AND THE AUTOMORPHISMS OF BOUNDED SYMMETRIC DOMAINS

WILHELM KAUP
Mathematisches Institut der Universität
Auf der Morgenstelle 10
D-72076 Tübingen, Germany

Abstract. We describe bounded symmetric domains in complex Banach spaces and their biholomorphic automorphisms in terms of the underlying JB*-triple structures. Of particular interest in this context is the square root of a certain Bergman operator - we give a new description of this square root as exponential of a hermitian operator which gives better norm estimates. We do not give full proofs. These will appear elsewhere.

1. Preliminaries

Let D be a bounded domain in a complex Banach space U. By definition, a function $f: D \to U$ is called *holomorphic* if for every $a \in D$ the Fréchet derivative $f'(a) \in \mathcal{L}(U)$ exists as a continuous linear operator on U. A bijection $g: D \to D$ is called *biholomorphic* or an *automorphism of* D if g and g^{-1} are holomorphic. Denote by $G := \mathrm{Aut}(D)$ the group of all biholomorphic automorphisms of D. Then D is called *symmetric* if to every $a \in D$ there is an automorphism $s = s_a \in G$ with $s^2 = \mathrm{id}$ having a as isolated fixed point. Then it is known that G acts transitively on D and that there is a complex Banach space E (uniquely determined up to an isometric isomorphism) such that D is biholomorphically equivalent to the open unit ball of E (compare [14] for details).

A complex Banach space E is called a *JB*-triple* if there exists a continuous ternary operation $(x, y, z) \mapsto \{xyz\}$ from E^3 to E (called the *Jordan triple product on* E) such that the following axioms are satisfied

(J_1) $\{xyz\}$ is symmetric bilinear in the outer variables x, z and conjugate linear in the inner variable y

(J_2) $L\{xyz\} = \{(Lx)yz\} - \{x(Ly)z\} + \{xy(Lz)\}$

(J_3) L is hermitian and has spectrum ≥ 0

(J_4) $\|L\| = \|a\|^2$

for all $a, x, y, z \in E$ and $L \in \mathcal{L}(E)$ defined by $L(z) := \{aaz\}$.

It is known [8] that for every x, y, z in a JB*-triple the following estimate holds

$$\|\{xyz\}\| \leq \|x\| \cdot \|y\| \cdot \|z\| \,. \tag{1.1}$$

The Jordan triple product on a JB*-triple is uniquely determined by the underlying Banach space structure and its open unit ball is symmetric. On the other hand,

S. González (ed.), Non-Associative Algebra and Its Applications, 204–214.

every complex Banach space E with symmetric open unit ball D (i.e. the group $G := \mathrm{Aut}(D)$ acts transitively on D) can be given the structure of a JB*-triple. In this sense we say that a given complex Banach space *is a JB*-triple* or it *is not a JB*-triple*.
For instance, every C*-algebra A is a JB*-triple. The triple product then is given by

$$\{xyz\} = (xy^*z + zy^*x)/2. \tag{1.2}$$

In particular, for every locally compact topological space S the the Banach space $\mathcal{C}_0(S)$ of all continuous complex-valued functions f on S vanishing at infinity is a JB*-triple - here $\|f\| = \sup|f(S)|$ and $\{fgh\} = f\overline{g}h$.

For every $x, y \in E$ denote by $x \square y$ or $L(x, y)$ the linear operator $z \mapsto \{xyz\}$ on E. Then L may be considered as a positive-definite operator-valued hermitian form on E. Denote by $Q(x)$ the antilinear operator $y \mapsto \{xyx\}$ on E. This depends quadratically on x and satisfies the fundamental formula

$$Q(Q(x)y) = Q(x)Q(y)Q(x)\,.$$

Furthermore

$$B(x,y) := 1 - 2x \square y + Q(x)Q(y)$$

is called the *Bergman operator* (compare [16], [15]).

A linear subspace $F \subset E$ is called a *subtriple* if $\{FFF\} \subset F$ holds. Every closed subtriple is a JB*-triple itself. In case F is only an $\mathbb{R}$–linear subspace with this property we call it a *real subtriple*. For every $a \in E$ denote by C_a the smallest closed (complex) subtriple of E containing a. Also let R_a be the smallest closed real subtriple containing a. Then it is known that $C_a = R_a \oplus iR_a$ holds and that there is a unique locally compact subset $S \subset \mathbb{R}$ together with a triple isomorphism $h: C_a \to \mathcal{C}_0(S)$ such that $S > 0$, $S \cup \{0\}$ is compact and such that $h(a)(s) = s$ for all $s \in S$. With this isomorphism R_a corresponds to the subspace $\mathcal{C}_0(S, \mathbb{R})$ of all real-valued functions.

The element $e \in E$ is called *unitary* if $e \square e = \mathrm{id}$ holds. Then E is a Jordan algebra with unit e in the product $x \circ y = \{xey\}$ and $x \mapsto x^* := \{exe\}$ defines an antilinear algebra involution. In particular $V := \{x \in E : x^* = x\}$ is a closed real Jordan subalgebra with $E = V \oplus iV$. More generally, the element $a \in E$ is called *invertible* in E if the antilinear operator $Q(a)$ is a bijection (and hence an $\mathbb{R}$-linear homeomorphism) of E and then $a^{-1} := Q(a)^{-1}a$ is called the *inverse* of a in E. It is obvious that the set of all invertible elements of E is open (may be empty) and that the mapping $x \mapsto x^{-1}$ is a real analytic mapping of period 2 there. Later on we will generalize the symbol a^{-1} for a more general type of invertibility.

The element $e \in E$ is called a *tripotent* if $\{eee\} = e$ holds. Every tripotent e induces a decomposition (called Peirce decomposition)

$$E = E_1 \oplus E_{1/2} \oplus E_0$$

where $E_k := E_k(e)$ is the k-eigenspace of $e \square e$ in E. The finite set $\mathcal{E} = \{e_1, \ldots, e_r\} \subset E$ is called an *orthogonal system of tripotents* if $e_i \neq 0$ and $\{e_i e_j e_i\} = \delta_{ij} e_i$ holds for all i, j. Then the linear span $H := \mathbb{C}e_1 \oplus \ldots \oplus \mathbb{C}e_r \subset E$ is a subtriple and

$H \square H \subset \mathcal{L}(E)$ consists of commuting operators. For every linear form $\lambda: H \to \mathbb{C}$ consider the root space

$$E_\lambda := \{x \in E : h \square e(x) = \lambda(h)x \quad \text{for all} \quad h \in H\}$$

where $e := e_1 + \ldots + e_r$. Then for every $\lambda, \mu, \nu \in H^* := \mathcal{L}(H, \mathbb{C})$

$$\{E_\lambda E_\mu E_\nu\} \subset E_{\lambda - \mu + \nu}$$

is an immediate consequence of (J_2) and furthermore

$$E = \bigoplus E_\lambda$$

where the finite sum is taken over all roots $\lambda \in H^*$ (i.e. $E_\lambda \neq 0$ - notice that every root takes on $\mathcal{E}$ only the values $0, 1/2$ or 1). It is clear that the dual basis $\lambda_1, \ldots, \lambda_r$ of H^* (i.e. $\lambda_i(e_j) = \delta_{ij}$) consists of roots. On the other hand, if we put $\lambda_0 := 0 \in H^*$, it is known that every root λ of $\mathcal{E}$ is of the form $\lambda = (\lambda_i + \lambda_j)/2$ for some $0 \leq i, j \leq r$, compare [16]. If we write E_{ij} instead of E_λ for this λ we have the Peirce decomposition with respect to $\mathcal{E}$ as

$$E = \bigoplus_{0 \leq i \leq j \leq r} E_{ij} \ .$$

For every $a = \sum \alpha_k e_k \in H$ and $\alpha_0 := 0$ the operator $a \square e = \sum \alpha_k e_k \square e_k$ leaves every E_{ij} invariant and is the multiplication by $(\alpha_i + \alpha_j)/2$ there. In the same way the Bergman operator $B(a, a)$ is the multiplication by $(1 - \alpha_i \overline{\alpha_i})(1 - \alpha_j \overline{\alpha_j})$ on E_{ij}.

2. The odd functional calculus

Let a be an arbitrary element in a JB*-triple E and denote for every field $\mathbb{K}$ by $\mathbb{K}^-[t] \subset \mathbb{K}[t]$ the subspace of all odd polynomials. For every $p \in \mathbb{C}^-[t]$ then $p(a) \in E$ is defined by $p(a) := q(a \square a)a = q(Q(a))a$ where $q \in \mathbb{C}[t]$ satisfies $p(t) = tq(t^2)$. In this way we get a well defined map $p: E \to E$ which is real-analytic polynomial (but not holomorphic if $p \neq 0$). In the special case $E = \mathbb{C}$ for instance the induced mapping is just the $\mathbb{T}$-equivariant extension of $p: \mathbb{R} \to \mathbb{C}$ to $\mathbb{C}$, where $\mathbb{T} := \{t \in \mathbb{C} : |t| = 1\}$. As usual $\mathbb{T}$-equivariance means $p(tz) = tp(z)$ for all $t \in \mathbb{T}$. Further examples are the odd powers $a^{2n+1} = (a \square a)^n a$ and in particular $a^3 = \{aaa\}$. The subspaces R_a and C_a defined in section 1 are just the closures of $\mathbb{R}^-[a]$ and $\mathbb{C}^-[a]$ in E.

The element $a \in E$ is called *(von Neumann) regular* if $a \in Q(a)E$, otherwise *singular*. Let us call the real number $s \geq 0$ a *singular value of* a if a is not contained in $(Q(a) - s^2)E$. The set $\mathsf{Sg}(a)$ of all singular values of a is called the *singular set of* a. In the special case of a rectangular matrix $a \in \mathbb{C}^{p \times q}$ the singular set $\mathsf{Sg}(a)$ consists as usual of all $s > 0$ such that s^2 is an eigenvalue of the square matrix aa^*. It can be shown that $\mathsf{Sg}(a) \subset \mathbb{R}$ always is a compact subset that does not change when computed with respect to any closed subtriple of E containing a. Furthermore, the $\mathbb{T}$-orbit $\mathbb{T}\mathsf{Sg}(a)$ is the set of all $h(a)$ where h runs over all non-zero triple homomorphisms $h: C_a \to \mathbb{C}$ and it also is the set of all $s \in \mathbb{C}$ such that the restriction of the $\mathbb{R}$-linear operator $(Q(a) - s^2)$ to C_a is not invertible.

Replacing the antilinear operator $Q(a)$ by the linear operator $a \square a$ we get the notion of triple spectrum: We call $s \in \mathbb{C}$ a *spectral value of* a if a is not contained in $((a \square a) - s^2)E$ or equivalently if the restriction of $((a \square a) - s^2)$ to C_a is not invertible. The set $\mathsf{Sp}(a)$ of all spectral values of a is called the *spectrum* of a. It can be shown that $\mathsf{Sp}(a) = \mathsf{Sg}(a) \cup -\mathsf{Sg}(a)$ holds and that $\mathsf{Sp}(a)$ is the closure of $\mathsf{Sp}'(a)\backslash\{0\}$ in $\mathbb{R}$ where $\mathsf{Sp}'(a) := \mathsf{Sp}(a) \cup \{0\}$. Always $\mathsf{Sp}(a) = -\mathsf{Sp}(a)$ is a compact subset of $\mathbb{R}$ and $\mathsf{Sp}(a) \neq \emptyset$ holds if $a \neq 0$. In [17] the spectrum $\mathsf{Sp}(a,b)$ has been defined for elements a, b of an arbitrary Jordan triple system. For $b = a$ in our case always $\mathsf{Sp}(a,a) = \mathsf{Sp}'(a)^2$ holds.

It can be shown that

$$\|p(a)\| = \sup \left|p(\mathsf{Sp}(a))\right| \tag{2.1}$$

holds for every $a \in E$ and for every odd polynomial p. This together with the Weierstrass Approximation Theorem allows the following extension of the polynomial functional calculus from above: Let $f: \mathsf{Sp}(a) \to \mathbb{C}$ be an odd continuous function. Choose a sequence (p_n) of odd polynomials converging uniformly on $\mathsf{Sp}(a)$ to f and define $f(a) := \lim p_n(a)$. Then it is clear that (2.1) also holds for f in place of p and we have

Lemma 2.2. *Let $S \subset \mathbb{R}$ be a compact subset with $S = -S$ and let $f: S \to \mathbb{C}$ be an odd continuous function. Then for $A := \{a \in E : \mathsf{Sp}(a) \subset S\}$ the induced mapping $f: A \to E$ is continuous.*

Proof. The mapping $f: A \to E$ is a uniform limit of continuous mappings $p_n: A \to E$. □

Suppose, the odd continuous function $f: \mathsf{Sp}(a) \to \mathbb{C}$ has a representation of the form $f(t) = g(t^2)t$ for a continuous function g on $\mathsf{Sp}(a)^2$ (i.e. $f'(0)$ exists). Then approximating g uniformly by a sequence (g_n) in $\mathbb{C}[t]$ we derive

$$f(a) = g(a \square a)a \tag{2.3}$$

where $g(a \square a) = \lim g_n(a \square a) \in \mathcal{L}(E)$.

For every $0 < r \leq +\infty$ denote by $I_r \subset \mathbb{R}$ and $D_r \subset E$ the open balls of radius r about the origin. For every $a \in D_r$ then $\mathsf{Sp}(a) \subset I_r$ holds. Therefore every odd continuous function $f: I_r \to \mathbb{C}$ induces a continuous mapping $f: D_r \to E$ (which we also will denote by $\tilde{f}$ for better distinction). In case f is differentiable in $0 \in \mathbb{R}$ also $\tilde{f}$ is differentiable in $0 \in E$ and $\tilde{f}'(0) = f'(0)\mathrm{id}_E$. We do not know which f give rise to a differentiable mapping $\tilde{f}$ (compare [1]).

Proposition 2.4. *For every odd real-analytic function $f: I_r \to \mathbb{C}$ the induced mapping $f: D_r \to E$ is also real-analytic.*

Proof. Write $f(t) = g(t^2)t$ with a real-analytic function g. The holomorphic functional calculus in the Banach algebra $\mathcal{L}(E)$ implies that $a \mapsto g(a \square a)$ defines a real analytic mapping $D_r \to \mathcal{L}(E)$. The statement follows with (2.3). □

The element $a \in E$ is called *algebraic* if $p(a) = 0$ holds for some nontrivial odd polynomial p, or equivalently, if a has finite spectrum $\mathsf{Sp}(a)$. Then C_a has finite dimension and a is a finite linear combination of pairwise orthogonal tripotents in C_a. In general, $a = 0$ may be the only algebraic element in E. On the other hand, in every JBW*-triple the set of all algebraic elements is dense. Fortunately, every JB*-triple E is a subtriple of a JBW*-triple (for instance the bidual of E, compare. [6],[3],[10] for details).

Let us fix the element $a \in E$ for a moment and put for simplicity $C := C_a$, $R := R_a$ and $S := \mathsf{Sp}(a)$. Consider in $\mathcal{C}(S, \mathbb{R})$ the subspace $\mathcal{C}^-(S, \mathbb{R})$ of all odd functions and let $\mathcal{C}_o^+(S, \mathbb{R})$ be the subspace of all even functions f such that $f(0) = 0$ if $0 \in S$. It is clear that both subspaces are isometrically isomorphic to $\mathcal{C}_o(S^+, \mathbb{R})$ for $S^+ := \{t \in S : t > 0\}$. It is known (compare [14]) that there exists a unique isometric $\mathbb{R}$-linear isomorphism from R onto $\mathcal{C}^-(S, \mathbb{R})$ such that $a \in R$ corresponds to the function $a(s) \equiv s$ on S. Let us identify both spaces in this way. Then for every $x, y, z \in R$ we have $\{xyz\} = xyz$ (usual product of functions) and also $\mathsf{Sp}(x) = x(S)$. For every odd continuous function $f : \mathsf{Sp}(x) \to \mathbb{R}$ the element $f(x) \in R$ is nothing but the composition of functions $f \circ x$.

Lemma 2.5. *Let V be the $\mathbb{R}$-linear span of $R \square R$ in $\mathcal{L}(E)$. Then $L = 0$ holds for every $L \in V$ with $L(R) = 0$.*

Proof. Let $\boldsymbol{E} = E^{**}$ be the bidual of E and denote by $\boldsymbol{R}$ the closure of R in $\boldsymbol{E}$ with respect to the topology $w^* := \sigma(\boldsymbol{E}, E^*)$. Then with $R \square R \subset \mathcal{L}(E)$ also $\boldsymbol{R} \square \boldsymbol{R}$ is a commutative set of operators in $\mathcal{L}(\boldsymbol{E})$. Fix $L \in V$ with $L(R) = 0$ and an $\varepsilon > 0$. There are $a_i, b_i \in R$ with $L = \sum a_i \square b_i$ and L extends to a w^*-continuous operator (compare [3]) on $\boldsymbol{E}$ with $L(\boldsymbol{R}) = 0$. The algebraic elements in $\boldsymbol{R}$ are (norm) dense in $\boldsymbol{R}$. Therefore there exist pairwise orthogonal tripotents $e_1, \ldots, e_r$ in $\boldsymbol{R}$ and $c_1, \ldots, c_r \in \mathbb{R}$ such that

$$M := \sum_{i=1}^{r} c_i M_i \quad \text{with} \quad M_i := e_i \square e_i$$

satisfies $\|L - M\| < \varepsilon/2$.

$$|c_i| = \|(L - M)(e_i)\| < \varepsilon/2$$

implies by means of the Peirce decomposition with respect to $e_1, \ldots, e_r$ that $\|M\| < \varepsilon/2$ and hence $\|L\| < \varepsilon$ holds. This implies $L = 0$ since $\varepsilon > 0$ was arbitrary. □

Proposition 2.6. *$R \square R$ is a closed $\mathbb{R}$-linear subspace of $\mathcal{L}(E)$ and there exists an isometric $\mathbb{R}$-linear bijection $\tau : R \square R \to \mathcal{C}_o^+(S, \mathbb{R})$ with $\tau(f \square g) = fg$ for all $f, g \in R$. Moreover, every $L \in R \square R$ has a unique representation $L = f \square f - g \square g$ with $f, g \in R$ and $f \square g = 0$.*

Proof. Using [5] p. 102 and a limit argument it follows that for every $f, g \in R \cong \mathcal{C}^-(S, \mathbb{R})$ the operator $f \square g$ only depends on the product $fg \in \mathcal{C}_o^+(S, \mathbb{R})$, i.e. τ is well defined by $f \square g \mapsto fg$, compare also [21], Lemma 2.1. Fix $h \in \mathcal{C}_o^+(S, \mathbb{R})$ and define $v \in \mathcal{C}^-(S, \mathbb{R})$ by $v(t) := |h(t)|^{1/2}$ for $t \in S^+$ and odd extension to S. Then there is a unique $u \in \mathcal{C}^-(S, \mathbb{R})$ with $h = uv$, i.e. $h = \tau(u \square v)$ and τ is surjective. Consequently to every $M, N \in R \square R$ there is $P \in R \square R$ with $\tau(M + N) = \tau(P)$. But $L := M + N - P$ vanishes on R. This implies $L = 0$ by (2.5) and hence $M + N = P$. Therefore $R \square R$ is an $\mathbb{R}$-linear subspace of $\mathcal{L}(E)$. Since the elements $L \in R \square R$ are hermitian and satisfy $\|L\| = \|L|_R\|$ the mapping τ is an $\mathbb{R}$-linear isometry. □

For every $h \in \mathcal{C}_o^+(S, \mathbb{R})$ we denote the operator $\tau^{-1}(h) \in R \square R$ by $h \square 1$ or also by $h(a) \square 1$ although neither h nor 1 are in R. As an example we will consider later the even function $h(t) = \ln(1-t^2)$ on $\mathsf{Sp}(a)$ in case $\|a\| < 1$ and obtain the hermitian operator $\ln(1-a^2)\square 1 \in R\square R$.

It should be pointed out that $R\square R$ in general is not a subalgebra of $\mathcal{L}(E)$ although $C_o^+(S,\mathbb{R})$ has the structure of an algebra. The square $(x\square x)^2$, $x \in R$, in general is not even hermitian. Also the operator $\exp(L) \in \mathsf{GL}(E)$, $L \in R\square R$, is not necessarily hermitian (see f.i. (4.5)). Nevertheless the norm $\|\exp(L)\|$ coincides with the spectral radius $\rho(\exp(L))$ (compare [4] p. 54 (7)). From (2.5) we get immediately

Lemma 2.7. *$R\square R$ is a subalgebra of $\mathcal{L}(E)$ if and only if $(x\square x)^2 = x^3\square x$ holds for every $x \in R$.*

Let us assume that 0 is not contained in the spectrum $S := \mathsf{Sp}(a)$ of the element $a \in E$, i.e. a is regular. Then the odd function $f(t) = t^{-1}$ is continuous on S and we get an element $a^{-1} := f(a)$ in $C_a \subset E$. This satisfies $a = Q(a)^2 a^{-3}$ and hence is strongly regular (i.e. $a \in Q(a)^2 E$). Therefore a^{-1} coincides with the generalized inverse (Moore-Penrose inverse) of a in the sense of [7]. Denote by E^{-1} the set of all regular elements of E. This set is not open in E in general and may even consist of the origin in E alone. Also, the mapping $x \mapsto x^{-1}$ is not continuous on E^{-1} in general (compare also (2.2)).

3. The automorphism group

Let E be a JB*-triple and let $D \subset E$ be the open unit ball. Then the group $G = \mathsf{Aut}(D)$ of all biholomorphic automorphisms of E is a real Banach Lie group in the topology of locally uniform convergence in D. The subgroup $K := \{g \in G : g(0) = 0\}$ is the group of all (surjective) linear isometries of the Banach space E (restricted to D) and coincides also with the group of all linear triple automorphisms of E (for this and the following compare [14]). It is an algebraic subgroup of $\mathsf{GL}(E)$ in the sense of [9] (over $\mathbb{R}$). The Lie algebra $\mathfrak{g}$ of G can be identified with the space $\mathfrak{aut}(D)$ of all complete holomorphic vector fields on E. Here a holomorphic vector field on D is a differential operator $X = f(z)\frac{\partial}{\partial z}$ where $f: D \to E$ is a holomorphic function. By definition, X is complete on D if for every $z \in D$ the differential equation

$$\frac{\partial h_t}{\partial t} = f(h_t) \tag{3.1}$$

has a solution $h_t(z) \in D$ to the initial value $h_0(z) = z$ for all real t. Then $\{h_t : t \in \mathbb{R}\}$ is a one parameter subgroup of G. Instead of h_t we write $\exp(tX)$.

The Lie algebra $\mathfrak{g} = \mathfrak{aut}(D)$ has a decomposition as direct sum of closed linear subspaces

$$\mathfrak{g} = \mathfrak{k} \oplus \mathfrak{p} \tag{3.2}$$

where $\mathfrak{k}$ is the Lie subalgebra of all triple derivations of E and $\mathfrak{p}$ is the Lie triple system of all constant-quadratic vector fields $X^\alpha := \alpha - \{z\alpha z\}\frac{\partial}{\partial z}$, $\alpha \in E$. By definition, a linear operator $\delta: E \to E$ is called a derivation if

$$\delta\{xyz\} = \{(\delta x)yz\} + \{x(\delta y)z\} + \{xy(\delta z)\}$$

holds for every $x, y, z \in E$. Every derivation of E is automatically continuous [2] and every $\delta = i(\alpha\square\alpha)$ is a derivation by (J_2).

The analogue of the decomposition (3.2) on the group level is

$$G = KP = PK \qquad \text{with} \qquad P := \exp(\mathfrak{p}). \tag{3.3}$$

P is a closed real-analytic submanifold of G and $\exp: \mathfrak{p} \to P$ is bianalytic. To every $a \in D$ there is a unique $p \in P$ with $p(0) = a$. We denote this automorphism p by g_a in the following. There is a unique vector field $X^\alpha = (\alpha - \{z\alpha z\})\frac{\partial}{\partial z} \in \mathfrak{p}$ with $g_a = \exp(X^\alpha)$. The elements a and α generate the same closed subtriple of E and they are related by $a = \tanh(\alpha)$ and $\alpha = \tanh^{-1}(a)$. In particular $g_0 = \mathrm{id}$ and $g_{-a} = g_a^{-1}$ for all $a \in D$.

As a consequence of [14] p. 132 every $g \in G$ has a representation

$$g(z) = a + \lambda(1 + z\,\square\, a)^{-1}z, \quad z \in D \tag{3.4}$$

where $a := g(0)$ and $\lambda := g'(0)$. On the other hand there is a unique representation

$$g = g_a k$$

with $k \in K$ and hence

$$\lambda = \lambda_a k$$

for $\lambda_a := g_a'(0)$. It is known that

$$\lambda_a = B(a,a)^{1/2} \in \mathrm{GL}(E)$$

where the square root is understood in the sense of the holomorphic functional calculus on $\mathcal{L}(E)$ - notice that the Bergman operator

$$\begin{aligned} B(a,a) &= 1 - 2a\,\square\, a + Q(a)^2 \\ &= (1 - a\,\square\, a)^2 + (a\,\square\, a)^2 - a^3\,\square\, a \end{aligned}$$

has positive spectrum. An alternative description is given in the following

Theorem 3.5. *For every $a \in D$ the operator $\lambda_a \in \mathrm{GL}(E)$ also satisfies*

$$\lambda_a = \exp(\ln(1 - a^2)\,\square\, 1) .$$

Proof. For $\alpha := \tanh^{-1}(a) \in E$ consider the one parameter group $h_t := \exp(tX^\alpha)$ where $X^\alpha = (\alpha - \{z\alpha z\})\frac{\partial}{\partial z} \in \mathfrak{p}$. Since h_t solves (3.1), i.e.

$$\frac{\partial h_t}{\partial t} = \alpha - (h_t\,\square\,\alpha)h_t$$

we get by differentiation that $h_t'(0)$ solves

$$\frac{\partial h_t'(0)}{\partial t} = -2(a_t\,\square\,\alpha)h_t'(0)$$

to the initial value $h_0'(0) = \mathrm{id}$ where $a_t := h_t(0) = \tanh(t\alpha)$. This equation has the explicit solution (compare also [20] p. 233 and [11])

$$h_t'(0) = \exp(A(t)) \ \text{ with } \ A(t) := -2\int_0^t (a_s\,\square\,\alpha)ds .$$

For $t = 1$ we have $a_1 = a$, $h_1'(0) = \lambda_a$ and

$$A(1) = -2\int_0^1 (\alpha \tanh(s\alpha)\square 1)ds = (\ln\cosh(\alpha)^{-2}) \square 1$$
$$= (\ln(1-\tanh(\alpha)^2)) \square 1 = \ln(1-a^2) \square 1 .$$ □

Corollary 3.6. *For every $a \in D$ and $r := \|a\|$ we have*

(*i*) $$\|\lambda_a\| \le 1 \quad \text{and} \quad \|\lambda_a^{-1}\| = \frac{1}{1-r^2}$$

(*ii*) $$\|\lambda_a - \mathsf{id}\| \le \|\lambda_a^{-1} - \mathsf{id}\| = \frac{r^2}{1-r^2} .$$

Proof. The hermitian operator $\theta := \ln(1-a^2)\square 1$ has spectrum $\ge \ln(1-r^2)$ and ≤ 0 because of [14] Corollary 3.5. For every real s the norm of $\exp(s\theta)$ coincides with its spectral radius by [4] p. 64 (7). This implies easily (*i*) and hence also (*ii*). □

For applications of (3.6.ii) compare [11].

The considerations so far also apply to the dual symmetric manifold $\widetilde{D}$ of D. This is a symmetric complex Banach manifold (the canonical generalization of the compact dual in finite dimensions) containing the Banach space E as a canonical coordinate chart in such a way that the Lie group $\widetilde{G}$ of all biholomorphic isometries of $\widetilde{D}$ has

$$\widetilde{\mathfrak{g}} := \mathfrak{k} \oplus \widetilde{\mathfrak{p}} \quad \text{with} \quad \widetilde{\mathfrak{p}} := i\mathfrak{p}$$

as Lie algebra (compare [12] for details). For every $a \in E$ and $\alpha := \tan^{-1}(a) \in E$ put $\widetilde{g}_a := \exp(\widetilde{X}^\alpha)$ where $\widetilde{X}^\alpha := (\alpha + \{z\alpha z\})\frac{\partial}{\partial z} \in \widetilde{\mathfrak{p}}$. Then $\widetilde{g}_a \in \widetilde{G}$ and $\widetilde{g}_a(0) = a \in E \subset \widetilde{D}$ for all $a \in E$.

The proof of (3.5) works in the same way in the dual situation and gives

Proposition 3.7. *For every $a \in E$ the derivative $\widetilde{\lambda}_a := \widetilde{g}_a'(0) = B(a,-a)^{1/2} \in$ $\mathsf{GL}(E)$ also satisfies*

$$\widetilde{\lambda}_a = \exp(\ln(1+a^2)\square 1) .$$

This implies immediately $\|\widetilde{\lambda}_a\| = 1 + r^2$ for all $a \in E$ and $r := \|a\|$ - a result already essentially proved by a direct argument in [18] in connection with the computation of the curvature of $\widetilde{D}$.

Corollary 3.8. *For every $a \in E$ and $r := \|a\|$ we have*

(*i*) $$\|\widetilde{\lambda}_a^{-1}\| \le 1 \quad \text{and} \quad \|\widetilde{\lambda}_a\| = 1 + r^2$$

(*ii*) $$\|\widetilde{\lambda}_a^{-1} - \mathsf{id}\| \le \|\widetilde{\lambda}_a - \mathsf{id}\| = r^2 .$$

4. The structure group

Let E be a JB*-triple with open unit ball D. As before denote by $G := \mathsf{Aut}(D)$ the biholomorphic automorphism group of D. Then the isotropy subgroup $K := \{g \in G : g(0) = 0\}$ coincides with the group $\mathsf{GL}(D)$ of all linear isometries of E and also with the automorphism group

$$\mathsf{Aut}(E) := \{g \in \mathsf{GL}(E) : g\{xyz\} = \{(gx)(gy)(gz)\} \quad \forall\ x, y, z \in E\}$$

of the triple product. Therefore $K \subset \mathsf{GL}(E)$ is a (real) algebraic subgroup of degree ≤ 3 in the sense of [9]. Actually, using the equivalent equations

$$g\{x(g^{-1}y)z\} = \{(gx)y(gz)\}$$

we see that K is already algebraic of degree ≤ 2. Denote by $\mathsf{Str}(E)$ the group of all $g \in \mathsf{GL}(E)$ with the following property: There is an operator $h \in \mathsf{GL}(E)$ such that

$$g\{x(hy)z\} = \{(gx)y(gz)\}\} \tag{4.1}$$

holds for all $x, y, z \in E$. Then h is uniquely determined by g and satisfies $h\{x(gy)z\} = \{(hx)y(hz)\}$ (compare [14] p. 524). We denote it by $g^\star$. It is easily verified that $g \mapsto g^\star$ defines an antiholomorphic antiautomorphism of period 2 of $\mathsf{Str}(E)$ with

$$K = \{g \in \mathsf{Str}(E) : gg^\star = \mathrm{id}\}\ .$$

We call $\mathsf{Str}(E)$ the *structure group* of E. It can be given the structure of a complex Lie group in the following way: Denote by E^- the Banach space E with the conjugate complex structure. Then $\mathsf{Str}(E)$ may be identified with the algebraic subgroup of all $(g, h) \in \mathsf{GL}(E) \times \mathsf{GL}(E^-)$ satisfying (4.1) - which essentially is the automorphism group of the Jordan pair (E, E^-) associated with the JB*-triple E (compare [16] for details).

The Lie algebra $\mathfrak{s}$ of $\mathsf{Str}(E)$ can be identified with the space of all operators $\lambda \in \mathcal{L}(E)$ satisfying the following property: There is an operator $\sigma \in \mathcal{L}(E)$ such that

$$\lambda\{xyz\} = \{(\lambda x)yz\} - \{x(\sigma y)z\} + \{xy(\lambda z)\}$$

holds for all $x, y, z \in E$. Again, σ is uniquely determined by λ and will be denoted by λ^*. The antilinear involution $\lambda \mapsto \lambda^*$ of the complex Lie algebra $\mathfrak{s}$ gives a decomposition

$$\mathfrak{s} = \mathfrak{k} \oplus \mathfrak{h} \tag{4.2}$$

into $(-1)-$ and $(+1)-$eigenspaces. There $\mathfrak{h}$ is the space of all $\lambda \in \mathcal{L}(E)$ satisfying (J_2), i.e. $\mathfrak{h} = \mathcal{H}(E)$ is the space of all hermitian operators on E. In particular, $\mathfrak{s}$ contains all operators $a \square b$ with $a, b \in E$ and $(a \square b)^* = b \square a$. Furthermore $\mathfrak{k} = i\mathfrak{h}$ is the space of all triple derivations of E and is just the Lie algebra of the Lie subgroup $K \subset \mathsf{Str}(E)$.

In analogy to the setting $P := \exp(\mathfrak{p}) \subset G$ let us consider $H := \exp(\mathfrak{h}) \subset \mathsf{Str}(E)$. All operators in H have spectrum in $\mathbb{R}^+$ and it is clear that $H \subset \mathsf{Str}(E)$ is a real analytic submanifold and that $\exp : \mathfrak{h} \longrightarrow H$ is bianalytic. The question occurs whether also a polar decomposition for $\mathsf{Str}(E)$ exists like (3.3). Clearly, every $k \in K$ acts on $\mathfrak{k}$ by $\lambda \mapsto k\lambda k^{-1}$ and hence in the same way on $\mathfrak{h} = i\mathfrak{k}$. Consequently, $k\exp(\lambda)k^{-1} = \exp(k\lambda k^{-1}) \in H$ holds for every $k \in K$ and $\lambda \in \mathfrak{h}$. Therefore

$$\Gamma := KH = HK$$

is an open subgroup of $\mathsf{Str}(E)$ and hence a complex Lie group by itself. We do not know whether $\Gamma \neq \mathsf{Str}(E)$ may occur (compare the analogues result for C*-algebras in [19], 4.1.21 Corollary).

For every x, y in the open unit ball D of E the Bergman operator $B(x,y)$ is in the connected identity component of Γ and $B(x,y)^* = B(y,x)$. In case x, y are invertible in E the operator $g := Q(x)Q(y)$ is in $\mathsf{Str}(E)$ and $g^* = Q(y)Q(x)$. Furthermore the structure group $\mathsf{Str}(E)$ leaves the open subset of all invertible elements of E invariant and has open orbits there. In case E is a C*-algebra with unit for every invertible $a \in E$ the left multiplication operator $L(a)$ defined by $x \mapsto ax$ is in the group Γ (use the polar decomposition of a) and $L(a)^* = L(a^*)$.

Example 4.3. Let $E := \mathcal{C}(S)$ for a compact topological space S. Then $\mathsf{Str}(E) = \Gamma$ is the set of all operators $f(s) \mapsto c(s)f(\varphi(s))$ where $c \in E$ does not vanish on S and $\varphi: S \to S$ is a homeomorphism.

Let again $D \in E$ be the open unit ball and let $G := \mathsf{Aut}(D)$ be the biholomorphic automorphism group of D. Let us call $\Gamma \subset \mathsf{Str}(E)$ the *restricted structure group* of E. For every $a \in D$ the automorphism $g_a \in P \subset G$ has derivative $\lambda_a := g'_a(0) \in H$ by (3.5). Hence for every $g \in G$ and every $a \in D$, $b := g(a)$ there is a representation $g = g_b k g_a^{-1}$ with $k \in K$, i.e. $g'(a) = \lambda_b k \lambda_a^{-1}$ proving (compare also [14])

Lemma 4.4. *For every $g \in \mathsf{Aut}(D)$ and for every $a \in D$ the derivative $g'(a) \in \mathsf{GL}(E)$ is contained in the restricted structure group Γ of E.*

In general the elements of $H \subset \Gamma$ are not hermitian operators on E:

Example 4.5. Suppose that E is irreducible (i.e. not a direct sum $E' \oplus E''$ of nontrivial triple ideals) and that $a \in E$ is an algebraic element that is not proportional to a tripotent in E. Then for every real $s \neq 0$ the power $\lambda_a^s \in H$ (in particular for $s = 2$ the Bergman operator $B(a,a)$) is not hermitian.

Proof. By assumption there is representation $a = \sum_1^r \alpha_k e_k$ for an orthogonal system $\{e_1 \dots e_r\}$ of tripotents in E with real coefficients satisfying $0 < \alpha_1 < \dots < \alpha_r$ and $r \geq 2$. Denote by E_{ij} the corresponding Peirce spaces. Then E irreducible implies $E_{12} \neq \{0\}$ by [13] p. 471. The operator λ_a leaves the subtriple $E_{11} \oplus E_{12} \oplus E_{22}$ invariant. Without loss of generality we may therefore assume in the following that $E = E_{11} \oplus E_{12} \oplus E_{22}$. Then $e := e_1 + e_2$ is a unitary element of E. Fix a real $s \neq 0$ and put $\theta := s\ln(1-a^2)\square 1$. Assume that $g := \lambda_a^s = \exp(\theta)$ is an hermitian operator on E. For $\beta_k := (1-\alpha_k)^s$ and $b := \beta_1 e_1 + \beta_2 e_2$ the hermitian operator $\lambda := (b\square e) - g$ satisfies $\lambda(e_1) = \lambda(e_2) = 0$. On E_{12} the operator λ is just multiplication by

$$\frac{1}{2}\left((1-\alpha_1^2)^{s/2} - (1-\alpha_2^2)^{s/2}\right)^2 > 0\,,$$

i.e. $\lambda \neq 0$ vanishes in e and has spectrum ≥ 0 thus contradicting the following Lemma. ☐

Lemma 4.6. *Let E be an arbitrary JB*-triple. Suppose $e \in E$ is a unitary element and $\lambda \in \mathcal{L}(E)$ is an hermitian operator with $\lambda(e) = 0$. Then $\lambda = 0$ holds if λ has spectrum ≥ 0.*

Proof. $V := \{x \in E : x^* = x\}$ with $x^* := \{exe\}$ is a real form of E invariant under the operator $i\lambda \in \mathcal{L}(E)$. Therefore, if we assume that λ has spectrum $S \geq 0$, we conclude that $iS \subset \mathbb{C}$ is invariant under complex conjugation, i.e. $S = -S \geq 0$ and hence $S = \{0\}$. □

References

1. Arazy J., Barton T.J., Friedman Y.: Operator differentiable functions. Integral Equations and Operator Theory **13**, 461-487 (1990)
2. Barton T.J., Friedman Y.: Bounded derivations of JB*-triples. Quart. J. Math. Oxford **41**, 255-268 (1990)
3. Barton T.J., Timoney R. M.: On biduals, preduals and ideals of JB*-triples. Math. Scand. **59**, 177-191 (1986)
4. Bonsall F.F., Duncan J.: Numerical Ranges of Operators on Normed Spaces and of Elements of Normed Algebras. Cambridge Univerisity Press 1971
5. Braun R., Kaup W., Upmeier H.: On the automorphisms of circular and Reinhardt domains in complex Banach spaces. manuscripta math. **25**, 97-133 (1978)
6. Dineen S.: Complete holomorphic vector fields on the second dual of a Banach space. Math. Scand. **59**, 131-142 (1986)
7. Fernández A., Garcia E., Sánchez E., Siles M.: Strong regularity and generalized inverses in Jordan systems. Comm. in Algebra. **20**, 1917-1936 (1992)
8. Friedman Y., Russo B.: The Gelfand-Naimark theorem for JB*-triples. Duke Math. J. **53**, 139-148 (1986)
9. Harris L.A., Kaup W.: Linear algebraic groups in infinite dimensions. Illinois J. Math. **21**, 666-674 (1977)
10. Horn G.: Characterization of the predual and ideal structure of a JBW*-triple. Math. Scand. **61**, 117-133 (1987)
11. Isidro J.M., Kaup W.: Determining Boundary Sets of Bounded Symmetric Domains. manuscripta math. **81**, 149-159 (1993)
12. Kaup W.: Algebraic Characterization of Symmetric Complex Banach Manifolds. Math. Ann. **228**, 39-64 (1977)
13. Kaup W.: Über die Klassifikation der symmetrischen Hermiteschen Mannigfaltigkeiten unendlicher Dimension I, II. Math. Ann. **257**, 463-483 (1981); **262**, 503-529 (1983)
14. Kaup W.: A Riemann mapping theorem for bounded symmetric domains in complex Banach spaces. Math. Z. **183**, 503-529 (1983)
15. Loos O.: Bounded symmetric domains and Jordan pairs. Mathematical Lectures. Irvine: University of California at Irvine 1977
16. Loos O.: Jordan Pairs. Lecture Notes in Mathematics Vol. 460. Berlin-Heidelberg-New York: Springer 1975
17. Loos O.: Properly algebraic and spectrum-finite ideals in Jordan systems. Math. Proc. Camb. Phil. Soc. **114**, 149-161 (1993)
18. Mellon P.: Holomorphic curvature of infinite dimensional symmetric complex Banach manifolds of compact type. Ann. Acad. Sci. Fennicae. **18**, 299-306 (1993)
19. Sakai S.: C*-Algebras and W*-Algebras. Berlin-Heidelberg-New York: Springer 1971
20. Stachó L: On the spectrum of inner derivations in partial Jordan triples. Math. Scand. **66**, 242-248 (1990)
21. Stachó L: On the algebraic classification of bounded circular domains. Proc. R. Ir. Acad. **91**, 219-238 (1991)

QUADRATIC DIFFERENTIAL EQUATIONS ON GRADED STRUCTURES

MICHAEL K. KINYON
Indiana University South Bend
South Bend, IN 46634 USA

Abstract. In this note, an example is constructed which illustrates one of the major results of the contribution of N. Hopkins [1] to these proceedings.

1. Background

Let V be a real finite dimensional vector space and let $p : V \to V$ be a polynomial mapping. Thus for each $X \in V$, $p(X) = \sum_{j \geq 0} p_j(X)$ where each p_j is homogeneous of degree j, i.e., $p_j(aX) = a^j p_j(X)$ for all $a \in \mathbf{R}$, $X \in V$.

For $P \in V$, we consider the associated initial value problem

$$\dot{Z} = p(Z)$$

$$Z(0) = P.$$

We denote the corresponding solution by $Z(t, P)$.

Of particular interest is the case where p itself is homogeneous of degree, say, m. For then corresponding to p, there exists an m-linear mapping $\tilde{p} : V \times \cdots \times V \to V$ such that $\tilde{p}(X, \ldots, X) = p(X)$ for all $X \in V$; in particular $\tilde{p}(X_1, \ldots, X_m) = \frac{1}{m!} p^{(m)}(0) \cdot (X_1, \ldots, X_m)$ where $p^{(m)}$ is the m-th derivative of p. As is customary [5] we shall not notationally distinguish between p and $\tilde{p}$. Thus corresponding to p, there exists an *m-ary algebra* $A = (V, p)$.

A special case of considerable interest is the *quadratic* case $m = 2$. Here we allow ourselves a further abbreviation $XY := p(X, Y)$ and we write the corresponding differential equation as $\dot{Z} = Z^2$. This allows us to say that the equation $\dot{Z} = Z^2$ *occurs* in the algebra A.

We refer the interested reader to the literature for detailed studies of the nonassociative approach to polynomial differential equations in general, and quadratic differential equations occurring in algebras in particular (see [2] and [5] for more extensive bibliographies).

For an m-ary algebra $A = (V, p)$, the *automorphism group* is defined as usual to be

$$Aut\ A = \{\phi \in GL(V) : \phi p(X) = p(\phi X) \ \ \forall X \in V\}.$$

The Lie algebra of this Lie group is the *derivation algebra*

$$Der\ A = \{D \in gl(V) : Dp(X) = 2p'(X) \cdot DX \ \ \forall X \in V\}.$$

S. González (ed.), Non-Associative Algebra and Its Applications, 215–218.

For $P \in A$, we let $A(P)$ denote the subalgebra of A generated by P. It is well-known and easily observed that the solution trajectory to $\dot{Z} = p(Z)$ through the initial point P remains in $A(P)$ for all $t \in \mathbf{R}$ for which the solution exists [4].

We now focus on the quadratic case $\dot{Z} = Z^2$.

For $\phi \in Aut\ A$ and $P \in V$, the decomposition of (the complexification of) $A(P)$ relative to the eigenspaces of ϕ can provide information about the qualitative behavior of the trajectory through P. Of interest is the case where P is an eigenvector for ϕ corresponding to the eigenvalue -1. For then $A(P)$ has a natural $\mathbf{Z}_2$-grading $A(P) = A_0(P) \oplus A_1(P)$ where $A_i(P) = \{Q \in A(P) : \phi Q = (-1)^i Q\}$. Conversely, if $A = A_0 \oplus A_1$ is a $\mathbf{Z}_2$-grading of A (that is $A_i A_j \subseteq A_{i+j \pmod 2}$), then the mapping $\phi : A \to A$ defined by $\phi(Z_0 + Z_1) = Z_0 - Z_1$ is an automorphism of A satisfying $A^2 = I$.

In her contribution to these proceedings [1], Nora Hopkins addressed the following issue: suppose that a trajectory passes through the (-1)-eigenspaces of two automorphisms at distinct times. What can one say about the qualitative behavior of the trajectory? She obtained the following answer [1, Theorem 3.2].

Theorem 1: *Let $\dot{Z} = Z^2$ occur in A and suppose there exist $\phi_1, \phi_2 \in Aut\ A$ with $\phi_1 P = -P$ and $\phi_2 Z(t_0, P) = -Z(t_0, P)$ for some $t_0 \neq 0$. If $(\phi_1\phi_2)|_{A(P)}$ has finite order, then $Z(t, P)$ is periodic.*

Remark: As an examination of the proof of this theorem shows, the quadratic nature of the right hand side of the differential equation manifests itself solely in the identity $Z(t, -P) = -Z(-t, P)$. In fact, the theorem holds for all differential equations $\dot{Z} = p(Z)$ where p is a polynomial (or even analytic) mapping satisfying $p(-X) = p(X)$ for $X \in V$; thus p consists of only even terms.

In the remainder of this note, we construct an example (homogeneous, quadratic) of the phenomenon described in this theorem. We begin this construction in the next section with a refined decomposition of the algebra A and of the corresponding differential equation.

2. The Construction

Suppose that $\phi_1, \phi_2 \in Aut\ A$ each satisfy $\phi_i^2 = I$, and let $A = A_0^1 \oplus A_1^1$ and $A = A_0^2 \oplus A_1^2$ be the corresponding $\mathbf{Z}_2$-gradings. In order to simplify the subsequent construction, we make the following assumption: there exist subspaces A_{00}, A_{01}, A_{10}, and A_{11} such that

$$A = A_{00} \oplus A_{01} \oplus A_{10} \oplus A_{11}$$

$$A_0^1 = A_{00} \oplus A_{01},\ A_1^1 = A_{10} \oplus A_{11}$$

$$A_0^2 = A_{00} \oplus A_{10},\ A_1^2 = A_{01} \oplus A_{11}$$

With this assumption and the grading rules $A_j^i A_k^i \subseteq A_{j+k \pmod 2}^i$, we find that relative to the complete decomposition into four subspaces, the algebra A is $\mathbf{Z}_2 \times \mathbf{Z}_2$-graded, that is

$$A_{ij} A_{kl} \subseteq A_{i+k, j+l \pmod 2}.$$

Thus for $Z = Z_{00} + Z_{01} + Z_{10} + Z_{11}$, we find that the quadratic differential equation $\dot{Z} = Z^2$ decomposes as follows:

$$\begin{aligned}
\dot{Z}_{00} &= Z_{00}^2 + Z_{01}^2 + Z_{10}^2 + Z_{11}^2 \\
\dot{Z}_{01} &= 2Z_{00}Z_{01} + 2Z_{10}Z_{11} \\
\dot{Z}_{10} &= 2Z_{00}Z_{10} + 2Z_{01}Z_{11} \\
\dot{Z}_{11} &= 2Z_{00}Z_{11} + 2Z_{01}Z_{10}
\end{aligned}$$

Next we make our example specific and assume that each subspace A_{ij} is one-dimensional. Thus $A \cong \mathbf{R}^4$ as a vector space. Simplifying our subscripts, this means that the quadratic system has the form

$$\begin{aligned}
\dot{x}_1 &= a_1x_1^2 + a_2x_2^2 + a_3x_3^2 + a_4x_4^2 \\
\dot{x}_2 &= 2b_1x_1x_2 + 2b_2x_3x_4 \\
\dot{x}_3 &= 2c_1x_1x_3 + 2c_2x_2x_4 \\
\dot{x}_4 &= 2d_1x_1x_4 + 2d_2x_2x_3
\end{aligned}$$

For this system, the (matrix representations of the) corresponding automorphisms are

$$\phi_1 = \begin{bmatrix} 1 & 0 & 0 & 0 \\ 0 & 1 & 0 & 0 \\ 0 & 0 & -1 & 0 \\ 0 & 0 & 0 & -1 \end{bmatrix} \quad \text{and} \quad \phi_2 = \begin{bmatrix} 1 & 0 & 0 & 0 \\ 0 & -1 & 0 & 0 \\ 0 & 0 & 1 & 0 \\ 0 & 0 & 0 & -1 \end{bmatrix}.$$

Now the difficulty with applying Theorem 1 in practice that one must establish the existence of a trajectory connecting the two (-1)-eigenspaces. With ten free parameters, this is a daunting task. Instead we try a different route, using the following theorem from [2]. As before $Z(t, P)$ denotes the (unique) solution to $\dot{Z} = Z^2$, $Z(0) = P$.

Theorem: *Let $D \in Der\ A$. Then $Z(t, P) = e^{tD}P$ if and only if $DP = P^2$.*

This theorem was used to construct quadratic systems exhibiting periodic and quasiperiodic trajectories in [2] [3]. Here we use it to construct a trajectory connecting the (-1)-eigenspaces of ϕ_1 and ϕ_2.

Set

$$D = \begin{bmatrix} 0 & 0 & 0 & 0 \\ 0 & 0 & 1 & 0 \\ 0 & -1 & 0 & 0 \\ 0 & 0 & 0 & 0 \end{bmatrix}.$$

We force D to be a derivation of the algebra A. Routine calculations put restrictions on the coefficients (structure constants), yielding the following system.

$$\begin{aligned}\dot{x}_1 &= a_1x_1^2 + a_2(x_2^2 + x_3^2) + a_4x_4^2 \\ \dot{x}_2 &= 2b_1x_1x_2 + 2b_2x_3x_4 \\ \dot{x}_3 &= 2b_1x_1x_3 - 2b_2x_2x_4 \\ \dot{x}_4 &= 2d_1x_1x_4.\end{aligned}$$

Next we solve the pair of equations $DP = P^2$ and $\phi_1 P = -P$ (or more accurately, we force a solution to exist). Since P is a (-1)-eigenvector for ϕ_1, we can normalize it so that in coordinates $P = (0,1,0,r)^t$ for some $r \in \mathbf{R}$. Now $DP = (0,0,-1,0)^t$, while $P^2 = (a_2 + a_4r^2, 0, -2b_2r, 0)^t$. Thus we require $2b_2r = 1$ so that $r = 1/2b_2$, and $0 = a_2 + a_4r^2 = a_2 + a_4/4b_2^2$. With these stipulations, the first equation in our system reduces to

$$\dot{x}_1 = a_1x_1^2 + a_2(x_2^2 + x_3^2 - 4b_2^2x_4^2),$$

and the rest of the system remains unchanged.

Now $Z(t,P) = e^{tD}P$ and we observe that $Z(\pi/2, P) = (0,0,-1,1/2b_2)$. Thus $\phi_2 Z(\pi/2,P) = -Z(\pi/2,P)$. Therefore the conditions of Theorem 1 are satisfied and the solution $Z(t,P) = e^{tD}P$ is periodic (as we also observe from the solution formula itself). This completes the construction.

Finally, at the referee's suggestion, we make an observation. Since the algebra has an infinite automorphism group, our example could be seen as a bit unsatisfactory. However, it is clear from continuous dependence of solutions on parameters that any system that is close to the given one and retains the finite automorphism group will also satisfy the hypotheses of Hopkins' theorem.

References

1. N. C. Hopkins, Quadratic differential equations in graded algebras, these Proceedings.
2. M. K. Kinyon and A. A. Sagle, Quadratic dynamical systems and algebras, to appear in J. Diff. Eq.
3. M. K. Kinyon and A. A. Sagle, Automorphisms and derivations of ordinary differential equations and algebras, to appear in Rocky Mountain Math. J.
4. L. Markus, Quadratic differential equations and nonassociative algebras, in "Contributions to the Theory of Nonlinear Oscillations", Vol. V, L. Cesari, J.P. LaSalle, and S. Lefschetz (eds.), Princeton Univ. Press, Princeton, 1960, pp. 185-213
5. S. Walcher, "Algebras and Differential Equations", Hadronic Press, Palm Harbor, 1991

A GENERALIZATION OF NOVIKOV RINGS

ERWIN KLEINFELD
Division of Mathematical Sciences
The University of Iowa
Iowa City, Iowa 52242 U.S.A.

and

HARRY F. SMITH
Department of Mathematics
Statistics and Computing Science
University of New England
Armidale, N.S.W. 2351 AUSTRALIA

Abstract. Generalizations of theorems on simple Novikov algebras by E. I. Zel'manov and J. M. Osborn to a subvariety in the join of associative and Novikov are obtained.

Key words: algebras, Novikov algebras

In [1, 2] a class of rings was introduced that satisfy the identities

$$x(yz) = y(xz) \tag{1}$$

and

$$x(yz - zy) = (xy)z - (xz)y\ . \tag{2}$$

The latter identity can be written in associator form as

$$(x, y, z) = (x, z, y)\ . \tag{3}$$

In subsequent publications [4, 5] these rings have been called Novikov rings. We have in mind replacing Equation 1 by the more general identity

$$(w, x, yz) = y(w, x, z)\ . \tag{4}$$

In order to clarify this we shall call rings satisfying Equation 1 and Equation 3 strongly Novikov and rings satisfying Equation 3 and Equation 4 weakly Novikov.

Simple, finite dimensional strongly Novikov algebras over a field of characteristic zero have been classified by E. I. Zel'manov [7]. They must be associative and commutative fields. We are able to generalize this result to weakly Novikov rings. However the conclusion in that case is that they must be associative rings, for weakly Novikov rings are within the variety of associative rings. A similar comment applies to the main result due to J. M. Osborn [5].

Lemma 1. *If R is strongly Novikov then it is weakly Novikov. Moreover weakly Novikov rings are a subclass of associative rings whereas strongly Novikov rings are not.*

S. González (ed.), Non-Associative Algebra and Its Applications, 219–222.

Proof. Note that $(w,x,yz) = (wx)(yz) - w\{x.yz\} = y\{wx.z\} - y\{w.xz\} = y(w,x,z)$, using only Equation 4 and the definition of the associator. Thus strongly Novikov rings are weakly Novikov. In an asociative ring that satisfies Equation 1 we would have the identity $[x,y]z = 0$. It is easy to see that 2×2 rational matrices do not satisfy this identity, so that strongly Novikov rings are not a subclass of the variety of associative rings. This concludes the proof.

Henceforth we assume that R is a weakly Novikov ring.

Definition. *The right nucleus N or R is defined as $N = \{n \in R | (R,R,n) = 0\}$.*

Lemma 2. *The right nucleus N of R is an ideal of R. The associator ideal A of R is the set of all finite sums of associators. Moreover $NA = 0$.*

Proof. Let $n \in N$, and $x,y,z \in R$. Then $(x,y,zn) = z(x,y,n) = 0$, using Equation 4. Thus N is a left ideal. It follows from Equation 3 that n is also in the middle nucleus, in other words that $(R,n,R) = 0$. At this point we can use the Teichmueller identity, which holds in every ring: $(xy,n,z) - (x,yn,z) + (x,y,nz) = x(y,n,z) + (x,y,n)z$. However 4 of these associators are already known to be zero, since $-(x,yn,z) = -(x,z,yn) = 0$. Thus also $(x,y,nz) = 0$. This shows that N is a right ideal as well, hence an ideal. But then Equation 4 implies

$$N(R,R,R) = 0\ . \tag{5}$$

The associator ideal A is known to be the set of all finite sums of associators and left multiples of associators in an arbitrary ring. But then Equation 4 reduces this to finite sums of associators in R. Consequently Equation 5 implies $NA = 0$. This completes the proof.

Lemma 3. $N \subseteq \{x(yz) - y(xz)\}$.

Proof. We observe that $(w,x,yz) = y(w,x,z) = y(w,z,x)$ using Equation 3 and Equation 4. Thus

$$(w,yz,x) = y(w,z,x)\ . \tag{6}$$

By a combination of Equation 4 and Equation 6 we deduce that $(a,b,x(yz)) = x(a,b,yz) = (a,xb,yz) = y(a,xb,z) = y\{x(a,b,z)\} = y(a,b,xz) = (a,b,y(xz))$. But then $(a,b,x(yz) - y(xz)) = 0$, completing the proof.

Lemma 4. *If $a,b,x,y,z \in R$ and $p = (a,x(b,y,z))$, then the 12 permutations which map $\{a,b\}$ on itself and $\{x,y,z\}$ on itself all fix p.*

Proof. It follows from Equation 3 that the permutation (yz) fixes p. Since $p = (a,x,by.z - b.yz) = by.(a,x,z) - b.y(a,x,z) = (b,y,(a,x,z))$, using only Equation 4 and the definition of associator, we see that the permutation $(ab)(xy)$ also fixes p. By combining these two results we see that $p = (a,x,(b,y,z)) = (a,x(b,z,y)) = (b,z,(a,x,y)) = (b,z,(a,y,x)) = (a,y,(b,z,x)) = (a,y,(b,x,z))$. Thus (xy) also fixes p. But then (ab) fixes p. At this point we have enough generators to see that all 12 permutations fix p. This concludes the proof.

Definition. *Let $J_1 = A, \dots J_{n+1} = \Sigma(R, R, J_n)$.*

Lemma 5. *The J_n are all ideals and $J_n \subseteq J_{n+1}$. If R is semiprime and not associative then $J_n \neq 0$ for all positive integers n.*

Proof. Assume we have already shown that J_n is an ideal. It follows from Equation 4 that $R(R, R, J_n) = (R, R, RJ_n)$, so that J_{n+1} is a left ideal. However we get from the Teichmueller identity that $(R, R, J_n)R = (R, R, J_nR) - (R, RJ_n, R) + (RR, J_n, R) - R(R, J_n, R)$, so that induction and Equation 3 J_{n+1} is a right ideal as well and consequently an ideal. Thus we have proved the first part. Suppose that R is semiprime and not associative. Then $J_1 \neq 0$. Let n be the smallest positive integer such that $J_n = 0$. Define $I = N \cap A$. Since $NA \subseteq I^2$ it follows from Lemma 2 that $I^2 = 0$, so that semiprime now implies $I = 0$. However $J_n = 0$ implies $N \subseteq J_{n-1}$. Since it is obvious that $A \subseteq J_{n-1}$, we have $I \subseteq J_{n-1}$ so that $J_{n-1} = 0$. This is a contradiction, since n was supposed to be the smallest positive integer such that $J_n = 0$. Thus no such positive integer exists and the proof is complete.

Theorem 1. *If R is prime and not associative then $N = 0$, and R must be strongly Novikov.*

Proof. Since $A \neq 0$ and $NA = 0$ as a consequence of Lemma 2, then prime implies $N = 0$. But then Lemma 3 yields $x(yz) - y(xz) \in N$, and so R must be strongly Novikov. This completes the proof of the theorem.

Corollary: Simple, finite dimensional, weakly Novikov algebras over a field of characteristic zero must be associative, hence total matrix rings over division algebras.

Proof. If R satisfies the hypothesis and is not associative then it must be strongly Novikov as a consequence of the theorem. But then [7] – Theorem 1 implies R must be a field, a contradiction. Thus R must have been associative to begin with. This concludes the proof of the corollary.

References

1. A. A. Balinskii and S. P. Novikov, Poisson brackets of hydrodynamic type, Frobenius algebras and Lie algebras, Dokl. Akad. Nauk SSSR **283** (1985), 1036–1039; English Transl. Soviet Math. Dokl. **32** (1985), 228–231.
2. I. M. Gelfand and I. Ya. Dorfman, Hamiltonian operators and related algebraic structures, Funktsional Anal. I. Prolozhen, **13** No. 4 (1979), 13–30; English Transl. Funct. Anal. and Appl. **13** (1979), 248–262.
3. E. Kleinfeld, On rings satisfying $(x, y, z) = (y, x, z)$, Algebras Groups Geom., **4** (1987), 129–138.
4. J. M. Osborn, Novikov algebras, Nova J. Algebra Geom., **1** (1992), 1–14.
5. J. M. Osborn, Simple Novikov algebras with an idempotent, Comm. Algebra, **20** No. 9 (1992), 2729–2753.

6. J. M. Osborn and E. I. Zelmanov, Nonassociative algebras related to Hamiltonian operators in the formal calculus of variations, to appear.
7. E. I. Zelmanov, On a class of local translation invariant Lie algebras, Soviet Math. Dokl. 35 (1987), 216–218; English Transl. Soviet Math. Dokl. 35 (1987), 216–218.

AUTOMORPHISMES ET DÉRIVATIONS DANS LES ALGÈBRES DE BARKER

BEN YAKOUB L'MOUFADAL
Faculté Des Sciences
Département des Mathématiques
B. P. 2121 - Tétouan - Maroc

Abstract. Le but de cette note est de caractériser certains J-Automorphismes et J-dérivations d'une classe de sous-algèbres A de l'algèbre des matrices triangulaires à coefficients dans une algèbre non associative R. Soit ψ un J-Automorphisme de A qui laisse fixe E_{ii} et applique E_{ij} dans le noyau de A. (E_{ij} étant la matrice ayant 1 dans la position (i,j) et 0 ailleurs), alors $\psi \in Aut(A)$ et $\psi = I_U \circ \phi^{\#}$ où I_U est la conjugaison par l'élément inversible $U \in N(A)$, $\phi \in Aut(R)$ et $\phi^{\#}((u_{ij})) = (\phi(u_{ij}))$. De même toute J-dérivation de A qui envoie E_{ij} dans $N(A)$ est une dérivation de A qui est la somme d'une dérivation intérieure et d'une dérivation $d^{\#}$ de A telle que $d^{\#}((u_{ij})) = (d(u_{ij}))$ où d est une dérivation de R.

Mots clés : Algèbre de Barker, Automorphisme, Dérivation, Graphe, Matrice triangulaire, Noyau.

1. Introduction

Pendant ces dernières années, plusieurs chercheurs ont étudié les automorphismes et les dérivations dans les algèbres $T_n(K)$ des matrices triangulaires à coefficients dans K, par exemple Barker et Kezlan ont montré dans [3] que tout K-Automorphisme de $T_n(K)$ est intérieur où K est un anneau commutatif intègre. Dans un article très récent [7] Kezlan a généralisé ce résultat au cas où K est un anneau commutatif. Dans [1] Barker s'est intéressé à certaines sous-algèbres de $T_n(K)$ où K est un corps commutatif, ces sous-algèbres sont déterminées à partir de graphes ([1], [2] et [6]). Dans [8] Jondrup a montré que si R est une algèbre simple artinienne de dimension finie sur son centre K, alors tout K-automorphisme (resp. dérivation) de $T_n(R)$ est intérieur. Pour le cas non associatif, Benkart et Osborn [5] ont étudié les automorphismes et les dérivations de $M_n(R)$ l'algèbre de matrices à coefficients dans une algèbre non associative R.

Dans ce travail on va contribuer à étudier les automorphismes et les dérivations de certaines sous-algèbres de $T_n(R)$ où R est une algèbre non associative, ces sous-algèbres seront appelées les algèbres de Barker (définition 2. 2)

2. Algèbres de Barker

2.1. Préliminaires sur les Graphes ([1], [2], [6])

On appelle graphe tout couple $G = (V(G), E(G))$ dont le premier élément est un ensemble et le second est un sous-ensemble de $V(G) \times V(G)$. Les éléments de $V(G)$

S. González (ed.), *Non-Associative Algebra and Its Applications*, 223–228.

sont appelés les sommets de G et les éléments de $E(G)$ sont appelés les arêtes de G. Un graphe partiel P d'un graphe G est le graphe $P = (V(P), E(P))$ avec $V(P) = V(G)$ et $E(P) \subseteq E(G)$. Soient $T = (V(T), E(T))$ le graphe tel que $V(T) = \{1, 2, ..., n\}$ $(n > 1)$ et $E(T) = \{(i,j) : i \leq j\}$ et Γ un graphe partiel de T. On dit que Γ est connexe si et seulement si pour tout $i \leq j$, il existe une suite $i = j_1, j_2, \ldots, j_p = j$ telle que $(j_q, j_{q+1}) \in E(\Gamma)$ ou bien $(j_{q+1}, j_q) \in E(\Gamma)$ pour tout $q = 1, 2, \ldots, p$. Γ est dit transitif si et seulement si $(i,j) \in E(\Gamma)$ et $(j,k) \in E(\Gamma)$ implique $(i,k) \in E(\Gamma)$. En suite, on dit que Γ vérifie la propriété d'interpolation si et seulement si $(i,j) \in E(\Gamma)$, $(i,k) \in E(\Gamma)$ avec $j < k$ alors $(i,k) \in E(\Gamma)$.

2.2. Algèbres de Barker

Soient K un anneau commutatif et R une K-algèbre unitaire non nécessairement associative, Γ un graphe partiel de T qui est connexe, transitif et vérifie la propriété d'interpolation. On note : $T_n(\Gamma, R)$ la K-algèbre de matrices triangulaires supérieures $M = (m_{st})$ à coefficients dans R telles que : si $(i,j) \notin E(\Gamma)$ on a $m_{ij} = 0$. L'algèbre $T_n(\Gamma, R)$ sera appelée l'algèbre de Barker associée à Γ et R.

3. Automorphismes des algèbres de Barker

Soit R une algèbre non associative unitaire de caractéristique $\neq 2$, on définit $N(R)$ le noyau de R par : $N(R) = \{a \in R : (a,x,y) = (x,a,y) = (x,y,a) = 0 \,\forall\, x,y \in R\}$ où $(a,b,c) = (ab)c - a(bc)$. On vérifie facilement que $N(R)$ est une algèbre associative et si a est un élément inversible de $N(R)$, l'application I_a de R dans R définie par $I_a(x) = axa^{-1}$ est un automorphisme de R. Tout automorphisme de R est un automorphisme de R^+ (R^+ étant R muni du produit de Jordan : $xoy = \frac{1}{2}(xy + yx)$ pour tout $x, y \in R$). Soit $A = T_n(\Gamma, R)$ une algèbre de Barker associée à Γ et R. Si f est une application de R dans R on définit l'application $f^{\#}$ de A dans A par : $f^{\#}((u_{ij})) = (f(u_{ij}))$ pour tout $(u_{ij}) \in A$. Dans A^+ il n'existe que les quatre produits non nuls suivants :

$$(aE_{ii}) \circ (bE_{ii}) = (aob)E_{ii} \;;\; (aE_{ii}) \circ (bE_{ij}) = \tfrac{1}{2}abE_{ij}$$
$$(aE_{ij}) \circ (bE_{jj}) = \tfrac{1}{2}abE_{ij} \;;\; (aE_{ij}) \circ (bE_{jk}) = \tfrac{1}{2}abE_{ik}.$$

pour tout $a, b \in R$ (où E_{ij} est la matrice ayant 1 dans la position (i,j) et 0 ailleurs). Notons dans la suite $Aut(A)$ le groupe des automorphismes de A, $J\text{-}Aut(A) = Aut(A^+)$ et $(J\text{-}Aut(R))^{\#} = \{\phi^{\#}$ tel que $\phi \in J\text{-}Aut(R)\}$.

En utilisant quelques techniques de Benkart et Osborn dans [5] on démontre facilement le lemme suivant :

Lemme 3.1 *Soient A une algèbre de Barker et $\psi \in J\text{-}Aut(A)$ tel que $\psi(E_{ii}) = E_{ii}$ pour tout $1 \leq i \leq n$. On a pour tout $a, b \in R$:*

0/ $\psi(aE_{ii}) = \psi_i(a)E_{ii}$ avec $\psi_i \in J\text{-}Aut(R)$.

1/ $\psi(E_{ij}) = \alpha_{ij}E_{ij}$ avec $\alpha_{ij} \in R$.

2/ $\psi(aE_{ij}) = \psi_i(a)\alpha_{ij}E_{ij} = \alpha_{ij}\psi_j(a)E_{ij}$.

3/ $\psi_i(a)\alpha_{ij} = \alpha_{ij}\psi_j(a)$.

4/ $\alpha_{ik} = \alpha_{ij}\alpha_{jk}$ $(i \leq j \leq k)$.

5/ $\psi^{-1}(E_{ij}) = \beta_{ij}E_{ij}$ *et* $\psi_i(\beta_{ij})\alpha_{ij} = 1 = \alpha_{ij}\psi_j(\beta_{ij})$.

6/ $\alpha_{ij} \in N_m(R) = \{a \in R : (x,a,y) = 0 \text{ pour tout } x, y \in R\}$.

7/ $\psi_i(a)[\psi_i(b)\alpha_{ij}] = \psi_i(ab)\alpha_{ij}$.

8/ $[\alpha_{ij}\psi_j(a)]\psi_j(b) = \alpha_{ij}\psi_j(ab)$.

9/ $\psi_i(a)[\psi_i(b)\alpha_{ij}] = [\alpha_{ij}\psi_j(a)]\psi_j(b)$.

Proposition 3.2 *Soient A une algèbre de Barker et $\psi \in J\text{-}Aut(A)$ tel que : $\psi(E_{ii}) = E_{ii}$ pour tout $1 \leq i \leq n$. Il existe $U, V \in A$ avec $U.V = 1$ et $R_V \circ L_U \circ \psi \in (J\text{-}Aut(R))^\#$, R_V et L_U étant les opérateurs de multiplication à droite et à gauche respectivement.*

Preuve. Soient $U = E_{11} + \alpha_{1,2}E_{2,2} + \ldots + \alpha_{1n}E_{nn}$ et $V = E_{1,1} + \psi_2(\beta_{1,2})E_{2,2} + \ldots + \psi_n(\beta_{1n})E_{nn}$. $U.V = 1$ d'après la propriété (5) du lemme précédent, l'application $\omega = R_V \circ L_U$ vérifie

- $(\omega \circ \psi)(aE_{1,1}) = \psi_1(a)E_{1,1}$

- $(\omega \circ \psi)(aE_{ii}) = \omega(\psi_i(a)E_{ii}) = [(\alpha_{1i}\psi_i(a))\psi_i(\beta_{1i})]E_{ii} \overset{(3)}{=} [(\psi_1(a)\alpha_{1i})\psi_i(\beta_{1i})]E_{ii} \overset{(6)}{=} [\psi_1(a)(\alpha_{1i}\psi_i(\beta_{1i}))]E_{ii} \overset{(5)}{=} \psi_1(a)E_{ij}$

- $(\omega \circ \psi)(aE_{ij}) = \omega(\alpha_{ij}\psi_j(a)E_{ij}) = [[\alpha_{1i}(\alpha_{ij}\psi_j(a))]\psi_j(\beta_{1j})]E_{ij} \overset{(6)+(4)}{=} [(\alpha_{1j}\psi_j(a)\psi_j(\beta_{1j})]E_{ij} \overset{(3)}{=} [(\psi_1(a)\alpha_{1j})\psi_j(\beta_{1j})]E_{ij} \overset{(6)+(5)}{=} \psi_1(a)E_{ij}$.

On en déduit que : $\omega o \psi = R_V o L_U o \psi = \psi_1^\#$ ce qui complète la preuve. □

Théorème 3.3 *Soient A une algèbre de Barker et $\psi \in J\text{-}Aut(A)$ tel que :*
i/ $\psi(E_{ii}) = E_{ii}$ pour tout $1 \leq i \leq n$.
ii/ $\psi(E_{ij}) \in N(A)$ pour tout $1 \leq i \leq j \leq n$.
On a : $\psi = I_U \circ \phi^\#$ avec $\phi \in Aut(R)$ et $I_U(X) = U.X.U^{-1}$ pour tout $X \in A$ où U est inversible dans $N(A)$.

Preuve. Dans ce cas on a : $\alpha_{ij} \in N(R)$ pour tout $1 \leq i \leq j \leq n$ Donc $U \in N(A)$ avec $U.V = 1 = V.U$, en effet : on a $U.V = 1$ d'après la proposition précédente de plus on a :

$$\begin{aligned}
\psi_j(\beta_{1j})\alpha_{1j} &\overset{(5)}{=} (\psi_1(\beta_{1j})\alpha_{1j})(\psi_j(\beta_{1j})\alpha_{1j}) \\
&\overset{(6)}{=} \psi_1(\beta_{1j})[\alpha_{1j}(\psi_j(\beta_{1j})\alpha_{1j})] \\
&= \psi_1(\beta_{1j})[(\alpha_{1j}\psi_j(\beta_{1j})\alpha_{1j}] \\
&\overset{(5)}{=} \psi_1(\beta_{1j})\alpha_{1j} \\
&\overset{(5)}{=} 1.
\end{aligned}$$

D'où $U.V = 1 = V.U$. Si on multiplie à droite l'équation (7) du lemme 3.1 par $\psi_j(\beta_{ij})$ on déduit que $\psi_i \in Aut(R)$ pour tout $1 \leq i \leq n$. Donc $V \in N_m(A)$ et l'application $\omega = R_V \circ L_U$ de la proposition précédente est un automorphisme de A, en effet si $X, Y \in A$ on a :

$$\begin{aligned} \omega(XY) &= U(XY)V && = U[X(VU)Y]V \\ &= U[(XV)(UY)]V && = [(UXV)(UY)]V \\ &= [(UXV)[(UY)(VU)]]V && = [(UXV)[(UYV)U]]V \\ &= [[(UXV)(UYV)]U]V && = (UXV)(UYV) \\ &= \omega(X)\omega(Y). \end{aligned}$$

D'où $\psi = \omega^{-1} \circ \psi_1^{\#} \in Aut(A)$, de plus $\psi^{-1} \in Aut(A)$ implique $\beta_{ij} \in N(R)$ c'est à dire $V \in N(A)$. Donc U est inversible dans $N(A)$ avec $U = V^{-1}$, $\omega = I_{U^{-1}}$ et $\psi = I_U o \psi_1^{\#}$. □

Corollaire 3.4 *Si R est une algèbre alternative et $\psi \in J\text{-}Aut(A)$ tel que $\psi(E_{ii}) = E_{ii}$ pour tout $1 \leq i \leq n$ alors $\psi = I_U o \phi^{\#}$ où U est inversible dans $N(A)$ et $\phi \in Aut(R)$.*

Preuve. On a $\alpha_{ij} \in N(R)$ pour tout $1 \leq i \leq n$ car R est alternative le reste se découle du théorème 3.3. □

4. Dérivations dans les algèbres de Barker

Soit R une algèbre non associative, une application linéaire D de R dans R est une dérivation si : $D(ab) = D(a)b + aD(b)$ pour tout $a, b \in R$. Si $a \in N(R)$ l'application ad_a de R dans R donnée par : $ad_a(x) = ax - xa$ est une dérivation de R et si la caractéristique de R est $\neq 2$ on vérifie que tout dérivation de R est une dérivation de R^+. En utilisant les techniques de ([5] paragraphe 5) on démontre les facilement les résultats suivantes :

Lemme 4.1 *Soit A une algèbre de Barker, si D est une dérivation de A^+ alors $D(E_{ii}) \in N(A)$ pour tout $1 \leq i \leq n$.*

Lemme 4.2 *Soient A une algèbre de Barker et D une dérivation de A^+ il existe $U \in N(A)$ tel que : $(D - ad_U)(E_{ii}) = 0$ pour tout $1 \leq i \leq n$.*

Preuve. D'après le lemme précédent on a $D(E_{ii}) \in N(A)$. Soit $U = D(E_{1,1})E_{1,1} + \ldots + D(E_{nn})E_{nn} \in N(A)$ et on vérifie que $ad_U(E_{ii}) = D(E_{ii})$ pour tout $1 \leq i \leq n$. □

Lemme 4.3 *Soient A une algèbre de Barker et D une dérivation de A^+ telle que $D(E_{ii}) = 0$ pour tout $1 \leq i \leq n$. Alors D induit des transformations D_{ij} de R dans R telles que : $D(aE_{ij}) = D_{ij}(a)E_{ij}$. De plus D_{ii} est une dérivation de R^+ pour tout $1 \leq i \leq n$.*

Lemme 4.4 *Soient A une algèbre de Barker et D une dérivation A^+ telle que : $D(E_{ii}) = 0$ pour tout $1 \leq i \leq n$. Alors $D(E_{ij}) \in N_m(A)$ pour tout $1 \leq i \leq n$.*

Lemme 4.5 *Soient A une algèbre de Barker et D une dérivation de A^+ telle que : $D(E_{ij}) \in N(A)$ pour tout $1 \leq i < j \leq n$. Il existe $V \in N(A)$ tel que $(D - ad_V)(E_{ij}) = 0$ pour tout $1 \leq i \leq j \leq n$ où $E_{ij} \in A$.*

Preuve. D'après le lemme 4.2 on peut supposer que $D(E_{ii}) = 0$ pour tout $1 \leq i \leq n$, donc D induit une dérivation D_1 sur B^+ où :

$$B = \{M \in T_{n-1}(R) : \begin{pmatrix} 0 & 0 \\ 0 & M \end{pmatrix} \in A\}$$

et par induction il existe $V \in N(B)$ tel que $(D_1 - ad_V)(E_{ij}) = 0$ pour tout $2 \leq i \leq j \leq n$. On peut donc supposer $D(E_{1,1}) = 0$ et $D(Eij) = 0$ pour tout $2 \leq i \leq j \leq n$. Soit p minimal tel que $E_{1p} \in A$, on a $D(E_{1p}) = D_{1p}(1)E_{1p}$ et posons $W = D_{1p}(1)E_{1,1}$ on vérifie facilement que l'on a :
i/ $ad_W(E_{1,1}) = D(E_{1,1})$.
ii/ $ad_W(E_{ij}) = 0 = D(E_{ij})$ pour tout $2 \leq i \leq j \leq n$ où $E_{ij} \in A$.
iii/ $ad_W(E_{1p}) = D_{1p}(1)E_{1p} = D(E_{1p})$. Donc $D' = D - ad_W$ vérifie $D'(E_{1,1}) = 0$, $D'(E_{ij}) = 0$ pour tout $2 \leq i \leq j \leq n$ où $E_{ij} \in A$ et $D'(E_{1p}) = 0$. Soit $q > p$ tel que $E_{1q} \in A$ on a : $D'(E_{1q}) = 2D'(E_{1p} o E_{pq}) = 2[D'(E_{1p}) \circ E_{pq} + E_{1p} o D'(E_{pq})] = 0$. D'où $D'(E_{ij}) = 0$ pour tout $1 \leq i \leq j \leq n$. Enfin, il existe $W \in N(A)$ avec $(D - ad_W)(E_{ij}) = 0$ pour tout $1 \leq i \leq j \leq j \leq n$ où $E_{ij} \in A$. □

Lemme 4.6 *Soient A une algèbre de Barker et D une dérivation de A^+ avec $D(E_{ij}) = 0$ pour tout $1 \leq i \leq j \leq n$ où $E_{ij} \in A$. Il existe d une dérivation de R telle que : $D = d^{\#}$.*

Preuve. Comme dans le lemme précédent D induit une dérivation D_1 dans B^+ où

$$B = \{M \in T_{n-1}(R) : \begin{pmatrix} 0 & 0 \\ 0 & M \end{pmatrix} \in A\}.$$

Par induction $D_1 = d^{\#}$ où d est une dérivation de R. D'autre part, soit $p \in \{2, \ldots, n\}$ tel que $E_{1p} \in A$ on a : $D(aE_{1p}) = 2D(E_{1p} o (aE_{pp})) = 2E_{1p} o (d(a)E_{pp}) = d(a)E_{1p} = D_{1p}(a)E_{1p}$, et $D(aE_{1p}) = 2D(aE_{1,1} \circ E_{1p}) = 2D_{1,1}(a)E_{1,1} \circ E_{1p} = D_{1,1}(a)E_{1p}$. Donc $D_{1,1} = d$ et $D = d^{\#}$. □

Théorème 4.7 *Soient A une algèbre de Barker et D une dérivation de A^+ telle que $D(Eij) \in N(A)$ pour tout $1 \leq i \leq j \leq n$ où $E_{ij} \in A$. Il existe $U \in N(A)$ et d une dérivation de R tels que : $D = ad_U + d^{\#}$.*

Preuve. C'est une conséquence immédiate des lemmes 4.5 et 4.6. □

Corollaire 4.8 *Soient R une algèbre alternative et $A = T_n(\Gamma, R)$ une algèbre de Barker associée à Γ et R on a*

$$Der(A^+) = ad_{N(A)} + (Der(R))^{\#} = Der(A).$$

Preuve. D'après le lemme 4.4 on a : $D(E_{ij}) \in N(A) = N_m(A)$ le reste découle du théorème 4.7. □

References

1. G. P. Barker, *Automorphism Groups of Algebras of Triangular Matrices :* Linear. Algebra. Appl : 121 207-215 1989.
2. G. P. Barker, *Automorphisms of Triangular Matrices over Graphs.* (Préprint).
3. G. P. Barker et T. P. Kezlan, *The Automorphism Group of Matrices in Current Trends in Matrix Theory (F. Uhlig and R. G. Grone. Eds...)* North-Holland - New-York 1987.
4. G. P. Barker et T. P. Kezlan, *Automorphisms of Algebras of Triangular Matrices,* Arkiv. Math (à paraître).
5. G. M. Benkart et J. M. Osborn, *Derivation and Automorphisms of nonassociative Matrix Algebra,* Trans. Amer. Math. Soc. 263 N^0, 2 411-430 1981.
6. L. Ben Yakoub, *Sur le théorème de Skolem-Noether,* Thèse de troisième cycle 1990 - Rabat - Maroc.
7. T. P. Kezlan, *A note on Algebra Automorphism of Triangular Matrices over commutative Rings,* Linear Algebra Appl : 135 181-184 1990.
8. S. Jondrup, *Automorphisms of Upper Triangular Matrices Rings,* Arch. Math. Vol. 49 497-502 1987.
9. S. Jondrup, *The Group of Automorphisms of Certain Subalgebras of Matrix Algebra,* J. Algebra, 141 106-114 1991.
10. S. Jondrup, *Automorphisms and Derivations of Upper Triangular Matrix Rings :* (Preprint).

LATTICE ISOMORPHISMS OF JORDAN ALGEBRAS OVER ARBITRARY FIELDS

JESÚS LALIENA*
Departamento de Matemáticas, Universidad de La Rioja. Spain

Abstract. We study Jordan algebras M whose lattice of subalgebras is isomorphic to the lattice of subalgebras of a Jordan matrix algebra, $J = H(D_n, J_A)$, where D is either a quadratic extension field (if $n \geq 2$), a central division quaternion algebra (if $n \geq 3$) or a central division Cayley-Dickson algebra (if $n = 3$). We prove that M is also a Jordan matrix algebra of the same kind as J.

In [1]J. A. Anquela proved that a Jordan algebra over an algebraically closed field F, such that $charF \neq 2$, with lattice of subalgebras isomorphic to the lattice of subalgebras of a finite dimensional semisimple Jordan algebra J over F is isomorphic to J.

We are going to study Jordan algebras whose lattice of subalgebras is isomorphic to the lattice of subalgebras of a Jordan matrix algebra, $H(D_n, J_A)$, where D is either a central quadratic extension field (if $n \geq 2$), a central division quaternion algebra (if $n \geq 3$) or a division Cayley-Dickson algebra (if $n = 3$).

In the following we will consider finite dimensional Jordan algebras over F, a field with $charF \neq 2$. In fact, it is known (see [1]) that Jordan algebras with the same lattice of subalgebras as a finite dimensional one are also finite dimensional.

Let J, M be Jordan algebras. $L(J)$ and $L(M)$ will denote their lattices of subalgebras. A lattice isomorphism or L-isomorphism between J and M is a one-to-one map ψ from $L(J)$ onto $L(M)$ such that

$$\psi(A \vee B) = \psi(A) \vee \psi(B), \qquad \psi(A \cap B) = \psi(A) \cap \psi(B)$$

for all $A, B \leq J$, where $A \vee B$ is the subalgebra of J generated by A and B.

We will use the following notation

$< X >$ means vector subspace of J over F generated by $X \subseteq J$.

(X) means subalgebra of J spanned by $X \subseteq J$.

$l(J)$ is the length of the longest chain of subalgebras of J.

Let D_n be the algebra of $n \times n$ matrices with entries in D. $H(D_n, J_A)$ will denote $\{(x_{ij}) \in D_n : A^{-1}(j(x_{ij}))^t A = (x_{ij})\}$ with $n \geq 2$, A is a diagonal matrix whose diagonal entries are $(a_1, \ldots, a_n)$ with a_i, a_i^{-1} in the nucleus of D and $j(a_i) = a_i$, D an alternative algebra with involution j and where t denotes the tranposed matrix.

Also we will use in the following the clasification of Jordan algebras with length 2. It was proved in [4] and appears also in [1].

---

* Partially supported by the DGICYT (Ps. 90-0129) and by the DGA (PCB-6/91)

S. González (ed.), Non-Associative Algebra and Its Applications, 229–234.

1. Algebras with length 4

Let e_{ij} i,j=1,..,n be the usual matrix units in F_n. If $i = j$ we will denote e_{ii} by e_i.

Lemma 1.1. *Let $J = H(D_2, J_A)$ Jordan algebra such that D is a division associative algebra over F with involution j and A a diagonal matrix such that $A =$ diag $\{1, \nu\}$. If $l(J) = 4$, then D is a quadratic extension field with j the usual involution.*

Proof. First we will prove that $j(x) \neq x$ for every $x \in D - F$. Suppose $x \in D - F$ and $j(x) = x$. Then we have the following chain of subalgebras

$$0 \leq (e_1) \leq (e_1, e_2) \leq F(x).e_1 \oplus (e_2) \leq F(x).e_1 \oplus F(x).e_2 \leq J$$

and so $l(J) \geq 5$, that is a contradiction. Therefore, if $j(x) = x$, it follows that $x \in F$, and so $\nu \in F$ and $xj(x) \in F$ for every $x \in D - F$.

Now let $x \in D - F$ and $K = F(x)$ extension field of F. Since $H(K_2, J_A) \leq H(D_2, J_A)$ and $l(H(K_2, J_A)) \geq 4$, we have that $D = K$. We suppose that there exists $y \in K$ such that $\{1, x, y\}$ are linearly independent over F. Then we have the following chain of subalgebras

$$0 \leq (e_1) \leq (e_1, e_2) \leq (e_1, e_{12} + \nu^{-1} e_{21}) \leq (e_1, e_{12} + \nu^{-1} e_{21}, x e_{12} + \nu^{-1} j(x) e_{21}) \leq J$$

and so $l(J) \geq 5$, that is a contradiction. Therefore $K = F(x)$ is two dimensional and j such that $j(\lambda 1 + \mu x) = \lambda 1 - \mu x$.•

Now let J be Jordan algebra of a symmetric bilinear form , f, over a vectorial space V over the base field F with dimension of V finite and greater than one. It is easy to prove that J has a subalgebra isomorphic to $F \oplus F$ if and only if there exists a nonzero idempotent in J different of 1, and this is equivalent to prove that there exists $x \in V$ such that $f(x, x) = 1$.

Lemma 1.2. *Let J be a Jordan algebra of a nondegenerate symmetric bilinear form, f, over a vectorial space V over the base field F with dimension of V equal to 3. Suppose that there exists $x \in V$ such that $f(x, x) = 1$ and also that the bilinear form is represented by the diagonal matrix $A = diag\{1, \alpha, \beta\}$. Then*

i)$J \cong H((F \oplus F)_2, J_I)$, and $F \oplus F$ with the exchange involution, if and only if $\frac{-\alpha}{\beta} \in F^2$

ii) $J \cong H((K)_2, J_B)$, and K with the usual involution, if and only if $\frac{-\alpha}{\beta} \notin F^2$

Proof. Let $\{x, z, u\}$ be an orthogonal basis of V such that the associated matrix to f in those basis is $A =$ diag $\{1, \alpha, \beta\}$. If $-\alpha/\beta = \gamma^2$ with $\gamma \in F$, then

$$\{1/2 + 1/2x, 1/2 - 1/2x, z + \gamma u, z - \gamma u\}$$

have the same multiplication table as the following basis of $H((F \oplus F)_2, J_I)$ over F

$$\{(1,1)e_1, (1,1)e_2, \alpha(1,0)e_{12} + \alpha(0,1)e_{21}, (0,1)e_{12} + (1,0)e_{21}\}$$

If $\frac{-\alpha}{\beta} \notin F^2$, let $K = F(c)$ quadratic extension field of F with c a root of $X^2 + \frac{\alpha}{\beta}$. We consider

$$\{e_1, e_2, e_{12} + \beta^{-1} e_{21}, c e_{12} - \beta^{-1} c e_{21}\},$$

basis of $H(K_2, J_C)$ over F with C = diag $\{1, \beta\}$. It has the same multiplication table as the basis of J over F given by

$$\{1/2 + 1/2x, 1/2 - 1/2x, u, z\}.\bullet$$

Proposition 1.3. *Let $J = H(K_2, J_A)$ such that $K = F(x)$ is a quadratic extension field with the usual involution and $A = diag\{1, \nu\}$ with $\nu \in F - \{0\}$. Let M be Jordan algebra such that $L(J)$ and $L(M)$ are isomorphic. Then $M \cong H(L_2, J_B)$ where L is a quadratic extension field of F with the usual involution and $B = diag\{1, \beta\}$, $\beta \in F - \{0\}$. Moreover if ϕ is L-isomorphism between $L(J)$ and $L(M)$, then $\phi < e_i > = < y_i >$ with y_i i= 1,2 nonzero orthogonal idempotents in M, such that $y_1 + y_2 = 1_M$*

Proof. J is Jordan algebra of a nondegenerate symmetric bilinear form over a vectorial space V over F with dimension of V equal to 3. From Proposition (3,1) in [1]we have that M is either a Jordan algebra of a nondegenerate symmetric bilinear form over a vectorial space with dimension 3 or $M \cong H(D_2, J_C)$ with D an associative, not commutative, finite dimensional division algebra over F, with involution which fixes exactly F, being C= diag$\{1, \gamma\}$, $\gamma \in F - \{o\}$. If $M \cong H(D_2, J_C)$, since $l(J) = l(M) = 4$, we have, using Lemma 1.1, that D is a quadratic extension field of F. If J is a Jordan algebra of a symmetric bilinear form, using Lemma 1.2 we have that either $M \cong H((F \oplus F)_2, J_D)$ or $M \cong H(L, J_E)$ with L a quadratic extension field of F and D = diag $\{1, \alpha\}$, E= diag $\{1, \delta\}$, $\alpha, \delta \in F - \{0\}$. Proposition (3,6) in [4]prove that Jordan algebras with lattice of subalgebras isomorphic to $L(H((F \oplus F)_2, J_I)$ are isomorphic to it. So $M \cong H(L, J_E)$ with L a quadratic extension field.

Now we consider $J_1 = < e_1, e_2, e_{12} + \nu^{-1}e_{21} >$. It is a Jordan algebra of a nondegenerate symmetric bilinear form. We know using Lemma (2,2) in [1]that $\phi(J_1)$ is also a Jordan algebral of a nondegenerate symmetric bilinear form. From Lemma (2,1) in [1]we have that every subalgebra of J_1 with length 2 contains the subalgebra $F.1_{J_1}$, therefore $\phi(F.1_{J_1})$ is contained in every subalgebra of $\phi(J_1)$ with length two. So $\phi(J_1) = < y >$ with y the identiy element in $\phi(J_1)$.

Let us consider also $J_2 = < 1_J, e_{12} + \nu^{-1}e_{21}, xe_{12} + \nu^{-1}xe_{21} >$. It is a Jordan algebra of a nondegenerate symmetric bilinear form. If J_2 is a division algebra, J_2 has a unique minimal subalgebra that is $F.1_J$ (see [4]). In this case $\phi(J_2)$ has also a unique minimal subalgebra and so $\phi(J_2)$ is a division algebra or a nilpotent algebra because of Lemma (2,3) in [4]. But $F.1_J \geq (e_1, e_2) \cong F \oplus F$, and from the clasification of Jordan algebras with length two we have that $\phi(e_1, e_2) \cong F \oplus F$. Since $\phi(F.1_J) \geq \phi(e_1, e_2)$, $\phi(F.1_J)$ is not nilpotent and so $\phi(J_2)$ is division algebra with identity element in the unique minimal subalgebra, that is, in $\phi(F.1_J)$. If J_2 is not a division algebra, then, from Lemma (2,2) in [1], $\phi(J_2)$ is a Jordan algebra of a nondegenerate symmetric bilinear form. So, we can prove, as before with $\phi(J_1)$, that $\phi(F.1_J) = < y >$ with y the identity element in $\phi(J_2)$. Since $M = \phi(J_1) \vee \phi(J_2)$, we can show, using the following linearization of the Jordan identity

$$2(xz)(tx) + x^2(tz) = 2((xz)t)x + (x^2t)z \tag{1.1}$$

and induction on the length on the generators of M, that $y = 1_M$. Moreover since $\phi(e_1,e_2) \cong F \oplus F$ and $\phi(F.1_J) =< y >$ it follows that $\phi < e_1 >=< y_1 >$, $\phi < e_2 >=< y_2 >$ with y_1, y_2 orthogonal idempotents such that $y_1 + y_2 = y = 1_M$.•

2. Jordan matrix algebras

Here we study Jordan algebras with the same lattice of subalgebras that a Jordan matrix algebra $H(D_n, J_A)$, where D is either a quadratic extension field of F, a central quaternion algebra over F, or a central division Cayley-Dickson algebra ($n = 3$) over F, and where $A = diag\{1, a_2, ..., a_n\}$ with $a_i \in F - \{0\}$.

Theorem 2.1. *Let $J = H(K_n, J_A)$ with $n \geq 3$, K a quadratic extension field of F and $A = diag\{1, a_2, ..., a_n\}$ $a_i \in F - \{0\}$. If M is a Jordan algebra such that $L(J)$ and $L(M)$ are isomorphic, then $M \cong H(L_n, J_B)$ with L a quadratic extension field of F and $B = diag\{1, b_2, ..., b_n\}$ $b_i \in F - \{0\}$*

Proof. i) Let $D = H(F_n, J_A)$ and $\phi : L(J) \longrightarrow L(M)$ L-isomorphism. From Theorem (2,1) in [1]and his proof we know that $\phi(D) = H(F_n, J_B)$ with $B = diag\{1, b_2, ..., b_n\}$, $b_i \in F - \{0\}$, and moreover $\phi < e_i >=< y_i >$ i=1,...,n with y_i connected orthogonal idempotents in M such that $y_1 + ... + y_n = 1_{\phi(D)}$ and $\phi < e_1, e_j, e_{1j} + a_j^{-1} e_{j1} >=< y_1, y_j, w_j >$ with w_j invertible element which connects y_1, y_j and $w_j^2 = b_j^{-1}(y_1 + y_j)$

ii)We are going to see that $y_1 + ... + y_n = 1_M$. Let $K = F(x)$ and we consider $C_{1j} =< e_1, e_j, e_{1j} + a_j^{-1} e_{j1}, xe_{1j} - a_j^{-1} xe_{j1} >$ with $1 \neq j \leq n$. We have that $C_{1j} \cong H(K_2, J_{A_j})$ with $A_j =$ diag $\{1, a_j\}$. We know because of Proposition 1, that $\phi(C_{1j}) \cong H((L_j)_2, J_{B_j})$ with $L_j = F(d_j)$ a quadratic extension field, $B_j =$ diag $\{1, b_j\}$ because $< y_1, y_j, w_j > \leq \phi(C_{1j})$ and the proof of Lemma 1.2, and moreover $1_{\phi(C_{1j})} = y_1 + y_j$ with $\phi < e_1 >=< y_1 >, \phi < e_j >=< y_j >$. We will show that $y_k.\phi(C_{1j}) = 0$ if $1, j \neq k \leq n$. Let $\phi(C_1 j) =< y_1, y_j, w_j, z_j >$ with $z_j^2 \in F.1_{\phi(C_{1j})}, z_j w_j = 0, y_1 w_j = y_j w_j = 1/2 w_j, y_1 z_j = y_j z_j = 1/2 z_j$. We have that $\phi(C_{1j} \vee < e_k >) = \phi(C_{1j}) \vee < y_k >$ is semisimple algebra with length 5 and such that $\phi(C_{1j}) \vee < y_k >= \phi(C_{1j}) \oplus < y_k >$ because of Lemmas (4,1), (4,2) in [1]. So $y_k.\phi(C_{1j}) = 0$. From Proposition 1 we know $1_{\phi(C_{1j})} = y_1 + y_j$. Also, since $J = (e_i, e_{1i} + a_i^{-1} e_{i1}, xe_{1i} - xa_i^{-1} e_{i1} : i = 1, ..., n)$ we have that $M = (y_i, w_i, z_i : i = 1, .., n)$. From these last three results, that is, $y_k.\phi(C_{1j}) = 0, 1_{\phi(C_{1j})} = y_1 + y_j, M = (y_i, w_i, z_i : i = 1, .., n)$, using induction on the length on the generators of M and the linearization of the Jordan identity (1.1) we can prove that $y_1 + ... + y_n = 1_M$.

iii) Now from the Coordinatization Theorem (see [2]) we have that $M \cong H(D_n, J_C)$ where D is an alternative algebra with involution j and $C =$ diag $\{1, c_2, .., c_n\}$. But if $M = \sum_{i,j=1}^n M_{ij}$ is the Peirce decomposition of M relative to y_i we have that $H(D_2, J_H) \cong M_{11} + M_{1j} + M_{jj} \geq (y_1, y_j, w_j, z_j)$ and we recall that $(y_1, y_j, w_j, z_j) \cong H((L_j)_2, J_{B_j})$ with $L_j = F(d_j)$ a quadratic extension field. Since $M = (y_i, w_i, z_i : i = 1, .., n)$ it follows that $L_i = L_j = D$ and $C = B$.•

Theorem 2.2. *Let $J = H(Q_n, J_A)$ $n \geq 3$, Q a central division quaternion algebra over F and $A =$ diag $\{1, a_2, .., a_n\}$, $a_i \in F - \{0\}$. If M is a Jordan algebra such that $L(J)$ and $L(M)$ are isomorphic, then $M \cong H(\bar{Q}_n, J_B)$, $\bar{Q}$ a central division quaternion algebra over F and $B =$ diag $\{1, b_2, .., b_n\}$, $b_i \in F - \{0\}$.*

Proof. i) Let $E = H(F_n, J_A)$ and $\phi : L(J) \longrightarrow L(M)$ L-isomorphism. From Theorem (2,1) in [1]and his proof we know that $\phi(E) \cong H(F_n, J_B)$ with $B =$ diag $\{1, b_2, .., b_n\}, b_i \in F - \{0\}$, and moreover $\phi < e_i > = < y_i >$ $i = 1, ..., n$ with y_i connected orthogonal idempotents in M such that $y_1 + ... + y_n = 1_{\phi(E)}$.

ii) We are going to see that $y_1 + ... + y_n = 1_M$. Let $\{1, c_1, c_2, c_3\}$ be the usual basis of Q over F. We know that the extension fields $K_i = F(c_i)$ $i = 1, 2, 3$ are quadratic extension fields. So $\phi(H((K_i)_n, J_A)) = H((L_i)_n, J_B)$ because of Theorem 2.1, where $L_i = F(d_i)$ is also quadratic extension field of F and $y_1 + ... + y_n$ is the identity element in $H((L_i)_n, J_A)$ for $i = 1, 2, 3$ (see ii) in proof of Theorem 2.1). Since $J = (H((K_i)_n, J_A) : i = 1, 2, 3)$ it follows that $M = (H((L_i)_n, J_A) : i = 1, 2, 3)$. Now using the linearization of the Jordan identity (1.1) and induction on the length on the generators of M we can prove that $y_1 + y_2 + .. + y_n = 1_M$. From the Coordinatization Theorem (see [3]) we have that $M \cong H(D_n, J_B)$ with D alternative algebra with involution, j, and $B =$ diag $\{1, b_2, ..., b_n\}$.

iii) Now using Theorem (2,2) of Chapter III in [3]and because $\phi : L(J) \longrightarrow L(M)$ is L-isomorphism such that $\phi(H(F_n, J_A)) = H(F_n, J_B)$ we have that the following lattices of subalgebras are L-isomorphic

$X = \{$ subalgebras of $Q\}$ $\qquad Y = \{J_1 \leq J : H(F_n, J_A) \leq J_1\}$
$Z = \{M_1 \leq M : H(F_n, J_B) \leq M_1\}$ $\qquad U = \{D_1 \leq (D, j) : F.1_D \leq D_1\}$.

So, because of Theorem 2.1, we know that the proper subalgebras of M containing $H(F_n, J_B)$ are isomorphic to $H(K_n, F_B)$ where K is a quadratic extension field with the usual involution. Therefore the symmetric elements of (D, j) are in F. We are going to prove that $\{$ subalgebras of (D, j) containing $F\} = \{$ subalgebras of $D\}$. Let N be subalgebra of D. If there exists $a \in N$ such that $a^2 \neq 0$, then $F \leq N$ because a^2 is symmetric element. If $a^2 = 0$ for every $a \in N$, then since j is involution and the symmetric elements are in F we have that $a = \lambda.1 + b$ with $b \neq 0$, $j(b) = -b$ and $\lambda \in F$. But $a^2 = 0$ implies $\lambda = 0$ and so $j(a) = -a$. Let $P = (a) \vee F \cdot 1_D$, it is a subalgebra of (D, j). Since X, Y, Z, U are isomorphic lattices, and $P \leq (D, j)$ such that $F.1_D \leq P$ and $l(P) = 2$, we have that $H(P_n, J_B) = \phi(H(E_n, J_A)$ with E a quadratic extension field of F and subalgebra of Q. But, from Theorem 2.1 we know that P is also a quadratic extension field. Therefore we have a contradiction and so $a^2 \neq 0$ for some $a \in N$ and $\{$ subalgebras of (D, j) containing $F\} = \{A : A \leq D\}$.

iv) Hence we have that $L(Q)$ and $L(D)$ are L-isomorphic. Alternative algebras L-isomorphic to Q have been studied in [3]. So we have that D is either nilpotent, purely inseparable extension field with dimension p^2 and $p = charF$ or associative division algebra with dimension m^2 and m the dimension of the maximal subfields of D. But Q has subalgebras that are quadratic extension fields of F. So, because of Theorem 2.1 and because X, Y, Z, U are isomorphic lattices, it follows that D has subalgebras that are also quadratic extension fields of F. Therefore D can not be nilpotent and if D is purely inseparable extension field of F, it follows that dimension of D is 4 with $charF = 2$, contradiction, because we have supposed at the begining that $charF \neq 2$. Therefore D is an associative division algebra with dimension 4 and thus D is a central quaternion division algebra.•

Theorem 2.3. *Let $J = H(C_3, J_A)$ with C a central division Cayley-Dickson algebra over F and $A = \text{diag}\ \{1, a_2, a_3\}$ $a_i \in F - \{0\}$. If M is a Jordan algebra such that $L(J), L(M)$ are L-isomorphic, then $M \cong H(\bar{C}_3, J_B)$ with $\bar{C}$ a central division Cayley-Dickson algebra over F and $B = \text{diag}\ \{1, b_2, b_3\}$, $b_i \in F - 0$.*

Proof. As in the proof of Theorem 2.2 we can see, using Theorem (2,1) in [1], that y_1, y_2, y_3 are connected orthogonal idempotents in M with $< y_i > = \phi < e_i >$ and if $E = H(F_3, J_A)$ then $\phi(E) = H(F_3, J_B)$ with $B = \text{diag}\ \{1, b_2, b_3\}$ and $y = y_1 + y_2 + y_3 = 1_{\phi(E)}$.

Now we will prove that $y = 1_M$. Let $\{1, c_1, ..., c_7\}$ be the usual basis of C over F (see [5]). We have that $Q_0 = (1, c_1, c_2, c_3)$, $Q_1 = (1, c_1, c_4, c_5)$, $Q_2 = (1, c_2, c_4, c_6)$, $Q_3 = (1, c_3, c_4, c_7)$ are central division quaternion algebras. So from Theorem 2.2 and his proof it follows that $\phi(H(Q_i)_3, J_A) \cong H((\bar{Q}_i)_3, J_B)$ where $\bar{Q}_i$ are division quaternion algebras and $y_1 + y_2 + y_3$ is the identity element in $\phi(H((Q_i)_3, J_A))$ for $i = 0,1,2,3$. Since $J = (H((Q_i)_3, J_A) : i = 0,1,2,3)$, we have that $M = (H((\bar{Q}_i)_3, J_B) : i = 0,1,2,3)$. Now using the linearization of the Jordan identity (1.1) we can prove, by induction on the length on the generators of M, that $y = 1_M$.

Therefore from the Coordinatization Theorem (see [3]) $M = H(D_3, J_B)$ with D alternative algebra with involution and $B = \text{diag}\ \{1, b_2, b_3\}$. From Theorem (2,2) of Chapter III in [2] and because $\phi : L(J) \longrightarrow L(M)$ is L-isomorphism such that $\phi(H(F_n, J_A)) = H(F_n, J_B)$ we have that the following lattices are L-isomorphic

$$X = \{\text{ subalgebras of } C\} \qquad Y = \{J_1 \leq J : H(F_n, J_A) \leq J_1\}$$
$$Z = \{M_1 \leq M : H(F_n, J_B) \leq M_1\} \qquad U = \{D_1 \leq (D, j) : F.1_D \leq D_1\}.$$

So from Theorems 2.1 and 2.2, we know that the proper subalgebras of M containing $H(F_n, J_B)$ are isomorphic either to$H(K_n, J_B)$ or to $H(Q_n, J_B)$ where K is a quadratic extension field and Q a central division quaternion algebra with the usual involution. Therefore the symmetric elements of (D, j) are in F.

Now we want to prove that { subalgebras of (D, j) containing $F\}$ = { subalgebras of $D\}$. We can use for it the same proof as in iii) of Theorem 2.2.

Thus $X = L(C)$ and $U = L(D)$ are isomorphic. From [3] we have that D is either a purely inseparable extension field of F with dimension p^3 with $p = charF$ or D is a central division Cayley-Dickson algebra over F. But C has subalgebras that are quadratic extension fields of F and so, because of Theorem 2.1 and because X, Y, Z, U are isomorphic it follows that D has subalgebras that are quadratic extension fields. So if D is a purely inseparable extension field we have that $p = 2$, contradiction. So D is a central division Cayley-Dickson algebra.•

References

1. J.A. Anquela. Lattice definability of semisimple Jordan algebras.Comm. Algebra 19 (1991), 1409-1427.
2. N. Jacobson. *Structure and representations of Jordan algebras*. American Mathematical Society. Providence, Rhode Island 1968.
3. J.A. Laliena. Lattice isomorphisms of Alternative algebras. J. Algebra 128 (1990), 335-355.
4. J.A. Laliena. Lattice isomorphisms of Jordan algebras.Nonassociative algebraic models. Nova Science Publishers, Inc. New York. (1992), 195-211.
5. K.A. Zhevlakov; A.M. Slinko; I.P. Shestakov; A.I. Shirshov. *Rings that are nearly associative.* Academic Press, New York 1982.

MULTIBARIC ALGEBRAS

LOPEZ-SANCHEZ, JESUS and RODRIGUEZ SANTA MARIA, EMILIA
Departamento de Matematica Aplicada (Biomatematica)
Facultad de Biologia
Universidad Complutense
28040-Madrid (SPAIN)

Abstract. Most of the baric algebras arising in population genetic models are baric algebras whose weight homomorphism is uniquely determined. However, some algebras with genetic realization have defined two or more weight homomorphisms. In the present work we generalize the concept of train algebra for multibaric algebras and we consider the bibaric train algebras of rank 3 and 4. In particular we show that for these ranks, being Jordan is equivalent to being power associative. Furthermore we give a characterization of bibaric train algebras of rank 3 based on the properties of a refined Peirce decomposition. Also, some results about train algebras of rank 4 are presented.

1. Algebras with genetic realization and bibaric algebras

It is well known the existence of algebras with genetic realization which admit more than one natural basis. For example, the gametic algebra for simple Mendelian segregation has infinite natural basis. However, the weight homomorphism induced over the algebra by these basis is the same [4]. We now consider the following example: Let $A = \langle e_1, e_2, e_3, e_4 \rangle_R$ be the commutative 4-dimensional algebra over R with multiplication table: $e_1^2 = e_1$, $e_2^2 = e_2$, $e_1 \cdot e_3 = \frac{1}{2}e_3$, $e_2 \cdot e_3 = \frac{1}{2}e_3$, $e_2 \cdot e_4 = \frac{1}{2}e_4$ and the other products are zero. The two basis $B_1 = (e_1, e_1 + e_2, e_1 + e_3, e_1 + e_4)$ and $B_2 = (e_2, e_2 + e_1, e_2 + e_3, e_2 + e_4)$ are natural, but the weights induced over A by B_1 and B_2 are different.

Definition 1.1. An algebra A over a field K is called a *bibaric* algebra when it admits two different weight homomorphisms $\omega_1 : A \to K$ and $\omega_2 : A \to K$. We denote these algebras by (A, ω_1, ω_2) .

Definition 1.2. A n-dimensional commutative algebra over R or C has a double genetic realization when it admits two natural basis such that the weights induced are different.

A first question is proposed: Has every bibaric algebra with genetic realization a double genetic realization ? This example shows that the answer is negative: Let $A = \langle e_1, e_2, e_3 \rangle_R$ be a commutative 3-dimensional algebra with multiplication table: $e_1^2 = e_1$, $e_1 \cdot e_2 = \frac{1}{3}e_3$, $e_2^2 = e_2 - 2e_3$, $e_2 \cdot e_3 = \frac{1}{2}e_3$ and the other products are zero. A is bibaric since the weights defined by $\omega_1(e_1) = 1$, $\omega_1(e_2) = \omega_1(e_3) = 0$ and $\omega_2(e_2) = 1$, $\omega_2(e_1) = \omega_2(e_3) = 0$ are different. Furthermore, the elements e_1, $e_1 + \frac{1}{3}e_2$, $e_1 + \frac{1}{2}e_3$ form a natural basis, being ω_1 the weight induced. We say in this case that the algebra A has a genetic realization with respect to ω_1. However, it has not a genetic realization with respect to ω_2 because it has not an idempotent e with $\omega_2(e) = 1$. When the weight homomorphisms ω_1 and ω_2 are equivalent, it

S. González (ed.), Non-Associative Algebra and Its Applications, 235–240.

is, when exits an algebra isomorphism $f : A \to A$ such that $\omega_1 = \omega_2 \circ f$, we have:

Proposition 1.3. *Let (A, ω_1, ω_2) be a n-dimensional commutative bibaric algebra over R or C such that ω_1 is equivalent to ω_2. If A has one genetic realization with respect to ω_1 then the algebra has a double genetic realization and admits two natural basis which induce the weights ω_1 and ω_2 with identical multiplication constants.*

The proof is very easy. Let $(a_i)_{1\leq i\leq n}$ be a natural basis which induces ω_1, then $\big(f(a_i)\big)_{1\leq i\leq n}$ is also a natural basis which induces ω_2 and it has identical multiplication constants that $(a_i)_{1\leq i\leq n}$.

Conversely, when A has two natural basis $(a_i)_{1\leq i\leq n}$ and $(b_i)_{1\leq i\leq n}$ which induce the weights ω_1 and ω_2 respectively and its multiplication constants are identical it is easy to prove that the isomorphism f defined by $f(a_i) = b_i$ is an algebra isomorphism such that $\omega_2 \circ f = \omega_1$.

However, the commutative 3-dimensional algebra $A = \langle e_1, e_2, e_3\rangle_R$ with multiplication table: $e_1^2 = e_1$, $e^2 = e_2$, $e_1 \cdot e_2 = \frac{1}{2}e_3$, $e_2 \cdot e_3 = \frac{1}{4}e_3$, $e_3^2 = e_3$ and with the other products zero, provides an example of an algebra with a double genetic realization whose weights are not equivalent.

Remark: These results can be easily extended to multibaric algebras in general, that is, to algebras which admit a number $k > 1$ of weight homomorphisms different (*k-baric* algebras).

2. B-train algebras

An important class of baric algebras is formed by the train algebras. In these algebras the weight homomorphism is unique. A natural generalization of the train algebra concept for bibaric algebras is:

Definition 2.1. A n-dimensional commutative bibaric algebra (A, ω_1, ω_2) is called a *b-train* algebra when the coefficients of the rank polynomial of A are only functions of $\omega_1(x)$ and $\omega_2(x)$. That is, the rank equation has the form:

$$x^r + p_1\big(\omega_1(x), \omega_2(x)\big)x^{r-1} + \ldots + p_{r-1}\big(\omega_1(x), \omega_2(x)\big)x = 0 \quad \forall \in A$$

where $p_i\big(\omega_1(x), \omega_2(x)\big)$ are homogeneous polynomials of degree i in $\omega_1(x)$ and $\omega_2(x)$.

A b-train algebra has associate three subalgebras which are train algebras: The algebra $A_1 = ker\,\omega_2$ with weight $\omega_1|_{ker\,\omega_2}$ whose rank equation is

$$x^r + p_1\big(\omega_1(x), 0\big)x^{r-1} + \ldots + p_{r-1}\big(\omega_1(x), 0\big)x = 0;$$

the algebra $A_2 = ker\,\omega_1$ with weight $\omega_2|_{ker\,\omega_1}$ whose rank equation is

$$x^r + p_1\big(0, \omega_2(x)\big)x^{r-1} + \ldots + p_{r-1}\big(0, \omega_2(x)\big)x = 0;$$

and the algebra $A_3 = \{x \in A | \omega_1(x) = \omega_2(x)\}$ with weight $\omega = \omega_i|_{A_3}$ whose rank equation is

$$x^r + \gamma_1\omega(x)x^{r-1} + \ldots + \gamma_{r-1}\omega(x)^{r-1}x = 0$$

where $\gamma_i\omega(x)^i = p_i\left(\omega(x), \omega(x)\right)$, for $i = 1, \ldots, r-1$.

If we call I and I_i $(i = 1, 2, 3)$ the sets of idempotent elements of A and A_i respectively, we have $I = I_1 \cup I_2 \cup I_3$. Furthermore, it is easy to prove the uniqueness of the weights ω_1 and ω_2 of the a b-train algebra.

The above definition can be generalized to any *k-baric* algebra. In this case, the number of associate subalgebras is $2^k - 1$.

3. B-train algebras of rank 3

In the following we assume that K has characteristic $\neq 2$. The rank of a b-train algebra in necessarily greater than 2 and the unique form of the rank equation for a b-train algebra of rank 3 is

$$x^3 - (\omega_1(x) + \omega_2(x))x^2 + \omega_1(x)\omega_2(x)x = 0 \tag{1}$$

Its associate subalgebras have the rank equations:

$A_1: \ x^3 - \omega_1(x)x^2 = 0$
$A_2: \ x^3 - \omega_2(x)x^2 = 0$
$A_3: \ x^3 - 2\omega(x)x^2 + \omega(x)^2 x = 0$

These subalgebras are the only train algebras of rank 3 which are Jordan algebras [2]. In our case it results:

Theorem 3.1. *Let A be a b-train algebra of rank 3. Then A is a Jordan algebra.*

Proof. We apply the usual linearization method. From (1) we have for any $x, y \in A$:

$$\begin{aligned} x^2y + 2x(xy) = 2(\omega_1(x) + \omega_2(x))xy + (\omega_1(y) + \omega_2(y))x^2 - \\ \omega_1(x)\omega_2(x)y - (\omega_1(x)\omega_2(y) + \omega_1(y)\omega_2(x))x \end{aligned} \tag{2}$$

Multiplying (2) by x, replacing y by xy in itself, and taking the difference between these equations, we obtain

$$x(x^2y) - x^2(xy) = (\omega_1(y) + \omega_2(y))(x^3 - (\omega_1(x) + \omega_2(x))x^2 + \omega_1(x)\omega_2(x)x) = 0$$

On other hand, the subalgebras A_i are characterized by the following Peirce decomposition and properties [2]: $A_i = Ke_i \oplus U^i_{\frac{1}{2}} \oplus U^i_0$, for $i = 1, 2$; where e_i is an idempotent of A_i and $U^i_j = \{x \in A_i | e_i x = jx\}$; furthermore, $\left[U^i_{\frac{1}{2}}\right]^2 \subseteq U^i_0$, $U^i_{\frac{1}{2}} \cdot U^i_0 \subseteq U^i_{\frac{1}{2}}$, $\left[U^i_0\right]^2 = \langle 0 \rangle$ and $x^3 = 0$ for any $x \in ker\,\omega_1 \cap ker\,\omega_2$. $A_3 = Ke_3 \oplus U^3_{\frac{1}{2}} \oplus U^3_1$ and on satisfy $\left[U^3_{\frac{1}{2}}\right]^2 \subseteq U^3_1$, $U^3_{\frac{1}{2}} \cdot U^3_1 \subseteq U^3_{\frac{1}{2}}$, $[U^3_1] = \langle 0 \rangle$ and $x^3 = 0$ for any $x \in ker\,\omega_1 \cap ker\,\omega_2$.

In order to obtain a refined Peirce decomposition for the b-train algebras of rank 3 we first prove that it is possible to find two orthogonal idempotents.

Proposition 3.2. *Let (A, ω_1, ω_2) be a b-train algebra of rank 3. Then: i) If $e_1 \in A_1$ and $e_2 \in A_2$ are idempotents, $(e_1e_2)^2 = 0$. ii) There are idempotents $\overline{e}_1 \in A_1$ and $\overline{e}_2 \in A_2$,such that $\overline{e}_1\overline{e}_2 = 0$.*

Proof. If $e_1e_2 \neq 0$ we consider the elements $x = e_1 + e_2 + \lambda(e_1e_2)$, $\lambda \in K$. Since they satisfy the equation $x^3 \quad 2x^2 \mid x = 0$, it results $(e_1 + e_2)(e_1e_2) = e_1e_2$ and $2(e_1 + e_2)((e_1 + e_2)(e_1e_2)) - 3(e_1 + e_2)(e_1e_2) + e_1e_2 + (e_1e_2)^2 = 0$. ¿From the above identities we get $(e_1e_2)^2 = 0$. Now, we define $\overline{e}_1 = e_1 - e_1e_2$ and $\overline{e}_2 = e_2 - e_1e_2$. Then $\overline{e}_1\overline{e}_2 = e_1e_2 - (e_1 + e_2)(e_1e_2) = 0$.

Theorem 3.3. *Let (A, ω_1, ω_2) be a bibaric algebra. Then A is a b-train algebra of rank 3 if and only if the following conditions are satisfied: i) A has two orthogonal idempotents e_1 and e_2, and the Pierce decomposition with respect to them* *is* $A = Ke_1 \oplus Ke_2 \oplus U^{12}_{\frac{1}{2}\frac{1}{2}} \oplus U^{12}_{\frac{1}{2}0} \oplus U^{12}_{0\frac{1}{2}}$ *where* $U^{12}_{ij} = \{x \in ker\,\omega_1 \cap ker\,\omega_2 \mid e_1x = ix, e_2x = jx\}$. *ii)* $\left[U^{12}_{\frac{1}{2}\frac{1}{2}}\right]^2 = \left[U^{12}_{\frac{1}{2}0}\right]^2 = \left[U^{12}_{0\frac{1}{2}}\right]^2 = \langle 0\rangle$, $\left[U^{12}_{\frac{1}{2}\frac{1}{2}}\right]\left[U^{12}_{\frac{1}{2}0}\right] \subseteq U^{12}_{0\frac{1}{2}}$, $\left[U^{12}_{\frac{1}{2}\frac{1}{2}}\right]\left[U^{12}_{0\frac{1}{2}}\right] \subseteq U^{12}_{\frac{1}{2}0}$, $\left[U^{12}_{\frac{1}{2}0}\right]\left[U^{12}_{0\frac{1}{2}}\right] \subseteq U^{12}_{\frac{1}{2}\frac{1}{2}}$. *iii)* $x^3 = 0$ *for any* $x \in ker\,\omega_1 \cap ker\,\omega_2$.

Proof. From the Peirce decomposition of A_i, $i = 1, 2$, : $A_i = Ke_i \oplus (ker\,\omega_1 \cap ker\,\omega_2)$ we have

$$A = Ke_1 \oplus Ke_2 \oplus (ker\,\omega_1 \cap \ker\,\omega_2) = Ke_1 \oplus Ke_2 \oplus U^{12}_{\frac{1}{2}\frac{1}{2}} \oplus U^{12}_{\frac{1}{2}0} \oplus U^{12}_{0\frac{1}{2}} \oplus U^{12}_{00}$$

¿From the proposition (3.2), we can choose e_1 and e_2 orthogonal idempotents. Since $e_1 + e_2$ is a idempotent of A_3 and $A_3 = K(e_1 + e_2) \oplus U^3_1 \oplus U^3_{\frac{1}{2}}$ we obtain $U^{12}_{00} = \langle 0\rangle$. If $x \in \left[U^{12}_{\frac{1}{2}\frac{1}{2}}\right]^2$, then $x \in \left[U^1_{\frac{1}{2}}\right]^2 \subseteq U^1_0$ and $x \in \left[U^2_{\frac{1}{2}}\right]^2 \subseteq U^2_0$, and thus $x \in U^{12}_{00}$. Therefore, $\left[U^{12}_{\frac{1}{2}\frac{1}{2}}\right]^2 = \langle 0\rangle$. The other relations are similarly shown.
Conversely, if the conditions *i)*, *ii)* and *iii)* are fulfilled, for any element $x = \omega_1(x)e_1 + \omega_2(x)e_2 + u + v + w$ of $A = Ke_1 \oplus Ke_2 \oplus U^{12}_{\frac{1}{2}\frac{1}{2}} \oplus U^{12}_{\frac{1}{2}0} \oplus U^{12}_{0\frac{1}{2}}$ on have

$$x^2 = \omega_1(x)^2e_1 + \omega_2(x)^2e_2 + (\omega_1(x) + \omega_2(x))u + \omega_1(x)v + \omega_2(x)w \\ +2uv + 2uw + 2vw$$

$$x^3 = \omega_1(x)^3e_1 + \omega_2(x)^3e_2 + (\omega_1(x)^2 + \omega_2(x)^2 + \omega_1(x)\omega_2(x))u + \omega_1(x)^2v \\ +\omega_2(x)^2w + 2(\omega_1(x) + \omega_2(x))(uv + uw + vw)$$

and on verify that $x^3 - (\omega_1(x) + \omega_2(x))x^2 + \omega_1(x)\omega_2(x)x = 0$.

We observe that the idempotent elements of the subalgebras A_i may be state as: $I_1 = \{e_1 + u + 2uv | u \in U^{12}_{\frac{1}{2}\frac{1}{2}}, v \in U^{12}_{\frac{1}{2}0}\}$, $I_2 = \{e_2 + u + w + 2uw | u \in U^{12}_{\frac{1}{2}\frac{1}{2}}, v \in U^{12}_{0\frac{1}{2}}\}$ and $I_3 = \{e_1 + e_2 + v + w - 2vw | v \in U^{12}_{\frac{1}{2}0}, w \in U^{12}_{0\frac{1}{2}}\}$. Furthermore, it is easy to prove that all idempotents in I_3 can be expressed as the sum of two orthogonal idempotents in I_1 and I_2 , respectively.

We can associate to A three integers (m, p, q) which are related to a refined Peirce decomposition of A by

$$m = \dim U^{12}_{\frac{1}{2}\frac{1}{2}},\ p = \dim U^{12}_{\frac{1}{2}0} \quad \text{and} \quad q = \dim U^{12}_{0\frac{1}{2}}.$$

When $p = q = 0$, that is if $A = Ke_1 \oplus Ke_2 \oplus U^{12}_{\frac{1}{2}\frac{1}{2}}$, then $(e_1 + e_2)(\lambda e_1 + \mu e_2 + u) = \lambda e_1 + \mu e_2 + u$ so $e_1 + e_2$ is the unity element 1 of algebra. One easily can see that all elements in A satisfy the equation $x^2 - (\omega_1(x) + \omega_2(x))x + \omega_1(x)\omega_2(x)1 = 0$. Linearizing this equation, result

$$xy = \tfrac{1}{2}((\omega_1(x) + \omega_2(x))y + (\omega_1(y) + \omega_2(y))x) - \tfrac{1}{2}(\omega_1(x)\omega_2(y) + \omega_1(y)\omega_2(x))1$$

and thus all 2+m-dimensional algebras of rank 3 of type $(m, 0, 0)$ are isomorphics.

When m=q=0, that is when $A = Ke_1 \oplus Ke_2 \oplus U^{12}_{\frac{1}{2}0}$, A_2 has exactly one idempotent, e_2, and the multiplication in A is given by

$$xy = \tfrac{1}{2}(\omega_1(x)y + \omega_1(y)x) - \tfrac{1}{2}(\omega_1(x)\omega_2(y) + \omega_1(y)\omega_2(x) - 2\omega_2(x)\omega_2(y))e_2$$

Again, all algebras of type $(0,p,0)$ are isomorphics. The algebras of type $(0,p,0)$ and $(0,0,p)$ are isomorphics if one swaps e_1 for e_2.

It is known that every train algebra of rank 3 satisfies a train equation in the plenary powers x, x^2 and $(x^2)^2$, [1], but the contrary is not true [3]. However, for the bibaric algebras we have:

Proposition 3.4. *Let (A,ω_1,ω_2) be a bibaric algebra. Then are equivalents: i) A is a b-train algebra with rank equation (1). ii) A satisfies the b-train equation for the first three plenary powers*

$$(x^2)^2-(\omega_1(x)^2+\omega_2(x)^2+\omega_1(x)\omega_2(x))x^2+(\omega_1(x)^2\omega_2(x)+\omega_1(x)\omega_2(x)^2)x=0 \quad (3)$$

Proof. 1. Replacing y by x^2 in (2) and multiplying (1) by x, (3) is easily obtained. 2. Taking the square of (3) and replacing x by x^2 in that equation, and taking the difference between the resulting equations, on get $(x^2)^2-(\omega_1(x)+\omega_2(x))x^3+\omega_1(x)\omega_2(x)x^2=0$. Now, if we take the difference between (3) and the last equation, the equation (2) is obtained.

4. Train algebras of rank 4

Before to consider b-train algebras of rank 4, we establish some results about train algebras of rank 4.

Proposition 4.1. *Let (A,ω) be a train algebra of rank 4. If A is power-associative, its rank equation has one of the following expressions: $x^4-\omega(x)x^3=0$ (4), $x^4-2\omega(x)x^3+\omega(x)^2x^2=0$ (5), $x^4-3\omega(x)x^3+3\omega(x)^2x^2-\omega(x)^3x=0$ (6).*

Proof. If A is a train algebra of rank 4 its equation has the form $x^4+\alpha\omega(x)x^3+\beta\omega(x)^2x^2-(1+\alpha+\beta)\omega(x)^3x=0$ with $\alpha,\beta\in K$. The linearization of the above equation provides

$$\begin{aligned}x^3y+x(x^2y)+2x(x(xy))=&-\alpha\omega(y)x^3-\alpha\omega(x)(x^2y+2x(xy))-2\beta\omega(x)\omega(y)x^2\\&-2\beta\omega(x)^2xy+3(1+\alpha+\beta)\omega(x)^2\omega(y)x+(1+\alpha+\beta)\omega(x)^3y \qquad (7)\end{aligned}$$

We start multiplying (7) by x, replacing y by xy in (7) and considering the difference between the resulting equations. Following if we set $y=x$ and we assume that the algebra is power-associative then the resulting equation is $\omega(x)^2((\alpha^2+\alpha-2\beta)x^3+(\alpha\beta+3\alpha+5\beta+3)\omega(x)x^2-(\alpha^2+\alpha\beta+4\alpha+3\beta+3)\omega(x)^2x)=0$. As by hipothesis the algebra is of rank 4, the only possible values for α and β are: i) $\alpha=-1,\beta=0$; ii) $\alpha=-2,\beta=0$; iii) $\alpha=-3,\beta=3$.

However, it is easy to find train algebras of rank 4 satisfying (4), (5) or (6) which are not power-associatives.

It is known that not all power-associative train algebras of rank 4 are Jordan algebras [2]. However, we have:

Theorem 4.2. *Let (A,ω) be a train algebra of rank 4 whose rank equation is (4) or (6). Then, are equivalent: i) A is power-associative. ii) A is a Jordan algebra. (In the last case it is required that the characteristic of K be also different of three.)*

Proof. Suppose that A is a power-associative algebra satisfying (4). We linearize the identities $(x^2)^2=\omega(x)x^3$, $x^2x^3=\omega(x)^2x^3$ and $x^3x^3=\omega(x)^3x^3$ to obtain

$$4x^2(xy) = \omega(y)x^3 + \omega(x)(x^2y + 2x(xy)) \tag{8}$$
$$x^2(x^2y) + 2x^2(x(xy)) + 2x^3(xy) = 2\omega(x)\omega(y)x^3 + \omega(x)^2(x^2y + 2x(xy)) \tag{9}$$
$$2x^3(x^2y) + 4x^3(x(xy)) = 3\omega(x)^2\omega(y)x^3 + \omega(x)^3(x^2y + 2x(xy)) \tag{10}$$

If we replace y by xy in (9), we get after some calculations

$$2x^3(x(xy)) = \omega(x)^2\omega(y)x^3 - \tfrac{1}{4}\omega(x)(x^2(x^2y) + 4x(x(x(xy)))) + \tfrac{1}{4}\omega(x)^2(3x^2(xy) + 6x(x(xy))) \tag{11}$$

On the other hand, when we replace x by x^2 in (8) and we have in mind that $(x^2)^3 = \omega(x)^3x^3$, we can deduce:

$$4x^3(x^2y) = \omega(y)\omega(x)^2x^3 + \omega(x)^2x^3y + 2\omega(x)x^2(x^2y) \tag{12}$$

Now, we substitute (11) and (12) in (10) to obtain

$$x^2(x^2y) = 4x(x(x(xy))) + \omega(x)\omega(y)x^3 - \omega(x)(x^3y + 3x^2(xy) + 6x(x(xy))) + 2\omega(x)^2(x^2y + 2x(xy)) \tag{13}$$

When we substitute (13) in (9), we get:

$$x^3(xy) + 2x(x(x(xy))) = \tfrac{1}{4}\omega(x)\omega(y)x^3 + \tfrac{1}{4}\omega(x)(2x^3y + 5x^2(xy) + 10x(x(xy))) - \tfrac{1}{2}\omega(x)^2(x^2y + 2x(xy)) \tag{14}$$

But the linearization of $(x^2)^2 = x^4$ provides the identity

$$4x^2(xy) = x^3y + x(x^2y) + 2x(x(xy)) \tag{15}$$

and replacing y by xy in (15) it can deduce

$$x^3(xy) + 2x(x(x(xy))) = \tfrac{3}{4}\omega(y)\omega(x)x^3 + \tfrac{1}{4}\omega(x)(4x^2(xy) - x(x^2y) + 6x(x(xy))) \tag{16}$$

Taking the difference between (14) and (16) we obtain:

$$x^2(xy) + x(x^2y) + 2x^3y + 4x(x(xy)) = 2\omega(y)x^3 + 2\omega(x)(x^2y + 2x(xy)) \tag{17}$$

Finally, using (8) and (15) in (17), on get $x^2(xy) = x(x^2y)$.

The proof for the algebra satisfying (6) is similar.

5. B-train algebras of rank 4

The proofs for the following results are analogs to proposition 4.1 and the theorem 4.2; however the calculations are more tedious and are omitted here.

Proposition 5.1. *If (A, ω_1, ω_2) is a power-associative b-train algebra of rank 4, its rank equation is one of the following equations*

$$x^4 - (\omega_1(x) + \omega_2(x))x^3 + \omega_1(x)\omega_2(x)x^2 = 0$$
$$x^4 - (2\omega_1(x) + \omega_2(x))x^3 + (\omega_1(x)^2 + 2\omega_1(x)\omega_2(x))x^2 - \omega_1(x)^2\omega_2(x)x = 0$$
$$x^4 - (2\omega_2(x) + \omega_1(x))x^3 + (\omega_2(x)^2 + 2\omega_2(x)\omega_1(x))x^2 - \omega_2(x)^2\omega_1(x)x = 0$$

Theorem 5.2. *Let (A, ω_1, ω_2) be a b-train algebra of rank 4. Then are equivalent: i) A is power-associative. ii) A is a Jordan algebra.*

References

1. I.M.H. Etherington. Commutative train algebras of ranks 2 and 3. J. London Math. Soc. **15**, 136-149 (1.940).
2. M. Ouattara. Sur les T-Algebres de Jordan. Lin. Alg. Appl. **144**, 11-21 (1.991).
3. S. Walcher. Algebras which satisfy a train equation for the first three plenary powers. Arch. Math. **56**, 547-551 (1.991).
4. A. Wörz-Busekros. Algebras in genetics. Lecture Notes in Biomath. 36, Berlin-Heidelberg-New York. (1.980).

THE BERNSTEIN PROBLEM IN MATHEMATICAL GENETICS AND BERNSTEIN ALGEBRAS

YU. I. LYUBICH
Technion, 32000 Haifa, Israel

Abstract. The Bernstein problem is to explicitly describe all quadratic evolucionary operators V such that $V^2 = V$. It is equivalent to the description of all stochastic Bernstein algebras. We present our results in this direction.

All references below are taken from the list in book [1].

In works [BerS22], [BerS23], [BerS23a], [BerS24] S.N.Bernstein posed and partially investigated an important problem concerning mathematical expression of fundamental laws of biological heredity. Let us to formulate this problem in more modern terms.

Let $\{1, 2, \cdots, n\}$ be a set of all distinct biological types of individuals in an infinetely large population. Let $x = (x_i)_{i=1}^n$ be a probability distribution of types in some generation, so all the $x_i \geq 0$ and $s(x) \equiv \sum_i x_i = 1$. Every such point $x \in \mathbb{R}^n$ is called a *state* of the population. The set of all state is the basic symplex $\Delta^{n-1} \subset \mathbb{R}^n$. Let us denote by $p_{ik,j}$ the probability that the type j appears in the next generation from parents whose types are i and k. Obviously, $p_{ik,j} \geq 0$ and $\sum_j p_{ik,j} = 1$; moreover, $p_{ik,j} = p_{ki,j}$ assuming that the biological characters of types are the same for males and females.

Suppose that this population is reproduced at random and without selection. Then its state $x' = (x'_j)_{j=1}^n$ in the generation will be

$$x'_j = \sum_{i,k} p_{ik,j} x_i x_k \quad (1 \leq j \leq n). \tag{1}$$

These formulas define a mapping $V : \Delta^{n-1} \to \Delta^{n-1}$ called the *evolutionary operator* (*e.o.* for short) of the given population. Therefore, $x' = Vx$ for the *equilibria states*.

The Bernstein Problem. *To describe explicitly all e.o. such that for all states in a generation the corresponding states in the next generation are equilibria.*

In other words, an explicit form of V such that $V^2 = V$ is required. The latter property is the *Stationarity Principle* in Bernstein's terminology. If it is valid then V (or the population) is called *Bernstein* or $\equiv$ *stationary* [LyuY71]. The simplest examples are: the *unit* e.o., $Vx = x$, and the *constant* one, $Vx = c$ where $c \in \Delta^{n-1}$. Some nontrivial examples appear only if $n \geq 3$.

S. González (ed.), Non-Associative Algebra and Its Applications, 241–244.

Example 1. There is the classical *Hardy-Weinberg* e.o. for $n = 3$:

$$x'_j = x_j^2 + x_j x_3 + \frac{1}{4}x_3^2 = (x_j + \frac{1}{2}x_3)^2 \qquad (j = 1, 2),$$

$$x'_3 = 2x_1x_2 + x_1x_3 + x_2x_3 + \frac{1}{2}x_3^2 = 2(x_1 + \frac{1}{2}x_3)(x_2 + \frac{1}{2}x_3).$$

The coefficients come here from the Mendelian law for a pair of genes A, a. The types 1, 2, 3 in this population are just the genotypes AA, aa, Aa. This e.o. is Bernstein because it preserves the values $p = x_1 + \frac{1}{2}x_3$ and $q = x_2 + \frac{1}{2}x_3$. (It is biologically clear since p, q are probabilities of the genes A, a.)

Example 2. There is the *Bernstein quadrille* e.o. for $n = 4$:

$$x'_1 = (x_1 + x_3)(x_1 + x_4), \quad x'_2 = (x_2 + x_3)(x_2 + x_4),$$

$$x'_3 = (x_1 + x_3)(x_2 + x_3), \quad x'_4 = (x_1 + x_4)(x_2 + x_4).$$

This is Bernstein because it preserves the values $x_1 + x_3, x_1 + x_4, x_2 + x_3, x_2 + x_4$. These ones can also be interpreted as probabilities of some genes [LyuY71].

Bernstein solved his problem for $n = 3$. In [LyuY71] the Bernstein problem was restricted to the case of a *stationary gene structure (s.g.s.)* in a sense suggested by Examples 1,2. Namely, a linear form $f(x)$ is called *invariant* if $f(Vx) = f(x)$. An e.o. V is called *regular* (or *conservative* in a later terminology) if V can be expressed through invariant linear forms. By our definition, a population has *s.g.s.* if its e.o. is conservative. Every conservative e.o. in Bernstein but the converse is not true even for $n = 3$ [LyuY71]. I believe that the Bernstein problem out of s.g.s. makes no genetical sense, so it becomes purely mathematical since the Bernstein population is two-level only in the case of s.g.s. [1].

The Bernstein problem in the conservative case ($\equiv$ *s.g.s.*) was completely solved in works [LyuY71], [LyuY73], [LyuY74a], [LyuY77a] by analyzing some algebraic structure behind it. For the first step, any given Bernstein e.o. V extends by (1) to the whole space $\mathbf{R}^n$ and then $V^2x = s^2(x)Vx$ [LyuY71]. Being quadratic every e.o. clearly generates an algebra structure in $\mathbf{R}^n$ such that $Vx = x^2$ ($p_{ik,j}$ are the structure constants of this non-associative commutative algebra). Then the previous identity takes the form $(x^2)^2 = s(x)^2x^2$ (appeared explicitly in [LyuY74a]). (Note that $s : \mathbf{R}^n \to \mathbf{R}$ is an algebra homomorphism [LyuY71]). Algebras with such an identity are also called *Bernstein*. A basic structure theorem for them was established in [LyuY71] (Theorem 4.1). The conservative algebras are exactly such that $A/AnnA$ is unit, i.e. $x^2 = s(x)x$. This fact plays a principal role in our analysis of s.g.s. [LyuY73], [LyuY74a] which also includes some convexity arguments based on the probabilistic properties of $p_{ik,j}$. These ones mean that the algebra is *stochastic* [LyuY77c]. Accordingly, the Bernstein problem sounds as *'to explicitly describe all stochastic Bernstein algebras'*.

After the above quoted works an extensive study of general Bernstein algebras has been carried out ([HolP75], [LyuY77b], [LyuY80], [LyuY84], [GriA87],...).

Note that there is another class of Bernstein algebras, so-called *exceptional* ones (see below) for which the Bernstein problem was completely solved [LyuY76b] (cf. [LyuY75] where just the case of exceptional conservative was considered).

Let us explain the above mentioned algebraic notions and facts. By the basic structure theorem, every idempotent e defines a decomposition $A = \mathbb{R}e \oplus U \oplus W$ with $U = ImL_e, W = KerL_e$ where $L_e y = 2ey$ is a projection in $Kers$. In this context

$$x^2 = s^2 e \oplus (su + 2uw + w^2) \oplus u^2 \tag{2}$$

for $x = se \oplus u \oplus w$ which $u \in U$ and $w \in W$. Moreover, some identities ($u^3 = 0, u(uw) = 0$ etc.) are fulfilled. The pair of integer (m, δ) where $m = dimU + 1$ and $\delta = dimW = n - m$ does not depend on e and it is called the *type* of the algebra. The estimate $dimJ \leq m$ for the subspace of invariant linear forms and the criterion "A *is conservative* $\Leftrightarrow dimJ = m$" holds [LyuY71]. By the way, the class of conservative algebras can be defined by identity: $x^2 y = \sigma(x)xy$ [LyuY77b]. In terms of (2) this means $uw = 0, w^2 = 0$. If $A^2 = A$ and $\delta > \frac{(m-1)(m-2)}{2}$ then A is conservative [LyuY74]. A Bernstein algebra A is *exceptional* if $dimA/A^2 = \delta$ or equivalently, $u^2 = 0$ [LyuY77b]. (In general, $dimA/A^2 \leq \delta$).

Every Bernstein algebra of type (m, δ) with $m \leq 2$ or $\delta \leq 1$ is conservative or exceptional. In particular, it is true for $n \leq 4$. Thus, the Bernstein problem is solved for the types $(2, n-2)$ and $(n-1, 1)$, in particular, for $n \leq 4$ (cf. [LyuY73a], [LyuY75a]).

Now let us formulate our main results in the Bernstein problem. For simplicity we suppose that the e.o. V is *normal* in the following sense: a) there is no j such that $x'_j = 0$; b) there is no a pair j_1, j_2 such that x'_{j_1} and x'_{j_2} are proportional; c) there is no a pair j_1, j_2 such that all x'_j-s depend only on $x_{j_1} + x_{j_2}, x_i (i \neq x_{j_1}, x_{j_2})$. In the conservative case the normality provides a canonical identification of the abstract types 1,2,...,n with genotypes like AA, aa, Aa ([1], Section 4.2). Moreover, every e.o. (not only Bernstein) can be reduced to a normal one by a *normalization* [LyuY74a] and [1], Sections 3.9, 4.2).

Theorem 1. *Let a stochastic Bernstein algebra be conservative and normal. Then its e.o.* $Vx = x^2$ *has one of the following two forms.*

1. Elementary gene structure (e.g.s.). *In this case*

$$x'_i = p_i^2 + 2p_i \sum_{k:k \neq i} \theta_{ik} p_k \quad (1 \leq i \leq m); x'_j = 2\theta_j p_{i_j} p_{k_j} \quad (m+1 \leq j \leq n)$$

where $p_1, \cdots, p_m$ *are linear invariant forms,*

$$p_i(x) = x_i + \sum_{j=m+1}^{n} \pi_{ij} x_j \quad (1 \leq i \leq m).$$

The matrix (π_{ij}) *is a column stochastic with exactly two nonzero entries* $\pi_{i_j j}, \pi_{k_j j}$ *in every column. All the pairs* (i_j, k_j) *are different.*
All θ_{ik} *are nonnegative and all* θ_j *are positive. The relations*

$$\theta_{i_j k_j} + \pi_{i_j j}\theta_j = \frac{1}{2}, \quad \theta_{k_j i_j} + \pi_{k_j j}\theta_j = \frac{1}{2}$$

hold and $\theta_{ik} = \theta_{ki} = \frac{1}{2}$ *for all other pairs* (i, k).

2. Nonelementary gene structure (n.e.g.s.). *In this case* $n = \alpha\beta$ *with* $\alpha > 1$ *and* $\beta > 1$ *and the coordinates can be labeled as* x_{ij}: $1 \leq i \leq \alpha, 1 \leq j \leq \beta$, *so that* $x'_{ij} = p_i p_j$ *where* p_i *and* p_j *are linear invariant forms,*

$$p_i = \sum_j x_{ij}, \quad q_j = \sum_i x_{ij}.$$

Conversely, every above described e.o. (algebra) is conservative and normal.

The hardy-Weinberg e.o. is a simplest example of e.g.s. ($n = 3, m = 2, \theta_{12} = \theta_{21} = 0, \theta_3 = 1, \pi_{13} = \pi_{23} = \frac{1}{2}$). The Bernstein quadrille e.o. is a simplest example of n.e.g.s. ($n = 4, \alpha = \beta = 2$). The above written general formulas can also be interpreted in terms of genes [LyuY71], [LyuY73], [LyuY73c].

Theorem 2. *Let a stochastic Bernstein algebra be exceptional and normal. Then it is unit and therefore it is conservative.*

Conjecture [1]. *Every stochastic Bernstein normal algebra is conservative.*

In view of stated results this conjecture is true for the types (m, δ) with $m \leq 2$ or $\delta \leq 1$, in particular, for $n \leq 4$. For some related conjectures see [1]. Some results in this direction were obtained by a combinatorial topology [LyuY77c], [LyuY79b].

Bernstein's works contain also several deep theorems like a necessity of the Mendelian law under the Stationarity Principle. His proofs are elementary but rather complicated. In [LyuY73b], [LyuY76a], [LyuY79a] all these theorems have been proved using Bernstein algebras.

Finally, let us another problem closely related to the Bernstein one.

Problem. Which Bernstein algebras are stochastic up to the choice of a basis ?

References

1. Yuri I. Lyubich, Mathematical Structures in Population Genetics. Biomathematics, v.22, Springer-Verlag, Berlin - Heidelberg, 1992.

LES ALGÈBRES DE MUTATION

C. MALLOL ET R. VARRO
Département de Mathématiques et Informatique Appliquées, Université de Montpellier III, BP 5043, 34032 Montpellier cedex 1, France

Abstract. The definition of ωM-algebra is introduced and we study a special class : the baric algebras with nilpotent kernels of index 2, named mutation algebras.

1. Introduction

Les idées qu'on développera par la suite, trouvent leur source dans le fait suivant : considérons une population isolée, panmictique et non soumise à la sélection naturelle. Si $a_1, \ldots, a_m$ sont des types géniques, μ_{ji} le taux de mutation par génération de a_i vers a_j ($\mu_{ji} \geq 0, \sum_{1 \leq k \leq m} \mu_{ki} = 1$) et $x_i(t)$ la fréquence de a_i à la t-ième génération, la fréquence de ce type à la génération suivante est donnée par $x_i(t+1) = \sum_{1 \leq j \leq m} \mu_{ij} x_j(t)$. Dans [6], on associe à cette population l'algèbre de base $\{e_1, \ldots, e_m\}$ avec $e_i e_j = \frac{1}{2}\left(\sum_{1 \leq k \leq m} (\mu_{ki} + \mu_{kj}) e_k\right)$.

1.1. DÉFINITION

On dit qu'une algèbre A est de *mutation* s'il existe une application linéaire $M : A \to A$ et une base $B = (e_i)_{i \in I}$, appelée de mutation, telle que $M(e_i) = \sum_{k \in I} \mu_{ki} e_k$, $\sum_{k \in I} \mu_{ki} = 1$, et $e_i e_j = \frac{1}{2} M(e_i + e_j)$.

Du fait que $\sum_{k \in I} \mu_{ki} = 1$, M est multiplicative ; de plus l'application $\omega : A \to K$, $\omega(e_i) = 1$, est une pondération et $xy = \frac{1}{2}(\omega(y) M(x) + \omega(x) M(y))$.

2. Le ωM-produit

Soient K un corps, car$(K) \neq 2$, A un K-espace vectoriel, $B = (e_i)_{i \in I}$ une base de A, $\omega : A \to K$ une forme linéaire et $M : A \to A$ une application linéaire, M et $\omega \neq 0$. Un calcul simple montre qu'il n'existe pas de structure d'algèbre sur A telle que $x^2 = M(x)$, pour tout $x \in A$. Cependant, nous avons :

2.1. THÉORÈME

Il y a équivalence entre :
(a) Se donner B, M et définir sur A la structure d'algèbre : $e_i e_j = \frac{1}{2} M(e_i + e_j)$.
(b) Se donner ω, M et munir A du produit $xy = \frac{1}{2}(\omega(y) M(x) + \omega(x) M(y))$.

Démonstration. Si (a) est vérifiée, soit ω la forme linéaire définie par $\omega(e_i) = 1$,

S. González (ed.), Non-Associative Algebra and Its Applications, 245–250.

$e_i \in B$; un calcul simple montre que $x^2 = \omega(x)M(x)$. Réciproquement, soit $\varepsilon \in A$, $\omega(\varepsilon) = 1$; si N est une base de $\ker(\omega)$ alors $B = \{\varepsilon\} \cup \{\varepsilon + x / x \in N\}$ est une base de A et $ee' = \frac{1}{2}M(e+e')$ pour tout e, $e' \in B$. ◇

2.2. Définition

On dit qu'une K-algèbre A est une ωM-algèbre s'il existe des applications linéaires non nulles $\omega : A \to K$ et $M : A \to A$ telles que le produit de A est donné par $xy = \frac{1}{2}(\omega(y)M(x) + \omega(x)M(y))$. ◇

2.3. Remarque

(a) D'après 2.1, tout espace vectoriel A avec une application linéaire non nulle $M : A \to A$ peut être muni d'une structure de ωM-algèbre si l'on se donne une forme linéaire $\omega \neq 0$ ou, ce qui est équivalent, une base de A comme en 2.1-a.
(b) Il faut noter que dans une ωM-algèbre, la forme linéaire ω n'est pas nécessairement une pondération, contrairement au cas des algèbres de mutation (cf. 2.6 et 2.7). ◇

2.4. Théorème

Soit A une algèbre. Les affirmations suivantes sont équivalentes :
(a) A est une ωM-algèbre. (b) Il existe un hyperplan E de A tel que $E^2 = \{0\}$.

Démonstration. (a) $\Longrightarrow$ (b) Il suffit de prendre $E = \ker(\omega)$. (b) $\Longrightarrow$ (a), soit $e \notin E$ et ω la forme linéaire définie par $\omega(\lambda e + y) = \lambda$ où $\lambda \in K$ et $y \in E$. Si M est l'application linéaire définie par $M(x) = 2ex - \omega(x)e^2$ on vérifie rapidement que $x^2 = \omega(x)M(x)$ pour tout $x \in A$. ◇

2.5. Remarque

Soit A une algèbre vérifiant 2.4.b. Le choix d'un élément $e \notin E$ équivaut à choisir une forme ω telle que $\ker(\omega) = E$ et $\omega(e) = 1$ et de ce fait, M est uniquement déterminée : en effet, si $\varepsilon \in A$, $\omega(\varepsilon) = 1$, $e - \varepsilon \in E$ donc $(e-\varepsilon)y = 0$, $y \in E$, d'où, $(e-\epsilon)^2 = 0$ et $e\varepsilon = \frac{1}{2}(e^2 + \varepsilon^2)$. Il découle que $2\varepsilon x - \omega(x)\varepsilon^2 = 2ex - \omega(x)e^2$.

L'application M peut s'écrire : $M = 2L_e - L_e \circ \mathcal{E}_\omega$ où L_e est la multiplication par e et $\mathcal{E}_\omega$ est le premier projecteur associé à la décomposition $Ke \oplus \ker(\omega)$, application définie par $\mathcal{E}_\omega(x) = \omega(x)e$. ◇

2.6. Proposition

Soit A une ωM-algèbre. Les affirmations suivantes sont équivalentes :
(a) $\omega \circ M = \omega$; (b) ω est une pondération.
Dans ces conditions, M est multiplicative, A est uniquement pondérée et sa structure de ωM-algèbre est unique.

Démonstration. (a) $\Longrightarrow$ (b) On sait que $x^2 = \omega(x)M(x)$ donc $\omega(x^2) = \omega(x)\omega(M(x))$; or, comme $\omega \circ M = \omega$, il en résulte $\omega(x^2) = \omega(x)^2$. Quant à (b) $\Longrightarrow$ (a), on a $\omega(M(x)) = 2\omega(ex) - \omega(x)\omega(e^2)$ donc si ω est une pondération, $\omega(M(x)) = \omega(x)$, car $\omega(e) = 1$. Pour la dernière affirmation, comme $x^2 = \omega(x)M(x)$ on a $M(x^2) = \omega(x)M^2(x) = \omega(M(x))M^2(x) = M(x)^2$. Pour le reste, tenant compte de 2.4, il suffit de démontrer que la pondération est unique ce qui se fait sans difficulté car $\ker(\omega)^2 = \{0\}$. ◇

2.7. Théorème

Soit A une algèbre. Les affirmations suivantes sont équivalentes :
(a) A est de mutation. (b) A est une ωM-algèbre avec $\omega \circ M = \omega$. (c) A admet une pondération ω telle que $\ker(\omega)^2 = \{0\}$.

Démonstration. Que (a) $\Longrightarrow$ (b) résulte de 1.1. Réciproquement, (b) $\Longrightarrow$ (a) car si A est une ωM-algèbre avec $\omega \circ M = \omega$, il suffit de choisir une base B de A formée d'éléments de poids 1. L'équivalence entre (b) et (c) découle des résultats 2.4 et 2.6. ◇

2.8. Remarque

(a) Les algèbres de mutation apparaissent dans [2] sous le nom d'algèbres $A_{T,\omega}$.
(b) Soient (A, ω) une algèbre pondérée et N l'idéal engendré par $\ker(\omega)^2$. Comme $N \subset \ker(\omega)$, A/N est pondérée par $\overline{\omega}(\overline{x}) = \omega(x)$ où $\overline{x}$ est la classe de x. La surjection canonique $A \to A/N$ est un morphisme d'algèbres pondérées et $\ker(\overline{\omega})^2 = \{0\}$. Toute algèbre pondérée se projete donc sur une algèbre de mutation.
(c) Si A est de mutation, elle est train-spéciale (car $\ker(\omega)^2 = \{0\}$) donc génétique. De plus, de $xy = \frac{1}{2}(\omega(y)M(x) + \omega(x)M(y))$ on établit facilement que $x^{[n+1]} = \omega(x)^{2^n-1}M^n(x)$ et $x^{n+2} = \frac{1}{2^n}\omega(x)^{n+1}(M^{n+1} + \sum_{1\leq k\leq n} 2^{n-k}M^k)(x)$. De ce fait, si A est de dimension finie et si r est le degré du polynôme minimal de M, on montre sans difficulté que A est une train-algèbre de rang $r+1$ et que A vérifie une train-équation pour les $r+1$ premières puissances pleines. Or, en combinant les puissances pleines et principales on peut établir une équation sans conditions sur la dimension et indépendante du polynôme minimal : $x^{[3]} = 2\omega(x)x^3 - \omega(x)^2x^2$. Cette équation (résultant du fait que $x^2 - \omega(x)x \in \ker(\omega)$) est vérifiée en outre par les algèbres de Bernstein-Jordan d'ordre 1. ◇

Désormais A désigne une algèbre de mutation.

Les algèbres de mutation forment une classe particulière d'algèbres non associatives. En effet, d'après [4] il y a équivalence entre : (a) A est de Jordan, (b) A est à puissances associatives, (c) $M^3 = 3M^2 - 2M$.

2.9. Proposition

Les énoncés suivantes sont équivalents : (a) A est associative. (b) A est alternative. (c) A est de Moufang. (d) $M^2 = 2M$ sur $\ker(\omega)$. (e) Il existe $\varepsilon \in A$ tel que $2x^3 - 3\omega(x)x^2 + \omega(x)^3\varepsilon = 0$ pour tout $x \in A$.
Dans ces conditions, A admet un unique idempotent.

Démonstration. Montrons (c) $\Longrightarrow$ (d) : par hypothèse on a $0 = (xzx)y - x(z(xy)) = \omega(x)[2\omega(zy)(M^2-2M)(x)+\omega(xy)(M^3-2M^2)(z)+\omega(zx)(M^3-4M)(y)]$. En prenant $\omega(x) \neq 0$, $\omega(z) \neq 0$, pour tout $y \in \ker(\omega)$, $(M^3 - 4M)(y) = 0$; de même, prenant $\omega(x) \neq 0$, $\omega(y) \neq 0$, on a $(M^3 - 2M^2)(z) = 0$ pour tout $z \in \ker(\omega)$. Ceci veut dire que $M^3 - 4M = M^3 - 2M^2 = 0$ sur $\ker(\omega)$ d'où le résultat. (d) $\Longrightarrow$ (e). Comme $M^2 = 2M$ sur $\ker(\omega)$, $M^2 - 2M$ est constante sur les éléments de poids 1. Il existe donc $\varepsilon \in A$, $\omega(\varepsilon) = 1$ (car $\omega \circ M = \omega$), tel que pour tout $x \in A$, $\omega(x) \neq 0$, $(M^2 - 2M)(\omega(x)^{-1}x) = -\varepsilon$. Il en résulte la relation $2x^3 - 3\omega(x)x^2 + \omega(x)^3\varepsilon = 0$, vérifiée aussi, de façon triviale, par les éléments de

$\ker(\omega)$. (e) $\Longrightarrow$ (a). On déduit de $2x^3 - 3\omega(x)x^2 + \omega(x)^3\varepsilon = 0$ que $M^2 - 2M$ est constante sur les éléments de poids 1, donc est nulle sur $\ker(\omega)$. Or pour tout $x, y, z, \in A$ on a : $x(yz) - (xy)z = \frac{1}{4}\omega(y)(M^2 - 2M)(\omega(x)z - \omega(z)x)$. Quant à la dernière affirmation, de $(M^2 - 2M)(\varepsilon) = -\varepsilon$, on a $\varepsilon^{[3]} = 2\varepsilon^2 - \varepsilon$; puis, si on remplace x par ε dans $2x^3 - 3\omega(x)x^2 + \omega(x)^3\varepsilon = 0$ on obtient $\varepsilon^3 = \frac{1}{2}(3\varepsilon^2 - \varepsilon)$ d'où $\varepsilon^4 = \frac{1}{4}(7\varepsilon^2 - 3\varepsilon)$. Or $\varepsilon^{[3]} = \varepsilon^4$ car A est associative, d'où $\varepsilon^2 = \varepsilon$. L'unicité résulte par l'application de la relation $2x^3 - 3\omega(x)x^2 + \omega(x)^3\varepsilon = 0$ à un autre idempotent non nul. ◇

2.10. Remarque

Si $x^2 = x$ alors $M(x) = x$. L'exemple qui suit montre que l'existence de points fixes de M n'est pas une condition suffisante pour l'existence d'idempotents : soit A l'algèbre de base $\{e_1, e_2\}$, $\omega(e_1) = \omega(e_2) = 1$ et $M(e_1) = 2e_1 - e_2$, $M(e_2) = e_1$. Si $x = \alpha_1 e_1 + \alpha_2 e_2$ alors $M(x) = (2\alpha_1 + \alpha_2)e_1 - \alpha_1 e_2$ donc $x^2 = (\alpha_1 + \alpha_2)[(2\alpha_1 + \alpha_2)e_1 - \alpha_1 e_2]$. On voit que $x^2 = x$ entraine $x = 0$. Les résultats qui suivent fournissent des conditions pour l'existence d'un idempotent. ◇

2.11. Proposition

A possède un idempotent si au moins une des conditions suivantes est remplie :
(a) $\ker(M - Id) \not\subset \ker(\omega)$; la réciproque est vraie. (b) M est diagonalisable ; la réciproque n'est pas vraie. (c) A est réelle de dimension finie et $\|M(e_i)\| = 1$, $i \in I$, où $\|\sum_{i\in I} \alpha_i e_i\| = \sum_{i\in I} |\alpha_i|$.

Démonstration. Montrons (b), soit $x \in A$, $x \neq 0$, tel que $M(x) = \lambda x$ avec $\lambda \in K$. Comme $M(x) - x = (\lambda - 1)x$, si $\lambda \neq 1$ on a $x \in Im(M - Id)$. Or, $Im(M - Id) \subset \ker(\omega)$ car $\omega \circ M = \omega$; donc si A admet une base de vecteurs propres de M, forcément 1 est valeur propre et $\ker(M - Id) \not\subset \ker(\omega)$, donc (a) est vérifiée. La réciproque n'est pas vraie : soit A l'algèbre de mutation de base $B = \{x, y\} \cup D$ avec $\omega(z) = 1$, $z \in B$ et $M(x) = 3x - 2y$, $M(y) = 2x - y$, $M(e) = e$, $e \in D$. Les éléments de D sont des idempotents et un calcul simple montre que la seule valeur propre de M est 1. Or, une base de $\ker(M - Id)$ est $\{x - y\} \cup D$ donc $\ker(M - Id) \neq A$. Pour ce qui est de (c), la condition $\|M(e_i)\| = 1$ équivaut à $\mu_{ki} \geq 0$ pour tout $k, i \in I$. On a $|\omega(x)| \leq \|x\|$ et $\|M(x)\| \leq \|x\|$ d'où $\|x^2\| \leq \|x\|^2$. L'application $x \to x^2$ est donc continue et applique le polyèdre compact contractile $\{x \in A/\omega(x) = \|x\| = 1\}$ sur lui-même. D'après le théorème du point fixe de Lefschetz, elle admet un point fixe. ◇

2.12. Remarque

(a) Si une algèbre satisfait les conditions de 2.11-c, elle contient une sous-algèbre gamétique : celle engendrée par les idempotents d'ordre 1. Du point de vue de la modélisation génétique ces conditions étant vérifiées, il en résulte que dans une population il y a toujours une sous-population qui se stabilise dès la première génération et sur laquelle la mutation n'a pas d'effet.
(b) Si A possède un idempotent e, l'ensemble des idempotents de A est $Ip(A) = \{e + x/x \in \ker(\omega), M(x) = x\}$. ◇

Nous finissons ce travail par quelques réflexions sur les endomorphismes et les dérivations.

3. Endomorphismes et dérivations

Soient A une algèbre de mutation, $e \in A$, $\omega(e) = 1$. Dans ce qui suit nous utiliserons constamment les "règles de calcul" : $2xy = \omega(y)M(x)+\omega(x)M(y)$, $x, y \in A$, $M(e) = e^2$; si $x^2 = 0$ alors $x \in \ker(\omega)$ et dans ce cas, $M(x) = 2ex$ et $xy = \omega(y)ex$.

Soit $End_\omega(A)$ l'ensemble formé par les applications f, linéaires et multiplicatives, telles que $\omega \circ f \neq 0$. ◇

3.1. Théorème

Soient $f, d : A \longrightarrow A$ deux applications linéaires.

(a) $f \in End_\omega(A)$ si et seulement si $\omega \circ f = \omega$ et $f \circ M = M \circ f$.

(b) $d \in Der_K(A)$ si et seulement si $\omega \circ d = 0$ et $d \circ M = M \circ d$.

Démonstration. Le résultat (b) a été prouvé en [5] (thm. 4.3.4), nous en donnons ici une démonstration différente. Soit d une dérivation. Supposons que $d(e) \in \ker(\omega)$ (*) : on a $d(M(e)) = d(e^2) = 2ed(e) = M(d(e))$. Puis, si $x \in \ker(\omega)$ on a : $d(M(x)) = 2d(ex) = 2(ed(x) + d(e)x) = 2ed(x)$ (1) d'où $\omega(d(M(x))) = 2\omega(d(x))$ (2) ; or, $x^2 = 0$, donc $2xd(x) = \omega(d(x))M(x) = 0$, d'où, en appliquant $\omega \circ d$, $\omega(d(x))\omega(d(M(x))) = 0$ d'où, par (2), $\omega(d(x)) = 0$. Enfin, de (1), $d(M(x)) = 2ed(x) = M(d(x))$ car $d(x) \in \ker(\omega)$. Il reste à montrer que la supposition (*) : si A admet un idempotent, de $e^2 = e$ on obtient $d(e) = 2ed(e)$ d'où $\omega(d(e)) = 2\omega(ed(e)) = 2\omega(d(e))$ donc $\omega(d(e)) = 0$. Si A n'admet pas d'idempotent, on pose $e^2 = e + x$ avec $x \in \ker(\omega)$ et $ex \neq 0$ (sinon e^2 serait un idempotent). On a $2ed(e) = d(e^2) = d(e) + d(x)$ donc $\omega(d(e)) = \omega(d(x))$; montrons que $\omega(d(x)) = 0$: on a $0 = d(x^2) = 2xd(x) = \omega(d(x))M(x) = 2\omega(d(x))ex = 0$, d'où $\omega(d(x)) = 0$. La réciproque découle de l'identité : $d(xy) - xd(y) - yd(x) = \frac{1}{2}[(d \circ M - M \circ d)(\omega(x)y + \omega(y)x) - \omega(d(x))M(y) - \omega(d(y))M(x))]$. ◇

3.2. Remarque

Le résultat précédent montre que les morphismes et les dérivations coincident sur $\ker(\omega)$. Par la suite nous préciserons le lien entre $End_\omega(A)$ et $Der_K(A)$ dans le cas où A admet un idempotent. Soient $e \in A$, $e \neq 0$, $e^2 = e$; dans ces conditions le projecteur $\mathcal{E}_\omega$ (2.5) est un élément de $End_\omega(A)$ et $L_{\mathcal{E}_\omega}$ agit comme une sorte de fonction caractéristique au sens de $\mathcal{E}_\omega \circ g = \mathcal{E}_\omega$ si $g \in End_\omega(A)$ et $\mathcal{E}_\omega \circ g = 0$ si $g \in Der_K(A)$. De plus :

3.3. Proposition

La translation $d \longrightarrow d + \mathcal{E}_\omega$, établit une bijection entre $Der_K(A)$ et $End_\omega(A)$.

Démonstration. Il suffit de voir que les conditions de 3.1 sont vérifiées. ◇

Soient $C(\omega, M) = \{h \in End_K(Ker(\omega))/h \circ M = M \circ h\}$ et $GL(\omega, M)$ les éléments inversibles de $C(\omega, M)$.

Nous noterons $Ker(M - Id) \Delta C(\omega, M)$ l'ensemble $Ker(M - Id) \times C(\omega, M)$ muni du produit semi-direct $(x, y)(y, g) = (x + h(y), h \circ g)$ et $Ker(M - Id) \bigtriangledown C(\omega, M)$ le même ensemble muni de l'opération $(x,y)(y,g) = (h(y) - g(x), [h,g])$, où $[h,g]$ est le crochet de Lie.

3.4. PROPOSITION

(a) $End_K(A) \simeq Ker(M - Id)\Delta C(\omega, M)$, isomorphisme de monoides.
(b) $Der_K(A) \simeq Ker(M - Id) \bigtriangledown C(\omega, M)$, isomorphisme d'algèbres de Lie.

Démonstration. Montrons (a) : Si $f \in End_K(A)$, $f(e) - e \in Ker(M - Id)$; par conséquent l'application $f \to (f(e) - e, f_{|Ker(\omega)})$ est bien définie et admet pour réciproque $(x,y) \to f$, avec $f(y) = h(y)$ si $y \in ker(\omega)$ et $f(e) = e + x$ ($x \in \ker(M - Id) \Longrightarrow e + x \in Ip(A)$). Enfin, si $f, g \in End_K(A)$, on a $((f \circ g)(e) - e,$ $f \circ g_{|Ker(\omega)}) = (f(e) - e + f(g(e)) - e), f \circ g_{|Ker(\omega)})$ donc $((f \circ g)(e) - e, f \circ g_{|Ker(\omega)}) =$ $(f(e) - e, f_{|Ker(\omega)})(g(e) - e, g_{|Ker(\omega)})$. ◇

3.5. COROLLAIRE

Les groupes $Aut_K(A)$ et $Ker(M - Id)\Delta Gl(\omega, M)$ sont isomorphes.

Remerciements.
Nous remercions le referee pour ses suggestions de rédaction.

4. Bibliographie

1. I.M.H. ETHERINGTON, Genetic algebras, Proc. Roy. Soc. Edinburgh, 59 : 242-258 (1939).
2. Y.I. LYUBICH, Classification de certains types d'algèbres de Bernstein, Vestinik Kharkovskogo Univ. Ser. Matematica i Mecanica, 177 : 86-94 (1979) (en russe).
3. A. MICALI et M. OUATTARA, Sur les algèbres de Jordan génétiques II, Cahiers Mathématiques Univ. Montpellier II, 38 : 77-107 (1989).
4. A. MICALI et M. OUATTARA, Sur les algèbres de Jordan génétiques, Ann. Sci.Univ. Blaise Pascal Clermont Ferrand II, 27 : 193-227 (1991).
5. M. OUATTARA, Algèbres de Jordan et algèbres génétiques, Cahiers Mathématiques Univ. Montpellier II, 37 (1988).
6. A. WORZ-BUSEKROS, Algebras in genetics, Lect. Notes in Biomathematics 36, Springer-Verlag, Berlin (1980).

NON ASSOCIATIVE GRADED ALGEBRAS*

DOLORES MARTÍN BARQUERO and CÁNDIDO MARTÍN GONZÁLEZ
Departamento de Matemática Aplicada, Universidad de Málaga.
Plaza de El Ejido s/n. 29013 Málaga.

Abstract. In this paper we deal with simple non necessarily associative n-graded algebras over a commutative unitary ring. No additional finiteness condition (chain conditions, etc) is needed for our main result : we give a functorial construction which provides kl-graded algebras from k-graded algebras (for arbitrary positive integers k and l). In particular it provides l-graded algebras from non-graded algebras. Furthermore, if the starting k-algebra is simple, the resulting kl-algebra is also simple. We prove that any simple n-graded algebra arises as the result of applying this functor to a certain k-graded algebra which is simple as ungraded algebra.

1. Introduction and preliminary definitions

The associative $\mathbb{Z}_2$-graded algebras were first studied by C. T. C. Wall from the structural viewpoint (see [4]). After this, some generalizations ([1], [3]) and new technics have been introduced in the theory of two-graded algebras (as for instance the link of the structure of the algebra with the triple system structure of its odd part, [2]). The graded Lie algebras, the irruption of the superalgebras, and the possibility of exploiting the graded algebras in the study of triple systems, have produced an arising interest on graded algebras. Let F be any commutative unitary ring and A a (not necessarily associative) F-algebra. Let $n \in \mathbb{Z}, n > 1$, we say that A is a $\mathbb{Z}_n$*-graded algebra* whenever $A = \bigoplus_{i=0}^{n-1} A_i$ with $\{A_i : i = 0, \ldots, n-1\}$ being a finite set of F-submodules of A such that $A_iA_j \subset A_{i+j}$ (sum in $\mathbb{Z}_n$). The F-submodules A_i of A will be called *homogeneous components* of A. If A and B are $\mathbb{Z}_n$-graded algebras over F, then an *homomorphism* $f : A \to B$ is an F-modules homomorphism such that $f(A_i) \subset B_i$ for all $i \in \mathbb{Z}_n$ and $f(xy) = f(x)f(y)$ for all $x, y \in A$. In case that $A \subset B$ and f is the inclusion map, then we say that A is a $\mathbb{Z}_n$*-graded subalgebra* of B. If besides we have that $AB \subset A$ and $BA \subset A$, we say that A is an $\mathbb{Z}_n$*-graded ideal* of B. The *right* and *left graded ideals* are defined in an obvious way. We say that a $\mathbb{Z}_n$-graded algebra A is simple when $A^2 \neq 0$ and A has no $\mathbb{Z}_n$-graded ideals other than $\{0\}$ and A itself. We say that an ideal I of a (not graded) algebra A is *minimal* when $I \neq 0$ and if J is an ideal of A with $J \subset I$ then $J = \{0\}$ or $J = I$. The obvious definitions for *minimal right ideal* (respectively *minimal left ideal* should also be mentioned as well as the fact that all of these definitions may also be given in a graded context.

* Supported by the DGICYT, Ps. 90-0119, and by the Plan andaluz para la consolidación de grupos de investigación y desarrollo tecnológico.

S. González (ed.), Non-Associative Algebra and Its Applications, 251–256.

2. Constructing graded algebras

In this paragraph we give a method for constructing a $\mathbb{Z}_{nm}$-graded algebras from a given $\mathbb{Z}_m$-graded algebra. In particular we can construct $\mathbb{Z}_n$-graded algebras from non-graded algebras. This method will provide us with the prototype of $\mathbb{Z}_n$-graded simple algebra, since one of the main results in this paper claims that essentially there are no more $\mathbb{Z}_n$-graded simple algebras than the described in this paragraph. We introduce previously some notations : if X is any F-module we define the F-modules monomorphism

$$e_i : X \to \overbrace{X \oplus \cdots \oplus X}^{n} \quad (0 \leq i \leq n-1), \; e_i : x \mapsto (\overbrace{0 \cdots 0}^{i} x, 0 \cdots 0).$$

Let A be any $\mathbb{Z}_k$-graded algebra $A = \oplus_{i=0}^{k-1} A_i$. We consider $B = \overbrace{A \oplus \cdots \oplus A}^{l}$, with the following $\mathbb{Z}_n$ grading ($n := lk$): $B = \bigoplus_{i=0}^{n-1} B_i$ with $B_{kq+r} = \{e_q(x_r) : x_r \in A_r\}$, $0 \leq r < k$ and the product $e_j(x_i)e_s(y_t) = e_p(x_i y_t)$ where $p = [j+s+c_k(i,t)](mod\ l)$ and $c_k(i,t) = 1$ if $i+t \geq k$ and $c_k(i,t) = 0$ otherwise. We shall denote this algebra by $Gr(A,l,k)$. The following result is easy to prove from the previous definitions:

Theorem 1 *If A is a $\mathbb{Z}_k$-graded algebra (simple as ungraded algebra), and $l \in \mathbb{Z}$, $l \geq 1$, then the $\mathbb{Z}_n$-graded algebra $B = Gr(A,l,k)$ (with $n = lk$) is simple.*

Lemma 1 *Let $i,j \in \{0,\dots,l-1\}$, $k = m.n$ and $r,s \in \{0,\dots,k-1\}$, $r = \alpha n + \beta$, $s = \alpha' n + \beta'$ with $0 \leq \beta, \beta' < n$. Then*

$$m[i+j+c_k(r,s)]_{\ mod\, l} + [\alpha+\alpha'+c_n(\beta,\beta')]_{\ mod\, m} = [m(i+j)+\alpha+\alpha'+c_n(\beta,\beta')]_{\ mod\, lm}.$$

The proof consists of an easy though tedious analysis of the formula in the following cases : (1) $0 \leq r+s < k$ and $0 \leq \beta+\beta' < n$. (2) $0 \leq r+s < k$ and $n \leq \beta+\beta' < 2n$. (3) $k \leq r+s < 2k$ and $0 \leq \beta+\beta' < n$. (4) $k \leq r+s < 2k$ and $n \leq \beta+\beta' < 2n$.

Theorem 2 *Let be $A = Gr(B,l,k)$ a kl-graded algebra and $B = Gr(C,m,n)$ a mn-graded algebra with $k = mn$ then A is isomorphic to the kl-graded algebra $Gr(C,lm,n)$.*

Let $e_i : B \to A$ and $f_j : C \to Gr(C,lm,n)$ be defined by

$$e_i(x) = (\overbrace{0\dots,0}^{i}, x, 0\dots,0), \quad f_j(z) = (\overbrace{0\dots,0}^{j}, z, 0\dots,0)$$

for $i = 0,\dots,l-1$, $j = 0,\dots,lm-1$. We define

$$\Theta : A \to \overbrace{C \oplus \dots \oplus C}^{lm}, \quad e_i(x_r) \mapsto f_{im+\alpha}(a_\beta)$$

where $x_r = (\overbrace{0,\dots,0}^{\alpha}, a_\beta, 0, \dots, 0)$ for some $a_\beta \in C$.

It is easy to prove that Θ is an homomorphism of lk-graded F-modules, let us see that Θ is a homomorphism of algebras, that is to say , $\Theta(e_i(x_r)e_j(y_s)) = \Theta(e_i(x_r)\Theta(e_j(y_s))$ for x_r as before and

$$y_s = (\overbrace{0,\ldots,0}^{\alpha'}, b_{\beta'}, 0, \ldots, 0) \tag{1}$$

for some $b'_\beta \in C$. In fact

$$e_i(x_r)e_j(y_s) = e_{[i+j+c_k(r,s)]\ mod\, l}(x_r y_s)$$

$$\Theta(e_i(x_r)e_j(y_s)) = f_{m[i+j+c_k(r,s)]\ mod\, l+[\alpha+\alpha'+c_n(\beta,\beta')]\ mod\, m}(a_\beta b'_\beta). \tag{2}$$

On the other hand

$$\Theta(e_i(x_r))\Theta(e_j(y_s)) = f_{mi+\alpha}(a_\beta)f_{mj+\alpha'}(b_{\beta'}) =$$

$$f_{[(i+j)m+\alpha+\alpha'+c_n(\beta,\beta')]\ mod\, lm}(a_\beta b'_\beta). \tag{3}$$

By the previous lemma the results of (2) and (3) coincide, hence Θ is a homomorphism of algebras. We prove now that Θ is an isomorphism . If we have $\Theta(e_i(x_r)) = 0$ then $f_{im+\alpha}(a_\beta) = 0$ according to the definition this implies that $a_\beta = 0$ hence $x_r = 0$, as a consequence $e_i(x_r) = 0$ and then Θ is a monomorphism. To see that Θ is an epimorphism, we take any element $f_t(a_\beta) \in Gr(C, lm, n)$. As we have $t = mi + \alpha$ with $0 \leq \alpha < m$, we define

$$x_r = (\overbrace{0,\ldots,0}^{\alpha}, a_\beta, 0, \ldots, 0)$$

which is an element in B and $e_i(x_r) \in A$. Then it follows easily that $\Theta(e_i(x_r)) = f_t(a_\beta)$.

Corollary 1 *Let be $A \cong Gr(B, l, k)$ a kl-graded algebra and $B \cong Gr(C, l', k')$ a $l'k'$-graded algebra with $k = l'k'$ then A is isomorphic to the kl-graded algebra $Gr(C, ll', k')$.*

Theorem 3 *If A as a simple $\mathbb{Z}_n$-graded algebra , and $I \neq 0$ a (non graded) proper ideal of A then*

- *1) $I \cap A_i = \{0\}$ for all $i \in \mathbb{Z}_n$.*
- *2) There is a subset $S \subset \mathbb{Z}_n$ such that $I \cap (\oplus_{i\in S} A_i) \neq \{0\}$ and for any $S' \subset \mathbb{Z}_n$ with Card$(S') <$ Card(S), we have $I \cap (\oplus_{i\in S'} A_i) = \{0\}$.*
- *3) Fix $k_0 \in S$, then*

$$A_{k_0+j} \subset I + \bigoplus_{i\in S-\{k_0\}} A_{i+j}$$

for all $j \in \mathbb{Z}_n$. If $A_0 \neq 0$, the set S is 2) may be chosen containing 0.

If I is a proper ideals such that $I \neq \{0\}$, and $I \cap A_i \neq \{0\}$ for a certain i, then define $I_j := A_j \cap I$ for all $j \in \mathbb{Z}_n$. It is clear that $\oplus_{j\in\mathbb{Z}_n} I_j \neq \{0\}$ is a graded ideal of A hence it coincides with A. Consequently $A_j \subset I$ for all j and so $I = A$ a contradiction. In order to prove 2), we must realize that there is a natural number $k > 1$ and elements $l_1, \ldots, l_k \in \mathbb{Z}_n$ such that :

- (I) $I \cap (A_{l_1} \oplus \cdots \oplus A_{l_k}) \neq \{0\}$.
- (II) For any $p \in N, 0 < p < k$ and any $m_1, \ldots, m_p \in \mathbb{Z}_n$, we have $I \cap (A_{m_1} \oplus \cdots \oplus A_{m_p}) = \{0\}$.

Indeed : as we have $I \cap A_i = \{0\}$ for all i, we consider the set

$$X = \{k \in N : k > 1, I \cap (A_{l_1} \oplus \cdots \oplus A_{l_k}) \neq 0,\ \exists l_1, \ldots, l_k \in \mathbb{Z}_n \text{ all different } \}$$

which is non empty because $n \in X$. If we choose k as the least number in X, then clearly k sastisfies the required properties (I) and (II) above. Now define $S := \{l_1, \ldots, l_k\}$ and the asserts 1) and 2) are clearly satisfied by S. To see 3), we define

$$I_{k_0+j} := A_{k_0+j} \cap (I + \bigoplus_{i \in S - \{k_0\}} A_{i+j})$$

It is easy to see that for $j = 0$, $I_{k_0} \neq 0$ and that $\oplus_i I_i$ is a graded ideal of A. Then each $I_{k_0+j} = A_{k_0+j}$ implying 3). Finally suppose that $0 \notin S$. As we have $A_0 \neq 0$ we can define $T = S - k_0 := \{i - k_0 : i \in S\}$ we have

$$A_0 \subset I + \bigoplus_{i \in T - \{0\}} A_i$$

hence if $0 \neq a_0 \in A_0$ then $a_0 = x + \sum_{i \in T-\{0\}} a_i$ with $0 \neq x \in I$ and $a_i \in A_i$. Therefore $x \in I \cap (\oplus_{i \in T} A_i)$ that is to say $I \cap (\oplus_{i \in T} A_i) \neq \{0\}$. As $Card(T) = Card(S)$, the remaining is trivial.

Theorem 4 *Let A be a simple $\mathbb{Z}_n$-graded F-algebra, and $I \neq 0$ a (non graded) proper ideal of A. Then there is a natural number $0 < k < n$ such that $n = kl$, and a set $\{\theta_{i+rk,i+sk} : 0 \leq i \leq k-1, 0 \leq r, s \leq l-1\}$ of F-modules isomorphisms $\theta_{i+rk,i+sk} : A_{i+sk} \to A_{i+rk}$ such that*

- *1) $\theta_{\alpha\alpha} = Id$, $\theta_{\alpha\beta}\theta_{\beta\gamma} = \theta_{\alpha\gamma}$ for all $\alpha, \beta, \gamma \in \mathbb{Z}_n, \alpha = i+rk, \beta = i+sk, \gamma = i+tk$.*
- *2) $\theta_{\alpha,\beta+\gamma}(x_\beta x_\gamma) = x_\beta \theta_{\alpha-\beta,\gamma}(x_\gamma) = \theta_{\alpha-\gamma,\beta}(x_\beta)x_\gamma$ for all α, β, γ with $\alpha \equiv \beta + \gamma \bmod k$ and $(x_\beta, x_\gamma) \in A_\beta \times A_\gamma$*
- *3) Defining $B_j := \sum_{i=0}^{k-1} A_{i+kj}$ for $j = 0, \ldots, l-1$ (with $kl = 0$), the map $B_0 \to B_j$ given by*

$$(x_0, \ldots, x_{k-1}) \mapsto (\theta_{kj,0}(x_0), \ldots \theta_{k-1+kj,k-1}(x_{k-1}))$$

is a F-modules isomorphism.

We fix a set S as given in the previous theorem, and $k_0, k_1 \in S$ with $k_0 < k_1$. Define $h := k_1 - k_0$ and $k := gcd(n, k_1 - k_0)$. For any integers r, i we have $A_{i+rh} = A_{k_0+(i-k_0+rh+)}$ and by 3) of the previous theorem we have $A_{i+rh} \subset I \cap (\bigoplus_{p \in S-k_0} A_{p+i-k_0+rh})$.Then for any $x_{i+rh} \in A_{i+rh}$ we have $x_{i+rh} = z + \sum_{p \in S-\{k_0\}} x_{p+i-k_0+rh}$ with $x_{p+i-k_0+rh} \in A_{p+i-k_0+rh}$ and $z \in I$. Moreover if $x_{i+rh} = z' + \sum_{p \in S-\{k_0\}} x'_{p+i-k_0+rh}$ with $z' \in I$ and each $x'_{p+i-k_0+rh} \in A_{p+i-k_0+rh}$ then $0 = (z - z') + \sum_{p \in S-\{k_0\}} (x_{p+i-k_0+rh} - x'_{p+i-k_0+rh})$ hence $z - z' \in I \cap (\oplus_{p \in S-\{k_0\}} A_{p+i-k_0+rh})$ and this intersection being 0 by the previous theorem, implies $z = z'$ and therefore $x_{p+i-k_0+rh} = x'_{p+i-k_0+rh}$ for each p. In this way we can define $\theta_{i+(r+1)h,i+rh} : A_{i+rh} \to A_{i+(r+1)h}$ mapping x_{i+rh} to $x_{i+(r+1)h}$. It is easily seen this is an F-modules homomorphism. Let us put now $\alpha := i+rh$. We want to see that $\theta_{\alpha+h,\alpha}$ is an isomorphism, take $\theta_{\alpha+h,\alpha}(x_\alpha) = 0$, then $x_\alpha = z + \sum_{p\in S-\{k_0,k_1\}} x_{p+\alpha-k_0}$ with $z \in I$ and $x_{p+\alpha-k_0} \in A_{p+\alpha-k_0}$. Then $z \in (I \cap (\oplus_{j\in S-\{k_1\}} A_{p-k_0+\alpha} = 0$ and consequently each $x_{p+\alpha-k_0} = 0$. The epimorphic character of $\theta_{\alpha+h,\alpha}$ follows from the relations

$$A_{\alpha+h} \subset I \cap (A_\alpha \oplus (\oplus_{p\in S-\{k_0,k_1\}} A_{p+\alpha-k_0}))$$
$$A_\alpha \subset I \cap (A_{\alpha+k} \oplus (\oplus_{p\in S-\{k_0,k_1\}} A_{p+\alpha-k_0}))$$

(see 3) of Theorem 3). Now we give a step forward by defining $\theta_{i+sh,i+rh} := \theta_{i+sh,i+(s-1)h} \cdots \theta_{i+(r+1)h,i+rh}$ in case that $s > r$, $\theta_{i+sh,i+rh} := \theta^{-1}_{i+rh,i+sh}$ if $s < k$ and $\theta_{i+rh,i+rh} := Id$ for any r, i. Now the assert 1) is straightfoward. To see 2) we define $\beta := i + sh$ and $\gamma := i + th$, and then we prove that

$$\theta_{\beta+\gamma+h,\beta+\gamma}(x_\beta x_\gamma) = x_\beta \theta_{\gamma+h,\gamma}(x_\gamma)$$

for all $(x_\beta, x_\gamma) \in A_\beta \times A_{-}g$. Indeed, we have

$$x_\gamma = z + x_{\gamma+h} + \sum_{p\in S-\{k_0,k_1\}} x_{p+\gamma-k_0}$$

$$x_\beta x_\gamma = x_\beta z + x_\beta x_{\gamma+h} + \sum_{p\in S-\{k_0,k_1\}} x_\beta x_{p+\gamma-k_0}$$

hence $\theta_{\beta+\gamma+h,\beta+\gamma}(x_\beta x_\gamma) = x_\beta x_{\gamma+h} = x_\beta \theta_{\gamma+h,\gamma}(x_\gamma)$. In a similar way we should prove $\theta_{\beta+\gamma+h,\beta+\gamma}(x_\beta x_\gamma) = \theta_{\beta+h,\beta}(x_\beta) x_\gamma$ and the assert 2), follows easily from these identities. To prove 3), we recall that each $\theta_{\alpha,0}$ ($\alpha = i + rh$) is a F-modules isomorphism. We also remark the following : as $k := gcd(n, h)$ and $n = kl$, each number $i+rh$ with $0 \le i \le h-1$ can also be writen in the form $j+qk$ with $0 \le j \le k-1$ (and necessarily $0 \le q \le l-1$). Reciprocally each number $j + qk$ may be recovered as a certain $i + rh$ since the subgroups of $\mathbb{Z}_n$ generated by k and h coincide. Furthermore the numbers $i + rh$ and $i + sh$ when written in the new form are $j + qk$ and $j + tk$ and reciprocally. In this way all that we have proved may be translated to agree with the statement of the theorem.

Theorem 5 *Let A be a simple $\mathbb{Z}_n$-graded F-algebra. Then A is simple as ungraded algebra or is isomorphic to the $\mathbb{Z}_n$-graded algebra $Gr(B, n_1, n_2)$ with $n = n_1 n_2$ and B being a $\mathbb{Z}_{n_2}$-graded algebra which is simple as ungraded algebra.*

If A is not simple, we consider the F-modules B_j defined in the previous theorems. We endow B_0 with an algebra structure by means of the following product : take $x_p, x_q \in B_0$ with $(x_p, x_q) \in A_p \times A_q$. We put $p + q = ck + r$ with $0 \le r < k$ (division in $\mathbb{Z}$). Then define $x_p \cdot x_q := \theta^{-1}_{ck+r,r}(x_p x_q)$. Relative to this product, B_0 is a $\mathbb{Z}_k$-graded algebra. We have to distinguish the product of two elements $x, y \in B_0$ from that of the same elements in A. We shall denote by yuxtaposition the product in A and by a dot the product in the algebra B_0. To check that B_0 is simple we fix a $\mathbb{Z}_k$-graded ideal $J = J_0 \oplus \cdots \oplus J_{k-1}$ such that $J \ne 0$. Define now $J_{i+rk} := \theta_{i+rk,i}(J_i)$ for $0 \le i < k$ and $J' := \oplus_{i,r} J_{i+rk}$. It is straightforward that J' is a $\mathbb{Z}_n$-graded ideal of A and therefore $J' = A$ implying $J = B_0$. The $\mathbb{Z}_k$-graded algebra B_0 is simple. We now prove that A is isomorphic to $Gr(B_0, l, k)$: we define $f : A \to Gr(B_0, l, k)$ by $f(x_{i+rk}) = (e_r \circ \theta^{-1}_{i+rk,i})(x_i + rk)$ where we recall that e_r is the natural inclusion of B_0 into the r-th direct summnad of $Gr(B_0, l, k)$. By definition $f(A_i)$ agrees with the i-th homogeneous component of $Gr(B_0, l, k)$, hence f is a graded map. As f is a composition of F-modules homomorphisms, then f is it. Let us see that $f(xy) = f(x)f(y)$ for all $x \in A_{i+rk}$ and $y \in A_{j+sk}$.

$$f(x)f(y) = e_r\theta^{-1}_{i+rk,i}(x) \cdot e_s\theta^{-1}_{j+sk,j}(y) = e_q(\theta^{-1}_{i+rk,i}(x) \cdot \theta^{-1}_{j+sk,j}(y))$$

with $q = (r + s + c_k(i, j))(mod\ l)$. If we prove

$$\theta^{-1}_{i+ki,i}(x) \cdot \theta^{-1}_{j+sk,j}(y)) = \theta_{(i+j)\ mod\,k + kq, (i+j)\ mod\,k}(xy)$$

then we should have

$$f(x)f(y) = e_q\theta_{(i+j)\ mod\,k + kq, (i+j)\ mod\,k}(xy) = f(xy)$$

because $xy \in A_{(i+j)\ mod\,k + kq}$. So, we must prove the before said relation. Define $\alpha := (i + j)\ _{mod\,k}$, then

$$\theta_{\alpha,i+j}(\theta^{-1}_{i+ki,i}(x)\theta^{-1}_{j+sk,j}(y)) = \theta^{-1}_{i+ki,i}(x)\theta_{\alpha-i,j}(\theta^{-1}_{j+sk,j}(y)) =$$

$$\theta^{-1}_{i+ki,i}(x)\theta_{\alpha-i,j+ks}(y) =$$

$$\theta_{i,i+ki}(x)\theta_{\alpha-i,j+ks}(y) = \theta_{\alpha,i+j+k(r+s)}(xy) = \theta_{\alpha,i+j}(\theta_{i+j,i+j+k(r+s)}(xy))$$

implying that $\theta^{-1}_{i+ki,i}(x)\theta^{-1}_{j+sk,j}(y)) = \theta^{-1}_{i+j+k(r+s),i+j}(xy)$ hence

$$\theta^{-1}_{i+j,\alpha}\left[\theta^{-1}_{i+ki,i}(x)\theta^{-1}_{j+sk,j}(y)\right] = \theta_{\alpha,i+j+(r+s)k}(xy)$$

$$\theta^{-1}_{i+ki,i}(x) \cdot \theta^{-1}_{j+sk,j}(y) = \theta_{\alpha+qk,\alpha}(xy)$$

So we have proved $A \cong Gr(B_0, l, k)$ for some k such that $0 < k < n$. If $k = 1$ then B_0 is a simple (ungraded) algebra and the result is proved. We can finish the proof argueing by induction on n. In fact if $n = 2$ the result follows from what we have already proved, if the result is true for all the positive integers m such that $m < n$ then we have $B_0 \cong Gr(B, l', k')$ with $k = l'k'$ and then by corollary 1 $A \cong Gr(B, n_1, n_2)$ with $n_1 = ll'$ and $n_2 = k'$.

References

1. Castellón A and Cuenca J. A. and Martín C. 'Applications of Ternary H^*-algebras to associative H^*-superalgebras'. To appear in *Algebra, Groups and Geometries.*
2. Cuenca, J.A, García A, and Martín C. 'Prime two-graded algebras with nonzero socle'. Preprint.
3. Cuenca J. A. and Martín C. 'Jordan two-graded H^*-algebras'. In Nonassociative Algebraic Models. S. Gonzalez and H. C. Myung Eds. pp. 91–99.
4. Wall, C.T.C.:1963, 'Graded Brauer Groups',*J. reine angew*, Vol. no. **213**, pp. 187–199.

TRACIAL ELEMENTS FOR NONASSOCIATIVE H*-ALGEBRAS

J. MARTINEZ MORENO
Departamento de Análisis Matemático
Universidad de Granada. 18071-Granada, Spain

Abstract. We introduce the concept of tracial element for a nonassociative H^*-algebra A with zero annihilator which extends the analogous one for associative H^*-algebras by Saworotnow and Friedell. We show that the set of tracial elements enjoys many of the properties of the associative case. The canonical associative bilineal form on A given by $< a, b >:= (a/b^*)$ for all a, b in A, is continuously linearizable with respect to the tracial norm if and only if A has an operator-bounded approximate unit.

1. Introduction

Let H be a complex Hilbert space, the associative H^*-algebra $\mathcal{HS}(H)$ of all Hilbert-Schmidt operators on H contains a $*$-invariant ideal which is a very interesting algebra in the operator theory on Hilbert spaces. It is known as the trace-class of operator on H, $(TC(H))$. The trace-class was introduced by Schatten as of all possible products FG with F and G in $\mathcal{HS}(H)$. $TC(H)$ has a canonical norm $|\cdot|$ for which it is a Banach $*$-algebra and, if we forget the product, $(TC(H), |\cdot|)$ becomes in a natural way the dual of the Banach space of compact operators on H, as well as the unique complete predual of the space of bounded linear operators on H. From an algebraic point of view $TC(H)$ is a primitive Banach algebra with dense socle. All of the above comments can be seen in Schatten's book [51].

Since the theory of structure for complex associative H^*-algebras with zero annihilator, developed by Ambrose in [1], reduces at essence the theory of such algebras to the particular case of H^*-algebras of the type $\mathcal{HS}(H)$, it is reasonable to think that an abstract concept of tracial-element might be introduced on an arbitrary complex associative H^*-algebra with zero annihilator in such a way that in the particular case where the H^*-algebra is $\mathcal{HS}(H)$, it reconstructs the algebra $TC(H)$. The reasonable conjecture above expounded were verified successfully by P. P. Saworotnow and J. C. Friedell in [50] and [47].

The abstract theory of Saworotnow-Friedell about elements of trace in associative H^*-algebras with zero annihilator is in the base of subsequent development of the concept of "Hilbert-module" over an associative H^*-algebra with zero annihilator, which was also introduced by Saworotnow [45], and studied with fullness later (see [2, 7, 13, 14, 17, 26, 27, 33, 34, 35, 36, 37, 38, 40, 46, 48, 49, 54, 55]).

In the present moment where the theory of nonassociative H^*-algebras has reached a state of appreciable maturity as much in your general aspect (see [11, 15, 16, 20, 21, 24, 44, 57]) as in your aplication at family classes of nonassociative algebras defined by identities (see [3, 4, 5, 8, 9, 10, 12, 19, 22, 28, 29, 31, 39, 42, 52, 53, 56, 58, 59, 60]), it seems reasonable to consider the possibility to give life to the ideas

S. González (ed.), Non-Associative Algebra and Its Applications, 257–268.

of Saworotnow-Friedell about tracial-elements in this nonassociative environment.

This is the purpose of this paper in which we introduce the concept of element of trace in an H^*-algebra A with zero annihilator (Definition 2.1) which coincides with the analogous by Saworotnow-Friedell in the associative case and, as it will be shown, such concept has a behavior very similar to the classical concept in its context. Explicitly if we denoted by $\tau c(A)$ the set of all tracial elements of A, $\tau c(A)$ is a subspace of A containing A^2 so automatically an ideal which is $*$-invariant, and for convenient norm, $\|\cdot\|_t$, $\tau c(A)$ becomes a dual Banach space in such a way that the product is separately w^*-continuous (Theorem 2.1).

Automorphisms and derivations over an H^*-algebra with zero annihilator preserve the ideal of the tracial elements, and seen as operators in such ideal are w^*-continuous (theorems 2.2 and 2.3)

Since any H^*-algebra with zero annihilator is the l^2-sum of a family of topologically simple H^*-algebras [21], we study the ideal of the tracial elements of a such l^2-sum and we prove that, at essence, such ideal is the l^1-sum of the ideals of the tracial elements of every sumand H^*-algebra (Theorem 2.4). This result is further on of divertissement and it is useful in the developement of our paper, so for example, starting from it we obtain that an H^*-algebra A is topologically simple if and only if $\tau c(A)$ is a prime algebra if and only if $\tau c(A)$ has not non trivial direct sumands (Corollary 2.1).

For an H^*-algebra A with zero annihilator, $\tau c(A)$ is a semiprime algebra with known extended centroid (Proposition 2.1). ¿From this fact united to results in [44] and above result we obtain that every homomorphism from any complete normed complex algebra into $\tau c(A)$ with dense range is continuous.

Section 3 is devoted to study the relation between the complete normed algebra $\tau c(A)$ and the problem of the possibility of linearizing continuously the canonical associative symmetric bilinear form $<,>$ on the H^*-algebra A given by $<a,b>=(a/b^*)$ for all a,b in A. To be more precise the problem can be formulated as follows: given an H^*-algebra A with zero annihilator, to find a Banach space $(X,|\cdot|)$ such that $A^2\subseteq X\subseteq A$ and for which there exists a continuous linear funtional for the bar one norm, f, satisfying $<a,b>=f(ab)$ for all a,b in A. Our main result (Theorem 3.1) assures that the Banach space $(\tau c(A),\|\cdot\|_t)$ is one solution to this problem if and only if A has an "operator-bounded" approximate unit, so having of more general nonassociative context where the trace introduced in [50] has meaning. In this respect we also point out that the canonical form $<,>$ on A is linearizable (without continuity) if and only if A has an approximate unit (Proposition 3.1).

2. The algebra of the tracial elements for an H*-algebra

We recall that an H^*-*algebra* is a complex nonassociative algebra A with an algebra involution $*$ (called the H^*-algebra involution), which is a Hilbert space relative to an inner product $(./.)$ satisfying the so called H^*-*axiom*, namely:

$$(ab/c)=(a/cb^*)=(b/a^*c) \text{ for all } a,b,c \text{ in } A$$

It is well known, [21; Proposition 2 (i)], that the product of every H^*-algebra is continuous for the topology of the Hilbert norm associated to the inner product, so by changing the inner product by a suitable positive multiple if necessary, every H^*-algebra becomes a complete normed algebra, a fact that we will assume in what follows. Moreover, if A has zero annihilator ($a \in A, aA = Aa = 0 \Rightarrow a = 0$), then the topology of the Hilbert norm is the unique complete normable topology on A making its product continuous ([43; Remark 2.8.i)]). As a consequence isomorphisms betwen H^*-algebras with zero annihilator are automatically continuous. Also the H^*-algebra involution of any H^*-algebra with zero annihilator is continuous, in fact it is isometric ([23; Proposition 2 (ix)]).

In what follows all H^*-algebras A we cosider will be assumed to have zero annihilator. As it is usual we denote by L_a and R_a the operators of left and right multiplication by a in A.

If A,B are algebras with an algebra involution $*$ and F is a linear mapping from A into B, F^* will denote the mapping $a \to (F(a^*))^*$ from A into B. In the particular case that A, B are H^*-algebras and F is continuous, $F^\bullet : B \to A$ will denote the hilbertian adjoint operator of F.

Definition 2.1. An element a in an H^*-algebra A is a *tracial element* if the set of real numbers $\{|(b/a)| : b \in A \text{ and } \|L_b\| \leq 1\}$ is bounded. $\tau c(A)$ will denote the set of all tracial elements of A and for an element a in $\tau c(A)$ we will define the *tracial norm* of a by $\|a\|_t := Sup\{|(b/a)| : b \in A, \|L_b\| \leq 1\}$.

Clearly $\tau c(A)$ is a subspace of A and the tracial norm is a seminorm on $\tau c(A)$. Moreover for all a in $\tau c(A)$ we have

$$\|a\|^2 \leq \|L_a\|\,\|a\|_t \leq \|a\|\,\|a\|_t$$

from which we have $\|a\| \leq \|a\|_t$, hence $\|\cdot\|_t$ is a norm on $\tau c(A)$.

For an element a in an H^*-algebra A we will write $p(a) := \|L_a\|$. Since A has zero annihilator, we have that p is a norm on A related with the previous one by the inequality $p \leq \|\cdot\|$. From this it follows the inclusion $(A,p)' \subseteq (A, \|\cdot\|)'$ for the dual spaces. Since $*$ is isometric and $\|L_{x^\bullet}\| = \|L_x^\bullet\| = \|L_x\|$ for all x in A, it follows that $\tau c(A)$ is $*$-invariant and $\|a^*\|_t = \|a\|_t$ for all a in $\tau c(A)$. Now the mapping $a \to \hat{a}$ from $(\tau c(A), \|\cdot\|_t)$ into $(A,p)'$ given by $\hat{a}(x) := (x/a^*)$ for all x in A is clearly linear and isometric. Using the inclusion $(A,p)' \subseteq (A, \|\cdot\|)'$ and inclusion $(A,p)' \subseteq (A, \|\cdot\|)'$ andchet representation theorem we have that, given f in $(A,p)'$ there exist a unique a in A such that $f(x) = (x/a^*)$ for all x in A. This says to us that a^*, and therefore a, is in $\tau c(A)$ and $f = \hat{a}$.

Therefore we have proved there exists a canonical identification between $(\tau c(A), \|\cdot\|_t)$ and $(A,p)'$. In such identification the operation of an element a in $(\tau c(A), \|\cdot\|_t)$ over an element x in (A,p) is given by $a(x) = (x/a^*)$. In what follows we will consider $(\tau c(A), \|\cdot\|_t)$ as the dual Banach space of the completed of (A,p).

Theorem 2.1. *Let A be an H^*-algebra. Then $\tau c(A)$ is a $*$-invariant ideal of A containing A^2. $\|\cdot\|_t$ is an algebra norm on $\tau c(A)$ under which $\tau c(A)$ becomes a dual Banach space and satisfies:*

i) $\|ab\|_t \leq \|a\|\,\|b\|$ for all a, b in A.

ii) $\|a\| \leq \|a\|_t$ for all a in $\tau c(A)$

iii) $\|a^*\|_t = \|a\|_t$ for all a in $\tau c(A)$.

Moreover for every b in A the mappings $a \to ba$ and $a \to ab$ from $\tau c(A)$ into $\tau c(A)$ are w^*-continuous. In particular the product of $\tau c(A)$ is separately w^*-continuous.

Proof. If a, b are in A, then Cauchy-Schwarz inequality and H^*-axiom give:

$$|(c/ab)| = |(cb^*/a)| = |(L_c(b^*)/a)| \leq \|L_c(b^*)\| \, \|a\| \leq \|L_c\| \, \|b\| \, \|a\|.$$

From which it follows that ab is in $\tau c(A)$ and $\|ab\|_t \leq \|a\| \, \|b\|$, which proves i). This property together with $\tau c(A)$ is a subspace of A permits to assure that A^2 is in $\tau c(A)$, hence $\tau c(A)$ is an ideal of A.

The rest of the proof it follows easily of previous commentaries to the Theorem and the following Lemma applied to the case where $A = B$ and F is the operator L_a (resp. R_a) taking into account $L_a^{*\bullet} = R_a$ (resp. $R_a^{*\bullet} = L_a$) and R_a (resp.L_a) is continuous on (A, p).

Lemma 2.1. *Let A, B be H^*-algebras, F a continuous linear mapping from A into B, and assume that $F^{*\bullet}$ is continuous as mapping from (B, p) into (A, p) with norm M. Then $F(\tau c(A))$ is contained in $\tau c(B)$ and F, seen as mapping from $(\tau c(A), \|\cdot\|_t)$ into $(\tau c(B), \|\cdot\|_t)$, is w^*-continuous with norm M.*

Proof. We denote by G the mapping $F^{*\bullet}$ and consider the unique continuous linear extension of G to the completed of (B, p) and (A, p). Since both mappings have the same transpose map, G^t, it follows that G^t is w^*-continuous with norm equal to the norm of G ([30; Chapter 3, §12]). Now, taking into account the canonical identifications between $(\tau c(A), \|\cdot\|_t)$ (resp. $(\tau c(B), \|\cdot\|_t)$) and the dual space of the completed of (A, p) (resp. (B, p)), we have

$$(b/G^t(a)^*) = G^t(a)(b) = a(G(b)) = (G(b)/a^*) = (b/G^\bullet(a^*))$$

for all b in B and a in $\tau c(A)$. From which we have that $G^{\bullet *}$ carries $\tau c(A)$ into $\tau c(B)$ and, seen as a mapping between such spaces, coincides with G^t. To conclude the proof is sufficient to observe that $G^t = (F^{*\bullet})^t = (F^{*\bullet})^{\bullet *} = F$.

The next result assures that the set of tracial elements of an H^*-algebra A is determined by the algebraic structure of A.

Theorem 2.2. *If A, B are H^*-algebras and Φ is an isomorphism from A onto B, then $\Phi(\tau c(A)) = \tau c(B)$. Moreover Φ seen as isomorphism from $\tau c(A)$ onto $\tau c(B)$ is w^*-continuous and $\|\Phi\|_t \leq \|\Phi\|^2$.*

Proof. As we know Φ is a bicontinuous isomorphism and using the H^*-axiom we obtain $L_{\Phi^{*\bullet}(b)} = \Phi^{*\bullet} L_b \Phi$ for all b in B. Therefore

$$p(\Phi^{*\bullet}(b)) = \|L_{\Phi^{*\bullet}(b)}\| \leq \|\Phi^{*\bullet}\| \, \|L_b\| \, \|\Phi\| = \|\Phi\|^2 p(b)$$

for all b in B. Hence $\Phi^{*\bullet} : (B, p) \to (A, p)$ is continuous with norm less or equal that $\|\Phi\|^2$. Now, Lemma 2.1 gives $\Phi(\tau c(A)) \subseteq \tau c(B)$ and the second part in the statement. Similar argument applied on Φ^{-1} concludes the proof.

A particular consequence of the above theorem is that the automorphisms of any H^*-algebra A preserve the tracial elements and seen as automorphisms on $\tau c(A)$ are w^*-continuous. The next theorem says that the derivations have the same properties.

Theorem 2.3. *If D is a derivation on an H^*-algebra A, then $\tau c(A)$ is invariant by D and, seen D as derivation on $\tau c(A)$ is w^*-continuous and satisfies $\|D(a)\|_t \leq 2\|D\|\|a\|_t$ for all a in $\tau c(A)$.*

Proof. By [57; Theorem 3] D is continuous and by [20; Theorem 2.1] $D^{**} = -D$. Now apply similar argument in Theorem 2.2 taking into account $L_{D(b)} = DL_b - L_b D$.

If $\{A_i\}_{i \in I}$ is a family of H^*-algebras, then the l^2-sum, $B = \overset{l^2}{\oplus} A_i$, of such family is in a natural way a new H^*-algebra by defining the product and the involution pointwise, so it is interesting the calculation of $\tau c(B)$ in terms of the $\tau c(A_i)$. This is the message of the next theorem.

Theorem 2.4. *$(\tau c(B), \|\cdot\|_t)$ is, in a canonical way, totally isomorphic to $\overset{l^1}{\oplus} ((\tau c(A_i), \|\cdot\|_{t_i})$. Such canonical isomorphism is w^*-bicontinuous.*

Proof. Firstly observe that if for every $i \in I$ F_i is a continuous linear operator on $(A_i, \|\cdot\|)$ and the family $\{\|F_i\|\}$ is bounded , then the mapping $\{a_i\} \to \{F_i(a_i)\}$ defines a continuous linear operator F on $(B, \|\cdot\|)$ with norm $\|F\| = Sup\{\|F_i\| : i \in I\}$. Now if $\{a_i\}$ is in B, from inequalities $\sum \|a_i\|^2 < +\infty$ and $p(a_i) = \|L_{a_i}\| \leq \|a_i\|$ it follows that $\{a_i\}$ is an element of the c_0-sum, $\overset{c_0}{\oplus} (A_i, p)$, of the normed spaces (A_i, p) and the family $\{\|L_{a_i}\|\}_{i \in I}$ is bounded. Therefore the mapping Φ from (B, p) into $\overset{c_0}{\oplus} (A_i, p)$ given by $\Phi(\{a_i\}) = \{a_i\}$ is well defined and is isometric, because the observation in the beginning of the paragraph (choose $F_i = L_{a_i}$ and observe that $F = L_{\{a_i\}}$). On the other hand, since the c_{00}-sum, $\overset{c_{00}}{\oplus} (A_i, p)$, is dense in $\overset{c_0}{\oplus} (A_i, p)$ and it is contained in the image of Φ, is follows that Φ has dense range. Hence $\hat{\Phi} : (\hat{B}, p) \to \overset{c_0}{\oplus} (\hat{A_i}, p)$, the unique linear extension of Φ to respective completed spaces, is an isometric linear biyection. Therefore its transpose map $\Phi^t : \overset{l_1}{\oplus} (\tau c(A_i), \|\cdot\|_{t_i}) \to (\tau c(B), \|\cdot\|_t)$ is also an isometric linear map which is w^*-bicontinuous [30; Corollary, p. 308]. Finally from definition of transpose map we easily obtain that $\hat{\Phi}^t(\{a_i\}) = \{a_i\}$ for all $\{a_i\}$ in $\overset{l^1}{\oplus} (\tau c(A_i), \|\cdot\|_{t_i})$, so the mapping $\{a_i\} \to \{a_i\}$ from $(\tau c(B), \|\cdot\|)$ onto $\overset{l^1}{\oplus} ((\tau c(A_i), \|\cdot\|_{t_i})$ is the wanted natural isomorphism.

If A is an algebra with a norm for which the product of A is continuous and P is a dense subalgebra of A, then the closure in A of any ideal of P is an ideal of A. In particular if A is an H^*-algebra and Q is an ideal of $\tau c(A)$, then $\bar{Q}$ is an ideal of A. Hence if A is *topologically simple* (nonzero product and no nonzero proper closed ideals) then $\tau c(A)$ is a prime algebra. On the other hand, if P is a closed ideal of A, its orthogonal is also a closed ideal of A and both are $*$-invariant [21; Proposition 2 (5)], this together with the above theorem permit to assure that if $\tau c(A)$ has not non trivial diret sumands, then A is topologically simple. Thus we have proved the following corollary.

Corollary 2.1. *An H^*-algebra A is topologically simple if and only if $\tau c(A)$ is a prime algebra, if and only if $\tau c(A)$ has not non trivial direct sumands.*

The extended centroid $C(A)$ of a prime algebra A is a field extension of the base field for A. It was introduced in [32] for the associative case and later in [23] for the general case and, colloquially speaking, A can be enlarged to obtain a prime algebra over $C(A)$ with the property that its extended centroid coincides with $C(A)$. A prime algebra with the property that its extended centroid coincides with the base field, is called *centrally closed* and a such algebra enjoies of perfections which facilitate the knowledge of the same (see for example [25] and [18]). The concept of extended centroid has been carried to the more general case of semiprime algebras [6], and to this reference the reader we send for its precise definition. Our next result assure that if A is an H^*-algebra then $\tau c(A)$ is a semiprime algebra, and includes a precise determination of the extended centroid of $\tau c(A)$.

Throughout the paper C will denote the field of complex numbers.

Proposition 2.1. *Let A be a H^*-algebra. Then $\tau c(A)$ is a semiprime algebra with extended centroid equal C^I where I denotes the cardinal of the family of the minimal closed ideals of A. In particular if A is topologically simple, then $\tau c(A)$ is a prime centrally closed algebra.*

Proof. Since $\tau c(A)$ is a dense ideal of A and A is a normed semiprime algebra [21; Proposition 1 (ii)], $\tau c(A)$ is an essential ideal of A. So by [15; Theorem 3] $\tau c(A)$ is a semiprime algebra such that $C(\tau c(A)) = C(A)$. The rest of the proof it follows from the description of $C(A)$ given in [15; Proposition 4].

Given a complete norm algebra A we will say that A has the *α property* if every homomorphism from any complete norm algebra into A with dense range is continuous. Every H^*-algebra A has the α property and below we will prove that $\tau c(A)$ has the same property. Previously we enunciate the following Proposition of easy proof using the Closed Graph Theorem.

Proposition 2.2. *Let C be a complete normed algebra and assume there exists a separating family of homomorphisms densely value from C into complete normed algebras which have the α property. Then C has the α property.*

Theorem 2.5. *Let A be an H^*-algebra. Then every homomorphism from any complete norm complex algebra into $\tau c(A)$ with dense range is continuous.*

Proof. First let us assume that A is topologically simple. By Proposition 2.1 $\tau c(A)$ is a prime centrally closed algebra . Moreover on $\tau c(A)$ there exists a continuous associative symmetric bilinear form given by $< a, b >= (a/b^*)$ for all a, b in $\tau c(A)$. Now by [44; Proposition 2] we have that $\tau c(A)$ has the α property.

If A is not topologically simple, by [21; Theorem 1] and Theorem 2.4, $\tau c(A) = \overset{l^1}{\oplus} \tau c(A_i)$, where $\{A_i\}$ is the family of the minimal closed ideals of A which are topologically simple H^*-algebras. Now since for all i, $\tau c(A_i)$ has the α property and the family of proyections $\{\pi_i : \tau c(A) \to \tau c(A_i)\}$ is a separating family of surjective homomorphisms, we have that $\tau c(A)$ has the α property because of Proposition 2.2.

3. Existence of a trace function

Every H^*-algebra A has a canonical associative symmetric bilinear form $<,>$ given by

$$< a, b >:= (a/b^*) \text{ for all } a, b, c \text{ in } A.$$

The first problem of which we tackle is the possibility of linearizing the above canonical form, namely: is there a linear form f on A such that $< a, b >= f(ab)$ for all a, b in A ?. The answer is given by the next proposition.

We recall that a left approximate unit ($l.a.u.$) in a normed algebra B is a net $\{u_i\}_{i\in I}$ in B such that the net $\{u_i a\}_{i\in I}$ converges to a for all a in B. Analogously it is defined right approximate units ($r.a.u.$). A net which is $l.a.u.$ and $r.a.u.$ is said to be an approximate unit ($a.u.$).

In the sequel, $BL(H)$ will denote the Banach algebra of all bounded linear operators on a prefixed Hilbert space H.

Lemma 3.1. *Let A be an H^*-algebra and let C be a convex subset of A. If A has a l.a.u. $\{u_i\}_{i\in I}$ with u_i in C for all i in I, then A has also an a.u. $\{v_j\}_{j\in J}$ with v_j in C for all j in J.*

Proof. Using [41; Theorem 2.1.5] it is easy to show that the closures of convex subsets of $BL(H)$ for the strong-$*$ operator topology and for the strong operator topology are the same, and clearly a net $\{u_i\}_{i\in I}$ is a *l.a.u.* if and only if the net of operators $\{L_{u_i}\}_{i\in I}$ converges in the strong operator topology to the identity operator I_A in A. Therefore if $\{u_i\}_{i\in I}$ satisfies the hypothesis of lemma, then I_A is in the closure for the strong operator topology of convex set $\{L_c : c \in C\}$, hence there exists a net $\{v_j\}_{j\in J}$ in C such that the nets $\{L_{v_j}(a)\}_{j\in J}$ and $\{L^\bullet_{v_j}(a)\}_{j\in J}$ converge, in the norm topology, to a for all a in A. Finally, since $L^\bullet_x = R^*_x$ for all x in A and $*$ is isometric, we have $\{L^\bullet_{v_j}(a)\}_{j\in J}$ converges in the norm topology to a if and only if $\{R_{v_j}(a)\}_{j\in J}$ converges in the norm topology to a, and the proof is concluded.

Proposition 3.1. *Let A be an H^*-algebra. Then the following assertions are equivalent:*

i) The canonical associative symmetric bilinear form on A is linearizable.

ii) A has l.a.u.

iii) A has a.u.

Proof. *i)⇒ii).* Assume *ii)* is not true. In view of the proof of above lemma, the identity operator I_A is not in the closure for weak operator topology of $\{L_a : a \in A\}$. Therefore, by Hahn-Banach Theorem, there exists a weak continuous linear functional φ in $BL(A)$ such that

$$\varphi(I_A) \neq 0 \text{ and } \varphi(L_a) = 0 \text{ for all } a \text{ in } A. \tag{3.1}$$

But by the well-known caracterisation of continuous linear funtionals for the weak operator topology in $BL(A)$, there are n in $\mathbb{N}$ and $a_1, \ldots, a_n, b_1, \ldots, b_n$ in A such that $\varphi(F) = \sum_{k=1}^{n}(F(a_k)/b_k)$ for all F in $BL(A)$. So the conditions (3.1) can be reformulate as

$$\sum_{k=1}^{n}(a_k/b_k) \neq 0 \text{ and } \sum_{k=1}^{n}(L_a(a_k)/b_k) = 0 \text{ for all } a \text{ in } A.$$

Hence, using the H^*-axiom and the fact that $*$ is isometric, we can assure there are $2n$ elements $a_1, \ldots, a_n, b_1, \ldots, b_n$ in A such that $\sum_{k=1}^{n} a_k b_k^* = 0$ and $\sum_{k=1}^{n}(a_k/b_k) \neq 0$. This is a contradiction with assumption *i)*.

The implication *ii)⇒iii)* it follows from Lemma 3.1.

iii)⇒i). If $\{u_i\}_{i\in I}$ is an *a.u.* in A, then for any natural number n and for any $a_1, \ldots, a_n, b_1, \ldots, b_n$ in A we have

$$\sum_{k=1}^{n}(a_k/b_k^*) = lim \sum_{k=1}^{n}(a_k/u_i b_k^*) = lim \sum_{k=1}^{n}(a_k b_k/u_i),$$

so $lim(a/u_i)$ exists for all a in A^2 and the mapping $g : a \to lim(a/u_i)$ from A^2 in C is linear. Any linear extension of g to the totality of A linearize the canonical associative symmetric bilinear form on A.

The situation arises enough restrictive if you want a continuous linearization relative to hilbertian norm as the following proposition shows.

Proposition 3.2. *For an H^*-algebra A, the following assertions are equivalent:*

i) The canonical associative symmetric bilinear form on A is continuously linearizable.

ii) A has unit.

iii) There exists a w^*-continuous linear functional f on $\tau c(A)$ satisfying $(a/b^*) = f(ab)$ for all a, b in A.

Proof. *i)⇒ii).* By assumption there exists a continuous linear funtional f on A such that there exists a continuous linear funtional f on A such thatchet representation theorem there is u in A such that $(a/b^*) = f(ab) = (ab/u) = (a/ub^*)$ for all a, b in A. Hence u is a left unit in A. Since u^* is a right unit in A we have that u is unit in A.

ii)⇒iii). If u is the unit of A, then the mapping $a \to (a/u)$ from $\tau c(A)$ in C satisfies the conditions in *iii)*.

iii)⇒i). Lef f be the linear funtional satisfying *iii)*. Seen f as an element of the closure of (A, p) in its bidual, there exists a sequence $\{u_n\}$ in A such that

1) $\{u_n\}$ is a Cauchy sequence in (A, p), and

2) $\{(u_n/a^*)\}$ converges to $f(a)$ for all a in $\tau c(A)$.

By 1) $\{L_{u_n}\}$ is a Cauchy sequence in $BL(A)$, so if T is the limit of such sequence, then $\{(u_n/ab)\} = \{(L_{u_n}(b^*)/a)\}$ converges to $(T(b^*)/a)$ for all a, b in A. But by 2) $\{(u_n/ab)\}$ olso converges to $f(b^*a^*)$ equal to (b^*/a) by assumption. Therefore $(Tb^*/a) = (b^*/a)$ for all a, b in A, hence $T = I_A$. Thus we have proved that I_A is in the closure for the norm operator topology of the set $L_A := \{L_a : a \in A\}$. Now using $L_a^{\bullet *} = R_a$ we can assure that I_A is in the closure of $\{R_a : a \in A\}$, from which we have $||I_A - R_b|| < 1$ for suitable b in A. It says that R_b is invertible in the Banach algebra $BL(A)$. The inequalities

$$||a|| = ||R_b^{-1} R_b(a)|| \leq ||R_b^{-1}|| \, ||ab|| \leq ||R_b^{-1}|| \, ||L_a|| \, ||b||,$$

for all a in A, prove that the mapping $a \to L_a$ from A into L_A is an homeomorphism and as a consequence L_A is norm-closed in $BL(A)$. This prove that I_A belongs to L_A. Finally if u is such that $I_A = L_u$, then u is a left unit in A. Writing $f(a) = (a/u)$ for all a in A we have f is a continuous linear funtional linearizing the canonical associative symmetric bilinear form in A.

If H is an infinite dimensional complex Hilbert space, then the H^*-algebra $\mathcal{HS}(H)$ has an approximate unit but has not unit. The same is true for every infinite dimensional associative H^*-algebra, so in view of the above proposition it seems interesting to study the possibility of linearizing continuously the canonical form on any H^*-algebra A, now giving up that the linear form on A works on the whole of A and the cotinuity has to be referred to some complete norm whose topology is stronger than the Hilbert norm topology. So the problem is to find a complete normed space $(X, |\cdot|)$ such that $A^2 \subseteq X \subseteq A$ and for which there exists a continuous linear funtional for the bar one norm which linearizes the canonical form on A. Our main result assures that the Banach space $(\tau c(A), \|\cdot\|_t)$ is one solution to this problem if and only if A has an operator-bounded approximate unit.

A *l.a.u.* $\{u_i\}_{i\in I}$ (resp. *a.u.*) in a normed algebra is said *operator-bounded* (*l.o.b.a.u.*) (resp. *o.b.a.u.*) if $\|L_{u_i}\| \leq M$ for all i in I.

Theorem 3.1. *For an H^*-algebra A, the following assertions are equivalent:*

i) There exists a continuous linear funtional (with norm less or equal M) $f : (\tau c(A), \|\cdot\|_t) \to C$ satisfiying $f(ab) = (a/b^*)$ for all a, b in A.

ii) A has l.o.b.a.u. (by M).

iii) A has o.b.a.u. (by M).

Proof. *i)⇒ii).* Assume that *ii)* is not true. Then I_A is not in the closure for the strong operator topology of MB_{L_A} (B_{L_A} equal to the unit ball of L_A), therefore, by Hahn-Banach Theorem, there exists a strongly continuous linear functional φ on $BL(A)$ such that

$$Sup\{|\varphi(T)| : T \in MB_{L_A}\} < |\varphi(I_A)|. \tag{3.2}$$

According to what we have seen in the proof of Proposition 3.1, $\varphi(F) = \sum_{k=1}^{n}(F(a_k)/b_k)$ for convenients n in $\mathbb{N}$ and nonzero $a_1, \ldots, a_n, b_1, \ldots, b_n$ in A. Thus, replacing φ in (3.2) and applying that *i)* is true we obtain:

$$M\|\sum_{k=1}^{n} b_k a_k^*\|_t = Sup\{|\sum_{k=1}^{n}(a/b_k a_k^*) : \|L_a\| \leq M\} =$$

$$Sup\{|\sum_{k=1}^{n}(L_a(a_k)/b_k)| : \|L_a\| \leq M\} = Sup\{|\varphi(T)| : T \in MB_{L_A}\} < |\varphi(I_A)| =$$

$$|\sum_{k=1}^{n}(a_k/b_k)| = |\sum_{k=1}^{n} f(a_k b_k^*)| \leq M\|\sum_{k=1}^{n} a_k b_k^*\|_t = M\|\sum_{k=1}^{n} b_k a_k^*\|_t.$$

Which is a contradiction.

ii)⇒iii). Let $\{u_i\}_{i\in I}$ be a *l.o.b.a.u.* by M. Since the set $\{a \in A : \|L_a\| \leq M\}$ is a convex set containig to $\{u_i\}_{i\in I}$, it suffices to apply Lemma 3.1.

iii)⇒i). Let $\{u_i\}_{i\in I}$ be an *o.b.a.u.* by M. For arbitraries n in $\mathbb{N}$ and $a_1, \ldots, a_n$, $b_1, \ldots, b_n$ in A, we have

$$|\sum_{k=1}^{n}(a_k b_k/u_i)| \leq \|\sum_{k=1}^{n} a_k b_k\|_t \|L_{u_i}\| \leq M\|\sum_{k=1}^{n} a_k b_k\|_t.$$

Hence

$$|\sum_{k=1}^{n}(a_k/b_k^*)| = lim|\sum_{k=1}^{n}(a_k/u_i b_k^*)| = lim|\sum_{k=1}^{n}(a_k b_k/u_i)| \leq M\|\sum_{k=1}^{n} a_k b_k\|_t,$$

what permit to assure that $\sum_{k=1}^{n} a_k b_k \to \sum_{k=1}^{n} (a_k/b_k^*)$ defines a funtion from A^2 into C which is linear and $\|\cdot\|_t$-continuous with norm less or equal to M. The Hahn-Banach Theorem conclude the proof.

The purpose of our paper has been to make a new way in the theory of nonossociative H^*-algebras. As it is natural we exemplify the ideas which we have expounded. In this sense we have to tell that the associative theory (see [47; Theorem 1] and [50; Theorem]) remains fully collected into ours. However contrary to what happens in the associative case we have built an example of Lie H^*-algebra A such that A^2 is not dense in $(\tau c(A), \|\cdot\|_t)$. The fact that A^2 is dense in $\tau c(A)$ is a desirable situation implying the uniqueness of the trace-form which appears in Theorem 3.1. Also we know an example of a semisimple noncommutative Jordan H^*-algebra satisfying the equivalent conditions in Proposition 3.1 but not the ones in Theorem 3.1. Unfortunately the algebra in our example is not topologically simple.

Other interesting questions to answer are:

a) Is every derivation on $\tau c(A)$ continuous ?

b) Is it possible to extend every continuous derivation on $\tau c(A)$ to the totality of A?

4. Acknowledgements

The most part of my previous production in the H^*-algebras field has been effected in collaboration with M. Cabrera and A. Rodríguez. If moreover it is bored in mind the geographical and human proximity which I own respect these colleagues, it would be unimaginable to think that the ideas developed in this paper would not have been checked with they. In fact this check has been placed and the work has been very benefited with their suggestions and commentaries. In consequence the author would like to express their gratitude.

References

1. Ambrose, W.: 1945, Structure theorems for a special class of Banach algebras, *Trans. Amer. Math. Soc.* Vol. **57**, pp. 364-386.
2. Amson, J. C. and Reddy, N. G.: 1990, A Hilbert algebra of Hilbert-Schmidt quadratic opertator, *Bull. Austral. Math. Soc.* Vol. **41**, pp. 123-134.
3. Balachandran, V. K.: 1969 , Simple L^*-algebras of classical type, *Math. Ann.* Vol. **180**, pp. 205-219.
4. Balachandran, V. K. and Swaminathan, N.: 1972, Real L^*-algebras, *J. Pure. Appl. Math.* Vol. **3**, pp. 1224-1246.
5. Balachandran, V. K.: 1986, Real L^*-algebras, *J. Funtional Analysis*, Vol. **65**, pp. 64-75.
6. Baxter, W. E. and Martindale 3rd, W. S.: 1979, Central closure of semiprime nonassociative rings, *Comm. Algebra*, Vol. **7**, pp. 1103-1132.
7. Cabrera, M., El Marrakchi, A., Martínez, J. and Rodríguez, A.: 1993, Bounded Differential operator on Hilbert modules and derivations of structurable H^*-algebras, *Comm. Algebra*, Vol. **8**, pp. 2905-2945.
8. Cabrera, M., El Marrakchi, A., Martínez, J. and Rodríguez, A.: A Allison-Kantor-Koecher-Tits construction for Lie H^*-algebras, to appear in *J. Algebra*.
9. Cabrera, M., Martínez, J. and Rodríguez, A.: 1988, Mal'cev H^*-algebras, *Math. Proc. Camb. Phil. Soc.*, Vol. **103**, pp. 463-471.

10. Cabrera, M., Martínez, J. and Rodríguez, A.: 1988, Nonassociative real H^*-algebras, *Publicacions Matemtiques*, Vol. **32**, pp. 267-274.
11. Cabrera, M., Martínez, J. and Rodríguez, A.: 1989, A note on real H^*-algebras, *Math. Proc. Camb. Phil. Soc.*, Vol. **105**, pp. 131-132.
12. Cabrera, M., Martínez, J. and Rodríguez, A.: 1992, Structurable H^*-algebras, *J. Algebra*, Vol. **147**, pp. 19-62.
13. Cabrera, M., Martínez, J. and Rodríguez, A.: 1992, Hilbert modules over H^*-algebras in relation with Hilbert ternary rings. Proceedings of the workshop on Nonassociative algebraic models, *Nova Science Publishers, INC.*, New York, pp. 33-44.
14. Cabrera, M., Martínez, J. and Rodríguez, A.: Hilbert modules revisited: orthonormal bases and Hilbert-Schmidt operators, to appear in *Glasgow Math. J.*
15. Cabrera, M. and Rodríguez, A.: 1990, Extended centroid and central closure of semiprime normed algebras: a first approach, *Comm. Algebra*, Vol. **18**, pp. 2293-2326.
16. Cabrera, M. and Rodríguez, A.: 1992, Nonassociative ultraprime normed algebras, *Quart. J. Math. Oxford*, Vol. **43**, pp. 1-7.
17. Castellón, A. Cuenca, J. A. and Martín, C.: 1992, Ternary H^*-algebras, *Bollettino U. M. I.*, Vol. **6-B**, pp. 217-228.
18. Cobalea, M. A. and Fernández, A.: 1988, Prime noncommutative Jordan algebras and central closure, *Algebras, Groups and Geometries*, Vol. **5**, pp. 129-136.
19. Cuenca, J. A., García, A. and Martín, C.: 1990, Structure theory for L^*-algebras, *Math. Proc. Cab. Phil. Soc.*, Vol. **107**, pp. 361-365.
20. Cuenca, J. A. and Rodríguez, A.: 1985, Isomorphisms of H^*-algebras, *Math. Proc. Camb. Phil. Soc.*, Vol. **97**, pp. 93-97.
21. Cuenca, J. A. and Rodríguez, A.: 1987, Structure theory for noncommutative Jordan H^*-algebras, *J. Algebra*, Vol. **106**, pp. 1-14.
22. Cuenca J. A. and Sánchez, A.: Structure theory for real noncommutative Jordan H^*-algebras, to appear in *J. Algebra.*
23. Erickson, T. S., Martindale, W. S. 3rd and Osborn, J. M.: 1975, Prime nonassociative algebras, *Pacific J. Math.*, Vol. **60**, pp. 49-63.
24. Fernández, A. and Rodríguez, A.: 1986, A Wedderburn theorem for nonassociative complete normed algebras, *J. London Math. Soc.*, Vol. **33**, pp. 328-338.
25. Filippov, V. T.: 1977, Theory of Mal'tsev algebras, *Algebra i Logika*, Vol. **16**, pp. 101-108.
26. Giellis, G. R.: 1971, Trace-class for a Hilbert module, *Proc. Amer. Math. Soc.*, Vol. **29**, pp. 63-68.
27. Giellis, G. R.: 1972, A Characterization of Hilbert modules, *Proc. Amer. Math. Soc.*, Vol. **36**, pp. 440-442.
28. de la Harpe, P.: 1971, Classification des L^*-bres
29. de la Harpe, P.: 1971, Classification des L^*-bresparables, *C. R. Acad. Sci. Paris Ser. A*, Vol. **272**, pp. 1559-1561.
30. de la Harpe, P.: 1972, Classical Banch-Lie algebras and Banach-lie Groups of operators in Hilbert space, *Lecture Notes in Math.*, Vol. **285**, *Springer-Verlag*, New York.
31. Horv th, J.: 1966 Topological Vector Spaces and Distributions, Vol. I, *Addison-Wesley*, London.
32. Kaplansky, I.: 1948, Dual rings, *Ann. Math.*, Vol. **49**, pp. 689-701.
33. Martindale, W. S. 3rd: 1969, Prime rings satisfying a generalized polynomial identity, *J. Algebra*, Vol. **12**, pp. 576-584.
34. Molnar, L.: 1991, A note on Saworotnow's representation theorem on positive definite functions, *Huston J. Math.*, Vol. **17**, pp. 89-99.
35. Molnar, L.: 1991, Reproducing kernel Hilbert A-modules, *Glasnik Mat.*, Vol. **25**, pp. 109-119.
36. Molnar, L.: 1992, A note on the strong Schwarz inequality in Hilbert A-modules, *Publ. Math. Debrecen*, Vol. **40**, pp. 323-325.
37. Molnar, L.: 1992, Modular bases in a Hilbert A-module, *Czechoslovak Math. J.*, Vol. **42**, pp. 649-656.
38. Molnar, L.: On A-linear operators on a Hilbert A-module, to appear in *Period. Math. Theory.*
39. Molnar, L.: On Saworotnow's Hilbert A-modules, preprint.
40. Neher, E.: 1993, Generators and Relations for 3-Graded Lie Algebras, *J. Algebra*, Vol. **155**, pp. 1-35.
41. Ozawa, M.: 1980, Hilbert B(H)-modules and stationary processes, *Kodai Math. J.*, Vol. **3**,

pp. 26-39.

42. Pedersen, G. K.: 1979, C^*-algebras and their automorphism groups, *Academic Press*, London. automorphism groups, *Academic Press*, London.rez de Guzm n, I.: 1983, Structure theorems for alternative H^*-algebras, *Math. Proc. Camb. Phil. Soc.*, Vol. **94**, pp. 437-446.
43. Rodríguez, A.: 1985, The uniqueness of the complete norm topology in complete normed nonassociative algebras, *J. Functional Analysis*, Vol. **60**, pp. 1-15.
44. Rodríguez, A.: Continuity of densely valued homomorphisms into H^*-algebras, to appear in *Quart. J. Math. Oxford.*
45. Saworotnow, P. P.: 1968, A generalized Hilbert space, *Duke Math.*, Vol. **35**, pp. 191-197.
46. Saworotnow, P. P.: 1970, Representation of a topological group on a Hilbert module, *Duke Math.*, Vol. **37**, pp. 145-150.
47. Saworotnow, P. P.: 1970, Trace-class and cetralizers of an H^*-algebra, *Proc. Amer. Math. Soc.*, Vol. **26**, pp. 100-104.
48. Saworotnow, P. P.: 1972, Generalized positive linear functionals on a Banach algebra with an involution, *Proc. Amer. Math. Soc.*, Vol. **31**, pp. 299-304.
49. Saworotnow, P. P.: 1981, Irreducible representations of a Banach algebra on a Hilbert module, *Houston J. Math.*, Vol. **7**, pp. 275-281.
50. Saworotnow, P. P. and Friedell J. C.: 1970, Trace-class for an arbitrary H^*-algebra, *Proc. Amer. Math. Soc.*, Vol. **26**, pp. 95-100.
51. Schatten, R.: 1970, Norm ideals of completely continuous operators, *Springer-Verlag*, New York.
52. Schue, J.R.: 1960, Hilbert space methods in the theory of Lie algebras, *Trans. Amer. Math. Soc.*, Vol. **95**, pp. 69-80.
53. Schue, J. R.: 1961, Cartan decomposition for L^*-algebras, *Trans. Amer, Math. Soc.*, Vol. **98**, pp. 334-349.
54. Smith, J. F.: 1972, The p-clases of a Hilbert module, *Proc. Amer. Math. Soc.*, Vol. **36**, pp. 428-434.
55. Smith, J. F.: 1974, The structure of Hilbert modules, *J. London Math. Soc.*, Vol. **8**, pp. 741-749.
56. Unsain, I.: 1972, Classification of the simple separable real L^*-algebras, *J. Diferential Geometry*, Vol. **7**, pp. 423-451.
57. Villena, A. R.: Continuity of derivations on H^*-algebras, to appear in *Proc. Amer. Math. Soc.*.
58. Viola Devapakkiam, C.: 1975, Hilbert space methods in the theory of Jordan algebras, *Math. Proc. Camb. Phil. Soc.*, Vol. **78**, pp. 293-300.
59. Viola Devapakkian, C. and Rema, P. S.: 1976, Hilbert space methods in the theory of Jordan algebras II, *Math. Proc. Camb. Phil. Soc.*, Vol. **79**, pp. 307- 319.
60. Zalar, B.: 1991, Continuity of derivations on Mal'cev H^*-algebras, *Math. Proc. Camb. Phil. Soc.*, Vol. **110**, pp. 455-459.

ON LATTICE ISOMORPHISM OF BERNSTEIN ALGEBRAS

C. MARTÍNEZ
Departamento de Matemáticas, Universidad de Oviedo
33.007 Oviedo, Spain

and

J. A. SÁNCHEZ-NADAL*
Departamento de Matemática Aplicada, Universidad de Zaragoza
50.009 Zaragoza, Spain

Abstract. In this paper Bernstein algebras and isomorphisms between their lattices of subalgebras are studied. The main result of the paper proves that if we have a lattice isomorphism between Bernstein algebras then it is always possible to define a new isomorphism between their lattices that keeps the nucleus.

1. Introduction

A Bernstein algebra A over an infinite field K, *char* $K \neq 2$ is a baric algebra (that is an algebra A with a nonzero homomorphism of algebras $\omega : A \longrightarrow K$, "weight homomorphism") satisfying that

$$(x^2)^2 = \omega(x)^2 x^2 \quad \forall\, x \in A.$$

Then $\operatorname{Ker}\omega$ is an ideal of codimension one, uniquely determined, because it is known that a Bernstein algebra has only one weight homomorphism. Therefore in a Bernstein algebra we have subalgebras of two types: ones of them are also Bernstein algebras and the other ones (those contained in $\operatorname{Ker}\omega$) are not Bernstein algebras. The first problem that we find in the study of lattice isomorphisms of Bernstein algebras is that the image of the nucleus of A, $\operatorname{Ker}\omega$, is not always the nucleus of A', $\operatorname{Ker}\omega'$.

So in [6], we study Bernstein algebras for which a lattice isomorphism not keeping nucleus can be defined. Now, we will use those results to define a new lattice isomorphism θ satisfying $\theta(\operatorname{Ker}\omega) = \operatorname{Ker}\omega'$.

2. Some previous results

It is known the existence of idempotent elements in every Bernstein algebra (all of them having weight equal to one). If e is an idempotent element, then $A = K(e)\dot{+}\operatorname{Ker}\omega$ and $\operatorname{Ker}\omega = U_e + Z_e$, where $U_e = \{x \in \operatorname{Ker}\omega / ex = \frac{1}{2}x\}$ and $Z_e = \{x \in \operatorname{Ker}\omega / ex = 0\}$. The pair $(1 + \dim U_e, \dim Z_e)$ is called "type of A" and it is independent of the particular idempotent element e.

* Partially supported by D.G.A. P. CB.-6/91.

S. González (ed.), Non-Associative Algebra and Its Applications, 269–274.

Products of elements in U_e and Z_e satisfy:

a) $U_eU_e \subseteq Z_e$, $U_eZ_e \subseteq U_e$, $Z_eZ_e \subseteq U_e$.

b) $u^3 = 0$, $u(uz) = 0$, $uz^2 = 0$ and $(uz)^2 = 0$ for every $u \in U_e$, $z \in Z_e$.

A Bernstein algebra A is called exclusive if $U_e^2 = 0$ for some idempotent element e. In this case, $U_e = U_f = U$ does not depend on the idempotent element e.

If was proved in [3], that if A and A' are two Bersntein algebras and $\theta : \mathrm{L}(A) \longrightarrow \mathrm{L}(A')$ is a lattice isomorphism, then $\dim A = \dim A'$, and $\dim S = \dim \theta(S)$ for every subalgebra S of A. It was also proved that in cases $\operatorname{type} A = (1+r, 0)$ or $\operatorname{type} A = (1,1)$, A and A' are isomorphic. (1)

In [4], it was proved that if $\theta : \mathrm{L}(A) \longrightarrow \mathrm{L}(A')$ is a lattice isomorphism, A and A' are Bernstein algebras with $\operatorname{type} A \neq (1,1)$, then there is an idempotent element e in A such that: $\theta(\langle e \rangle) = \langle e' \rangle$ with $(e')^2 = e' \in A'$. (2)

In a previous paper, [7], we studied those Bernstein algebras having a lattice isomorphism that doesn't keep the nucleus. The following results were proved:

Let A, A' be two Bernstein algebras, $\theta : \mathrm{L}(A) \longrightarrow \mathrm{L}(A')$ a lattice isomorphism with $\theta(\operatorname{Ker} \omega) \neq \operatorname{Ker} \omega'$. Then:

- A and A' are exclusive algebras. (3)
- $\operatorname{type} A = \operatorname{type} A' = (1+r, s)$ with $s = 0, 1$. (4)
- If $\operatorname{type} A \neq (1,1)$, then $\theta(A^2) = (A')^2$. (5)
- If $\operatorname{type} A = (1+n, 1)$ with $n \geq 1$, then there are bases $\{\bar{u}_1, ..., \bar{u}_r\}$ of U and $\{\bar{u}'_1, ..., \bar{u}'_r\}$ of U' satisfying $\theta(K(\bar{u}_i)) = K(\bar{u}'_i)$ for $i = 2, ..., r$, $\theta(K(\bar{u}_1)) = K(\bar{e}')$ with $(\bar{e}')^2 = \bar{e}'$ and $\theta^{-1}(K(\bar{u}'_1)) = K(\bar{e})$ with $\bar{e}^2 = \bar{e}$. Furthermore, if $A = K(e) \dot{+} U \dot{+} K(z)$ and $A = K(e') \dot{+} U' \dot{+} K(z')$ with $z \in Z_e$, $z' \in Z_{e'}$, then $uz \in U \setminus K(\bar{u}_2, ..., \bar{u}_r)$ $\forall u \in U \setminus K(\bar{u}_2, ..., \bar{u}_r)$ and in a similar way $u'z' \in U' \setminus K(\bar{u}'_2, ..., \bar{u}'_r)$ $\forall u' \in U' \setminus K(\bar{u}'_2, ..., \bar{u}'_r)$. (6)

3. The main Theorems

Now we will prove that if there is a lattice isomorphism between two Bernstein algebras A and A' , then there is also another one that keeps kernels of the weight homomorphism.

Lemma 1. *Let A and A' be Bernstein algebras, $\theta : L(A) \longrightarrow L(A')$ a lattice isomorphism of Berstein algebras with $\theta(\operatorname{Ker} \omega) \neq \operatorname{Ker} \omega'$. If $\dim U = 1$, then $\dim A = 2$ or $\dim A = 3$ and $A \cong A'$.*

Proof. The result about dimensions is consequence of a previous result (5) that assures, under these conditions, that $\dim Z = 0$ or 1.

If type of A is (2,0), then $A \cong A'$ by using (1).

If type of A is (2,1), then type A' is (2,1) by (4).

Then, by (6), $A = K(e) + K(u) + K(z)$, $A' = K(e') + K(u') + K(z')$ with $\theta(K(e)) = K(u')$, $\theta(K(u)) = K(e')$ and $uz = u$, $u'z' = u'$.

So $\langle u, z \rangle = K(u, z)$ is applied by θ in $K(e', x')$ where $x' \notin K(u')$.

Condition $e'x' \in K(e', x')$ assures that $x' \in K(z')$ and $(x')^2 = 0$.

Consequently $(z')^2 = 0$, and we can prove, in a similar way, that $z^2 = 0$.

It is clear now that A and A' are isomorphic. □

In the following we will suppose that $\dim U = r > 1$.

Lemma 2. *Let A, A' and θ be as in Lemma 1. Then there is a bijective application*

$$\begin{aligned}(\quad)' : K(\bar{u}_2, ..., \bar{u}_r) &\longrightarrow K(\bar{u}'_2, ..., \bar{u}'_r)\\ u &\longmapsto u'\end{aligned}$$

and an automorphism

$$\begin{aligned}\psi : K &\longrightarrow K\\ \alpha &\longmapsto \alpha^{\psi}\end{aligned}$$

such that: a) $\theta(K(u)) = K(u')$ for every $u \in K(\bar{u}_2, ..., \bar{u}_r)$ b) If $\theta(K(e)) = K(e')$ with $e^2 = e \in A$ and $(e')^2 = e' \in A'$ then $\theta(K(e + \alpha u)) = K(e' + \alpha^{\psi} u')$ $\forall \alpha \in K$ and $u \in K(\bar{u}_2, ..., \bar{u}_r)$.

Proof. By (1), A and A' are exclusive, so if $\bar{e}'$ is the idempotent given in (6), $\bar{e}' = e' + \bar{u}'$ with $0 \neq \bar{u}' \in U'$.

Then $\theta(K(e, \bar{u}_1)) = K(e', e' + \bar{u}') = K(e', \bar{u}')$ and we can take $\bar{u}_1$ satisfying that $\theta(K(e + \bar{u}_1)) = K(\bar{u}')$.

Also $\theta(K(e, \bar{u}_1)) = K(e', \bar{e}') = K(e', \bar{u}')$, that is, $\forall \alpha \in K \setminus \{1\}$, $\theta(K(e + \alpha \bar{u}_1)) = K(e' + \alpha^{\Gamma} \bar{u}')$ with $\alpha^{\Gamma} \in K$.

So we have a bijection $\Gamma : K \setminus \{1\} \longrightarrow K \setminus \{1\}$ with $0^{\Gamma} = 0$.

i) We define the bijection $(\quad)'$ in the following way:

If $u \in K(\bar{u}_2, ..., \bar{u}_r)$, then $\theta(K(u)) = K(u')$ where $u' \in K(\bar{u}'_2, ..., \bar{u}'_r)$ (since we know that $\theta(K(\bar{u}_2, ..., \bar{u}_r)) = K(\bar{u}'_2, ..., \bar{u}'_r)$). Also $\theta(K(e, u)) = K(e', u')$ and we can choose u' in a unique way satisfying $\theta(K(e + u)) = K(e' + u')$.

Consequently

$$\begin{aligned}(\quad)' : K(\bar{u}_2, ..., \bar{u}_r) &\longrightarrow K(\bar{u}'_2, ..., \bar{u}'_r)\\ u &\longmapsto u'\end{aligned}$$

(with u' the one above defined), is a bijection.

In the same way, we have $\psi_u : K \longrightarrow K$ a bijection by defining $\psi_u(\alpha)$ by:

$$\theta(K(e + \alpha u)) = K(e' + \psi_u(\alpha) u')$$

ii) It can be proved that $\psi_u = \psi$ doesn't depend on the choosen element u.

iii) Finally it is proved that ψ is a field automorphism. □

Observation. *With the above notacion, we have:*

(1) $\theta(K(\bar{u}_1 + \alpha u)) = K(e' + \bar{u}'_1 - \alpha^{\psi} u')$ $\forall \alpha \in K, \forall u \in K(\bar{u}_2, ..., \bar{u}_r)$.

(2) $\theta(K(e + \alpha \bar{u}_1 + \beta u)) = K((\alpha^{\psi} - 1)e' + \alpha^{\psi} \bar{u}'_1 - \beta^{\psi} u')$ $\forall \alpha, \beta \in K, \forall u \in K(\bar{u}_2, ..., \bar{u}_r)$.

Corollary 1. *The above bijection $(\quad)'$ is a semilineal bijective application of automorphism ψ.*

Theorem 1. *Let A and A' be Bernstein algebras, $\theta : L(A) \to L(A')$ a lattice isomorphism with $\theta(\operatorname{Ker}\omega) \neq \operatorname{Ker}\omega'$. If $\dim U = r > 1$ and there is $\bar{z} \in \operatorname{Ker}\omega \setminus U$ with $\bar{z}^2 = 0$ then there is a lattice isomorphism $\theta_1 : L(A) \to L(A')$ with $\theta_1(\operatorname{Ker}\omega) = \operatorname{Ker}\omega'$.*

Proof. I) $\exists\, \bar{z}' \in \operatorname{Ker}\omega' \setminus U'$ with $\theta(K(\bar{z})) = K(\bar{z}')$, and so $(\bar{z}')^2 = 0$.

II) $\exists\, \bar{e} \in A$, $\bar{e}^2 = \bar{e}$ and $\bar{z} \in Z_{\bar{e}}$ with $\theta(K(\bar{e})) = K(\bar{u}') \subseteq U'$ and by symmetry there is $\bar{e}' \in A'$, $\bar{z}' \in Z_{\bar{e}'}$, with $\theta^{-1}(K(\bar{e}')) = K(\bar{u}) \subseteq U$.

If e and e' are idempotent elements of A and A' respectively satisfying $\theta(K(e)) = K(e')$, (we know their existence by [7]), we can suppose $\bar{e} = e + \bar{u}_1$, $\bar{e}' = e' + \bar{u}_1'$.

If $\bar{u}$ and $\bar{u}'$ are the elements given in Lemma 2, then we can see that $\bar{u} = \bar{u}_1 + u$, where $u \in K(\bar{u}_2, ..., \bar{u}_r)$. Furthermore, $\bar{u}' = \bar{u}_1' + u'$ where u' is as in Lemma 2.

III) If S is a subalgebra of A not contained in A^2, then $\bar{z} \in S$.

IV) If $\tilde{u} \in U$ satisfies that $\tilde{u}\bar{z} = 0$, then $\tilde{u} = 0$.

V) If $v \in U$ is an element satisfying $v\bar{z} = v$, then $v \in K(\bar{u})$.

VI) If S is a subalgebra of A, not contained in A^2 and with $\theta(S) \not\subseteq \operatorname{Ker}\omega'$, then $\bar{u} \in S$.

VII) If $S \leq A$, $S \not\leq A^2$ and $S \not\leq \operatorname{Ker}\omega$, then $\bar{e} \in S$.

Applying by symmetry the above results to A', we can finally define:

$$\theta_1 : \mathrm{L}(A) \to \mathrm{L}(A') \quad \text{by :}$$

$\theta_1(S) = \langle \hat{x}_1, ..., \hat{x}_m \rangle$, where $S = \langle x_1, ..., x_m \rangle$ with $x_i \in A^2 \cup \{\bar{z}\}$ $i = 1, ..., m$, $\widehat{(\alpha e + \alpha_1 \bar{u}_1 + ... + \alpha_r \bar{u}_r)} = \alpha^{\psi} e' + \alpha_1^{\psi} \bar{u}_1' + ... + \alpha_r^{\psi} \bar{u}_r'$ and $\hat{\bar{z}} = \bar{z}'$.

In this way, θ_1 is a lattice isomorphism and $\theta_1(\operatorname{Ker}\omega) = \operatorname{Ker}\omega'$. □

Theorem 2. *Let A and A' be Bersntein algebras, $\theta : L(A) \longrightarrow L(A')$ a lattice isomorphism with $\theta(\operatorname{Ker}\omega) \neq \operatorname{Ker}\omega'$. Then there is $\theta_1 : L(A) \longrightarrow L(A')$ a lattice isomorphism with $\theta_1(\operatorname{Ker}\omega) = \operatorname{Ker}\omega'$.*

Proof. We know by (4) that, in this case, type $A = (1+r, 0)$ or type $A = (1+r, 1)$ by [5].

If type $A = (1 + r, 0)$ then $A \cong A'$ by (1) and we can construct a lattice isomorphism by using the isomorphism between A and A'.

If type $A = (1, 1)$ or type $A = (2, 1)$ then $A \cong A'$ by [9] and Lemma 1 respectively and we can use the same argument as before.

If type $A = (1+r, 1)$ with $r > 1$ and $\exists\, \bar{z} \in \operatorname{Ker}\omega \setminus U$ with $\bar{z}^2 = 0$, then the result follows from Theorem 1. So we can assume that $\forall x \in \operatorname{Ker}\omega$, $x^2 = 0 \Rightarrow x \in U$. The proof may be finished in the following steps:

I) If $B \leq \operatorname{Ker}\omega$, but $B \not\leq U$ with minimal dimension, then $B = K(u_1, ..., u_l, \bar{z})$ with $\bar{z}^2 = u_1$, $\bar{z}u_i = u_{i+1}$, $i = 1, ...l - 1$ and $\bar{z}u_l = 0$. Furthermore $u_1, ..., u_l \in K(\bar{u}_2, ..., \bar{u}_r)$ (with the notation of Lemma 2).

II) $\exists\, \bar{z}' \in \operatorname{Ker}\omega' \setminus U'$ with $\theta(\langle \bar{z} \rangle) = \langle \bar{z}' \rangle$ and satisfying the same properties than $\bar{z}$.

III) There is an element $\bar{e} \in A$ with $\bar{e}^2 = \bar{e}$, $\theta(K(\bar{e})) \subseteq U'$ and $\dim\langle \bar{e}, \bar{z} \rangle = 1 + \dim\langle \bar{z} \rangle$. By symmetry we have the same result for A'.

IV) If $S \leq \operatorname{Ker}\omega$, $S \not\leq U$ and $\dim S = \dim\langle \bar{z} \rangle$, then $S = \langle \bar{z} \rangle$.

V) If $S \leq A$, $S \not\leq \operatorname{Ker}\omega$, then $\bar{e} \in S$.

VI) It is possible to define

$$\theta_1 : \mathrm{L}(A) \to \mathrm{L}(A') \quad \text{by :}$$

$\theta_1(S) = \langle \hat{x}_1, ..., \hat{x}_m \rangle$, where $S = \langle x_1, ..., x_m \rangle$ with $x_i \in A^2 \cup \{\bar{z}\}$ $i = 1, ..., m$, $\widehat{(\alpha e + \alpha_1 \bar{u}_1 + ... + \alpha_r \bar{u}_r)} = \alpha^{\psi} e' + \alpha_1^{\psi} \bar{u}_1' + ... + \alpha_r^{\psi} \bar{u}_r'$ and $\hat{\bar{z}} = \bar{z}'$.

Then θ_1 is a lattice isomorphism and $\theta_1(\mathrm{Ker}\,\omega) = \mathrm{Ker}\,\omega'$. □

Corollary 2. *Let A and A' be Bersntein algebras, $\theta : L(A) \longrightarrow L(A')$ a lattice isomorphism with $\theta(\mathrm{Ker}\,\omega) \neq \mathrm{Ker}\,\omega'$. If type $A = (1+r, 1)$ and for every $z \in \mathrm{Ker}\,\omega \backslash U$ it is true that $\dim\langle z \rangle \geq r$, then $A \cong A'$.*

Example. *There are Bernstein algebras A and A' and $\theta : L(A) \longrightarrow L(A')$ a lattice isomorphism with $\theta(\mathrm{Ker}\,\omega) \neq \mathrm{Ker}\,\omega'$, but $A \not\cong A'$.*

In fact, let $A = K(\bar{e}) + K(\bar{u}_1, \bar{u}_2) + K(\bar{z})$ and $A' = K(\bar{e}') + K(\bar{u}_1', \bar{u}_2') + K(\bar{z}')$ ($char\ K \neq 3$) and multiplications given by:

A	$\bar{e}$	$\bar{u}_1$	$\bar{u}_2$	$\bar{z}$
$\bar{e}$	$\bar{e}$	$\frac{1}{2}\bar{u}_1$	$\frac{1}{2}\bar{u}_2$	0
$\bar{u}_1$	$\frac{1}{2}\bar{u}_1$	0	0	$\bar{u}_1$
$\bar{u}_2$	$\frac{1}{2}\bar{u}_2$	0	0	$2\bar{u}_2$
$\bar{z}$	0	$\bar{u}_1$	$2\bar{u}_2$	0

A'	$\bar{e}'$	$\bar{u}_1'$	$\bar{u}_2'$	$\bar{z}'$
$\bar{e}'$	$\bar{e}'$	$\frac{1}{2}\bar{u}_1'$	$\frac{1}{2}\bar{u}_2'$	0
$\bar{u}_1'$	$\frac{1}{2}\bar{u}_1'$	0	0	$\bar{u}_1'$
$\bar{u}_2'$	$\frac{1}{2}\bar{u}_2'$	0	0	$-\bar{u}_2'$
$\bar{z}'$	0	$\bar{u}_1'$	$-\bar{u}_2'$	0

It can be proved that A and A' are not isomorphic.

If $S \leq A$, $S \not\leq A^2$, then $\bar{z} \in S$ and, in the same way, if $S' \leq A'$, $S' \not\leq (A')^2$ then $\bar{z}' \in S'$.

So, for every $S \leq A$, we can take $x_1, ..., x_m \in A^2 \cup \{\bar{z}\}$ such that $S = < x_1, ..., x_m >$.

If $x = \bar{e} + \alpha\bar{u}_1 + \beta\bar{u}_2$, then we define $x' = (\alpha - 1)\bar{e}' + \alpha\bar{u}_1' - \beta\bar{u}_2'$, $\forall\, \alpha, \beta$.

If $x = \bar{u}_1 + \alpha\bar{u}_2$, then we define $x' = \bar{e}' + \bar{u}_1' - \alpha\bar{u}_2'$, $\forall\, \alpha$.

If $x = \bar{u}_2$, or $x = \bar{z}$, then we define $x' = \bar{u}_2'$ or $x' = \bar{z}'$ respectively.

In this way,

$$\theta : \mathrm{L}(A) \longrightarrow \mathrm{L}(A') \quad \text{defined by}$$

$\theta(S) = \langle x_1', ..., x_m' \rangle$, where $S = \langle x_1, ..., x_m \rangle$ and x_i' as above, is a lattice isomorphism with $\theta(\mathrm{Ker}\,\omega) \neq \mathrm{Ker}\,\omega'$. □

References

1. N. Jacobson, *Structure and Representations of Jordan Algebras*, Amer. Math. Soc. Colloq. Publ., Vol. 39, AMS, Providence, RI, 1968.
2. Bersntein S.: *Principe de stationarité et généralization de la loi de Mendel*, Comptes Rendus de l'Acad. Sci. Paris 177 (1923) 581-584.
3. Bertand M.: *Algèbres non-associatives et algèbres génétiques.* Memorial des Sciences Mathématiques, fascicule CLXII, Gauthier-Villars Editeur (1966).
4. Cortés T.: *Bernstein algebras: lattice isomorphisms and isomorphisms.* Nonassociative Algebraic Models. Santos González and Hyo Chul Myung. Ed. Nova Science Publish. (1992), 69-91.
5. González S.: *"One dimensional Subalgebras of a Bernstein Algebra".* To appear in Algebra and Logic.
6. Holgate P.: *Genetic algebras satisfying Bernstein's stationarity principle*, J. London. Math. Soc. 9 (2) (1975), 613-623.
7. Martínez C.-Sánchez-Nadal J. A. *Bernstein algebras with a lattice isomorphism that does not preserve the nucleus.* To appear in Comm. in Algebra.
8. Worz-Busekros A.: *Algebras in Genetics: Lecture Notes in Biomathematics*, vol 36, Springer-Verleq (1980).

BERNSTEIN ALGEBRAS WHOSE LATTICE IDEALS IS LINEAR

C. MARTINEZ and J. SETÓ
Departamento de Matemáticas. Universidad de Oviedo
33.007 Oviedo, Spain

Abstract. In this paper Bernstein algebras and their ideal lattices will be considered. Some general facts about ideals of a Bernstein algebra will be given and we will exhibit some ideal lattices of a Bernstein algebra in case exclusive and normal.

Finally, Bernstein algebra having a lineal ideal lattices will be characterized.

1. Introduction

So B will denote a finite dimensional algebra over a field K, char K = 2, I(B) denote the set of idempotent elements and ω the weight homomorphism. If e is an idempotent element, the Peirce decomposition associated to e is:

$B = Ke \dot{+} U \dot{+} Z$, $\dot{+}$ denotes, as usual, a direct sum of vector spaces and we will use $\oplus$ for direct sum of ideals. Let us assume $\dim B = n$, $\dim U = r$, $\dim Z = s$ $(n = r + s + 1)$.

We will also use the following notation:

K(A)	vectorial space spanned by A
B(A)	ideal generated by A in B
U^o	the largest ideal of B contained in U
Z^o	the largest ideal of B contained in Z
N^o	$= U^o \oplus Z^o$
a.bc	instead of a(bc)
$\mathcal{J}$	lattice ideals for B

Subalgebras lattices have been studied in many structures, to get information about the algebra. Also in Bernstein algebras some studies about the lattice of subalgebras have been made [see 2]. The lattice of ideals has been considered in less occasions, but in the Lie case has been studied by P. Benito [1].

Here we will study the lattice ideal of a Bernstein algebra and we will characterize totally those Bernstein algebra that have a linear lattice of ideals. We will prove that these are only two of such algebras and with small dimension (2 and 3 respectively). Furthermore both algebras are Jordan-Bernstein algebras.

2. Some general results about ideals of a Bernstein algebra

2.1.

Let I be an ideal of B, $I \subseteq Ker\omega$. Then there is $U_1 \subseteq U$, $Z_1 \subseteq Z$ such that $I = U_1 \dot{+} Z_1$ (one of them may be equal to zero).

S. González (ed.), Non-Associative Algebra and Its Applications, 275–278.

In fact, if $i = u + z$ is an element of I, with $u \in U$ and $z \in Z$, then $u = 2ue = 2ie \in I$ and so $z = i - u \in I$.

2.2.

If $I \triangleleft B$ and there is $e^2 = e \in I$, then $B^2 \subseteq I$ and consequently $I(B) \subseteq I$.

In fact, $e \in I$ implies that $u = 2eu \in I \ \forall u \in U$. So $U_e \subseteq I$ and consequently $U_e^2 \subseteq I$, that is, $B^2 = Ke \dot{+} U_e \dot{+} U_e^2 \subseteq I$. Now it suffices to remember that if f is other idempotent element, then $f = e + u + u^2$ for some $u \in U_e$.

2.3.

If I is an ideal of B, then $I \subseteq Ker\omega$ or $B^2 \subseteq I$.

2.4.

Since each ideal that is strictly contained in B^2 is also contained in $Ker\omega$, that is, I $Ker\omega \cap B^2 = U \dot{+} U^2$, we have in every Bernstein algebra the following ideals (that are not necessarily different or proper):

$U^o \subseteq U$, $\quad U \dot{+} U^2$, $\quad Ker\omega$, $\quad B^2$.

Let us notice that B is nuclear $\Leftrightarrow B = B^2 \Leftrightarrow Ker\omega = U \dot{+} U^2$.

$U^o = U \dot{+} U^2 \Leftrightarrow$ B is exclusive and $B = Ke \dot{+} U$.

So the lattice of ideals of B in this case is formed by U and all the subspaces of U.

We know that all the powers of $Ker\omega$ are also ideals of B.

2.5.

Consequently the ideals of B are exactly of two types:

a) Ideals contained in $Ker\omega$, that can be expressed as $I = U_1 \dot{+} Z_1$ with $U_1 \subseteq U$ and $Z_1 \subseteq Z$.

b) All the subspaces of B containing B^2.

Notice that we can have an ideal of $Ker\omega$ that is not ideal of B. For instance, if $B = Ke \dot{+} U \dot{+} Z$ is a trivial Bernstein algebra: $\forall u \in U, z \in Z$, $K(u+z)$ is a 1-dimensional ideal of $Ker\omega$, but it is not an ideal of B. That is, an ideal I of $Ker\omega$ is an ideal of B if and only if it is "homogeneous", that is:

$$x = u + z \in I \quad \Leftrightarrow \quad u, z \in I.$$

2.6.

The only ideals of B having a complement are those ideals contained en $Z - U^2$ and their complements.

We must note that every ideal contained in $Z - U^2$ has a complementary ideal (that contains B^2), but not every ideal of B containing B^2 has a complementary ideal.

Example: $B = Ke \dot{+} Ku \dot{+} Kz$ with $u^2 = z^2 = 0$ and $uz = u$.

Then $B^2 = Ke \dotplus Ku$ is an ideal containing B^2 that does not admit complementary.

2.7.

The ideals of dimension 1 in a Bernstein algebra are:

- Ke only if $U = 0$, in which case e is the only idempotent element of the trivial algebra of type (1, n-1).
- Kz with $z \in \mathrm{Ann}\,Ker\omega$.
- Ku with $u \in U^o$ (in case B nuclear, $u \in U^o \Leftrightarrow u \in \mathrm{Annker}\,\omega \Leftrightarrow Ku \trianglelefteq B$).

Notice that $u \in U^o$ does not implies always that $Ku \trianglelefteq B$. Let B be the Bernstein algebra $B = Ke \dotplus K(u_1, u_2) \dotplus Kz$ with $U^2 = 0$, $z^2 = u_1$, $zu_2 = u_1$ and $zu_1 = 2u_2$. Then $Ku \ntrianglelefteq B$. Even more, there are not ideals of dimension 1 contained in B and consequently B has not ideals of dimension 1.

2.8.

In general an ideal of dimension 2 does not contain an ideal of dimension 1. For instance: $B = Ke \dotplus K(u_1, u_2) \dotplus Kz$ as in the previous example is an algebra without ideals of dimension 1. So the 2-dimensional ideal $U = K(u_1, u_2)$ doesn't contain ideals of dimension 1.

However if the algebra is Jordan-Bernstein, then every ideal of dimension 2 contains an ideal 1-dimensional.

3. Bernstein algebras having lineal lattice of ideals

Let us suppose that $\mathcal{J}$ is totally ordered. Then B is nuclear, that is $B^2 = B$, because in other case we would have two ideals B^2 and $Ker\omega$ with $B^2 \not\subset Ker\omega$ and $Ker\omega \not\subset B^2$.

We know that in a nuclear Bernstein algebra, U^o is contained in the annhilitor of $Ker\omega$. So every subspace of contained in U^o is an ideal of B. Therefore

$\dim U^o = 0$ or $\dim U^o = 1$.

Considering $\overline{B} = {}^{B}/_{U^o}$. We know that $\overline{B}$ is also a nuclear Bernstein algebra and it is Jordan: $\overline{B} = Ke \dotplus \overline{U} \dotplus Z$ with $\overline{U}^o = 0$.

In $\overline{B}$ we may considerer the chains of subspaces:

$$\overline{U} \supset \overline{U}^3 \supset \ldots \supset \overline{U}^{2n-1} \supset \overline{U}^{2n+1} = 0$$

$$Z = \overline{U}^2 \supset \overline{U}^4 \supset \ldots \supset \overline{U}^{2m} = 0.$$

We also Know that every product of m elements in $Ker\overline{\omega}$ is in $\overline{U}^m$.

Let $n \in \mathbf{N}$ such that $\overline{U}^{2n-1} \neq 0 = \overline{U}^{2n+1}$. Clearly this implies that $\overline{U}^{2n+2} = 0$. If $\overline{U}^{2n} = 0$, the above implies that $\overline{U}^{2n-1} \subseteq \overline{U}^{o} = 0$. Consequently $\overline{U}^{2n} \neq 0$ and so $\overline{U}^{2n}.\overline{U} \subseteq \overline{U}^{2n+1} = 0$ and $\overline{U}^{2n}.Z = 0$. That is again $\overline{U}^{2n} \subseteq \mathrm{Ann}Ker\omega$ and $\dim \overline{U}^{2n} = 1$.

But we know that $\overline{U} \dotplus \overline{U}^2 \supset \overline{U}^2 \dotplus \overline{U}^3 \supset \ldots \supset \overline{U}^r \dotplus \overline{U}^{r+1} \supset \ldots$ is a chain of ideals of $\overline{B}$.

If $\dim \overline{U} - \overline{U}^3 \geq 2$, we can take $u_1, u_2 \in \overline{U}$ with $u_1 + \overline{U}^3$ and $u_2 + \overline{U}^3$ linearly independent. In this case $K(u_1) + \overline{U}^2 + \overline{U}^3$ and $K(u_2) + \overline{U}^2 + \overline{U}^3$ are two ideals of $Ker\omega$ not contained each in other.

Consequently $\dim \overline{U} = \dim \overline{U}^3 + 1$ and we may find a basis of $\overline{U}$ in the form: $\{\overline{u} = \overline{u}_1, \overline{u}_2, \ldots, \overline{u}_r\}$ where $\overline{u}_j \in \overline{U}^3$, $j = 2, 3, \ldots, r$.

But then $\overline{U}^2$ is generated by $\overline{u}^2$, $\overline{u}.\overline{u}_i \in \overline{U}^4$ and $\overline{u}_i.\overline{u}_j \in \overline{U}^6$, $2 \leq i,j \leq r$ and $\overline{U}^3$ by $\overline{u}^3 = 0$, $\overline{u}.\overline{u}\overline{u}_i \in \overline{U}^5$, $\overline{u}.\overline{u}_i\overline{u}_j \in \overline{U}^7$ and $\overline{u}^2\overline{u}_i$, $\overline{u}\overline{u}_i.\overline{u}_j$, $\overline{u}_i\overline{u}_j.\overline{u}_h \in \overline{U}^5$, that is, $\overline{U}^3 = \overline{U}^5 = 0$.
But $\overline{U}^3 = 0 = \overline{U}.\overline{U}^2 = \overline{U}.Z$, $\overline{B}$ is normal and $K\overline{u}' \dotplus Kz \lhd \overline{B}$ for every $\overline{u}' \in \overline{U}$. Consequently $\dim \overline{U} = 1$ and $\overline{B} = Ke \dotplus K\overline{u} \dotplus K\overline{u}^2$.

Hence, $B = Ke \dotplus Ku \dotplus Ku^2$ or $B = Ke \dotplus Ku$ with $u^2 = 0$ or $B = Ke \dotplus K(u, u_o) \dotplus Ku^2$ with $u_o^2 = u.u_o = 0$.

But in the last case $K(u_o)$ and $Ku \dotplus Ku^2$ are two ideals of B without containing relations in between. So this is impossible and we have finally only two possibilities:

1. $B = Ke \dotplus Ku$ with $u^2 = 0$ $(\dim U^o = \dim U = 1)$
2. $B = Ke \dotplus Ku \dotplus Ku^2$ $(\dim U^o = 0)$.

Notice that in both cases B is a Jordan Bernstein algebra.

References

1. P. Benito; *"Relaciones entre un álgebra de Lie y el retículo de sus ideales"* (D. Thesis). Univ. de Zaragoza, (1989).
2. T. Cortés; *"Classification of 4-dimensional Bernstein algebras"*; Comm. in Algebra 19 (5), (1991), 1429-1443.
3. S. González and C. Martínez; *"Idempotent elements in a Bernstein algebra"*; J. London Math. Soc. (2) 42, (1990), 430-436.
4. Y. Lyubich; *"Mathematical structures in Population Genetics"*; Springer Verlag, Berlin-Heidelberg, (1992).
5. C. Martínez; *"Free Nuclear Bernstein algebras"*; To appear in J. of Algebra.
6. A. Wörz-Busecros; *"Algebras in Genetics"*. Lecture notes in Biomathematics, vol 36, Springer Verlag, Berlin- Heidelberg (1980).

LOCAL ALGEBRAS

KEVIN MCCRIMMON *
Department of Mathematics
University of Virginia
Charlottesville, VA 22903

Abstract.
In joint work with Alain D'Amour, we have simplified and quadratified Zelmanov's characterization of prime Jordan triples and pairs. A key tool in our work is Loos' theory of the socle. Using Meyberg's local algebra, we can carry Anquela, Cortes, and Montaner's result that a primitive PI algebra has a nonzero socle back to primitive triples and pairs. This technique of using local algebras to pass information back and forth between systems and algebras is an important one, and we want to propagandize on its behalf.

Key words: Jordan algebra, Jordan triple, Jordan pair, local algebra, primitive, socle

1. The Philosophical Picture

With poetic licence, let me portray the general philosophy behind our work as follows. We can think of Jordan algebras as the civilized repository of all that is noble in Jordan theory, Jordan triples and pairs as barbaric domains in urgent need of refinement, and local algebras A_b as means of bringing rude Jordan triples J to the city, civilizing them, and sending them back to the countryside with their verneer of culture.

CIVILITAS		BARBARITAS
	Culture, Truth	
	$\longrightarrow\longrightarrow\longrightarrow$	
Jordan Algebras A	*Theorems*	Jordan Triples and Pairs J
	$\longleftarrow\longleftarrow\longleftarrow$	
	A_b	

I was born and bred in the city, but I've been living in the country for years now, and in fact the natives are very friendly and the scenery magnificent. Ottmar Loos might even have some comments about the effete ways of the city dwellers and the

* This research was partially supported by NSF Grant DMS-8903309.
The author also wishes to thank the Universities of Zaragoza and Oviedo for their support and hospitality.

S. González (ed.), Non-Associative Algebra and Its Applications, 279–284.

natural vigor of the barbarians, but for the purposes of this article let us portray the situation in oversimplified terms as a contrast of Couth versus Uncouth.

The first step in the process of civilizing a Jordan triple or pair J is to have him put on manners and convert him into an algebra so that he may gain entrance to the city. It is well-known that for any element $b \in J$ the homotope $J^{(b)}$ is a Jordan algebra, but he has many rough edges (a big radical) – he is still a *country bumpkin* or *yokel.* Kurt Meyberg long ago showed how to train him in etiquette: there is a natural radical whose quotient

$$A_b = J^{(b)}/Ker_J(b)$$

is refined enough to pass in polite society.

Efim Zelmanov showed [7] how a primitive Jordan triple or pair J of Clifford type could (1) be dressed up as a local algebra A_b (which was now a primitive algebra), (2) go to the city with his Clifford identity and purchase a lot of reduced idempotents, and then (3) return with his treasure back to the country, where the treasure could be converted into enough reduced tripotents or pairpotents for J to become a system of Classical Type and live happily ever after.

Our basic idea is to send J off to the city with an HPI (homotope polynomial identity) to buy a *socle* instead of reduced idempotents. Loos' incisive theory of the socle [4] shows this high-tech gadget can do the work of hundreds of reduced idempotents, avoiding scalar extension and complicated arguments.

$$\text{Primitive } A_b \quad \underset{HPI}{\overset{socle}{\rightleftarrows}} \quad \text{Primitive } J$$

Our approach to the classification of primitive Jordan systems involves two parts. The spirit (if not quite the letter) of Part I goes as follows:

J primitive HPI system
$\Longrightarrow$ A_b is a primitive PI algebra
$\Longrightarrow$ the algebra A_b has nonzero socle (by A-C-M)
$\Longrightarrow$ the system J has nonzero socle.

Part II proceeds as follows:

J primitive system of Clifford type
$\Longrightarrow$ J is primitive with a "degree ≤ 2" HPI
$\Longrightarrow$ J has nonzero socle of capacity ≤ 2 (by Part I)
$\Longrightarrow$ $J = Soc(J)$ is of Classical Type and Capacity ≤ 2 ,

so we can apply Loos' classification [3] of Jordan pairs of finite capacity.

2. The Socle versus Reduced Idempotents

While the socle is technically much more efficient than reduced idempotents, let us show that at heart they are really the same thing. The general version $Soc(J)$ is the span of the *simple elements* x (parts of *division pairpotents* (x,y), $P_x y = x$ with Peirce subalgebra $J_2(x,y) = (P_x J)^{(y)}$ a division algebra), whereas the split version (over a big algebraically closed field Φ) is the span of the *reduced elements* e (parts of *reduced pairpotents* (e,f) with Peirce subalgebra $J_2(e,f) = \Phi e$ a copy of the field determined by a *reduced idempotent* e).

It was Efim Zelmanov who showed us how to exorcise division algebras: any big algebraically closed field will serve as exorcist.

Theorem 2.1 (FDR's Theorem) *The only division system we have to fear over a big algebraically closed field Φ is Φ itself.*

(A field is **big** with respect to J if its *cardinality* is bigger by a step than J's *dimension*, $|\Phi| > 1 + dim\Phi(J)$). The reason this works is that (like the Jacobson radical)

any division algebra over a BIG field is algebraic

by *Amitsur's Resolvent Trick* [any element x has $x - \alpha 1$ nonzero, hence invertible, for all but at most one $\alpha \in \Phi$, and BY BIGNESS the $|\Phi| - 1 > dim_\Phi(J)$ inverses $(x - \alpha 1)^{-1}$ can't all be independent, so clearing denominators from a *linear* dependence relation $\sum_{i=1}^n \beta_i (x - \alpha_i)^{-1} = 0$ gives an *algebraic* dependence relation $p(x) = \sum_{i=1}^n \beta_i \prod_{j \neq i}(x - \alpha_j 1) = 0$], and

the only algebraic division algebra over an ALGEBRAICALLY CLOSED field Φ is Φ itself. □

3. The Local Algebra

We now introduce the hero of our story. Let J be a Jordan triple over an irrelevant ring of scalars Φ, and let b be an element of J. (For a Jordan pair V we take the polarized triple $J = J(V) = V^+ \bigoplus V^-$ of the pair.)

Definition 3.1 (Meyberg (6)) *The local algebra of J at b is $A_b = J^{(b)}/Ker_J(b)$ where the kernel of b is $Ker_J(b) = \{z \in J \mid P_b z = P_b P_z b = 0\}$. Here $Ker_J(b)$ is always an ideal of $J^{(b)}$, and coincides with the outer kernel $Outker_J(b) = \{z \in J \mid P_b z = 0\}$ of b in most cases (eg. if $\frac{1}{2} \in \Phi$, or if b is regular, or if J is nondegenerate or special).*

This is a special case of Loos and Neher's notion [5] of *subquotient*:

$Ker_J(b)$ is their $Ker(B)$ for $B = (b)$ the principal inner ideal generated by b, and A_b lives as a homotope inside the Jordan pair

$$(J, B)/(Ker(B), 0) = (A, B)$$

(we can actually form homotopes $A^{(c)}$ for all $c \in B$, not just $c = b$). This pair approach illuminates a general philosophical point, which we may state baldly as:

$A = J/Ker_J(b)$ has no intrinsic structure, its structure comes from a pairing with the inner ideal B (in particular, the element b).

For regular elements the local algebra has a nice description.

Example 3.2 *If b is regular (part of a pairpotent (c,b)), then the local algebra of J at b is just the Peirce subalgebra: $Ker_J(b) = J_0(c,b) \bigoplus J_1(c,b)$, $A_b \cong J_2(c,b)^{(b)}$ for J_i the Peirce spaces relative to (c,b). Note that if $J = M_{2,2}(\Phi)$ is the Jordan algebra of 2×2 matrices, and b,c the matrix units E_{12}, E_{21}, then $(b) = \Phi b$ and $(c) = \Phi c$ are inherently trivial inner ideals, but are nontrivially paired ($J_2(c,b)^{(b)} \cong \Phi$).* □

In the pair case the local algebra splits into a direct sum of two pieces.

Example 3.3 *If $J = J(V) = V^+ \bigoplus V^-$ is the polarized triple of a Jordan pair $V = (V^+, V^-)$ and $b = (b^+, b^-)$, then the local algebra of J at b is the direct sum $A_b = A^+_{b^-} \bigoplus A^-_{b^+}$ of algebras $A^\epsilon_{b^{-\epsilon}} = V^\epsilon / Ker_J(b^{-\epsilon})$.* □

For this reason, in pairs we usually take only local algebras at *homogeneous* elements $b = b^-$ or $b = b^+$.

4. Local-to-Global Inheritance

The main purpose of the local algebra A_b is to bring back valuables to the original system J. This raises the question: which local algebraic properties of A_b are inherited globally by J? Many important elemental properties do pass back to J. The following proof uses nothing more than the Fundamental Formula $P_{P(x)y} = P_x P_y P_x$ in Jordan triples and the definition of the local algebra.

Theorem 4.1 (Elemental Inheritance Theorem) *If J is nondegenerate and $\overline{z} \neq \overline{0}$ is trivial, reduced, regular, or simple in A_b, then so is the element $P_b z$ in J.*

PROOF. The element $P_b z$ is nonzero in J since $\overline{z} \neq \overline{0} \Longrightarrow z \notin Ker(b) = Outker(b)$ (by (3.1) and nondegeneracy of J) $\Longrightarrow P_b z \neq 0$. If $\overline{z}$ is trivial ($\overline{U}_{\overline{z}} A = \overline{0}$), reduced ($\overline{U}_{\overline{z}} A \subset \Phi\overline{z}$), regular ($\overline{U}_{\overline{z}} A \ni \overline{z}$), or simple ($\overline{0} \neq \overline{y} \in \overline{U}_{\overline{z}} A \Longrightarrow \overline{z} \in \overline{U}_{\overline{y}} A$) in A_b , then $P_b z$ is trivial ($P_z P_b J \subset Ker(b) \Longrightarrow P_{P(b)z} J = P_b P_z P_b J = 0$), reduced ($P_z P_b J \subset \Phi z + Ker(b) \Longrightarrow P_{P(b)z} J = P_b P_z P_b J \subset \Phi P_b z$), regular ($P_z P_b J + Ker(b) \ni z \Longrightarrow P_{P(b)z} J = P_b P_z P_b J \ni P_b z$), or simple ($0 \neq x = P_{P(b)z} a = P_b(P_z P_b a) = P_b y \Longrightarrow y = P_z P_b a \notin Ker(b) \Longrightarrow \overline{0} \neq \overline{y} \in \overline{U}_{\overline{z}} A \Longrightarrow \overline{z} \in \overline{U}_{\overline{y}} A \Longrightarrow z \in P_y P_b J + Ker(b) \Longrightarrow P_b z \in P_b P_y P_b J = P_{P(b)y} J = P_x J$) in J itself. □

From the fact that the socle is the span of the simple and trivial elements it follows that we can bring the socle back from A_b to J.

Theorem 4.2 (Socle Inheritance Theorem) *If some local algebra A_b of J has nonzero socle, then J itself has nonzero socle.* □

5. Global-to-Local Inheritance

So far we have seen elemental information flow from the local algebra A_b back to J. This passage needs to be a two-way street: we need properties to flow from J to A_b as well. In order for A_b to be able to buy a socle at the Anquela-Cortes-Montaner emporium, it needs to be primitive with PI. It will certainly be PI if J satisfies a *homotope PI* (a Jordan PI $f(x_1,\ldots,x_n,y)$ which is a Jordan product $f^{(y)}(x_1,\ldots,x_n)$ of the x's in the y-homotope of the free Jordan triple, such that $f(a_1,\ldots,a_n,b)=0$ for all a_i,b in J, ie. all homotopes $J^{(b)}$ satisfy the ordinary PI $f^{(b)}(x_1,\ldots,x_n)$). Thus the crux is to make sure that some A_b is primitive. This can be done for Jordan pairs, but we have been unable to overcome technical obstacles to carry the proof over to triples directly; instead, we employ subterfuge.

Recall the definition of primitivity for triples and pairs.

Definition 5.1 *A Jordan triple J is primitive if it is primitive at some b, where J is primitive at b if it has a b-primitizer K:*

(1) *K is a proper inner ideal in J*
(2) *K is b-modular with b-modulus c (ie. c is modulus for K in $J^{(b)}$)*
(3) *K supplements nonzero triple ideals: $0\neq I\lhd J\Longrightarrow I+K=J$.*

(A Jordan algebra is primitive if it is primitive at $b=1$).

Definition 5.2 *A Jordan pair V is primitive if it is ϵ-primitive at some $b^{-\epsilon}$ ($\epsilon=\pm1$), where V is ϵ-primitive at $b^{-\epsilon}$ if it has an ϵ-primitizer K^ϵ:*

(1) *K^ϵ is a proper inner ideal in V^ϵ*
(2) *K^ϵ is $(b^\epsilon,b^{-\epsilon})$-modular for some b^ϵ (modulus for K^ϵ in $V^{\epsilon(b^{-\epsilon})}$)*
(3) *K^ϵ supplements nonzero ϵ-ideals (the ϵ-components of pair ideals):*
$$0\neq I\lhd_\epsilon V^\epsilon\Longrightarrow I^\epsilon+K^\epsilon=V^\epsilon$$
(4) *V is core-free: $Q_{V^\epsilon}z^{-\epsilon}=Q_{V^\epsilon}Q_{z^{-\epsilon}}V^\epsilon=0\Longrightarrow z^{-\epsilon}=0$.*

Note that we propose *abandoning the condition that a primitizer be maximal among b-modular inner ideals*; the crucial property is the ideal supplementation property (3), not the maximality. (In the terminology of Anquela-Cortes-Montaner, we advocate using *proto-primitizers*.)

Theorem 5.3 (Global-Local Primitivity Inheritance Theorem) *If a Jordan pair V is $+$-primitive at b^- , then the local algebra $A^+_{b^-}$ is primitive.*

PROOF. Better not ask – see [2], [7]. □

The subterfuge we mentioned earlier consists in having Jordan triples J carried to the city on the backs of their sturdy brethren, the Jordan pairs $V(J)=(J,J)$.

Theorem 5.4 (Primitive SocleTheorem) *A primitive Jordan system satisfying a homotope PI has nonzero socle.*

PROOF IDEA. If V is a +-primitive Jordan pair with HPI and modulus (b^+, b^-), then by Primitivity Inheritance 5.3 $A^+_{b^-}$ is primitive PI, so by Anquela-Cortes-Montaner [1] it has nonzero socle; then $A_b = A^+_{b^-} \bigoplus A^-_{b^+}$ does too by (3.3), therefore by Socle Inheritance (4.2) so does V .

If J is a triple we do not have the analogue of 5.3, so we call in the associated Jordan pair $V(J) = (J, J)$ to carry us through. One can show that V is either (1) both + and - primitive, or (2) a subdirect sum $V \overset{=}{\sim} V_1 \bigoplus V_2$ of ϵ-primitive Jordan pairs $V_1 = V/I$ and $V_2 = V/I^*$ for I an ideal with $I \cap I^* = 0$. Then either (1) $A^+_{b^-}$ or (2) $(A_i)^\epsilon_{b^{-\epsilon}}$ is primitive with nonzero socle by (5.3) and Anquela-Cortes-Montaner. We need the following 2 basic facts about socles in pairs [4]:

(Fact (1)) : if $V = V(J) = (J, J)$ then $Soc(V) = (Soc(J), Soc(J))$
(Fact (2)) : if $I \lhd V$ then $Soc(I) = I \cap Soc(V)$.

In Case (1), $Soc(V) \neq 0$ implies $Soc(J) \neq 0$ by Fact (1). In Case (2), $0 \neq I \cong \pi_2(I) \lhd V_2 = V/I^*$ (by $I \cap I^* = 0$) and $Soc(V_2) \neq 0$ together imply (by primeness of V_2) that $0 \neq \pi_2(I) \cap Soc(V_2) = Soc(\pi_2(I))$ (by Fact (2)), therefore the same holds for the isomorphic copy I of $\pi_2(I)$, $0 \neq Soc(I) = I \cap Soc(V)$ (by Fact (2) again), in particular $Soc(V) \neq 0$, and again $Soc(V) \neq 0$ implies $Soc(J) \neq 0$ by Fact (1). □

Thus by hook or by crook, primitivity flows from a Jordan system J to a local algebra A_b , turns into a socle, and flows back to J.

Our Jordan systems have many other adventures on their way to the city and back, but we'll leave that for another story [2]. Our one example will suffice to indicate how local algebras can mediate between urbane city dwellers and their countrymen.

References

1. J. Anquela, T. Cortes, F. Montaner: *The structure of primitive quadratic Jordan algebras*, to appear in J. Alg.
2. A. D'Amour and K. McCrimmon: *Local algebras of Jordan systems*, to appear.
3. O. Loos: *Jordan Pairs*, Lecture Notes in Math. vol. 460, Springer-Verlag, Berlin- Heidelberg-New York (1975).
4. O. Loos: *On the socle of a Jordan pair*, Collect. Math. 40 (1989), 109-125 .
5. O. Loos and E. Neher: *Complementation of inner ideals in Jordan pairs*, to appear in J. Alg .
6. K. Meyberg: *Lectures on algebras and triple systems*, Lecture Notes , University of Virginia, Charlottesville (1972).
7. E. Zelmanov: *Prime Jordan triple systems I,II,III*, Siberian Math. J. 24 (1983), 23-37; 25 (1984), 50-61; 26 (1985), 71-82.

APPLICATIONS OF FOX DIFFERENTIAL CALCULUS TO FREE LIE SUPERALGEBRAS

ALEXANDER A. MIKHALEV* and ANDREJ A. ZOLOTYKH†
Department of Mechanics and Mathematics
Moscow State University
Moscow, 119899, RUSSIA

Abstract. We give some applications of Fox differential calculus to free Lie superalgebras. The main result is the theorem that the rank of an element is equal to the rank of the left ideal of the free associative algebra generated by its left Fox partial derivatives. As corollaries we have an algorithm to find the rank of an element and a criterion for an element to be primitive. The same results obtained for a system of elements.

Key words: Fox differential calculus, free Lie superalgebras, automorphisms, primitive elements

1. Introduction

Free differential calculus for free groups was introduced by R.H. Fox in [5], J. Birman in [2] gave a matrix criterion for an endomorphism of free group to be an automorphism. In [21] U.U. Umirbaev proved an analog of this result for free Lie algebras (Ch. Reutenauer in [15] also obtained such criterion; more general case was considered by V. Shpilrain in [17] and by U.U. Umirbaev in [23]). The general Jacobian problem for varieties of algebras was considered by A.V. Yagzhev, he solved it in [24] for the varieties of all algebras and of all anticommutative algebras. The Jacobian problem for free associative algebras was considered by W. Dicks and J. Lewin in [3] and by A.H. Schofield in [16]. O.G. Kharlampovich in [6] used Fox differential calculus in connection with Lyndon condition for Lie algebras.

In [7] G.P. Kukin gave a criterion for an element of the free Lie algebra L to be primitive (a is a primitive element of L if and only if $L/\mathrm{id}(a)$ is a free Lie algebra where $\mathrm{id}(a)$ is the ideal of L generated by a). U.U. Umirbaev in [22] gave a criterion for a system of elements of the free group to be primitive. In [18] V. Shpilrain gave a description of the rank of a homogeneous (relative to the length) element of the free Lie algebra (the rank of an element a of the length m is equal to the dimension of the linear space spanned by all its left Fox partial derivatives of degree $m-1$).

G.L. Feldman proved in [4] that two-generated Lie algebras of cohomology dimension one are free Lie algebras. Recently the authors together with U.U. Umirbaev [14] gave an example of a nonfree Lie algebra over a field of prime characteristic which cohomology dimension is equal to one. This example shows in our situation the difference between the rank of an element and the rank of the left ideal generated by its Fox partial derivatives in the case of prime characteristic.

* Partially supported by the grant of the American Mathematical Society
† Partially supported by the Russian foundation of fundamental researches, grant 93-011-1543

S. González (ed.), Non-Associative Algebra and Its Applications, 285–290.

2. Free Colour Lie (p-)superalgebras

Let G be an abelian group, K a field, char $K \neq 2$, $\varepsilon : G \times G \to K^*$ a skew symmetric bilinear form,

$$\varepsilon(g_1 + g_2, h) = \varepsilon(g_1, h)\,\varepsilon(g_2, h), \quad \varepsilon(g, h_1 + h_2) = \varepsilon(g, h_1)\,\varepsilon(g, h_2),$$

$$\varepsilon(g, h)\,\varepsilon(h, g) = 1$$

for all g, g_1, g_2, h, h_1, $h_2 \in G$,

$$G_- = \{g \in G \mid \varepsilon(g, g) = -1\}, \quad G_+ = \{g \in G \mid \varepsilon(g, g) = +1\}.$$

We say that a G-graded algebra $R = \bigoplus_{g \in G} R_g$ over K is *a colour Lie superalgebra* if

$$[x, y] = -\varepsilon(d(x), d(y))[y, x],$$

$$[x, [y, z]] = [[x, y], z] + \varepsilon(d(x), d(y))[y, [x, z]], \quad [v, [v, v]] = 0$$

with $d(v) \in G_-$ for G-homogeneous elements $x, y, z, v \in R$ where $d(a) = g$ for $a \in R_g$.

Let char $K = p > 2$. We say that a colour Lie superalgebra R over K is a *colour Lie p-superalgebra* if on homogeneous components R_g, $g \in G_+$, we have a mapping $x \to x^{[p]}$, $d(x^{[p]}) = pd(x)$, such that for all $z \in R$, $\alpha \in K$, $x, y \in R$, $d(x) = d(y) \in G_+$, the following conditions are satisfied:

$$(\alpha x)^{[p]} = \alpha^p x^{[p]}, \quad (\mathrm{ad}(x^{[p]}))(z) = [x^{[p]}, z] = (\mathrm{ad} x)^p(z),$$

$$(x + y)^{[p]} = x^{[p]} + y^{[p]} + \sum s_i(x, y)$$

where $is_i(x, y)$ is the coefficient on t^{i-1} in the polynomial $(\mathrm{ad}(tx + y))^{p-1}(x)$ (here $(\mathrm{ad} a)(b) = [a, b]$).

For convenience we use the notation $\varepsilon(a, b)$ instead of $\varepsilon(d(a), d(b))$.

If Q is a G-graded associative algebra over K then $[Q]$ denotes the colour Lie superalgebra with the operation [,] where $[a, b] = ab - \varepsilon(a, b)\,ba$ for G-homogeneous elements $a, b \in Q$.

If char $K = p > 2$ and $x^{[p]} = x^p$ for all $x \in Q$ with $d(x) \in G_+$, then $[Q]$ with the operation $[p]$ is a colour Lie p-superalgebra denoted by $[Q]^p$.

Let $X = \cup_{g \in G} X_g$ be a G-graded set, i.e. $X_g \cap X_f = \emptyset$ for $g \neq f$, $d(x) = g$ for $x \in X_g$, $A(X)$ the free G-graded associative K-algebra with unity, $F(X)$ the free nonassociative K-algebra. Let also I be the G-graded ideal of $F(X)$ generated by the homogeneous elements of the form

$$a \cdot b + \varepsilon(a, b)\, b \cdot a$$

and

$$(a \cdot b) \cdot c - a \cdot (b \cdot c) + \varepsilon(a, b)\, b \cdot (a \cdot c), \quad f \cdot (f \cdot f)$$

where $a, b, c, f \in F(X)$, $d(f) \in G_-$; then $L(X)$ is *the free colour Lie K-superalgebra* (i.e. every G-mapping φ of degree zero from X into any colour Lie K-superalgebra R with the same group G and form ε $(d(\varphi(x)) = d(x),\ x \in X)$ can be uniquely extended to a colour Lie superalgebra homomorphism $\overline{\varphi} : L(X) \to R$). It is clear that this universal property defines the free colour Lie superalgebra $L(X)$ uniquely up to a colour Lie superalgebra isomorphism. For $u \in F(X)$ we set $\widetilde{u} = u + I \in L(X)$.

Consider the subalgebra $[X]$ of $[A(X)]$ generated by X and the homomorphism $\pi : L(X) \to [X]$ such that $\pi(\widetilde{x}) = x$ for all $x \in X$. Then π is an isomorphism of colour Lie superalgebras. In the case when $\operatorname{char} K = p > 2$ let also $L^p(X)$ be the subalgebra of $[A(X)]^p$ generated by X. Then $L^p(X)$ is the free colour Lie p-superalgebra.

For more information on free Lie superalgebras see [1, 10, 11, 12].

3. Jacobian Matrices and Automorphisms

For any element a of the free associative algebra $A(X)$, $X = \{x_1, \ldots, x_n\}$, we have the unique presentation in the form $a = x_1 a_1 + x_2 a_2 + \cdots + x_n a_n + \alpha \cdot 1$, where $a_i \in A(X)$, $\alpha \in K$. We call the element a_i *the left Fox partial derivative of the element* a *by* x_i and denote $a_i = \dfrac{\partial a}{\partial x_i}$, $i = 1, \ldots, n$.

Theorem 1 ([13]) *Let* $X = \{x_1, \ldots, x_n\}$, $L(X)$ *be the free colour Lie superalgebra (in the case when* $\operatorname{char} K = p$ *let* $L^p(X)$ *be the free colour Lie p–superalgebra). Let also* $f_1, \ldots, f_n$ *be G-homogeneous elements of* $L(X)$ *(of* $L^p(X)$*),* $d(f_i) = d(x_i)$, $i = 1, \ldots, n$. *Then the endomorphism* $\varphi : L(X) \to L(X)$ $(\varphi : L^p(X) \to L^p(X))$ *where* $\varphi(x_i) = f_i$, $i = 1, \ldots, n$, *is an automorphism if and only if the matrix* $\left(\dfrac{\partial f_i}{\partial x_j}\right)$, $1 \le i, j \le n$, *is invertible over* $A(X)$.

4. Rank and Primitivity of Elements

We say that an element $a \in A(X)$ *is depending on* $x_i \in X$ if in the associative presentation of a we have with nonzero coefficient a monomial which contains x_i.

We say that a G-homogeneous element a of the free colour Lie superalgebra $L(X)$ has *the rank* k if k is the least number of free generators from $X = \{x_1, \ldots, x_n\}$ on which $\varphi(a)$ depends on (where φ is an automorphism of $L(X)$). We denote $k = \operatorname{rank}(a)$. By analogy we define the rank for a G-homogeneous element of the free colour Lie p-superalgebra $L^p(X)$.

Theorem 2 (A.A. Mikhalev, A.A. Zolotykh, 1993) *1. Let* $\operatorname{char} K = 0$, h *be a G-homogeneous element of* $L(X)$. *Then* $\operatorname{rank}(h)$ *is equal to the rank of the left ideal of* $A(X)$ *generated by the elements* $\left\{\dfrac{\partial h}{\partial x} \mid x \in X\right\}$ *(as a free left* $A(X)$*-module).*

2. In the case when $\operatorname{char} K = p > 2$ *the assertion 1 of the theorem takes place for any G-homogeneous element* h *of* $L^p(X)$.

As a corollary, we have an algorithm for finding the rank of an element h which is reduced to a construction of finite Gröbner basis of the left ideal of $A(X)$.

By analogy with the case of one element we define the rank of a system of G-homogeneous elements $a_1, \ldots, a_k$ of $L(X)$ (in the case when $\operatorname{char} K = p > 2$ we suppose that $a_1, \ldots, a_k \in L^p(X)$). Denote by $\{e_i \mid 1 \le i \le k\}$ the standard basis of the left $A(X)$-module $A(X)^k$.

Corollary *Let* char $K = 0$, *and let* $a_1, \ldots, a_k$ *be G-homogeneous elements of* $L(X)$ *(or* char $K = p$ *and* $a_1, \ldots, a_k \in L^p(X)$*). Then the rank of the system* $\{a_1, \ldots, a_k\}$ *is equal to the rank of the left* $A(X)$*-module generated by the elements*

$$\left\{\sum_{i=1}^{k} \frac{\partial a_i}{\partial x} e_i \mid x \in X\right\}.$$

We say that a G-homogeneous element h of $L(X)$ (of $L^p(X)$) is *primitive* if it can be embedded in a set of free generators of $L(X)$ (respectively, of $L^p(X)$).

Theorem 3 (A.A. Mikhalev, A.A. Zolotykh, 1993) *1. Let* char $K = 0$, h *be a G-homogeneous element of* $L(X)$*. Then the following conditions are equivalent:*
a) the element h *is primitive;*
b) there are elements $m_1, \ldots, m_n \in A(X)$ *such that*

$$\sum_{i=1}^{n} m_i \frac{\partial h}{\partial x_i} = 1.$$

2. In the case when char $K = p > 2$ *the assertion 1 of the theorem takes place for any G-homogeneous element* h *of* $L^p(X)$.

We say that a system of G-homogeneous elements $S \subseteq L(X)$ ($S \subseteq L^p(X)$) is *primitive* if S is a subset of some set of free generators of $L(X)$ (of $L^p(X)$).

Theorem 4 (A.A. Mikhalev, A.A. Zolotykh, 1993) *Let* char $K = 0$, $a_1, \ldots, a_k$ *be G-homogeneous elements of* $L(X)$*. Then the following conditions are equivalent:*
a) the system $\{a_1, \ldots, a_k\}$ *is primitive;*
b) there are elements $m_{ij} \in A(X)$, $i = 1, \ldots, n$; $j = 1, \ldots, k$, *such that for all* j, l

$$\sum_{i=1}^{n} m_{ij} \frac{\partial a_l}{\partial x_i} = \delta_{jl},$$

where δ_{jl} *is the Kronecker delta (i.e. the matrix* $\left(\frac{\partial a_j}{\partial x_i}\right)$ *is leftinvertible);*
2. In the case when char $K = p > 2$ *the assertion 1 of the theorem takes place for any system of G-homogeneous elements* $a_1, \ldots, a_k \in L^p(X)$.

Remark 1 The assertions of these theorems do not take place in the case when char $K = p > 2$ and $L(X)$ is the free Lie algebra. Indeed, let $X = \{x, y, z\}$, $h = x + [y, z] + (\operatorname{ad} x)^p(z)$. Then h is not a primitive element but the rank of above mentioned left ideal is equal to 1 (h is a primitive element of $L^p(X)$, but not of $L(X)$).

5. Example of a Nonfree Lie Algebra of Cohomology Dimension One

Remind that cohomology dimension of a Lie algebra L over a field K is equal to n if $H^{n+1}(L, M) = 0$ for any L-module M and there exists an L-module M_0 such that $H^n(L, M_0) \neq 0$.

Theorem 5 (A.A. Mikhalev, U.U. Umirbaev, A.A. Zolotykh, [14]) *Let K be a field,* $\operatorname{char} K = p > 2$, $X = \{x, y, z\}$, *$L(X)$ the free Lie algebra over K,*

$$h = x + [y, z] + (\operatorname{ad} x)^p(z),$$

$\operatorname{id}(h)$ *the ideal of $L(X)$ generated by h. Then the algebra $L(X)/\operatorname{id}(h)$ is not a free Lie algebra but cohomology dimension of $L(X)/\operatorname{id}(h)$ is equal to one.*

Remark 2 This example gives a contradiction with the claim of G.P. Kukin ([8, 9]) that Lie algebras of cohomology dimension one are free over an arbitrary field (i.e. the analog of Stallings-Swan theorem, see [19, 20]).

Remark 3 The algebra $L(X)/\operatorname{id}(\mathrm{h})$ is not a free Lie algebra, but its universal enveloping algebra is the free associative algebra of rank two.

References

1. Bahturin Yu.A., Mikhalev A.A., Zaitsev M.V. and Petrogradsky V.M., *Infinite dimensional Lie superalgebras*, Walter de Gruyter Publ., Berlin, New York, 1992.
2. Birman J.S., An inverse function theorem for free groups, Proc. Amer. Math. Soc. **41** (1973), 634–638.
3. Dicks W. and Lewin J., A Jacobian conjecture for free associative algebras, Commun. Algebra **10** (1982), 1285–1306.
4. Feldman G.L., Ends of Lie algebras, Uspekhi Matem. Nauk **38** (1983), no 1, 199–200. (= Russian Math. Surveys).
5. Fox R.H., Free differential calculus, I. Derivations in free group rings, Ann. Math. **57** (1953), 547–560.
6. Kharlampovich O.G., Lyndon condition for solvable Lie algebras, Izvestia Vuzov. Mat. 1984, no 9, 50–59.
7. Kukin G.P., Primitive elements of free Lie algebras, Algebra i Logika **9** (1970), no 4, 458–472.
8. Kukin G.P., On Lie algebras of cohomology dimension one, 5-th All-Union Symposium on the Theory of Rings, Algebras and Modules, Abstracts of reports, Novosibirsk, 1982, p. 86.
9. Kukin G.P., Lie algebras of cohomology dimension one, Plenary talk on the IV All-Union School "Lie algebras and their applications in mathematics and physics," Kazan, 1990, unpublished.
10. Mikhalev A.A., Subalgebras of free colour Lie superalgebras, Mat. Zametki **37** (1985), 653–661. (= Math. Notes, 356–360).
11. Mikhalev A.A., Free colour Lie superalgebras, Dokl. Akad. Nauk SSSR **286** (1986), 551–554. (= Soviet Math. Dokl. **33** (1986), 136–139).
12. Mikhalev A.A., Subalgebras of free Lie p-superalgebras, Mat. Zametki **43** (1988), 178–191. (= Math. Notes, 99–106).
13. Mikhalev A.A., On right ideals of the free associative algebra generated by free colour Lie (p-)superalgebras, Uspekhi Mat. Nauk **47** (1992), no 5, 187–188. (= Russian Math. Surveys).
14. Mikhalev A.A., Umirbaev U.U. and Zolotykh A.A., An example of a nonfree Lie algebra of cohomology dimension one, Uspekhi Matem. Nauk, to appear (= Russian Math. Surveys).
15. Reutenauer Ch., Applications of a noncommutative Jacobian matrix, J. Pure and Appl. Algebra **77** (1992), 169–181.

16. Schofield A.H., *Representations of rings over skew fields*, London Math. Soc. Lecture Notes Series, no 92, Cambridge University Press, 1985.
17. Shpilrain V., On generators of L/R^2 Lie algebras, Proc. Amer. Math. Soc., to appear.
18. Shpilrain V., On the rank of an element of a free Lie algebra, Proc. Amer. Math. Soc., to appear.
19. Stallings J.R., On torsion-free groups with infinitely many ends, Ann. Math. **88** (1968), no 2, 312–334.
20. Swan R.G., Groups of cohomological dimension one, J. Algebra **12** (1969), 585–610.
21. Umirbaev U.U., On Jacobian of Lie algebras, VI All-Union School on Varieties of Alg. Systems, Abstracts of reports, Magnitogorsk, 1990, 32–33.
22. Umirbaev U.U., A note on primitive elements of free groups, Proc. of the 5-th Siberian School "Algebra and Analysis," to appear.
23. Umirbaev U.U., Partial derivatives and endomorphisms of some relatively free Lie algebras, Sibirsk. Math. Zh., to appear.
24. Yagzhev A.V., On endomorphisms of free algebras, Sibirsk. Mat. Zh. **21** (1980), no 1, 181–192.

ON THE COHOMOLOGY FOR THE WITT ALGEBRA W(1, 1)

DANIEL K. NAKANO*
Northwestern University
Evanston, Illinois 60208 U.S.A

Abstract. In this paper the cohomology ring for the restricted enveloping algebra of $W(1,1)$ is computed. The procedure given depends on calculating the ordinary Lie algebra cohomology of a certain p-unipotent subalgebra of $W(1,1)$. For low primes this computation becomes tractable and formulas for the dimensions of the graded components of the cohomology ring are given.

1. Introduction

Let G be a connected reductive algebraic group scheme over an algebraically closed field, k, of characteristic $p > 0$ and let $F : G \to G$ be the Frobenius map with G_1 denoting the scheme-theoretic kernel of F. It is well known that there is a categorical equivalence between the representations for G_1 and the representations for the restricted enveloping algebra, $V(g)$, where $g = Lie\ G$. By using this correspondence Friedlander and Parshall [FP1] and Andersen and Jantzen [AJ] have shown that the cohomology ring, $H^*(G_1, k) \cong H^*(V(g), k)$, is isomophic to coordinate algebra of the nullcone for primes larger than the Coxeter number. For other types of restricted simple Lie algebras, namely the Lie algebras of Cartan type [BW], the structure of the cohomology ring still remains a mystery. In general the dimensions of the graded components of the cohomology ring are not even known.

In this paper a technique for computing the dimensions of the cohomology ring for the Witt algebra, $W(1,1)$, will be presented. This procedure is highly dependent on the calculation of the ordinary Lie algebra cohomology, $H^*(u, k)$, for a certain p-unipotent subalgebra, u, of $W(1,1)$. Unlike the classical Lie algebras, the dimensions of the Lie algebras of Cartan type depend on the given prime. In conversations with A.S. Dzhumadil'daev it seems as though this fact makes the general computation of $H^*(u, k)$ a formidable task. For low primes this computation is tractable and explicit formulas for the dimension of $H^*(V(W(1,1)), k)$ for $p = 5$ and 7 will be provided. An interesting sidelight of this calculation is how the representation theory of $W(1,1)$ is used to prove even and odd vanishing of the cohomology of simple modules corresponding to exceptional weights. The author would like to thank the organizers for their hospitality during the conference and for the opportunity to present the algebraic group techniques for Lie algebras of Cartan type, developed by Z. Lin and the author, during the meeting in Oviedo along with this application of these methods.

* supported by NSF grant DMS-9206284

S. González (ed.), Non-Associative Algebra and Its Applications, 291–295.

2. Main results

Let k be an algebraically closed field of characteristic $p \geq 5$ and let $A = k[x]/(x^p = 0)$. The Witt algebra, $W(1,1)$, can be realized as $Der\ A =< x^i \frac{d}{dx} : i = 0,1,...p-1 >$. Set $e_{i-1} = x^i \frac{d}{dx}$ for $i = 0,1,...p-1$. The Lie relations and the pth power operations are given by $[e_i, e_j] = (j-i)e_{i+j}$ for $-1 \leq i+j \leq p-2$ with 0 otherwise and $e_j^{[p]} = 0$ for $j \neq 0$; $e_0^{[p]} = e_0$. Let $b^+ =< e_i : i = 0,1,...p-2 >$, $u =< e_i : i = 1,2,...p-2 >$ and $t =< e_0 >$. By convention for a restricted Lie algebra, g, the restricted universal enveloping algebra will be denoted by $V(g)$.

Since b^+ is a completely solvable Lie algebra the simple $V(b^+)$ modules are one-dimensional and parameterized by the set of restricted weights: $X_1(T) = \{\lambda_j : j = 0,1,...p-1\}$. From [N] the simple $V(W(1,1))$ modules are also indexed by $X_1(T)$. In order to realize these modules consider for each $j = 0,1,...p-1$ the induced module $Z(j) = ind_{V(b^+)}^{V(g)} \lambda_j$. For $j = 2,3,..p-1$ the modules $L(j) = Z(j)$ are simple. In the case when $j = 0,1$ (exceptional weights) $Z(0)$ has a p-1 dimensional module, $L(1)$, as its head and the trivial module, $L(0) \cong k$, as its socle. Similarly, $Z(1)$ has the trivial module as its head and $L(1)$ as its socle.

Now let M be a $V(b^+)$ module. By applying the Lyndon-Hochschild-Serre spectral sequence we have

$$E_2^{p,q} = H^p(V(t), H^q(V(u), M)) \Rightarrow H^{p+q}(V(b^+), M).$$

Since $V(t)$ is semisimple this spectral sequence collapses and we have for all $n \geq 0$; $H^n(V(b^+), M) \cong H^n(V(u), M)^{V(t)}$. For computational purposes we will consider $G = Aut\ W(1,1)$. From [W] G is isomorphic to a semidirect product of T with U where T is a one-dimensional torus and U is a unipotent algebraic group scheme with $Lie\ G \cong b^+$ [LN1]. Let T_1 be the infinitesimal kernel of T under the Frobenius map. Then for all $n \geq 0$; $H^n(V(b^+), M) \cong H^n(V(u), M)^{T_1}$.

Let $\lambda \in X_1(T)$ and $u^\sharp$ be the dual space of u. Moreover, let $S^*(u^\sharp)$ be symmetric algebra on $u^\sharp$ and $H^*(u,k)$ be the ordinary Lie algebra cohomology of u. Since the fixed point functor under T_1 is exact we can compose this with the Ivanovskii spectral sequence [FP3, (1.3)] to produce another spectral sequence

$$E_1^{2p,q} = S^p(u^\sharp)^{(1)} \otimes H^q(u,k)_{-\lambda} \Rightarrow H^{2p+q}(V(u), \lambda)^{T_1}. \quad (2.1)$$

Note that (1) indicates that T_1 acts trivially on $S^p(u^\sharp)$. Implicitly we have used the isomorphism $(H^q(u,k) \otimes \lambda)^{T_1} \cong H^q(u,k)_{-\lambda}$. The first result provides a description of the $V(b^+)$ cohomology for low primes.

Proposition 2.1 Let $b^+ =< e_j : j = 0,1,...p-2 >$. Then there exists the following isomorphisms of vector spaces.

$$H^n(V(b^+), k) = \begin{cases} S^{\frac{n}{2}}(u^\sharp) \oplus S^{\frac{n-2}{2}}(u^\sharp) & n \text{ even } p = 5 \\ S^{\frac{n}{2}}(u^\sharp) \oplus S^{\frac{n-2}{2}}(u^\sharp) \oplus S^{\frac{n-4}{2}}(u^\sharp) & n \text{ even } p = 7 \\ 0 & n \text{ odd} \end{cases}$$

$$H^n(V(b^+),\lambda_1)=\begin{cases} S^{\frac{n-1}{2}}(u^\sharp)\oplus S^{\frac{n-3}{2}}(u^\sharp) & n \text{ odd } p=5\\ S^{\frac{n-1}{2}}(u^\sharp)\oplus S^{\frac{n-3}{2}}(u^\sharp)\oplus S^{\frac{n-5}{2}}(u^\sharp) & n \text{ odd } p=7\\ 0 & n \text{ even}\end{cases}$$

Proof: We should first remark that if $T=\{t_a : a\neq 0\}$ then the action of T on $u^\sharp$ is given by $t_a.e_j^\sharp = a^{-j}e_j^\sharp$. In order to calculate the ordinary Lie algebra cohomology $H^*(u,k)_{-\lambda}$ one can use the Koszul complex and the fact that the differentials must respect the action of T. This yields the following isomorphisms as vector spaces:

$$H^*(u,k)_{-\lambda}=\begin{cases} <\bar{1}>\oplus<\bar{e_2}^\sharp\wedge\bar{e_3}^\sharp> & p=5,\ \lambda=0\\ <\bar{e_1}^\sharp>\oplus<\bar{e_1}^\sharp\wedge\bar{e_2}^\sharp\wedge\bar{e_3}^\sharp> & p=5,\ \lambda=1\\ <\bar{1}>\oplus<\bar{e_2}^\sharp\wedge\bar{e_5}^\sharp-3e_3^\sharp\wedge\bar{e_4}^\sharp> & p=7\ \lambda=0\\ \oplus<\bar{e_2}^\sharp\wedge\bar{e_3}^\sharp\wedge\bar{e_4}^\sharp\wedge\bar{e_5}^\sharp> & \\ <\bar{e_1}^\sharp>\oplus<\bar{e_1}^\sharp\wedge\bar{e_2}^\sharp\wedge\bar{e_5}^\sharp> & p=7,\ \lambda=1\\ \oplus<\bar{e_1}^\sharp\wedge\bar{e_2}^\sharp\wedge\bar{e_3}^\sharp\wedge\bar{e_4}^\sharp\wedge\bar{e_5}> & \end{cases}$$

The differentials in the spectral sequence (2.1), d_r, have bidegree $(r,1-r)$. Since $H^q(u,k)_{-\lambda}$ is either zero for all odd q or zero for all even q it follows that $d_r=0$ for all $r\geq 1$. Hence, the spectral sequence stops at E_1 and

$$E_1^{2p,q}\cong E_\infty^{2p,q}\cong H^{2p+q}(V(b^+),\lambda_j)$$

for $j=0,1$. The statement of the proposition now follows by applying the computation of $H^*(u,k)_{-\lambda}$ provided above.◇

For notational convenience let $\delta_l=\lambda_0$ for l even and $\delta_l=\lambda_1$ for l odd. In the next theorem a procedure is presented for passing from the cohomological results for $V(b^+)$ to cohomological results for $V(W(1,1))$.

Theorem 2.2 Suppose that $H^n(V(b^+),k)=0$ for n odd and $H^n(V(b^+),\lambda_1)=0$ for n even. Then

$$H^n(V(g),k)\cong\begin{cases}\oplus_{l=0}^n H^l(V(b^+),\delta_l) & \text{for } n \text{ even}\\ 0 & \text{for } n \text{ odd}\end{cases}$$

$$H^n(V(g),L(1))\cong\begin{cases}\oplus_{l=0}^n H^l(V(b^+),\delta_l) & \text{for } n \text{ odd}\\ 0 & \text{for } n \text{ even}\end{cases}$$

Proof: The representation theoretic results provided above yield two short exact sequences:

$$0\to k\to Z(0)\to L(1)\to 0$$

$$0\to L(1)\to Z(1)\to k\to 0.$$

Given these two short exact sequences we have two long exact sequences in cohomology:

$$\to H^i(V(g),k)\to H^i(V(g),Z(0))\to H^i(V(g),L(1))\to H^{i+1}(V(g),k)\to \quad (2.2)$$

$$\to H^i(V(g),L(1))\to H^i(V(g),Z(1))\to H^i(V(g),k)\to H^{i+1}(V(g),L(1))\to$$

By applying Frobenius reciprocity for each $i = 0, 1, \ldots p-1$ we have $H^j(V(g), Z(i)) \cong H^j(V(b^+), \lambda_i)$. Now by using the assumption given in the statement of the theorem one can obtain two five term exact sequences (2.3) from the long exact sequence above (2.2):

$$\begin{aligned} 0 \to H^n(V(g), L(1)) \to H^{n+1}(V(g), k) \to H^{n+1}(V(b^+), \lambda_0) \\ \to H^{n+1}(V(g), L(1)) \to H^{n+2}(V(g), k) \to 0 \end{aligned} \tag{2.3a}$$

$$\begin{aligned} 0 \to H^m(V(g), k) \to H^{m+1}(V(g), L(1)) \to H^{m+1}(V(b^+), \lambda_1) \\ \to H^{m+1}(V(g), k) \to H^{m+2}(V(g), L(1)) \to 0 \end{aligned} \tag{2.3b}$$

for n odd and m even. Note that $H^0(V(g), L(1)) = Hom_{V(g)}(k, L(1)) = 0$ which implies that $H^1(V(g), k) = 0$ by the exact sequence (2.3a) (i.e. n=-1). From the exact sequence (2.3b) it follows that $H^2(V(g), L(1)) = 0$. By continuing this process one can easily see that $H^n(V(g), k) = 0$ for n odd and $H^n(V(g), L(1)) = 0$ for n even. Consequently, by using the exact sequences (2.3a-b) and the even and odd vanishing we can conclude that

$$\begin{aligned} H^{n+1}(V(g), k) \cong H^{n+1}(V(b^+), k) \oplus H^n(V(g), L(1)) \\ H^{m+1}(V(g), L(1)) \cong H^{m+1}(V(b^+), \lambda_1) \oplus H^m(V(g), k) \end{aligned} \tag{2.4}$$

for n odd and m even. By combining these two statements (2.4) we have

$$\begin{aligned} H^n(V(g), k) \cong H^n(V(b^+), k) \oplus H^{n-1}(V(b^+), \lambda_1) \oplus H^{n-2}(V(g), k) \\ H^m(V(g), L(1)) \cong H^m(V(b^+), \lambda_1) \oplus H^{m-1}(V(b^+), k) \oplus H^{m-2}(V(g), L(1)) \end{aligned}$$

for n even and m odd. The result now follows by using induction on n and m.◇

We can now present the following theorem which gives the dimensions of the cohomology groups $H^*(V(W(1,1)), k)$ for $p = 5$ and 7 by combining the results in Proposition 2.1 and Theorem 2.2.

Theorem 2.3 Let $g = W(1,1)$ and $p = 5$ or 7. Then there exists the following isomorphism as vector spaces:

$$H^n(V(g), k) = \begin{cases} S^{\frac{n}{2}}(u^\sharp) \oplus (\bigoplus_{i=1}^{3} S^{\frac{n-2}{2}}(u^\sharp)) & n \text{ even}, p = 5 \\ \oplus \bigoplus_{i=1}^{4} (\oplus_{j=2}^{\frac{n}{2}} S^{\frac{n-2j}{2}}(u^\sharp)) & \\ S^{\frac{n}{2}}(u) \oplus (\bigoplus_{i=1}^{3} S^{\frac{n-2}{2}}(u^\sharp)) & n \text{ even}, p = 7 \\ \oplus(\bigoplus_{i=1}^{5} S^{\frac{n-4}{2}}(u^\sharp)) \oplus \bigoplus_{i=1}^{6} (\oplus_{j=3}^{\frac{n}{2}} S^{\frac{n-2j}{2}}(u^\sharp)) & \\ 0 & n \text{ odd} \end{cases}$$

$$H^n(V(g), L(1)) = \begin{cases} (\bigoplus_{i=1}^{2} S^{\frac{n-1}{2}}(u^\sharp)) \oplus (\bigoplus_{i=1}^{4} S^{\frac{n-3}{2}}(u^\sharp)) & n \text{ odd}, p = 5 \\ \oplus \bigoplus_{i=1}^{4} (\oplus_{j=2}^{\frac{n-1}{2}} S^{\frac{n-1-2j}{2}}(u^\sharp)) & \\ (\bigoplus_{i=1}^{2} S^{\frac{n-1}{2}}(u^\sharp)) \oplus (\bigoplus_{i=1}^{4} S^{\frac{n-3}{2}}(u^\sharp)) & n \text{ odd}, p = 7 \\ \oplus(\bigoplus_{i=1}^{6} S^{\frac{n-5}{2}}(u^\sharp)) \oplus \bigoplus_{i=1}^{6} (\oplus_{j=3}^{\frac{n-1}{2}} S^{\frac{n-1-2j}{2}}(u^\sharp)) & \\ 0 & n \text{ even} \end{cases}$$

It is interesting to note that one can provide explicit polynomial expressions for the dimensions of the graded components of the cohomology for $p = 5$. These polynomials are given by $\dim_k H^n(V(g), k) = \frac{1}{6}(n+2)(n^2+n+3)$ for n even and $\dim_k H^n(V(g), L(1)) = \frac{1}{24}(n+1)(4n^2-n+21)$ for n odd.

References

[AJ] H. H. Andersen, J. C. Jantzen, Cohomology of induced representations for algebraic groups, *Math. Ann.* **269** (1984), 487-525.

[BW] R. E. Block, R. L. Wilson, Classification of restricted simple Lie algebras, *J. Algebra* **114** (1988), 115-259.

[FP1] E. M. Friedlander, B. J. Parshall, Cohomology of Lie algebras and algebraic groups, *Amer. J. Math.* **108** (1986), 225-253.

[FP2] E. M. Friedlander, B. J. Parshall, Support varieties for restricted Lie algebras, *Invent. Math.* **86** (1986), 553-562.

[FP3] E. M. Friedlander, B. J. Parshall, Geometry of p-unipotent Lie algebras, *J. Algebra* **109** (1987), 25-45.

[HolN] R. R. Holmes, D. K. Nakano, Block degeneracy and Cartan invariants for graded Lie algebras of Cartan type, *J. Algebra* **161** (1993), 155-170.

[LN1] Z. Lin, D. K. Nakano, Algebraic group techniques in the representation and cohomology theory of Lie algebras of Cartan type, *preprint*.

[LN2] Z. Lin, D. K. Nakano, Modular representation theory for Lie algebras of Cartan type, *preprint*.

[N] D. K. Nakano, Projective modules over Lie algebras of Cartan type, *Memoirs of AMS* **98** (**470**) (1992).

[W] R. L. Wilson, Automorphisms of graded Lie algebras of Cartan type, *Comm. Alg.* **3** (**7**) (1975), 591-613.

3-GRADED LIE ALGEBRAS AND JORDAN PAIRS

ERHARD NEHER*
Department of Mathematics, University of Ottawa
585 King Edward, PO Box 450 STN A
Ottawa, ON K1N 6N5, Canada

Abstract. The well-known Kantor-Koecher-Tits construction associates to every Jordan pair V a 3-graded Lie algebra $L = L_1 \oplus L_0 \oplus L_{-1}$ with $(L_1, L_{-1}) = V$. We use this to describe Lie algebras graded by 3-graded root systems: they are exactly the central extensions of Kantor-Koecher-Tits algebras of Jordan pairs covered by a grid.

1. 3-graded Lie algebras and Jordan pairs. All algebraic structures will be over a ring k containing $\frac{1}{2}$ and $\frac{1}{3}$. The two basic structures we will consider are 3-graded Lie algebras and (linear) Jordan pairs.

Recall ([L]) that a *(linear) Jordan pair* is a pair of k-modules $V = (V^+, V^-)$ together with two trilinear maps ($\sigma = \pm$)

$$\{\ldots\} : V^\sigma \times V^{-\sigma} \times V^\sigma \to V^\sigma : (x,y,z) \to \{xyz\} =: D(x,y)z$$

which have the following properties:

(1) $$\{xyz\} = \{zyx\}, \text{ and}$$

(2) $$[D(x,y), D(u,v)] = D(\{xyu\}, v) - D(u, \{yxv\})$$

for all $x, u \in V^\sigma$ and $y, v \in V^{-\sigma}$, $\sigma = \pm$.

A Lie algebra L over k will be called *3-graded* if it has a decomposition $L = L_1 \oplus L_0 \oplus L_{-1}$ such that $[L_i\, L_j] \subset L_{i+j}$, where $L_k = 0$ if $k \notin \{\pm 1, 0\}$, and $[L_1\, L_{-1}] = L_0$. There is a close connection between 3-graded Lie algebras and Jordan pairs.

First of all, to any 3-graded Lie algebra one can associate a Jordan pair (L_1, L_{-1}) by defining, for $\sigma = \pm 1$, trilinear maps

(3) $$\{\ldots\} : L_\sigma \times L_{-\sigma} \times L_\sigma \to L_\sigma : (x,y,z) \to \{x\,y\,z\} := D(x,y)z := [[x\,y]\,z].$$

One easily verifies that (1) and (2) are satisfied. Conversely, every Jordan pair arises as (L_1, L_{-1}) of some 3-graded Lie algebra, as we will explain now. For a Jordan pair $V = (V^+, V^-)$ and $(x,y) \in V$ put $\delta(x,y) = (D(x,y), -D(y,x))$. We denote by $\mathcal{D}$ the k-span of all maps $\delta(x,y)$. It follows from (2) that $\mathcal{D}$ is a subalgebra of the Lie algebra $gl(V^+) \times gl(V^-)$ consisting of derivations of V. On $\mathcal{K} = \mathcal{K}(V) = V^+ \oplus \mathcal{D} \oplus V^-$ we define a product by

(4) $$[x^+ \oplus a \oplus x^-, y^+ \oplus b \oplus y^-] = (a_+ y^+ - b_+ x^+) \oplus ([ab] + \delta(x^+, y^-) - \delta(y^+, x^-)) \oplus (a_- y^- - b_- x^-)$$

where $x^\sigma, y^\sigma \in V^\sigma$ and $a = (a_+, a_-)$, $b = (b_+, b_-) \in \mathcal{D}$. One can show that (4) defines a Lie algebra on $\mathcal{K}$. Since this construction originates from the fundamental papers [Ka1-3], [Ko] and [T], we call $\mathcal{K}$ the *Kantor-Koecher-Tits* algebra of V. It

* Partially supported by NSERC-operating grant A 8836.

S. González (ed.), Non-Associative Algebra and Its Applications, 296–299.

is easily checked that with $\mathcal{K}_1 = V^+$, $\mathcal{K}_0 = \mathcal{D}$, $\mathcal{K}_{-1} = V^-$, $\mathcal{K}$ becomes a 3-graded Lie algebra whose associated Jordan pair $(\mathcal{K}_1, \mathcal{K}_{-1})$ coincides with the given Jordan pair V.

If L is a 3-graded Lie algebra with associated Jordan pair V, the KKT-algebra of V is in general not isomorphic to L. But at least L is a central extension of $\mathcal{K}(V)$: $C = \{x \in L_0 \,|\, [xL_1] = 0 = [xL_{-1}]\}$ is a central ideal of L and $L/C \approx \mathcal{K}(V)$. Hence, modulo central extensions, Jordan 3-graded Lie algebras and Jordan pairs are the "same"[1].

2. 3-graded root systems and grids in Jordan pairs. Let R be a root system as defined in [N2], for example a finite root system in the usual sense ([Bo]). For $\alpha, \beta \in R$ we put $<\alpha, \beta> = 2\frac{(\alpha,\beta)}{(\beta,\beta)}$ where $(.,.)$ denotes the scalar product of the surrounding Euclidian space of R. We call (R, R_1) a *3-graded root system* if R has a disjoint decomposition $R = R_1 \cup R_0 \cup R_{-1}$ such that $R_{-1} = -R_1$, $R_0 = \{\alpha - \beta | \alpha, \beta \in R_1, <\alpha, \beta> \neq 0, \alpha \neq \beta\}$ and $(R_i + R_j) \cap R \subset R_{i+j}$, where $R_k = \emptyset$ if $k \notin \{\pm 1, 0\}$. One can show that a reduced root system can be 3-graded if and only if none of its irreducible components are isomorphic to E_8, F_4 or G_2.

For the precise definition of a grid, the reader is referred to [N1] or [N3]. We only recall here that a grid in a Jordan pair V is a special family of idempotents of V which have compatible Peirce decompositions. To every grid $\mathcal{G}$ one can associate a (up to isomorphism) unique 3-graded root system (R, R_1) and a bijection $R_1 \to \mathcal{G} : \alpha \mapsto e_\alpha$ such that $\{e^\sigma_\alpha e^{-\sigma}_\alpha e^\sigma_\beta\} = <\beta, \alpha> e^\sigma_\beta$ for all $\alpha, \beta \in R_1$ ([N3]). Every 3-graded root system is associated to some grid.

3. Lie algebras graded by root systems. Let R be a root system. A Lie algebra L over k is called R *-graded* if there exist submodules L^α, $\alpha \in R \cup \{0\}$, such that

(5.a) $L = \oplus_{\alpha \in R \cup \{0\}} L^\alpha$;

(5.b) for all $\alpha, \beta \in R \cup \{0\}$,
$$[L^\alpha L^\beta] \subset \begin{cases} L^{\alpha+\beta} & if \alpha + \beta \in R \cup \{0\} \\ \{0\} & if \alpha + \beta \notin R \cup \{0\} \end{cases};$$

(5.c) as Lie algebra, L is generated by $\cup_{\alpha \in R} L^\alpha$;

(5.d) for every $\alpha \in R$ there exists $0 \neq X_\alpha \in L^\alpha$ such that $H_\alpha := [X_{-\alpha} X_\alpha]$ operates on L^β ($\beta \in R \cup \{0\}$) by
$$[H_\alpha z_\beta] = <\beta, \alpha> z_\beta, \; (z_\beta \in L^\beta).$$

Remarks. 1) Lie algebras graded by finite root systems were first introduced by R. V. Moody and S. Berman [BM] in the setting of Lie algebras over fields of characteristic 0. It is easily seen that the definition given above is equivalent to the one in [BM] (or in [BZ]). The interest of Berman and Moody in these types of Lie algebras came from their close connection to the intersection matrix algebras of Slodowy. Later, it turned out that Lie algebras graded by root systems are

[1] One has to be careful here, the functor $L \to (L_1, L_{-1})$ does not give an equivalence of categories since arbitrary homomorphisms between Jordan pairs do not necessarily lift to Lie algebras homomorphisms of the corresponding KKT-algebras.

an essential ingredient for the structure theory of elliptic quasi-simple Lie algebras ([BGK] and [BGKN]). Another reason why R-graded Lie algebras are an interesting class of Lie algebras is that they generalize important classes of Lie algebras, see the examples in 2).

2) Let k be a field of characteristic 0. Every simple finite-dimensional Lie algebra over k which contains a non-zero toral subalgebra is graded by a root system [Se], e.g. every simple Lie algebra over an algebraically closed field of characteristic 0. Other examples are obtained as follows. Let g be the split semisimple Lie algebra over k with root system R and let K be an associative commutative k-algebra. Then $K \otimes g$ with the natural Lie algebra structure is R-graded, and so is every central extension of $K \otimes g$. For example, every untwisted affine Lie algebra is graded by a root system. In general, not every R-graded Lie algebra will be of this form. However, modulo central extensions, this is the case if R is of type D or E [BM].

3) Lie algebras graded by a root system are always perfect: $L = [L\, L]$.

As indicated in the examples, the class of Lie algebras graded by a root system is invariant under taking central extensions:

Proposition. *Let $\tilde{L} \to L$ be a central extension of perfect Lie algebras. Suppose that $i+j$ is invertible in k for all $i,j \in \{< \alpha, \beta > \in \mathbb{Z}; \beta \in R,\ \alpha \in R \cup \{0\}\}$ and $i \neq j$ (for a 3-graded root system this condition is fulfilled if $\frac{1}{2}$ and $\frac{1}{3} \in k$). Then $\tilde{L}$ is R-graded if and only if L is R-graded.*

Consequently, one has a 2-step program for classifying Lie algebras graded by root systems: first classify up to central extensions and then classify the central extensions of the models obtained in the first step.

4. Main Result. One can use the close relation between 3-graded Lie algebras and Jordan pairs (see **1.**) to classify Lie algebras graded by 3-graded root systems, up to central extensions.

Theorem. *Let (R, R_1) be a 3-graded root system. Then a Lie algebra L is R-graded if and only if L is a central extension of the Kantor-Koecher-Tits algebra of a Jordan pair covered by a grid.*

The proof of this theorem is an extension of the methods developed in [N4]. In one direction, if L is an R-graded Lie algebra put

$$L_i = \oplus_{\alpha \in R_i} L^\alpha \ (i = \pm 1)\ ,\ L_0 = L^0 \oplus (\oplus_{\alpha \in R_0} L^\alpha)\,.$$

Then one shows that $L = L_1 \oplus L_0 \oplus L_{-1}$ is a 3-grading of L (hence L is a central extension of the Kantor-Koecher-Tits algebra of the Jordan pair $V = (L_1, L_{-1})$). Moreover, for X_α as in (5.d) one proves that $\{(X_\alpha, -X_{-\alpha})\,;\ \alpha \in R_1\} \subset V$ is a grid which covers V and whose associated 3-graded root system is (R, R_1). For the other direction, it is enough to prove that the Kantor-Koecher-Tits algebra of a Jordan pair covered by a grid is R-graded. For details see [N5].

This theorem can be used to describe Lie algebras graded by 3-graded root systems, up to central extensions: one first classifies Jordan pairs covered by a grid (- a classification which can be obtained from the classification of Jordan triple systems covered by a grid [N1]) and then describes their Kantor-Koecher-Tits algebras. Details are also contained in [N5].

5. Concluding remarks. In [BM], S. Berman and R. V. Moody described Lie algebras over fields of characteristic 0 graded by simply-laced root systems $\neq A_1$, up to central extensions. At the conference, G. Benkart announced a classification of Lie algebras graded by non-simply-laced root systems which she had obtained in joint work with E. Zelmanov ([BZ]). They use the Kantor-Koecher-Tits construction for type C and a generalized Freudenhal-Tits construction for the other types (B_n, F_4 and G_2).

Both the work of Berman-Moody and of Benkart-Zelmanov cover types which are out of reach of our methods (E_8 for Berman-Moody, G_2 and F_4 for Benkart-Zelmanov). On the other side, these authors work over fields and can therefore make use of representation theory while we work over rings where representation theory does not seem to be available. Perhaps the more important difference is that their work is focused on the classification while our theorem above provides a general description of Lie algebras graded by 3-graded root systems.

References

[BM] Berman, S. and Moody V. R., *Lie algebras graded by finite root systems and the intersection matrix algebras by Slodowy,* Invent. Math. **108**, 323 - 347 (1992).

[BGK] Berman, S., Gao, Y., Krylіouk, I.,*Quantum tori and the structure of elliptic quasi-simple Lie algebras.* Preprint.

[BGKN] Berman, S., Gao, Y., Kryliouk, I. and Neher, E.,*The alternative quantum torus and the structure of elliptic quasi-simple Lie algebras of type A_2.* In preparation.

[BZ] Benkart, G and Zelmanov, E., *Lie algebras graded by root systems.* In preparation.

[Bo] Bourbaki, N., *Groupes et Algèbres de Lie,* Chap. VI, Masson, Paris 1981.

[Ka1] Kantor, I.L, *Classification of irreducible transitive differential groups,* Dokl. Akad. Nauk SSSR **158** (1964), 1271-1274.

[Ka2] Kantor, I.L., *Non-linear transformation groups defined by general norms of Jordan algebras,* Dokl. Akad. Nauk SSSR **172** (1967), 176-180.

[Ka3] Kantor, I.L., *Some generalizations of Jordan algebras,* Trudy Sem. Vektor. Tenzor. Anal. Vyp. **16** (1972), 407-499 (Russian).

[Ko] Koecher, M., *Imbedding of Jordan algebras into Lie algebras* I, Amer. J. Math. **89** (1967), 787-816; II, Amer. J. Math **90** (1968), 476-510.

[L] Loos, O., *Jordan pairs,* Lect. Notes Math. vol. 460, Berlin-Heidelberg: Springer 1975.

[N1] Neher, E.,*Jordan Triple Systems by the Grid Approach,* Lect. Notes Math. vol. **1280**, Berlin-Heidelberg: Springer 1987.

[N2] Neher, E.,*Systèmes de racines 3-gradués,* C. R. Acad. Sci. Paris, t. **310**, Série I. 687-690 (1990).

[N3] Neher, E., *3-graded root systems and grids in Jordan triple systems,* J. of Algebra **140**, 284-329 (1991).

[N4] Neher, E., *Generators and Relations for 3-Graded Lie algebras,* J. of ALgebra **155**, 1-35 (1993).

[N5] Neher, E., *Lie algebras graded by 3-graded root systems and Jordan pairs covered by grids,* preprint.

[Se] Seligman, G. B.,*Rational Methods in Lie Algebras,* Lecture Notes in Pure and Applied Mathematics vol. **17**, New York, Marcel Dekker 1976.

[T] Tits, J., *Une classe d'algèbres de Lie en relation avec les algèbres de Jordan,* Nederl. Akad. Wetensch. Proc. Ser. A **65**= *Indag.* Math. **24** (1962), 530-535.

SUPER-TRIPLE SYSTEMS, NORMAL AND CLASSICAL YANG-BAXTER EQUATIONS

SUSUMU OKUBO
Department of Physics and Astronomy
University of Rochester
Rochester, NY 14627

Abstract. The Yang-Baxter as well as classical Yang-Baxter equations have been recast as triple product equations, and some solutions of these equations are obtained.

Key words: Yang-Baxter equation, triple product systems

1. Introduction

1.1. YANG-BAXTER EQUATION

Let V be a finite-dimensional vector space over a field F, and let $e_1, e_2, \ldots, e_N$ with $N = \text{Dim } V$ be a basis of V. For scattering matrix elements $R^{ab}_{cd}(\theta)$ $(a,b,c,d = 1,2,\ldots,N)$ for a parameter θ, we introduce a linear mapping in $V \otimes V$ by

$$R(\theta)\ e_a \otimes e_b = \sum_{c,d=1}^{N} R^{dc}_{ab}(\theta)\ e_c \otimes e_d \quad . \tag{1.1}$$

For a tensor product space

$$V_n = V \otimes V \otimes \ldots \otimes V \qquad (n - \text{times}) \quad , \tag{1.2}$$

we then define linear mappings $R_{jk}(\theta)$ $(j < k)$ for $j,k = 1,2,\ldots,n$ similarly to Eq. (1.1) when it operates only to the jth and kth sub-spaces as

$$\begin{aligned} &R_{jk}(\theta) e_{a_1} \otimes e_{a_2} \otimes \ldots \otimes e_{a_n} \\ &= \sum_{c,d=1}^{N} R^{dc}_{a_j a_k}(\theta)\ e_{a_1} \otimes \ldots \otimes e_c \otimes \ldots \otimes e_d \otimes \ldots \otimes e_{a_n} \quad . \end{aligned} \tag{1.3}$$

The Yang-Baxter equation [1] is then the relation

$$R_{12}(\theta) R_{13}(\theta') R_{23}(\theta'') = R_{23}(\theta'') R_{13}(\theta') R_{12}(\theta) \tag{1.4a}$$

$$\theta' = \theta + \theta'' \tag{1.4b}$$

for $n = 3$.

When $R_{jk}(\theta) = R_{jk}$ is independent of the parameter θ, then we call the solution of Eqs. (1.4) to be a constant or θ-independent solution. As we will see shortly, the

S. González (ed.), Non-Associative Algebra and Its Applications, 300–308.

constant solution is not necessarily a special case of θ-dependent solutions of Eqs. (1.4) in which we set $\theta = 0$ or $\theta = \infty$. The θ-dependent Yang-Baxter equation occurs in exactly solvable one-dimensional quantum field theory, and in exactly solvable statistical mechanics, while the θ-independent case is important for discussions of Hopf algebra, quantum-group, and knot group [1] and [2].

It is of some interest also to consider the so-called classical Yang-Baxter equation where Eq. (1.4a) is now replaced by

$$[R_{12}(\theta), R_{13}(\theta')] + [R_{12}(\theta), R_{23}(\theta'')] + [R_{13}(\theta'), R_{23}(\theta'')] = 0 \quad . \tag{1.5}$$

The θ-independent solution of the classical Yang-Baxter equation is relevant to discussions of Lie bi-algebras.

The purpose of this note is first to rewrite both ordinary and classical Yang-Baxter equations as triple product equations and then to proceed to solve them for some θ-dependent solutions. The case for θ-independent solutions will be discussed elsewhere.

1.2. Super-Space

In order to maintain generality, we suppose that the underlying vector space V admits a Z_2-grading with

$$V = V_B \oplus V_F \tag{1.6}$$

and set

$$N_B = \text{ Dim } V_B \quad , \quad N_F = \text{ Dim } V_F \quad , \tag{1.7}$$

where the suffices B and F refer to bosonic and fermionic spaces. We also write

$$N = N_B + N_F \quad , \tag{1.8a}$$

$$N_0 = N_B - N_F \quad . \tag{1.8b}$$

Next, we introduce the parity (or signature) function $\sigma(x)$ in V by

$$\sigma(x) = \begin{cases} 0 \, , & \text{if } x \in V_B \\ 1 \, , & \text{if } x \in V_F \end{cases} \quad . \tag{1.9}$$

Hereafter, we consider only homogeneous elements of V, i.e. we have either $x \in V_B$ or $x \in V_F$.

The super-symmetric form $< x|y >$ in V is a bilinear form satisfying

$$\text{(i) } < x|y > = 0 \quad , \quad \text{if} \quad \sigma(x) \neq \sigma(y) \tag{1.10a}$$

$$\text{(ii) } < y|x > = (-1)^{xy} < x|y > \quad . \tag{1.10b}$$

Here and hereafter, we have, for simplicity, written

$$(-1)^{\sigma(x)\sigma(y)} = (-1)^{xy} \tag{1.11}$$

and assume that V always possesses a non-degenerate super-symmetric form. For a basis vectors $e_1, \, e_2, \, \ldots, \, e_N$ of V, g_{jk} defined by

$$g_{jk} = < e_j | e_k > \tag{1.12}$$

has its inverse tensor g^{jk} satisfying

$$\sum_{\ell=1}^{N} g_{j\ell} g^{\ell k} = \sum_{\ell=1}^{N} g^{k\ell} g_{\ell j} = \delta_j^k \quad . \tag{1.13}$$

Introducing e^j by

$$e^j = \sum_{k=1}^{N} g^{jk} e_k \quad , \quad e_j = \sum_{k=1}^{N} g_{jk} e^k \quad , \tag{1.14}$$

we have

$$\sigma(e_j) = \sigma(e^j) = \sigma_j \quad . \tag{1.15}$$

Moreover, any $x \in V$ can be expanded as

$$x = \sum_{j=1}^{N} e_j < e^j | x > = \sum_{j=1}^{N} < x | e_j > e^j \quad . \tag{1.16}$$

Any triple product such as $x\ y\ z$ defined in V is also assumed to satisfy the condition

$$\sigma(xyz) = \{\sigma(x) + \sigma(y) + \sigma(z)\} \pmod 2 \quad . \tag{1.17}$$

Finally, the scattering matrix elements $R_{cd}^{ab}(\theta)$ are required to obey the conservation law

$$R_{cd}^{ab}(\theta) = 0 \quad , \quad \text{if} \quad \sigma(a) + \sigma(b) \neq [\sigma(c) + \sigma(d)] \pmod 2 \quad . \tag{1.18}$$

2. Yang-Baxter Equations as Triple Product Equations

We introduce two θ-dependent triple products $[z, x, y]_\theta$ and $[z, x, y]_\theta^*$ in V by

$$\left[e^b, e_c, e_d\right]_\theta = \sum_{a=1}^{N} (-1)^{ab} R_{cd}^{ab}(\theta) e_a \quad , \tag{2.1}$$

$$[e^a, e_d, e_c]_\theta^* = \sum_{b=1}^{N} (-1)^{cd} R_{cd}^{ab}(\theta) e_b \tag{2.2}$$

or equivalently by

$$R_{cd}^{ab}(\theta) = (-1)^{ab} < e^a | [e^b, e_c, e_d]_\theta > \tag{2.3}$$

$$= (-1)^{cd} < e^b | [e^a, e_d, e_c]_\theta^* > \quad . \tag{2.4}$$

We note that the condition Eq. (1.17) is satisfied for these triple products in view of our ansatz Eq. (1.18).

From Eqs. (1.16), (2.3) and (2.4), we can rewrite Eq. (1.1) as

$$R(\theta)x \otimes y = (-1)^{xy} \sum_{j=1}^{N} [e^j, y, x]^*_\theta \otimes e_j$$

$$= \sum_{j=1}^{N} (-1)^{\sigma_j(x+y+\sigma_j)} e_j \otimes [e^j, x, y]_\theta \quad . \tag{2.5}$$

It is now then straightforward to see that the Yang-Baxter equation, Eq. (1.4a) becomes a triple product relation

$$\sum_{j=1}^{N} (-1)^{xz+\sigma_j z} [v, [u, e_j, z]_{\theta'}, [e^j, x, y]_\theta]^*_{\theta''}$$

$$= \sum_{j=1}^{N} (-1)^{uv+\sigma_j x} [u, [v, e_j, x]^*_{\theta'}, [e^j, x, y]^*_{\theta''}]_\theta \tag{2.6}$$

while the classical Yang-Baxter equation, Eq. (1.5) is rewritten as

$$(-1)^{xz} [u, [v, z, x]^*_{\theta'}, y]_\theta + (-1)^{yz} [u, x, [v, z, y]^*_{\theta''}]_\theta$$

$$+ \sum_{j=1}^{N} (-1)^{yz+u(y+z)+\sigma_j(x+u)} < u | [v, e_j, x]^*_{\theta'} > [e^j, x, y]^*_{\theta''}$$

$$= (-1)^{uv+y(x+z)} [v, [u, x, z]_{\theta'}, y]^*_{\theta''} + (-1)^{uv+z(x+y)} [v, z, [u, x, y]_\theta]^*_{\theta''}$$

$$+ \sum_{j=1}^{N} (-1)^{uv+(v+z)(x+y+\sigma_j)} < v | [u, e_j, z]_{\theta'} > [e^j, x, y]_\theta \quad . \tag{2.7}$$

The interesting case is when we have the identity

$$[z, x, y]^*_\theta = [z, x, y]_\theta \tag{2.8}$$

so that Eq. (2.6) becomes

$$\sum_{j=1}^{N} (-1)^{xz+\sigma_j z} [v, [u, e_j, z]_{\theta'}, [e^j, x, y]_\theta]_{\theta''}$$

$$= \sum_{j=1}^{N} (-1)^{uv+\sigma_j x} [u, [v, e_j, x]_{\theta'}, [e^j, z, y]_{\theta''}]_\theta \tag{2.9}$$

which has been studied in [3], [4] and [5]. Note that Eq. (2.9) is invariant under

$$u \leftrightarrow v \quad , \quad x \leftrightarrow z \quad , \quad \theta \leftrightarrow \theta'' \quad . \tag{2.10}$$

The condition Eq. (2.8) will be satisfied for all examples given in sections 3 and 4.

3. Ortho-Symplectic Super-Triple System

Let V be the vector space with the non-degenerate super-symmetric bilinear form $< x|y >$. For a given constant λ, suppose that a triple linear product $x\ y\ z$ satisfies the following axioms:

(i) $xyz + (-1)^{xy} yxz = 0$ (3.1)

(ii) $xyz + (-1)^{yz} xzy = 2\lambda < y|z > x - \lambda(-1)^{yz} < x|z > y - \lambda < x|y > z$ (3.2)

(iii) $uv(xyz) = (uvx)yz + (-1)^{(u+v)x} x(uvy)z + (-1)^{(u+v)(x+y)} xy(uvz)$ (3.3)

(iv) $< uvx|y > = -(-1)^{(u+v)x} < x|uvy >$. (3.4)

Actually, if $\lambda \neq 0$, then Eq. (3.4) is a consequence of other postulates. We call the vector space V with the triple product $x\ y\ z$ satisfying these axioms to be a ortho-symplectic super-triple system. For the special case of $V_F = 0$, it will reproduce the orthogonal triple system (OTS) defined in [3], [4], while the case of $V_B = 0$ will lead to the symplectic triple system (STS) of Yamaguchi and Asano [6] which is equivalent to the balanced ternary algebra of Faulker and Ferrar [7].

We can construct Lie-super triple systems from any ortho-symplectic super-triple system in the following two different ways. Let us introduce a new triple product $x \cdot y \cdot z$ in the same vector space V by

$$x \cdot y \cdot z = \sum_{j=1}^{N} (xye_j)e^j z + \frac{1}{3}\,\lambda(N_0 - 16)xyz \quad , \tag{3.5}$$

where $N_0 = N_B - N_F$ as in Eq. (1.8b). Generalizing the result of [4], we can prove [5] that $x \cdot y \cdot z$ is a Lie-super triple product, i.e. it satisfies

(i) $x \cdot y \cdot z + (-1)^{xy} y \cdot x \cdot z = 0$ (3.6a)

(ii) $(-1)^{xz} x \cdot y \cdot z + (-1)^{yx} y \cdot z \cdot x + (-1)^{zy} z \cdot x \cdot y = 0$ (3.6b)

(iii) $u \cdot v \cdot (x \cdot y \cdot z) = (u \cdot v \cdot x) \cdot y \cdot z + (-1)^{(u+v)x} x \cdot (u \cdot v \cdot y) \cdot z$
$\qquad + (-1)^{(u+v)(x+y)} x \cdot y \cdot (u \cdot v \cdot z)$. (3.6c)

However, it is possible for some cases to have $x \cdot y \cdot z = 0$ identically as we will show shortly.

The second way is a straightforward generalization of the method due to Yamaguchi and Asano [6] for STS who constructed a Lie triple system (LTS) in the space $V \oplus V$. For any pairs x_j, y_j, and $z_j \in V$ $(j = 1, 2)$ such that

$$\sigma(x_1) = \sigma(x_2) \equiv \sigma(x) \quad , \quad \sigma(y_1) = \sigma(y_2) \equiv \sigma(y) \quad ,$$
$$\sigma(z_1) = \sigma(z_2) \equiv \sigma(z) \quad , \tag{3.7}$$

we set

$$X = \begin{pmatrix} x_1 \\ x_2 \end{pmatrix} \quad , \quad Y = \begin{pmatrix} y_1 \\ y_2 \end{pmatrix} \quad , \quad Z = \begin{pmatrix} z_1 \\ z_2 \end{pmatrix} \tag{3.8}$$

and assign their parities oppositely now by

$$\sigma(X) = 1 - \sigma(x) \quad , \quad \sigma(Y) = 1 - \sigma(y) \quad , \quad \sigma(Z) = 1 - \sigma(z) \quad . \tag{3.9}$$

We introduce a new triple product $X\ Y\ Z$ by

$$W = XYZ = \begin{pmatrix} w_1 \\ w_2 \end{pmatrix} \tag{3.10}$$

with

$$w_1 = x_1 y_2 z_1 - x_2 y_1 z_1 - \lambda < x_1|y_2 > z_1 - \lambda < x_2|y_1 > z_1 + 2\lambda < x_1|y_1 > z_2 \quad , \tag{3.11a}$$

$$w_2 = x_1 y_2 z_2 - x_2 y_1 z_2 + \lambda < x_1|y_2 > z_2 + \lambda < x_2|y_1 > z_2 - 2\lambda < x_2|y_2 > z_1 \quad . \tag{3.11b}$$

We can show that these define a Lie-super triple system.

As a example of non-trivial ortho-symplectic super triple system, let $< x|y >$ be a super-symmetric bilinear form as in section 1b. Let J be a linear map in V, satisfying conditions

(i) $\sigma(Jx) = \sigma(x)$ (3.12a)

(ii) $J^2 = \lambda I_d$ (3.12b)

(iii) $< x|Jy > = - < Jx|y >$ (3.13c)

for any $x,\ y \in V$, where λ is a constant. Setting

$$\begin{aligned} xyz = {} & Jx < y|Jz > \ +Jz < x|Jy > \ +(-1)^{z(x+y)} Jy < z|Jx > \\ & +\lambda\{< y|z > x - (-1)^{z(x+y)} < z|x > y\} \quad , \end{aligned} \tag{3.14}$$

we can verify that this defines a ortho-symplectic super-triple system. Then, $x \cdot y \cdot z$ defined by Eq. (3.5) is easily calculated to yield

$$x \cdot y \cdot z = C\ x * y * z \tag{3.15}$$

where the constant C is given by

$$C = \frac{1}{3}\ \lambda(N_0 - 4) \tag{3.16}$$

and

$$\begin{aligned} x * y * z = {} & Jx < y|Jz > \ +(-1)^{z(x+y)} Jy < z|Jx > \ -2\ Jz < x|Jy > \\ & +\lambda\{< y|z > x - (-1)^{z(x+y)} < z|x > y\} \quad . \end{aligned} \tag{3.17}$$

Especially, if $N_0 = N_B - N_F = 4$, then we have $x \cdot y \cdot z = 0$. Other examples of similar kinds are found in [4] and [5].

4. Solutions of Yang-Baxter Equations

4.1. Solution of Normal Yang-Baxter Equation

Let V be the ortho-symplectic super-triple system. We seek a solution of Eq. (2.9) with an ansatz of

$$[z,x,y]_\theta = P(\theta)(-1)^{(x+y)z} xyz + Q(\theta) < x|y > z + R(\theta) < z|x > y + S(\theta)(-1)^{xy} < z|y > x \tag{4.1}$$

for functions $P(\theta)$, $Q(\theta)$, $R(\theta)$, and $S(\theta)$ of θ to be determined. Inserting Eq. (4.1) into Eq. (2.9) and assuming $x \cdot y \cdot z = 0$ identically, we find after some calculation that the solution for $P(\theta) \neq 0$ is given by

$$R(\theta)/P(\theta) = a + k\theta \quad , \tag{4.2a}$$

$$Q(\theta)/P(\theta) = \lambda - \frac{2a\lambda}{2(a-\lambda)+k\theta} \quad , \tag{4.2b}$$

$$S(\theta)/P(\theta) = -2\lambda - \frac{2\lambda a}{k\theta} \quad , \tag{4.2c}$$

where k is an arbitrary constant and we have set

$$a = -\frac{1}{6}\,\lambda(N_0 - 4) = -\frac{1}{6}\,\lambda(N_B - N_F - 4) \quad . \tag{4.3}$$

We note that except for the case of $N_0 = 4$, the solution is singular at $\theta = 0$. The solution generalizes the result of [4]. Moreover, it satisfies the crossing-symmetric property as well as the unitarity relation:

$$\frac{1}{P(\overline{\theta})}\,[z,x,y]_{\overline{\theta}} = -(-1)^{xz}\frac{1}{P(\theta)}\,[x,z,y]_\theta \quad , \tag{4.4a}$$

$$k\overline{\theta} = -k\theta - 2(a-\lambda) \quad . \tag{4.4b}$$

$$R_{12}(\theta)P_{12}R_{12}(-\theta)P_{12} = f(\theta)I_d \tag{4.5a}$$

$$f(\theta) = P(\theta)P(-\theta)\left[(k\theta)^2 - 4\lambda^2\right]\left[a^2 - (k\theta)^2\right]/(k\theta)^2 \quad , \tag{4.5b}$$

where P_{12} in Eq. (4.5a) is the permutation operator defined by

$$P_{12}\, x \otimes y \otimes z = y \otimes x \otimes z \quad . \tag{4.5c}$$

Therefore, it is quite likely that the solution may correspond to some exactly solvable quantum field theory.

4.2. Solution of Classical Yang-Baxter Equation

Here, we restrict ourselves for cases of $V_B = 0$ or $V_F = 0$. Then, it is convenient to rewrite Eq. (1.10b) as

$$< y|x > = \epsilon < x|y > \tag{4.6}$$

for either $\epsilon = +1$ (orthogonal) or $\epsilon = -1$ (symplectic). Moreover, we assume the validity of

$$< u|[z,x,y]_\theta > = < z|[u,y,x]_\theta > = < x|[y,u,z]_\theta > \quad . \tag{4.7}$$

Then, it is not hard to rewrite the classical Yang-Baxter equation, Eq. (2.7) as

$$\begin{aligned}&[u,[v,z,x]_{\theta'},y]_\theta+[u,x,[v,z,y]_{\theta''}]_\theta+\epsilon[[x,u,v]_{\theta'},z,y]_{\theta''}\\&\quad=[v,[u,x,z]_{\theta'},y]_{\theta''}+[v,z,[u,x,y]_\theta]_{\theta''}+\epsilon[[z,v,u]_{\theta'},x,y]_\theta\quad,\end{aligned}\tag{4.8}$$

which is invariant again under Eq. (2.10).

A class of solutions satisfying Eqs. (4.7) and (4.8) can be found as follows. Let $x\ y\ z$ be a θ-independent triple product, satisfying conditions:

(i) $< u|zxy > = < z|uyx > = < x|yuz >$ (4.9a)

(ii) $uv(xyz) = (uvx)yz - \epsilon x(vuy)z + xy(uvz)$ (4.9b)

Note that the condition Eq. (4.9b) is one of the Kantor's relation for the Kantor triple system [8]. Then,

$$[z,x,y]_\theta=\frac{1}{k\theta}\,xzy+\beta(\theta)<z|x>y\tag{4.10}$$

offers a solution of Eqs. (4.7) and (4.8), where k is an arbitrary constant and $\beta(\theta)$ is an arbitrary function of θ. The conditions Eqs. (4.9) are satisfied by several triple products. First for $\epsilon=-1$, any symplectic triple system (STS) satisifes them. Similarly for $\epsilon=1$, they can be satisfied by the orthogonal triple system (OTS), Jordan triple system, and Lie triple systems. The fact that Eqs. (4.7) and (4.8) are satisfied by both STS and OTS is easily seen from the result of [4]. For the Jordan triple system, let

$$xyz=x(yz)-y(xz)+(xy)z\quad,\tag{4.11a}$$

$$<x|y>\equiv\ \mathrm{Tr}(R_{xy})\tag{4.11b}$$

where $xy=yx$ is the standard Jordan product. Then, we have $< xy|z > = < x|yz >$, and it can be shown that the conditions Eqs. (4.7) and (4.8) are satisfied by Eqs. (4.11). If the underlying Jordan algebra is semi-simple, then the symmetric bilinear form $< x|y >$ defined by Eq. (4.11b) is non- degenerate also. As for the Lie-triple system, we note first the canonical construction [9], [10] of a Lie algebra with Lie-product $[x,y]$ and write:

$$[x,y,z]=[[x,y],z]\quad.\tag{4.12}$$

Suppose that the Lie algebra possesses a symmetric bilinear non-degenerate form $< x|y >$ satisfying

$$<[x,y]|z>\ =\ <x|[y,z]>\quad.\tag{4.13}$$

Then, again the conditions Eqs. (4.7) and (4.8) are satisfied. We note that Eq. (4.13) is automatically obeyed, provided that the Lie algebra is semi-simple.

Another solution is also obtained as follows. Let $< x|y >$ be a symmetric bilinear non-degenerate form ($\epsilon=1$) and identify

$$xyz=\ <y|z>x\tag{4.14a}$$

which then satisfies the conditions Eqs. (4.9). Therefore, the corresponding solution is rewritten as

$$[z,x,y]_\theta=\frac{1}{k\theta}\ <y|z>x+\beta(\theta)<z|x>y\tag{4.14b}$$

This is equivalent to the usual solution of

$$R_{12}(\theta) = \frac{1}{k\theta} P_{12} + \beta(\theta) I_d \quad . \tag{4.14c}$$

Acknowledgements

This work is supported in part by U.S. Department of Energy Grant No. DE-FG02-91ER40685.

References

1. M. Jimbo (editor), Yang-Baxter Equation in Integrable Systems, World Scientific, Singapore (1989).
2. L. H. Kauffman, Knots Theory and Physics, World Scientific, Singapore (1991).
3. S. Okubo, 'Triple Products and Yang-Baxter Equation [I], Quaternionic and Octonionic Ternary Systems', *Jour. Math. Phys.* (1993) pp. 3273–3291.
4. S. Okubo, 'Triple Products and Yang-Baxter Equation [II], Orthogonal and Symplectic Ternary Systems', *Jour. Math. Phys.* (1993) pp. 3292–3315.
5. S. Okubo, 'Super Triple Systems and Applications to Para-statistics and Yang-Baxter Equation', University of Rochester Report UR-1312 (June, 1993). To appear in the Proceeding of the 15th Montreal-Rochester-Syracuse-Toronto Meeting.
6. K. Yamaguchi and H. Asano, 'On the Freudenthal's Construction of Exceptional Lie Algebras', *Proc. Japan Acad.* **51** (1972), pp. 253–258.
7. J. R. Faulkner and J. C. Ferrar, 'On the Structure of Symplectic Ternary Algebras', Nederl. Akad. Wetensch. Proc. Ser. A., 75 = *Indag. Math.* **34** (1972) pp. 247–256.
8. I. L. Kantor, 'Models of Exceptional Lie Algebras', *Soviet Math. Dokl.* **14** (1973) pp. 254–258.
9. W. G. Lister, 'A Structure Theory of Lie Triple Systems', *Trans. Amer. Math. Soc.* **72** (1952) pp. 217–242.
10. K. Yamaguchi, 'On Weak Representation of Lie Triple System', *Kumamoto J. Sci. Ser.* **A8** (1968) pp. 170–114.

A CONJECTURE ON LOCALLY NOVIKOV ALGEBRAS

J. MARSHALL OSBORN*
University of Wisconsin
Madison, WI 53706

Abstract. In this paper we define a generalization of Novikov algebras that we call locally Novikov algebras, and we construct several classes of examples of these algebras. We conjecture that every simple locally Novikov algebra over an algebraically closed field of characteristic 0 belongs to one of the classes of examples that we construct here. These examples will be used in another paper ([12]) to construct new examples of Lie algebras of characteristic 0.

1. Introduction. An algebra is (left) Novikov if it satifies the two identities

$$(x, y, z) = (y, x, z), \tag{1.1}$$

$$(xy)z = (xz)y, \tag{1.2}$$

where $(x, y, z) = (xy)z - x(yz)$. These algebras were used to study problems connected with physics both by A. A. Belinskii and S. P. Novikov [1], and by I. M. Gelfand and I. Ya. Dorfman [4]. Other papers on these algebras are listed in the bibliography. Algebras satisfying (1.1) also arise in the study of affine structures on Lie groups (for example, see [5]). If A is a Novikov algebra, the algebra A^- obtained by taking A under the new product $[x, y] = xy - yx$ is a Lie algebra. For example the Virasoro algebra arises in this manner.

By way of motivation for the definition of locally Novikov algebras, we recall the following result which was we established in [10].

Theorem 1.1. *Let F be a field of characteristic 0 with the property that each element of F has an nth root in F for every positive integer n. Let A be a simple infinite dimensional Novikov algebra containing an element e with the properties that $e^2 = be$ for some $b \in F$ and that $A = \sum_\alpha A'_\alpha$ where $A'_\alpha = \{x \in A | (L_e - (\alpha + b))^n x = 0$ for some $n\}$. Then one of the following is true:*

(i) *A has a basis $\{x_i\}_{i \geq -1}$ where products are given by $x_i x_j = (j+1)x_{i+j}$.*

(ii) *A has a basis $\{x_\alpha\}$ where α ranges over an additive subgroup Δ of F, and products are given by $x_\alpha x_\beta = (\beta + b)x_{\alpha+\beta}$.*

(iii) A has a basis $\{x_{\alpha k}\}$ where α ranges over an additive subgroup Δ of F, and k ranges over the nonnegative integers, and where products are given by

$$x_{\alpha k} x_{\beta \ell} = (\beta + b)\binom{k+\ell}{k} x_{\alpha+\beta, k+\ell} + \binom{k+\ell-1}{k} x_{\alpha+\beta, k+\ell-1}. \tag{1.3}$$

* AMS Subject Classification - Primary: 17A30, 17A60, 17A65, 17A68, 17D25

S. González (ed.), Non-Associative Algebra and Its Applications, 309–313.

The Lie algebras coming from the Novikov algebras defined in parts (i) and (ii) of Theorem 2.1 have a one-dimensional torus, and the Lie algebras from part (iii) appear to have no torus in the usual sense. We were motivated to look for a generalization of Novikov algebras in order to get examples of Lie algebras which have tori of dimension greater than one.

If A is a nonassociative algebra and if $z \in A$, let $L_z = L(z)$ and $R_z = R(z)$ denote respectively left and right multiplication by z in A. An idempotent e in A is called a *right projection* if $R_e^2 = R_e$. Two idempotents e_1 and e_2 are called *orthogonal* if $e_1e_2 = 0 = e_2e_1$. Let $e_1, e_2, \ldots, e_n$ be pairwise orthogonal right projections in A, let $\alpha = (\alpha_1, \alpha_2, \ldots, \alpha_n)$ be an n-tuple of elements from the field F over which A is defined, and let

$$A_i = \{x \in A | (R(e_i) - I)x = 0\},$$

$$A_{i\alpha} = \{x \in A_i | (L(e_j) - (\alpha_j + \delta_{ij})I)x = 0 \text{ each } j\},$$

$$A'_{i\alpha} = \{x \in A_i | (L(e_j) - (\alpha_j + \delta_{ij})I)^m x = 0 \text{ for each } j, \text{ some } m = m(j)\}.$$

We call the algebra A *locally Novikov* if

(1) A satisfies (1.1);

(2) A contains a right identity element which is the sum of n mutually orthogonal right projections $e_1, e_2, \ldots, e_n$ called the *canonical idempotents* of A;

(3) For each i such that $1 \le i \le n$, A_i satisfies (1.2);

(4) $A_i = \sum_\alpha A'_{i\alpha}$;

(5) If C, D are nonzero right ideals of A_i, then $CD \neq 0$.

(6) If $x \in A_i$, and if $y \in A_j$ for $j \neq i$, then $yx = 0$ if and only if $e_j x = 0$.

In the next section we construct our classes of locally Novikov algebras, and in the final section we conjecture that we have constructed all simple locally Novikov algebras over an algebraically closed field of characteristic zero.

2. Examples of simple Locally Novikov Algebras. For each k with $1 \le k \le n$ let Δ_k be an additive subgroup of the field F, and let N_k be the group algebra over F with basis $\{x^\mu\}$ where μ ranges over Δ_k, and with the product defined by $x^\mu x^\nu = x^{\mu+\nu}$. Let Δ be the tensor product of $\Delta_1, \Delta_2, \ldots, \Delta_n$, and let B be the tensor product of $N_1, N_2, \ldots, N_n$. Then B has a natural associative product defined by

$$(x^{\alpha_1} \otimes \ldots \otimes x^{\alpha_n})(x^{\beta_1} \otimes \ldots \otimes x^{\beta_n}) = x^{\alpha_1+\beta_1} \otimes \ldots \otimes x^{\alpha_n+\beta_n}.$$

It will be convenient to use the abbreviated notation $x^\alpha = x^{\alpha_1} \otimes \ldots \otimes x^{\alpha_n}$, so that the product in B can be expressed by $x^\alpha x^\beta = x^{\alpha+\beta}$ for $\alpha, \beta \in \Delta$.

Let $(B)_1, (B)_2, \ldots (B)_n$ be n copies of B. We can make $A = \sum_{i=1}^n (B)_i$ into an algebra by defining

$$(x^\alpha)_i (x^\beta)_j = (\beta_i + \delta_{ij})(x^{\alpha+\beta})_j, \tag{2.1}$$

and extending by linearity. We claim that A is a locally Novikov algebra. To check Property (1) of the definition, we calculate that

$$((x^\alpha)_i, (x^\beta)_j, (x^\gamma)_k) = [(\beta_i + \delta_{ij})(x^{\alpha+\beta})_j](x^\gamma)_k - (x^\alpha)_i[(\gamma_j + \delta_{jk})(x^{\beta+\gamma})_k]$$

$$= (\beta_i + \delta_{ij})(\gamma_j + \delta_{jk})x^{\alpha+\beta+\gamma} - (\beta_i + \gamma_i + \delta_{ik})(\gamma_j + \delta_{jk})x^{\alpha+\beta+\gamma}$$

$$= -(\gamma_i + \delta_{ik} - \delta_{ij})(\gamma_j + \delta_{jk})x^{\alpha+\beta+\gamma}.$$

Since the coefficient in the last expression is left unchanged if i and j are interchanged, Property (1) of the definition is satisfied.

For each i such that $1 \leq i \leq n$, let $e_i = (x^0)_i$ where x^0 by a slight abuse of notation will denote $x^0 \otimes \ldots \otimes x^0$. It is easy to check that e_i acts as a right identity on $(B)_i$, and that it right annihilates $(B)_j$ for $j \neq i$. Then $e_1, e_2, \ldots e_n$ satisfy Property (2) of the definition. For Property (3), $(B)_i$ is a Novikov algebra since

$$[(x^\alpha)_i(x^\beta)_i](x^\gamma)_i = (\beta_i + 1)(x^{\alpha+\beta})_i(x^\gamma)_i = (\beta_i + 1)(\gamma_i + 1)(x^{\alpha+\beta+\gamma})_i,$$

and this expression is unchanged when β and γ are interchanged. It follows from (2.1) that a product in A_i can only be zero if the second factor is spanned by elements $(x^\beta)_i$ with $\beta_i = -1$. But this set of elements contains no right ideals of A_i, so Property (5) is satisfied. Finally Properties (4) and (6) are immediate from (2.1). Thus A is locally Novikov.

Let J be a subset of the integers between 1 and n such that for each $k \in J$ the group Δ_k is cyclic with generator 1. Then the subalgebra A_J of A is spanned by those elements $(x^{\alpha_1} \otimes \ldots \otimes x^{\alpha_n})_i$ where $\alpha_k \geq -\delta_{ki}$ for each $k \in J$. To see that A_J is a subalgebra, we look at (2.1) and suppose that the two terms on the left are in A_J. If $\alpha_k + \beta_k < 0$, then either (i) $\alpha_k = -1$, $\beta_k = 0$, and $i \neq j$; (ii) $\alpha_k = -1$, $\beta_k = 0$, and $i = j$; (iii) $\alpha_k = 0$ and $\beta_k = -1$, or (iv) $\alpha_k = -1 = \beta_k$. In the first case, $k = i \neq j$, and so the coefficient on the right side is $\beta_i + \delta_{ij} = 0$. In cases (ii) and (iii), $\alpha_k + \beta_k = -1$, which is allowed since $k = j$. Finally in case (iv), the coefficient is $\beta_i + \delta_{ij} = 0$. Thus A_J is a subalgebra, and it is immediate that A_J satisfies Properties (1) - (6).

Theorem 2.2 *The algebras A and A_J are simple*

Proof. Since A can be regarded as the special case of A_J when $J = 0$, we can do the proof just for A_J. Let C be a nonzero ideal of A_J, and let $w = \sum_{\alpha,i} c_{\alpha,i}(x^\alpha)_i \in C$ be nonzero. Multiplying on the right and left by the different idempotents e_j, we see from (2.1) that $c_{\alpha,i}(x^\alpha)_i \in C$ for each α and i. Thus we can choose $w = (x^\beta)_k$ for some β and k. If $k \notin J$, then for any α we have $w(x^\alpha)_k = (\alpha_k + 1)(x^{\alpha+\beta})_k \in C$, and $(x^\alpha)_k w = (\beta_k + 1)(x^{\alpha+\beta})_k \in C$. Thus $(x^{\alpha+\beta})_k \in C$ unless $\alpha_k = -1 = \beta_k$. In the latter case, $w(x^{\alpha+\epsilon_k})_k = (x^{\alpha+\epsilon_k+\beta})_k \in C$, and multiplying on the left by $(x^{-\epsilon_k})_k$ gives $(x^{\alpha+\beta})_k \in C$. We conclude that $(x^\gamma)_k \in C$ for any γ, or that $(B)_k \in C$.

If $k \in J$, then β_k is an integer ≥ -1, and we want to show first that we can take $\beta_k = 0$. If $\beta_k = -1$, then $w(x^{\epsilon_k})_i$ has the desired property. If $\beta_k \geq 1$ then repeated multiplication on the left by $(x^{-\epsilon_k})_i$ will lead to an element of the right form. When $\beta_k = 0$ we may multiply on the left by $(x^\alpha)_k$ to obtain $(x^{\alpha+\beta})_k \in C$ for any α. Hence $(B)_k \in C$ in this case also.

In particular, $e_k \in C$, from which it follows using (2.1) that $(x^\alpha)_i \in C$ for any $i \neq k$ and any α with $\alpha_k \neq 0$. Thus, $(B)_i \cap C \neq 0$ for each i, and the first part of the proof shows that $(B)_i \subset C$ for each i. ∎

The algebra A has other simple subalgebras besides the algebras A_J. If S is a set of F-linear identities on n variables, let A_S denote the subalgebra of A spanned by all root vectors whose roots satisfy S in the sense that substitutition of the n components of each root vector α of A_S into each identity of S gives 0. It is easy to see that A_S is a subalgebra. Letting $\Delta_i^{(S)} = \{\gamma_k | \gamma$ is a root of $A_S\}$, we shall establish

Theorem 2.3 *The algebra A_S is simple if $\Delta_i^{(S)} \neq 0$ for every i.*

Proof. Because the identities S are linear, A_S contains all of the elements e_i. Thus any nonzero ideal C of A_S must contain some basis element $(x^\beta)_i$. If $\beta_i \neq -1$, then $(x^{-\beta})_i(x^\beta)_i = (\beta_i + 1)e_i \in C$. If $\beta_i = -1$, take the product in the other order, and in either case we obtain $e_i \in C$. Hence, $(B)_i = (B)_i e_i \subset C$. Pick any $j \neq i$. By hypothesis there exists a root α of A_S such that $\alpha_i \neq 0$, and so $e_i(x^\alpha)_j \in C$ is a nonzero multiple of $(x^\alpha)_j$. By the first part of the proof, $(B)_j \subset C$. Since this is true for all $j \neq i$, $C = A_S$ and A_S is simple. ∎

Let J be a subset of $\{1, 2, \cdots, n\}$ with the property that for each $k \in J$ the group is a nonzero cyclic group with generator 1, and define $A_{SJ} = A_S \cap A_J$. Then A_{SJ}, being the intersection of subalgebras, is itself a subalgebra. It is possible that $\Delta_i^{(S)}$ is cyclic with generator 1 but that Δ_i is not cyclic. In this case we can replace A by its subalgebra of the same form by deleting the part of Δ_i which is not in the subgroup generated by 1. We have

Theorem 2.4. *The algebra A_{SJ} is simple if $\Delta_i^{(S)} \neq 0$ for every i, and if for each $k \in J$ there exists a root γ of A_{SJ} such that $\gamma_k = -1$ and $\gamma_j = 0$ for $j \in J$ and $j \neq k$.*

Proof. Again $e_i \in A_{SJ}$, so that any nonzero ideal C of A_{SJ} contains a basis element, say $w = (x^\beta)_i$. If $\beta_i = -1$, we can find a root η of A_{SJ} with $\eta_i = 1$, and the ith component of the root of $w(x^\eta)_i \in C$ will be zero. Thus we can suppose that $\beta_i \geq 0$ if $i \in J$. Further, since $A_{SJ} \subset A_J$, we see that β_k is a nonnegative integer for each $k \in J$ not equal to i. If $\beta_k = q > 0$, we can multiply w on the left q times by $(x^\gamma)_k$ where γ is the root whose exitence is hyppothesised in the statement of the theorem. In this way one obtains an element w' of C such that the kth component of the root of w' is 0. Doing this for each $k \in J$, we can assume that $w = (x^\beta)_i$ is chosen so that $\beta_k = 0$ for each $k \in J$. Then $(x^{-\beta})_i$ is in both A_J and A_S, and so it is in A_{SJ}. Hence, $(x^\beta)_i(x^{-\beta})_i = e_i \in C$.

By the first hypothesis of the theorem, there exists a root α of A_{SJ} such that $\alpha_i \neq 0$. Hence, for each $j \neq i$, $e_i(x^\alpha)_j$ is a nonzero element of $C \cap (B)_j$. Proceeding as in the first part of the proof, $e_j \in C$. We have shown that all of the idempotents e_j are in C, from which it follows that $C = A_{SJ}$, or that A_{SJ} is simple. ∎

3. The conjecture. The classes of examples that we have constructed seem to us to be the natural generalizations of the Novikov algebras that arise in parts (i) and (ii) of Theorem 1.1. The algebra of part (iii) does not seem to generalize to the case where the torus has dimension greater than one. This leads us to make the following

Conjecture on Locally Novikov Algebras: *Let L be a locally Novikov algebra over an algebraically closed field F of characteristic 0, and let the canonical idempotents of L be $e_1, \ldots, e_n$ where $n \geq 2$. Suppose further that the root spaces of A are finite dimensional. Then we conjecture that L is one of the algebras A, A_J, A_S, or A_{SJ} defined above.*

References

1. A. A. Balinskii and S. P. Novikov, *Poisson brackets of hydrodynamic type, Frobenius algebras and Lie algebras*, Dokl. Akad Nauk SSSR 283 (1985), 1036-1039; English transl. Soviet Math. Dokl. **32** (1985), 228-231.
2. V. P. Cherkashin, *Left-symmetric algebras with commuting right multiplications* (Russian), Vestnik Moscov. Univ. Ser. I Mat. Mekh. **1988**, No. 5, 47-50.
3. V. T. Filippov, *A class of simple nonassociative algebras*, Mat. Zametki, **45** (1989), 101-105.
4. I. M. Gelfand and I. Ya. Dorfman, *Hamiltonian operators and related algebraic structures*, Funktsional Anal. I. Prolozhen, **13** No. 4 (1979), 13-30; English Transl. Funct. Anal. and Appl., **13** (1979), 248-262.
5. H. Kim, *Complete left-invariant affine structures on nilpotent Lie groups*, J. Differential Geom. **24** (1986), 373-394.
6. E. Kleinfeld, *On rings satisfying* $(x, y, z) = (y, x, z)$, Algebras Groups Geom., **4** (1987), 129-138.
7. J. M. Osborn, *Novikov algebras*, Nova J. Algebra Geom., **1** (1992), 1-14.
8. J. M. Osborn, *Simple Novikov algebras with an idempotent*, Comm. Algebra, **20**, No. 9, (1992), 2729-2753.
9. J. M. Osborn, *Modules for Novikov algebras*, Proceedings of the II International Conference on Algebra, Barnaul, 1991.
10. J. M. Osborn, *Infinite dimensional Novikov algebras of characteristic zero*, to appear in J. Algebra.
11. J. M. Osborn, *Locally Novikov algebras*, in preparation.
12. J. M. Osborn, *New simple infinite dimensional Lie algebras of of characteristic 0*, in preparation.
13. J. M. Osborn and E. I. Zelmanov, *Nonassociative algebras related to Hamiltonian operators in the formal calculus of variations*, to appear.
14. E. I. Zelmanov, *On a class of local translation invariant Lie algebras*, Soviet Math. Dokl. **35** (1987), 216-218. English transl. Soviet Math. Dokl. **35** (1987), 216-218.

COALGEBRA, COCOMPOSITION AND COHOMOLOGY

ZBIGNIEW OZIEWICZ*
University of Wrocław, Institute of Theoretical Physics,
plac Maxa Borna 9, 50204 Wrocław, Poland, oziewicz@plwruw11.bitnet

EUGEN PAAL
University of Tallin, Department of Mathematics,
5 Ehitajate tee, Tallinn EE0108, Estonia, paal@ce.ttu.ee

and

JERZY RÓŻAŃSKI†
University of Wrocław, Institute of Theoretical Physics
plac Maxa Borna 9, 50204 Wrocław, Poland, rozanski@plwruw11.bitnet

Abstract. We present unified approach to the cohomology assignment for braided Lie coalgebra in a braided monoidal category and for n-ary cocomposition.

Key words: cohomology of n-ary cocomposition, cohomology of nonassociative coalgebra, monoidal category, braided coalgebra, permuted Lie coalgebra, Hecke Lie coalgebra
1991 Mathematics Subject Classification: 17A01, 17A45, 17B56

Foreword

Throughout this paper k is a field. Let L be a finite dimensional k-space and $L^* \equiv \text{lin}(L, k)$ be the dual k-space of L. The tensor product $\otimes$ means $\otimes_k$, T is the tensor algebra functor and $TL \equiv L^{\otimes}$ is a tensor algebra. A linear map $\Delta \in \text{lin}(L, L^{\otimes n})$ is said to be noncoassociative n-ary cocomposition on L, coalgebra for $n = 2$. Results proven for cocompositions give the valid results for compositions by categorical duality.

Let $\text{der}\, TL$ be a set of a (skew)derivations of degree $n-1$ (definition 1). We have a set bijection

$$\text{lin}(L, L^{\otimes n}) \ni \Delta \longleftrightarrow D_\Delta \in \text{der}\, TL, \quad D_\Delta | k \equiv 0 \quad \text{and} \quad D_\Delta | L \equiv \Delta.$$

Let $I \triangleleft TL$ be two sided ideal such that $D_\Delta I \subset I$. The cohomology assign to a triple $\{L, \Delta, I\}$, a differential algebra $\{TL/I, D_\Delta/I\}$.

Cohomology and homology theory of associative algebra was formulated by Hochschild in 1945. One year after, Claude Chevalley together with Samuel Eilenberg,

---

* In 1993-1994 at Universidad Nacional Autonoma de México, Facultad de Estudios Superiores Cuautitlán, oziewicz@redvax1.dgsca.unam.mx

† Supported by Polish Committee of Scientific Research KBN.

S. González (ed.), Non-Associative Algebra and Its Applications, 314–322.

inspired by Élie Cartan implicite considerations, gave the definition of the cohomology and homology groups of a Lie algebra (published in 1948). Independently the (co)homology theory of Lie algebra was defined by Jean-Luis Koszul (Thése 1950). A unified axiomatic treatment of both theories, Hochschild theory and Chevalley/Eilenberg/Koszul theory, with the help of the derived functors Tor (derived from $\otimes$) and Ext (derived from Hom) is presented by Henri Cartan and Samuel Eilenberg in the monograph *Homological Algebra* (1956).

The Cartan/Eilenberg axiomatic (co)homology theory does not make clear, why (co)homology theory should make sense for two varieties of algebra, associative and Lie, only? Already in 1948 Eilenberg pose the problem of an assignment of a cohomology to *any* nonassociative algebra. Yamaguti in 1963 formulated the cohomology for algebra introduced by Malcev in 1955 on the basis of the analytic loops. Cohomology of a Lie triple system has been elaborated by Harris in 1961 and by Yamaguti (1957-1969). The (co)homology assignment was extended to a graded algebra and graded Lie triple (Tilgner 1977) and make sense for Lie $\mathbb{Z}_2$-graded algebra and Lie multi-graded algebra (Bani and Tripathy 1984). One can look for a list of algebras for which a (co)homology assignment exists. To do this we ask what is special about associative and Lie algebra which allows a (co)homology assignment? The present note is motivated by this question.

We present an unified approach to the cohomology assignment for n-ary cocomposition and in particular to the cohomology of permuted and Hecke Lie coalgebras in a symmetric and braided monoidal categories.

Gurevich (1986 and 1993) invented Hecke braided Lie algebra and Majid (1993) in another way introduced a braided Lie algebra which is also a coalgebra with a counit. Our axioms of a Hecke Lie coalgebra are different. We postulate that a factor algebra TL/I for a quadratic ideal I generated by an eigenspace of a Hecke braid is a *differential* algebra for every Hecke Lie coalgebra. This means that a morphism in a Hecke braided monoidal category is defined in a more complicated way.

Graded derivation

Definition 1 *Let A be an associative unitary k-algebra, $A \otimes A \xrightarrow{m} A$. Let $\alpha, \beta \in$ alg$A$. An (α, β)-*derivation *of an algebra A is a linear map $D \in$ lin$A$ that satisfies Leibniz's rule*

$$D \circ m \equiv m \circ (D \otimes \alpha + \beta \otimes D): \quad A \otimes A \longrightarrow A.$$

The k-space of all (α, β)-derivations of A is denoted by $\operatorname{der}_{\alpha,\beta} A$.

Let A be $\mathbb{N}$-graded algebra, $A = \sum A_n$, $A_0 \equiv k$, $A_1 \equiv L$, and let $D \in \operatorname{der}_{\alpha,\beta} A$ be a homogeneous derivation. The definition 1 imply $|\alpha| = |\beta| = 0$, $\alpha 1 = \beta 1 = 1$ and $D1 = 0$. A homogeneous derivation $D_\Delta \in \operatorname{der}_{\alpha,\beta} A$, $D_\Delta | L \equiv \Delta$, can be viewed as a collection of linear maps $D \equiv \{D|A_n\}$,

$$D_\Delta | A_{n+1} = m^n \circ \sum_{k=0}^{n} \alpha^{\otimes k} \otimes \Delta \otimes \beta^{\otimes (n-k)}. \tag{1}$$

The formula (1) tells that a derivation D is determined by a values of α, β and of D on a generating space L. In case of the tensor algebra $A = TL$, those values may be chosen arbitrarily. If D^p is a derivation then $D^p = 0$ is equivalent to $D^p|L = 0$. We need a condition on α, $\beta, \in \operatorname{alg} A$ which for some $1 < p \in \mathbb{N}$ assure the implication $D \in \operatorname{der}_{\alpha,\beta} A \Longrightarrow D^p \in \operatorname{der}_{\alpha^p,\beta^p} A$. A condition we are looking for is

$$(D \otimes \alpha + \beta \otimes D)^p = D^p \otimes \alpha^p + \beta^p \otimes D^p, \quad 1 < p \in \mathbb{N}.$$

Let ξ, $\zeta \in k$. In what follows we assume that $\alpha, \beta \in \operatorname{alg} A$ are such that $\alpha|L = \xi \cdot \mathrm{id}_L$ and $\beta|L = \zeta \cdot \mathrm{id}_L$.[1]

Lemma 1 *The implication $D \in \operatorname{der}_{\xi,\zeta} A \Longrightarrow D^p \in \operatorname{der}_{\xi^p,\zeta^p} A$ holds iff $\xi\zeta = 0$ or if $\xi\zeta \neq 0$ and $(\zeta/\xi)^{|D|}$ is a nontrivial p^{th} root of unity, $\zeta^{p|D|} = \xi^{p|D|}$, $\zeta \neq \xi$. In particular:*

$$D \in \operatorname{der}_{\xi,\zeta} A \Longrightarrow D^2 \in \operatorname{der}_{\xi^2,\zeta^2} A \qquad \text{iff} \qquad \xi\zeta(\xi^{|D|} + \zeta^{|D|}) = 0.$$

$$D \in \operatorname{der}_{\xi,\zeta} A \Longrightarrow D^3 \in \operatorname{der}_{\xi^3,\zeta^3} A \qquad \text{iff} \qquad \xi\zeta(\xi^{2|D|} + \xi^{|D|}\zeta^{|D|} + \zeta^{2|D|}) = 0.$$

A $\mathbb{N}$-graded algebra A with $\beta_\zeta \in \operatorname{alg} A$ is also $\mathbb{Z}_p$-graded for $\zeta^p = 1$. The implication of lemma 1 for $p = 3$ has been related by Kerner (1992) to a cubic root of a translations.

Lemma 2 *Let $D \in \operatorname{der}_{\alpha,\beta} A$ be homogeneous, then*

$$D^2 = 0 \qquad \Longleftrightarrow \qquad D^2|L = 0 \quad \text{and} \quad D^2|A_2 = 0.$$

proof. The assertion follows from the recurrent formula

$$\begin{aligned} D^2|A_{n+1} &= m \circ \{(D^2|A_n) \otimes (\beta|L) + (\alpha^2|A_n) \otimes (D^2|L)\} + \\ & \quad m \circ \{[(\alpha \circ D)|A_n] \otimes [(D \circ \beta)|L] + [(D \circ \alpha)|A_n] \otimes [(\beta \circ D)|L]\}. \end{aligned}$$

Factor derivation

Let π be the epimorphism: $TL \to TL/I$. A k-linear map $A \in \operatorname{lin} TL$ factors to $A/I \in \operatorname{lin}(TL/I)$ iff $AI \subset I$ and then $\pi \circ A = (A/I) \circ \pi$. A linear map

$$\omega : \quad (TL) \otimes (TL) \longrightarrow TL$$

factors to a map $\omega/I : (TL/I) \otimes (TL/I) \longrightarrow TL/I$, iff

$$\omega(I \otimes TL + TL \otimes I) \subset I, \quad \text{and then} \quad \pi \circ \omega = (\omega/I) \circ (\pi \otimes \pi).$$

If a linear factor map A/I exists then exists a factor map $(A \circ \otimes)/I$ and

$$(A \circ \otimes)/I = (A/I) \circ (\otimes/I) : \quad (TL/I) \otimes (TL/I) \longrightarrow TL/I.$$

Let A and B be linear maps in $\operatorname{lin} TL$. If exist a linear factor maps, A/I and B/I, then exists a linear factor maps $(A \circ B)/I$, $\{\otimes \circ (A \otimes B)\}/I$, and

$$(A \circ B)/I = A/I \circ B/I,$$

$$\{\otimes \circ (A \otimes B)\}/I = (\otimes/I) \circ (A/I \otimes B/I) : \quad (TL/I) \otimes (TL/I) \longrightarrow TL/I.$$

[1] More general possibility is to assume spectral decomposition $\alpha|L = \sum \xi_i \alpha_i$, where $\xi_i \in k$ and $\{\alpha_i\}$ is a spectral set of idempotents.

Corollary 1 *The derivation $D \in \mathrm{der}_{\alpha,\beta} TL$ factors to the $(\alpha/I, \beta/I)$-derivation D/I of the factor algebra $D/I \in \mathrm{der}_{\alpha/I,\beta/I}(TL/I)$ iff $\alpha I \subset I$, $\beta I \subset I$ and $DI \subset I$.*

An ideal $I \triangleleft TL$ is said to be *homogeneous* if $I = \sum_{n>2}(L^{\otimes n} \cap I)$. For homogeneous ideal a factor algebra TL/I is $\mathbb{N}$-graded. In what follows we assume that I is a two sided homogeneous ideal in a tensor algebra TL. Then an algebra maps α and β factors to the factor-algebra maps α/I, $\beta/I \in \mathrm{alg}(TL/I)$.

A cocomposition $\Delta \in \mathrm{lin}(L, L^{\otimes n})$ determine unique homogeneous derivation $D_\Delta \in \mathrm{der}_{\xi,\zeta} TL$, such that $D_\Delta | k \equiv 0$ and $D_\Delta | L \equiv \Delta$ (1). A derivation D_Δ factors to $(D_\Delta / I) \in \mathrm{der}_{\xi,\zeta}(TL/I)$ iff $D_\Delta I \subset I$.

Cocommutative n-ary cocomposition

Let $R < L^{\otimes 2}$. A non associative algebra $L^{\otimes 2} \xrightarrow{m} L$, is said to be R-commutative if $R < \ker m$ (Michaelis 1980). Then exists $m/R : L^{\otimes 2}/R \to L$. Dualy for a coalgebra $\{L, m\}^* = \{L^*, \Delta\}$ we have $\mathrm{im}\Delta \subset R^\perp \equiv \{\alpha \in (L^{\otimes 2})^*, \alpha R = 0\}$. In this case $L \ni \ker \Delta = \ker(\Delta/S)$ iff S is in any *complement* to $R^\perp$ in $(L^{\otimes 2})^*$. This motivate the

Definition 2 *Let $R < L^{\otimes n}$. A n-ary cocomposition $\Delta \in \mathrm{lin}(L, L^{\otimes n})$ is said to be* R-cocommutative *if* $R \cap (\mathrm{im}\, \Delta) = 0$.

Let $P \in \mathrm{End}(L^{\otimes n})$ be idempotent such that $R \equiv \mathrm{im}\, P$. A tensor $P \circ \Delta$ is said to be P-cocommutator of a cocomposition $\{L, \Delta\}$. A P-cocommutative n-ary cocomposition is a triple $\{L, P, \Delta\}$ where $P \circ \Delta = 0$.

Jacobi condition for n-ary cocomposition

Let $D \in \mathrm{der}_{\alpha,\beta} TL$, be such that $D^p \in \mathrm{der}_{\alpha^p,\beta^p} TL$ and such that $DI \subset I$ for a homogeneous ideal $I \triangleleft TL$. Then

$$D^p/I = (D/I)^p = 0 \quad \Longleftrightarrow \quad \boxed{\mathrm{im}(D^p|L) \subset I}$$

The framed condition is said to be Jacobi condition for cocomposition $\Delta \equiv D|L$. In particular let $p = 2$ and let $\xi\zeta(\xi^{|D|} + \zeta^{|D|}) = 0$. Then $(D_\Delta/I)^2 \in \mathrm{der}_{\xi^2,\zeta^2}(TL/I)$ and therefore

$$(D_\Delta/I)^2 = 0 \quad \Longleftrightarrow \quad \boxed{\mathrm{im}(D_\Delta \circ \Delta) \subset I \cap L^{\otimes(2|D|+1)}} \qquad (2)$$

We proved the

Lemma 3 *Let Δ be n-ary cocomposition on L with a derivation $D_\Delta \in \mathrm{der}_{\xi,\zeta} TL$ (1). Let I be* homogeneous *ideal in a tensor algebra TL. A value of an assignment*

$$\{L, \Delta, I\} \longmapsto \{TL/I, D_\Delta/I\}, \quad \textit{with} \quad \ker \Delta = \ker(\Delta/I),$$

is $\mathbb{N}$-graded differential algebra iff the following conditions hold

(*i*) $\quad \xi^n\zeta + \zeta^n\xi = 0 \qquad (\Longleftrightarrow (D_\Delta)^2 \in \mathrm{der}_{\xi^2,\zeta^2} TL)$

(*ii*) $D_\Delta I \subset I \qquad (\Leftrightarrow \exists\, D_\Delta/I \in \text{der } \textit{and } \exists\, (D_\Delta)^2/I = (D_\Delta/I)^2 \in \text{der}),$

(*iii*) *cocommutativity:* $I \cap \operatorname{im}\Delta = 0 \in L^{\otimes n}, \quad (\Longleftrightarrow \ker\Delta = \ker(\Delta/I))$

(*iv*) *Jacobi condition:* $\operatorname{im}(D_\Delta \circ \Delta) \subset I \cap L^{\otimes(2n-1)} \quad (\Leftrightarrow (D_\Delta/I)^2 = 0).$

Let $P \in \operatorname{End}(L^{\otimes n})$ and $Q \in \operatorname{End}(L^{\otimes(2n-1)})$ be such that

$$(\operatorname{im}P) \cap (\ker P) = 0, \quad I \equiv \operatorname{gen}\{\operatorname{im}P\} \quad \text{and} \quad \ker Q \equiv I \cap L^{\otimes(2n-1)}. \tag{3}$$

In terms of the operators P, Q and Δ, for a given (ξ,ζ)-derivation (1) with the condition (i) of the lemma 3, we have three conditions which assure the cohomology assignment for n-ary cocomposition:
The existence of a factor-derivation $D_\Delta I \subset I \quad \Longleftrightarrow$

$$\boxed{Q \circ D_\Delta \circ P = 0} : \quad L^{\otimes n} \longrightarrow L^{\otimes(2n-1)}. \tag{4}$$

A cocommutativity

$$\boxed{P \circ \Delta = 0} \quad \Longrightarrow \quad (\operatorname{im}P) \cap (\operatorname{im}\Delta) = 0. \tag{5}$$

Jacobi condition

$$\boxed{Q \circ D_\Delta \circ \Delta = 0} : \quad L \longrightarrow L^{\otimes(2n-1)}. \tag{6}$$

Definition 3 *A tensor* $J \equiv Q \circ D_\Delta \circ \Delta$, *is said to be* coJacobiator *of a cocomposition* Δ.

One can consider the set of equations (4-6) from two points of view:

- Let a factor algebra TL/I be given. Then the equations (4-6) are identities for (possible empty) variety of n-ary cocomposition $\Delta : L \longrightarrow L \wedge L \equiv L^{\otimes 2}/(I \cap L^{\otimes 2})$. In this way we define in a next sections a permuted Lie coalgebra in a symmetric monoidal category as an example of variety of coalgebra possessing the cohomology assignment.
- Let n-ary cocomposition $\{L, \Delta\}$ be given. Then the equations (4-6) determine (possible empty) set of ideals in the tensor algebra TL, for each of which a factor algebra TL/I is a differential algebra.

The choice $I \equiv 0$ define a variety of n-ary coassociative cocomposition, which is analogous to a coassociative coalgebra, and possess the cohomology assignment. For this choice we are left with Jacobi condition only (2).

For $|D_\Delta| = 1$, $D_\Delta | L^{\otimes 2} = \Delta \otimes \xi\, \mathrm{id}_L + \zeta\, \mathrm{id}_L \otimes \Delta$, and $(D_\Delta)^2$ is a derivation iff $\xi\zeta(\xi + \zeta) = 0$. Three solutions leads to

$$\begin{aligned} &\text{if } \zeta = 0 && (D_\Delta)^2 | L = (\Delta \otimes \mathrm{id}_L) \circ \Delta, \\ &\text{if } \xi = 0 && (D_\Delta)^2 | L = (\mathrm{id}_L \otimes \Delta) \circ \Delta, \\ &\text{if } \xi + \zeta = 0 && (D_\Delta)^2 | L = (\Delta \otimes \mathrm{id}_L - \mathrm{id}_L \otimes \Delta) \circ \Delta. \end{aligned}$$

The last expression is the *coassociator* of a coalgebra $\{L, \Delta\}$. Therefore for $I \equiv 0$ Jacobi condition, $(D_\Delta)^2 = 0$, determine a variety of coassociative coalgebra

$$(\Delta \otimes \mathrm{id}_L - \mathrm{id}_L \otimes \Delta) \circ \Delta = 0.$$

The coassociativity is equivalent to nilpotency $(D_\triangle)^2 = 0$ and therefore a coassociative coalgebra $C \equiv \{L, \Delta\}$ possess a differential algebra $\{TL, D_\triangle\}$. A cohomology of $\{TL, D_\triangle\}$ is *by definition* the Hochschild scalar-valued cohomology $H^*(C, C)$ of a coassociative coalgebra C.

A Lie algebra is skew-symmetric and in this case I is a quadratic ideal generated by symmetric tensors. This can be generalized by considering an ideal generated by eigenspace of a braid operator which is a solution of the braid equation.

Braided monoidal category

Definition 4 (Yetter 1990) *A monoidal category is* pre-braided *if it is equipped with a natural transformation $\mathcal{B}$ from $\otimes$ to $\otimes^{opp}$, such that*

$$\begin{aligned} \mathcal{B}_{V\otimes W,U} &\equiv (\mathcal{B}_{V,U} \otimes \mathrm{id}_W) \circ (\mathrm{id}_V \otimes \mathcal{B}_{W,U}), \\ \mathcal{B}_{V,W\otimes U} &\equiv (\mathrm{id}_W \otimes \mathcal{B}_{V,U}) \circ (\mathcal{B}_{V,W} \otimes \mathrm{id}_U). \end{aligned}$$

A pre-braiding $\mathcal{B}$ consists of a family of morphisms $\mathcal{B} \equiv \{\mathcal{B}_{.,.}\}$, $\mathcal{B}_{W,V} \in \mathrm{lin}(W \otimes V, V \otimes W)$. Pre-braid hexagon,

$$(\mathcal{B}_{W,U} \otimes \mathrm{id}_V) \circ \mathcal{B}_{V,W\otimes U} = (\mathrm{id}_U \otimes \mathcal{B}_{V,W}) \circ \mathcal{B}_{V\otimes W,U}, \tag{7}$$

is a consequence of the hexagons of the definition 4 and of naturality (Joyal and Street 1985)

A *braided* monoidal category is a monoidal category with a distinguished *pair* of mutually invertible braidings $\mathcal{B} : \otimes \longmapsto \otimes^{opp}$ and $\mathcal{B}^{-1} : \otimes^{opp} \longmapsto \otimes$, such that $\mathcal{B} \circ \mathcal{B}^{-1} = \mathcal{B}^{-1} \circ \mathcal{B} = \mathrm{id}$. This means that $(\mathcal{B}_{A,B})^{-1} = (\mathcal{B}^{-1})_{B,A}$. The particular case of a braided monoidal category is when both braidings coincide $\mathcal{B} = \mathcal{B}^{-1}$. Then $\mathcal{B}$ is said to be a permutation, $\mathcal{B}^2 = \mathrm{id}$. This category was introduced by Mac Lane in 1971 under the name *symmetric* monoidal category. In symmetric monoidal category the hexagons of the definition 4 are no more independent, they are equivalent.

The morphisms in a pre-braided category are maps such that pre-braiding is natural (functorial). A maps $f \in \mathrm{lin}(X, Z)$, $g \in \mathrm{lin}(Y, W)$ are a morphisms if

$$\mathcal{B}_{Z,W} \circ (f \otimes g) = (g \otimes f) \circ \mathcal{B}_{X,Y}.$$

Definition 5 (Braided coalgebra) *Let $\mathcal{B} \in \mathrm{End}(L^{\otimes 2})$ be a braid operator, i.e. $\mathcal{B}$ is a solution of the braid equation (7). A coalgebra $\Delta \in \mathrm{lin}(L, L^{\otimes 2})$ is said to be* $\mathcal{B}$-braided *(or* $\mathcal{B}$-permuted *for $\mathcal{B}^2 = \mathrm{id}$) if Δ is a morphism in a braided monoidal category. This means that the following conditions hold*

$$\begin{aligned} \mathcal{B}_{L^{\otimes 2},L} \circ (\Delta \otimes \mathrm{id}_L) &= (\mathrm{id}_L \otimes \Delta) \circ \mathcal{B}_{L,L}, \\ \mathcal{B}_{L,L^{\otimes 2}} \circ (\mathrm{id}_L \otimes \Delta) &= (\Delta \otimes \mathrm{id}_L) \circ \mathcal{B}_{L,L}. \end{aligned} \tag{8}$$

If a braiding $\mathcal{B}$ is a twist, $\tau(v \otimes w) \equiv (w \otimes v)$, or if $\mathcal{B} = -\tau$ and $\mathrm{char} k = 2$, then the equations (8) are identities for any Δ. If a braiding $\mathcal{B}$ is a permutation, $\mathcal{B}^2 = \mathrm{id}$, then the equations (8) are equivalent to each other. If a braid is not a permutation, $\mathcal{B}^2 \neq \mathrm{id}$, then both equations (8) together could imply $\Delta \equiv 0$. In those cases *one* of the equations (8) holds only.

Example. Let $\{e_i\}$ be a basis of L and let a subspaces $(L^{\otimes 2})_i < L^{\otimes 2}$ and a permutation $\mathcal{S}$ be defined in terms of a set of commutation factors $\varepsilon_{ij} \in k$,

$$\begin{aligned}(L^{\otimes 2})_i &\equiv \operatorname{span}\{e_q \otimes e_p,\ \varepsilon_{ij} = \varepsilon_{qj}\varepsilon_{pj},\ \forall j, \text{ no sum}\},\\ \mathcal{S}(e_i \otimes e_j) &\equiv \varepsilon_{ij} \cdot e_j \otimes e_i, \quad \varepsilon_{ij}\varepsilon_{ji} = 1, \quad \text{no sums.}\end{aligned} \tag{9}$$

This determine an equivalence relation on basis vectors,

$$e_i \sim e_j \quad \Longleftrightarrow \quad (L^{\otimes 2})_i = (L^{\otimes 2})_j.$$

Let L_α be a subspace of L spaned by an equivalence class of basis vectors. In this way a set of the commutation factors $\{\varepsilon_{ij}\}$ determine a splitting of $L = \oplus L_\alpha$.

In particular let $\varepsilon_{ij} \equiv (-1)^{ij}$, then $(L^{\otimes 2})_i = \operatorname{span}\{e_q \otimes e_p,\ q + p = i \mod 2\}$, and L is $\mathbb{Z}/2\mathbb{Z}$-graded, $L = L_0 \oplus L_1$, $(L^{\otimes 2})_0 = L_0 \otimes L_0 + L_1 \otimes L_1$, etc.

Lemma 4 *Let a permutation $\mathcal{S}$ be as in (9). A coalgebra Δ is $\mathcal{S}$-permuted iff* $\operatorname{im}(\Delta|L_\alpha) \subset (L^{\otimes 2})_\alpha$, $\forall$ *equivalence class* α.

proof. Let $\{\varepsilon^i\}$ be a dual basis in L^*, $\varepsilon^i e_j \equiv \delta^i_j$. Let $\Delta^{pq}_i \equiv (\varepsilon^p \otimes \varepsilon^q)\Delta\, e_i$. Inserting (9) into morphism condition (8) we get

$$\Delta^{pq}_i(\varepsilon_{qj}\varepsilon_{pj} - \varepsilon_{ij}) = 0, \quad \forall j, \text{ no sums.}$$

The conditions $\varepsilon_{qj}\varepsilon_{pj} = \varepsilon_{ij}$ are similar to known in multigraded Lie algebra (Scheunert 1979). □

If a coproduct Δ is a $\mathcal{B}$-morphism then a linear map $D_\Delta|L^{\otimes 2}$ has the forms

$$\begin{aligned}D_\Delta(\mathcal{B}) &= \Delta \otimes \mathrm{id}_L - (\mathcal{B} \otimes \mathrm{id}_L) \circ (\mathrm{id}_L \otimes \mathcal{B}) \circ (\Delta \otimes \mathrm{id}_L) \circ \mathcal{B}^{-1}\\ &= (\mathrm{id}_L \otimes \mathcal{B}) \circ (\mathcal{B} \otimes \mathrm{id}_L) \circ (\mathrm{id}_L \otimes \Delta) \circ \mathcal{B}^{-1} - \mathrm{id}_L \otimes \Delta.\end{aligned}$$

Define

$$\begin{aligned}Q(\mathcal{B})|L^{\otimes 2} &\equiv \mathrm{id}_{L^{\otimes 2}} + \mathcal{B},\\ Q(\mathcal{B})|L^{\otimes 3} &\equiv \{\mathrm{id}_{L^{\otimes 3}} + \mathrm{id}_L \otimes \mathcal{B} + \mathcal{B}_{L^{\otimes 2},L}\} \circ (\mathrm{id}_{L^{\otimes 3}} + \mathcal{B} \otimes \mathrm{id}_L)\\ &= \{\mathrm{id}_{L^{\otimes 3}} + \mathcal{B} \otimes \mathrm{id}_L + \mathcal{B}_{L,L^{\otimes 2}}\} \circ (\mathrm{id}_{L^{\otimes 3}} + \mathrm{id}_L \otimes \mathcal{B})\\ &= (\mathrm{id}_{L^{\otimes 3}} + \mathcal{B} \otimes \mathrm{id}_L) \circ \{\mathrm{id}_{L^{\otimes 3}} + \mathrm{id}_L \otimes \mathcal{B} + \mathcal{B}_{L,L^{\otimes 2}}\}\\ &= (\mathrm{id}_{L^{\otimes 3}} + \mathrm{id}_L \otimes \mathcal{B}) \circ \{\mathrm{id}_{L^{\otimes 3}} + \mathcal{B} \otimes \mathrm{id}_L + \mathcal{B}_{L^{\otimes 2},L})\}.\end{aligned} \tag{10}$$

The last equalities are consequences of the braid equation (7).

Let $I_\lambda(\mathcal{B})$ be two sided ideal in TL generated by λ-eigenspace of a braid operator

$$I_\lambda(\mathcal{B}) \equiv \operatorname{gen}\{\ker(\lambda \cdot \mathrm{id}_{L^{\otimes 2}} - \mathcal{B})\} \quad \triangleleft\, TL. \tag{11}$$

Lemma 5 *Let $\mathcal{H}$ be a Hecke braid. Then an operator $Q(-\lambda^{-1}\mathcal{H})$ (10) is proportional to projector and* $\{I_\lambda(\mathcal{H})\} \cap L^{\otimes 3} = \ker\{Q(-\lambda^{-1}\mathcal{H})\}$.

Permuted Lie coalgebra

If $\mathcal{S}$ is a permutation, $\mathcal{S}^2 = \mathrm{id}_{L^{\otimes 2}}$, then

$$I(\pm\mathcal{S}) \equiv \mathrm{gen}\ker(\mathrm{id}_{L^{\otimes 2}} \mp \mathcal{S}) = \mathrm{gen}\,\mathrm{im}(\mathrm{id}_{L^{\otimes 2}} \pm \mathcal{S}),$$
$$\{I(\pm\mathcal{S})\} \cap L^{\otimes 3} = \ker\{Q(\mp\mathcal{S})\}.$$

Lemma 6 (Main) *Let $\mathcal{S} \in \mathrm{End}(L^{\otimes 2})$ be a permutation and let a coalgebra $\Delta \in \mathrm{lin}(L, L^{\otimes 2})$ be $\mathcal{S}$-permuted. Then $D_\Delta\, I(\mathcal{S}) \subset I(\mathcal{S})$.*

proof. As a direct consequence of a morphism condition (8) for a permutation $\mathcal{S}$ we have

$$(\mathrm{id}_{L^{\otimes 3}} + \mathcal{S}_{L,L^{\otimes 2}} + \mathcal{S}_{L^{\otimes 2},L}) \circ (\Delta \otimes \mathrm{id}_L - \mathrm{id}_L \otimes \Delta) =$$
$$\{(\mathrm{id}_L \otimes \mathcal{S}) \circ [(\mathcal{S} \circ \Delta) \otimes \mathrm{id}_L] + \Delta \otimes \mathrm{id}_L - \mathrm{id}_L \otimes \Delta\} \circ (\mathrm{id}_{L^{\otimes 2}} - \mathcal{S}).$$

For a permutation $\mathcal{S}$ we have an alternative expressions for Q (10),

$$Q(\mathcal{S}) = (\mathrm{id}_{L^{\otimes 3}} + \mathrm{id}_L \otimes \mathcal{S}) \circ (\mathrm{id}_{L^{\otimes 3}} + \mathcal{S}_{L,L^{\otimes 2}} + \mathcal{S}_{L^{\otimes 2},L}).$$

It follows that $Q(\pm\mathcal{S}) \circ D_\Delta \circ (\mathrm{id}_{L^{\otimes 2}} + \mathcal{S}) = 0.$ □

Lemma 7 *Let $\mathcal{S} \circ \Delta = -\Delta$. The coJacobiator of the definition 3, for a $\mathcal{S}$-permuted Lie coalgebra Δ possess an alternative forms*

$$\begin{aligned} J &= 4\{\Delta \otimes \mathrm{id}_L - \mathrm{id}_L \otimes \Delta - (\mathrm{id}_L \otimes \mathcal{S}) \circ (\Delta \otimes \mathrm{id}_L)\} \circ \Delta \\ &= 4\{\Delta \otimes \mathrm{id}_L - \mathrm{id}_L \otimes \Delta + (\mathcal{S} \otimes \mathrm{id}_L) \circ (\mathrm{id}_L \otimes \Delta)\} \circ \Delta \\ &= 4\{\mathrm{id}_{L^{\otimes 3}} - \mathrm{id}_L \otimes \mathcal{S} + (\mathcal{S} \otimes \mathrm{id}_L) \circ (\mathrm{id}_L \otimes \mathcal{S})\} \circ (\Delta \otimes \mathrm{id}_L) \circ \Delta. \end{aligned}$$

This expression is known for graded Lie algebra (Mitra Bani and Tripathy 1984) and coincide with Jacobi condition as given by Woronowicz (1989). An ideal $I(\mathcal{S})$ determine a variety of $\mathcal{S}$-permuted Lie coalgebra satisfying the identities $\mathcal{S} \circ \Delta = -\Delta$, $J = 0$ and morphism condition (8) as a variety of coalgebra possessing the cohomology assignment.

Hecke Lie coalgebra

For a Hecke braid $\mathcal{H}$ instead of the morphism conditions (8) for Δ, we assume that

$$Q(-\lambda^{-1} \cdot \mathcal{H}) \circ (\Delta \otimes \mathrm{id}_L - \mathrm{id}_L \otimes \Delta) \circ (q \cdot \mathrm{id}_{L^{\otimes 2}} + \lambda \cdot \mathcal{H}) = 0. \tag{12}$$

In a case of a permutation the condition (12) is a consequence of the a morphism condition (8), the lemma 6. This is not the case for a Hecke braid.

If a braid is not a permutation, then it seems that it would be better to consider a pair $\mathcal{B}$ and $\mathcal{B}^{-1} \neq \mathcal{B}$ on equal footing rather. This would leads to a pair of ideals $I(\mathcal{B})$ and $I(\mathcal{B}^{-1})$ and a pair of 'differential' factor algebras.

Woronowicz (1989) and Majid (1994) are showing an equivalence between the braid equation in an extended space $k \oplus L$, and Jacobi condition for a coalgebra $\{L, \Delta\}$. It seems to plausible to relate this equivalence to cohomology assignment.

References

1. Chevalley Claude and Samuel Eilenberg: 1948, 'Cohomology of Lie groups and Lie algebras', *Trans. AMS* **63** 85–124
2. Eilenberg Samuel: 1948, 'Extensions of general algebras', *Ann. Soc. Polon. Math.*, (Rocznik Polskiego Towarzystwa Matematycznego) **21** (1) 125-134
3. Gurevich Dmitri: 1986, 'The Yang-Baxter equations and a generlization of formal Lie theory', *Soviet Math. Doklady* **33** 758-762
4. Gurevich Dmitri: 1993, 'Hecke symmetries and braided Lie algebras', in *Spinors, Twistors, Clifford Algebras and Quantum Deformations* in series 'Fundamental Theories of Physics', edited by Zbigniew Oziewicz, Bernard Jancewicz and Andrzej Borowiec, Kluwer Academic Publishers, Dordrecht
5. Harris B.: 1961, 'Cohomology of Lie triple systems and Lie algebras with involutions', *Transactions of American Mathematical Society* **98** 148-162
6. Hochschild G.: 1945, 'On the cohomology groups of an associative algebra', *Annals of Mathematics* **46** 58-67
7. Joyal A. and R. Street: 1985, 'Braided monoidal categories', Macquarie Mathematics Reports 850067 (1985) and 860081 (1986)
8. Kerner Richard: 1992, '$\mathbb{Z}_3$-graded algebras and cubic root of the supersymmetry translations', *Journal of Mathematical Physics* **33** (1) 403–411
9. Koszul Jean-Louis: 1950, 'Homologie et cohomologie des algébras de Lie', Thése, *Bull. Soc. Math. France* **78** 1–63
10. Majid Shahn: 1993, 'Quantum and braided Lie algebras', preprint DAMTP/93-4, to appear in J. Geometry and Physics; 'Braided geometry: a new approach to q-deformations', preprint; 'Lie algebras and braided geometry', to appear in Proceedings of the XXIIth Conference on Differential Geometric Methods in Theoretical Physics, Ixtapa, Mexico, edited by Jaime Keller, 1994
11. Michaelis Walter: 1980, 'Lie coalgebras', *Advances in Mathematics* **38** 1–54
12. Mitra Bani and K.C. Tripathy: 1984, 'The cohomology of the generalized Lie algebras', *J. Math. Phys.* **25** 2550-2556
13. Scheunert M.: 1979, 'Generalized Lie algebras', *J. Math. Phys.* **20** 712
14. Tilgner Hans: 1977, 'A graded generalization of Lie triples', *Journal of Algebra*, **47** 190-196
15. Woronowicz Stanisław L.: 1989, 'Differential calculus on compact matrix pseudogroups (quantum groups)', *Commun. Math. Phys.* **122** 125-170
16. Yamaguti Kiyosi: 1957-58, 'On the Lie triple system and its generalization', *Journal Sci. Hiroshima University*, **A21** 155-160
17. Yamaguti Kiyosi: 1963, 'On the theory of Malcev algebras', *Kumamoto Journal Sci.* **A6** (1) 9-45
18. Yamaguti Kiyosi: 1969, 'On cohomology groups of general Lie triple systems', *Kumamoto J. Sci.*, **A8** (4) 135-146
19. Yetter David: 1990, 'Quantum groups and representations of monoidal categories', Math. Proc. Camb. Phil. Soc. **108** 261-290

MINIMAL POLYNOMIAL IDENTITIES OF BARIC ALGEBRAS

LUIZ ANTONIO PERESI *
Departamento de Matemática, Universidade de São Paulo
Caixa Postal 20570, São Paulo-SP, Brazil 01452-990

Abstract.
We construct minimal polynomial identities, i. e, identities of the lowest degree for nine classes of baric algebras. We use the technique of representing equations by matrices.

Key words: Nonassociative algebra, baric algebra, minimal identities

1. Introduction

A baric algebra is a pair (A,ω) consisting of a nonassociative algebra A over a field F and a nonzero algebra homomorphism $\omega : A \to F$. Classes of commutative baric algebras have been defined throught the literature by equations of the form $x^{\{n\}} - \omega(x)^{n-m}x^{(m)} = 0$, where $1 \leq m < n$ and $x^{\{n\}}$, $x^{(m)}$ denote powers of x with some distribution of parentheses. If a commutative algebra satisfies such an equation then it satisfies the identity $(x^{\{n\}}, y, x^{(m)}) = 0$, where the homomorphism ω does not appear. Here and so on (a,b,c) is the associator $ab.c - a.bc$, we use the term identity to mean polynomial identity and the term equation to mean any kind of identity.

Thus an interesting problem is to find identities of lowest degree not implied by the commutative law that hold for all algebras in such a class of commutative baric algebras. We call this type of identities minimal identities. In general such a class of algebras is not a variety of algebras since it is not closed by the operation of taking subalgebras. A more specific problem is then to determine the minimal variety of algebras containing the class, i. e., the variety defined by all minimal identities and commutativity.

This kind of problem has been studied by Costa [5], Alcalde-Burgueño-Mallol [6] and Correa-Hentzel-Peresi [7]. In this last paper we described a general method for 'processing identities' and then applied it to find the minimal variety that contains the class of Bernstein algebras, i. e., the class defined by equations $xy - yx = 0$ and $(x^2)^2 - \omega(x)^2x^2 = 0$. It was written with more emphasis in the method itself than applications. In the present paper we reverse this situation by applying the method to find minimal identities for nine classes of commutative baric algebras. For six of these classes we determine the minimal variety. In order to make the paper more understandable we include the full argument for the classes defined by $x^2 - \omega(x)x = 0$ and $x^3 - \omega(x)^2x = 0$. For the other classes we just state the results.

All algebras considered in the paper are commutative algebras over a field F. The

* Supported by the CNPq of Brazil.

S. González (ed.), Non-Associative Algebra and Its Applications, 323–329.

characteristic of F is zero or greater than the degree of the identities considered.

2. Degree two

Let (A,ω) be a baric algebra satisfying $x^2-\omega(x)x=0$. As clear A satisfies the Jordan identity $(x^2,y,x)=0$. On the other hand if we apply our method to search for identities of degree three we find none. Thus the minimal identities have degree four. To find all of them we proceed as follows.

Linearizing $x^2-\omega(x)x=0$ we obtain $f(a,b)=2ab-\omega(a)b-\omega(b)a=0$. Using this equation we obtain the eight possible equations of degree four:

$$\begin{array}{llll} f(a,b)c.d=0, & f(a,b).cd=0, & \omega(d)f(a,b)c=0, & f(ac,bd)=0, \\ \omega(cd)f(a,b)=0, & f(ac,b)d=0, & \omega(d)f(ac,b)=0, & f(ac.d,b)=0. \end{array} \tag{1}$$

These equations are expressed in terms of five association types. They are:

$$T_1.\,\omega(R)RR.R,\quad T_2.\,\omega(RR)RR,\quad T_3.\,\omega(RRR)R,\qquad T_4.\,(RR.R)R,\quad T_5.\,RR.RR.$$

(Here R has no meaning. It is just a convenient way to say how the parentheses and ω appear.) Any degree four equation satisfied by A is a consequence of equations (1) and commutativity. Since minimal identities are not consequence of the commutative law we need to know the equations it implies. We do this by applying commutativity to the association types. We obtain:

$$\begin{array}{lll} \omega(a)bc.d-\omega(a)cb.d=0, & \omega(ab)cd-\omega(ba)cd=0, & \omega(ab)cd-\omega(ab)dc=0, \\ \omega(abc)d-\omega(bac)d=0, & \omega(abc)d-\omega(cab)d=0, & (ab.c)d-(ba.c)d=0, \\ ab.cd-ba.cd=0, & ab.cd-cd.ab=0. & \end{array} \tag{2}$$

The minimal identities of A are the identities involving only types

T_4 and T_5 implied by equations (1) that are not consequence of equations (2).

Now, instead of working directly with the equations (1) and (2) we represent them by matrices and then perform calculations with these matrices. This technique was introduced by Hentzel [2] and was slightly modified in [7].

Let S_n denote the symmetric group and FS_n the corresponding group algebra. Clifton [3] gives an algorithm which associates to $\pi\in S_n$ a matrix A_π. We use the map $\pi\rightarrow A_\pi$ to construct an isomorphism between FS_n and a certain direct sum of matrix algebras. We call the direct summands (irreducible) representations although they are not representations in the usual sense for $n>4$. For $n=4$ we have five representations and the isomorphism is given by

$$\begin{array}{l} (12)\mapsto[1]\oplus\begin{bmatrix}1&0&-1\\0&1&-1\\0&0&-1\end{bmatrix}\oplus\begin{bmatrix}1&-1\\0&-1\end{bmatrix}\oplus\begin{bmatrix}1&-1&1\\0&-1&0\\0&0&-1\end{bmatrix}\oplus[-1] \\ (1234)\mapsto[1]\oplus\begin{bmatrix}-1&1&0\\-1&0&1\\-1&0&0\end{bmatrix}\oplus\begin{bmatrix}-1&0\\-1&1\end{bmatrix}\oplus\begin{bmatrix}1&-1&1\\1&0&0\\0&1&0\end{bmatrix}\oplus[-1]. \end{array} \tag{3}$$

Considering how the positions of the variables a, b, c, d are changed in the terms of an equation we can represent it by an element of the direct sum $FS_4\oplus\ldots\oplus FS_4$

(here the number of summands is the number of association types, i. e., five). For instance, the equation $f(a,b)c.d = 0$ expands to $2(ab.c)d - \omega(a)bc.d - \omega(b)ac.d = 0$ and is represent by $\begin{matrix} T_1 & T_4 \\ -I-(12) & \oplus\, 2I \end{matrix}$. Now using (3) we represent this equation by matrices. Representation number 2 for example gives

$$\overset{T_1}{\begin{bmatrix} -2 & 0 & 1 \\ 0 & -2 & 1 \\ 0 & 0 & 0 \end{bmatrix}} \oplus \overset{T_4}{\begin{bmatrix} 2 & 0 & 0 \\ 0 & 2 & 0 \\ 0 & 0 & 2 \end{bmatrix}}$$

It is crucial to keep the different association types separated.

It is too much work to calculate by hand all these matrices. So we use a computer program named CRUNCH. The input is a set of equations involving r different types. For each representation the program calculates a $m \times r$ block matrix of $k \times k$ matrices where m is the number of equations and k is the representation degree. The output is the row canonical form of this block matrix. The row canonical form obtained from the set of equations (2) is given in the left part of table I and that obtained when we consider together the sets (1) and (2) is given in the right part of table I. Since our interest is concentrated on association types T_4 and T_5 we have written only part of these matrices. Comparing we see that there are five new stairstep ones. They correspond to the minimal identities we are looking for.

Finally we write the minimal identities in polynomial form. Since representation 1 is the identity representation the identity given by it is

$$(x^2, x, x) = 0. \tag{4}$$

To find the identities given by the other representations we use the process described in [7]. As an example we obtain the identity appearing in representation 2. It is given by

$$\overset{T_4}{\begin{bmatrix} 0 & 0 & 0 \\ 0 & 0 & 0 \\ 0 & 1 & 0 \end{bmatrix}}.$$

The standard tableaus for this representation are $\tau_1 = \begin{matrix} 1 & 2 & 3 \\ 4 & & \end{matrix}$ $\tau_2 = \begin{matrix} 1 & 2 & 4 \\ 3 & & \end{matrix}$ $\tau_3 = \begin{matrix} 1 & 3 & 4 \\ 2 & & \end{matrix}$ Since the nonzero entry of the matrix appears in row 3 we consider the horizontal permutations I, (13), (14), (34), (134), (143) and the vertical permutations I, (12) of τ_3. Let $H = \{I+(13)+(14)+(34)+(134)+(143)\}\{I-(12)\}$. Since the position of the nonzero entry is $(3,2)$ we have to multiply H by the permutation that maps τ_3 to τ_2, i.e., (23). The identity is then given by type T_4 and $(12)H$ and is

$$(a,c,b)d + (b,cd,a) + (a,d,b)c = 0. \tag{5}$$

Representations 3 and 4 yield the identities

$$(b,c,a)d + (a,d,b)c = 0, \tag{6}$$

TABLE I
Row Canonical Form

Equations (2).

Representation 1

The rows contain only zeros.

Representation 2

T_4 T_5

1 1 2
1
1
1

Representation 3

T_4 T_5

1 2
1 2

Representation 4

T_4 T_5

1 −2
1 1
1
1
1

Representation 5

T_4 T_5
1
1

Equations (1) and (2).

Representation 1

T_4 T_5

1 −1

Representation 2

T_4 T_5
1 2
1
1
1
1

Representation 3

T_4 T_5
1
1
1
1

Representation 4

T_4 T_5
1
1
1
1
1
1

Representation 5

T_4 T_5
1
1

$$ac.bd - ad.bc = 0, \tag{7}$$

$$(a, cd, b) + (b, ad, c) + (c, bd, a) = 0. \tag{8}$$

=

Identity (6) implies (8), and a commutative algebra satisfies (4) and (5) if and only if it is a Jordan algebra. Since it is necessary to put commutativity, the Jordan law, identities (6) and (7) together to obtain the matrix in table I no further reduction is possible, any minimal identity is implied by these four identities, and thus we have

Theorem 1. *The minimal variety of algebras containing the class of commutative baric algebras which satisfy the equation $x^2 = \omega(x)x$ is defined by identities* $xy = yx$, $ac.bd = ad.bc$, $(a, c, b)d = (a, d, b)c$ *and* $(x^2, y, x) = 0$.

the

3. Degree three

As shown by Walcher [4] commutative baric algebras satisfying $x^3 = \omega(x)x^2$ are Jordan algebras. Now we have

Theorem 2. *Jordan algebras form the minimal variety of algebras which contains the class of commutative baric algebras defined by* $x^3 = \omega(x)x^2$.

Although our method of attacking the problem is general, for some classes of baric algebras the problem is not doable. The number of stairstep ones that give the minimal identities is too high. It is too much work to write down all these identities in polynomial form. In these cases we do not find the minimal variety containing the class. But we do find the degree n of a minimal identity and construct a set of identities which generate all minimal identities of type $[n-1,1]$. We say that an equation has type $[n-1,1]$ when it is expressed in the variables x and y and in each term the degree of x is $n-1$ and the degree of y is 1.

We find this situation when considering the class defined by $xy = yx$ and the equation $x^3 = \omega(x)^2 x$. The minimal identities have degree five and there are many of them. Clearly $(x^3, y, x) = 0$ is an identity and since its degree is five it is minimal. To find a set of generators for minimal identities of type $[4,1]$ we proceed as follows. Let $f(x) = x^3 - \omega(x)^2 x$, $p(y,x) = 2yx.x + x^2y - 2\omega(xy)x - \omega(x^2)y$, $g(a,b,c)$ the linearized form of $f(x)$ and $q(y,a,b)$ the linearized form of $p(y,x)$. Any algebra in this class satisfies the following equations:

$$
\begin{array}{llll}
f(x)x.y = 0, & f(x)y.x = 0, & f(x).xy = 0, & \omega(x)f(x)y = 0, \\
\omega(y)f(x)x = 0, & \omega(xy)f(x) = 0, & g(x^2,x,x)y = 0, & \omega(y)g(x^2,x,x) = 0, \\
p(y,x)x.x = 0, & p(y,x)x^2 = 0, & \omega(x)p(y,x)x = 0, & \omega(x)^2 p(y,x) = 0, \\
q(y,x^2,x)x = 0, & \omega(x)q(y,x^2,x) = 0, & q(yx,x,x)x = 0, & \omega(x)q(yx,x,x) = 0, \\
q(yx.x,x,x) = 0, & q(yx^2,x,x) = 0, & q(yx,x^2,x) = 0, & q(y,x^2,x^2) = 0, \\
q(y,x^3,x) = 0. & & &
\end{array}
\tag{9}
$$

These are all the equations of type $[4,1]$ that are consequence of $f(x) = 0$. They are expressed in terms of the following types:

$$
\begin{array}{llllll}
\omega(y)x^4, & \omega(x)(xy.x)x, & \omega(x)x^2y.x, & \omega(x)x^3y, & \omega(xy)x^3, & \omega(x^2)xy.x, \\
\omega(x^2)x^2y, & \omega(x^2y)x^2, & \omega(x^3)xy, & \omega(x^3y)x, & \omega(x^4)y, & \omega(y)x^2x^2, \\
\omega(x)xy.x^2, & (xy.x)x.x, & (x^2y.x)x, & x^3y.x, & x^4y, & (xy.x)x^2, \\
x^2y.x^2, & x^3.xy, & (xy.x^2)x, & x^2x^2.y. & &
\end{array}
$$

Now we consider (9) as a system of equations where the indeterminates are these types. Reducing the matrix of the system to row canonical form we obtain

$$
\begin{array}{ccccccccc}
2 & & & & & 1 & -3 & -1 & 1 \\
 & 1 & & & -1 & & & & \\
 & & 1 & & & & -1 & & \\
 & & & 1 & -1 & -2 & 2 & 1 & -1
\end{array}
$$

The columns of this matrix correspond to the last nine types. We write these identities in polynomial form:

$$
2(xy.x)x.x + x^2y.x^2 - 3x^3.yx - (xy.x^2)x + x^2x^2.y = 0, \tag{10}
$$

$$(x^2y,x,x)+(x,x,y)x^2=0, \tag{11}$$

$$(x^3,y,x)=0, \tag{12}$$

$$(x^2,x,x)y+(x^2,yx,x)+2(x^2,x,yx)+2(y,x,x)x^2=0. \tag{13}$$

Identities (11), (12) and (13) are independent and imply (10). Therefore, we may state
of

Theorem 3. *Let A be a commutative baric algebra satisfying $x^3-\omega(x)^2x=0$. Then A satisfies the identities (11), (12) and (13). Furthermore any minimal identity of type $[4,1]$ is a consequence of these identities.*

4. Degree four

We consider now classes of baric algebras defined by $xy=yx$ and one equation of degree four. We have the following results:

Theorem 4.

(i) *For the equation $x^2x^2=\omega(x)x^3$ the minimal variety is defined by $xy=yx$,*

$$(x^2x^2,x,x)+2(x,x^2,x^3)=0,\quad 3(x,x,x^2)x^2+2(x^3,x^2,x)+(x^3,x,x^2)=0.$$

(ii) *For the equation $x^2x^2=\omega(x)^2x^2$ the minimal variety is defined by $xy=yx$,*

$$(x^2x^2,y,x)-2(x^2,y,x)x^2=0,$$

$$(y,x^2x^2,x)+2(x^2,x,yx^2)+2y(x^2,x^2,x)+2(x,yx.x,x^2)=0.$$

(iii) *For the equation $x^4=\omega(x)x^3$ the minimal variety is defined by $xy=yx$,*

$$(x^3,x,x^2)+(x^3,x,x)x=0,$$

$$3(yx.x^2,x,x)+(x^2x^2,x,y)+2(x,x,(yx.x)x)+$$
$$(y,x,x^4)+(x,x,y)x^3+(x,yx,x^3)+2(x,y,x^3)x=0.$$

(iv) *For the equation $x^4=\omega(x)^2x^2$ the minimal variety is defined by $xy=yx$,*

$$(y,x,x)x^3+(y,x,x)x^2.x+(x^2,y,x^3)+(x,x,x^3y)+(x,x,x^2y)x=0,$$

$$2(x,x,yx)x^2+2(x,x^3,yx)+(y,x^2x^2,x)+(y,x^3,x)x+2(x,x,y)x^2.x+$$
$$2(x,x,yx.x)x+(y,x^3,x^2)+(x,x,x^3)y+2(x^2,y,x)x.x+(x,x,x^2y.x)=0.$$

Theorem 5.

(i) *For the equation $x^2x^2=\omega(x)^3x$ the minimal identities have degree 6 and those of type [5,1] are given by*

$$(x^2x^2,y,x)=0,$$

$$2(x^2,x,(x,x,y))+2(yx.x,x,x^2)+2(yx,x,x^3)+(x,yx.x,x^2)+(x^2,x^3,y)=0,$$

$$(x^2y,x^2,x)+(y,x^2,x^3)+2(x,x,y)x^3+4(x,x^2,yx.x)+$$

$$2(x,x,x^2).yx+2(x^3,yx,x)+(y,x^2,x)x^2=0,$$

$$10(yx,x^2,x)x+3(x^2,x^3,y)+6(yx,x,x^3)+5(x,x^2y,x^2)+$$
$$2(yx.x,x^2,x)+6(x^2,x,x^2y)=0.$$

(ii) *For the equation $x^4=\omega(x)^3x$ the minimal identities have degree 6 and those of type [5,1] are given by*

$$(x^4,y,x)=0,$$

$$2(y,x,x)x^3+2(y,x,x)x^2.x+4((y,x,x)x.x)x+2(x,x^4,y)=0,$$

$$(x,y,x^2x^2)+2(x,y,x^2)x^2+2(x^2,y,x^3)+2(x^3,x^2,y)+2((y,x,x)x.x)x+$$
$$4(x^3,yx,x)+3(y,x,x)x^2.x+3(x,x^2,yx.x)+3(x^3,x,x)y+(y,x,x)x^3+$$
$$(x,x^2,y)x.x+2(yx.x^2,x,x)+(y,x,x^2x^2)+(x^2,x,x).yx=0,$$

$$2(y,x,x)x^3+(x,x^2x^2,y)+3(x,x^2y,x^2)+3(x^2y.x,x,x)+$$
$$2(x,y,x^4)+2(yx,x^3,x)+10((y,x,x)x.x)x+(x^2,x^3,y)+$$
$$2(y,x^2,x)x.x+3(x,x^4,y)+(x,x,x^3y)+2(y,x^3,x)x=0.$$

Acknowledgements

The computer program CRUNCH used to obtain the results of this paper was written by Professor Irvin Roy Hentzel of Iowa State University. To reduce a set of identities to a subset of independent identities we used the software package ALBERT created by D. P. Jacobs, S. V. Muddana and A. J. Offutt of Clemson University.

References

1. J. M. Osborn, Varieties of algebras, Adv. Math. 8:163-369 (1972).
2. I. R. Hentzel, Processing identities by group representation, Computers in nonassociative rings and algebras (R. E. Beck and B. Kolman, Eds.), Academic Press, New York, pp. 14-40 (1977).
3. J. Clifton, A simplification of the computation of the natural representation of the symmetric group S_n, Proc. Amer. Math. Soc., 83:248-250 (1981).
4. S. Walcher, Bernstein algebras which are Jordan algebras, Arch. Math. 50:218-222 (1988).
5. R. Costa, Shape identities in genetic algebras, Linear Alg. Appl., to appear.
6. M. T. Alcalde, C. Burgueño, C. Mallol, Les Pol(n,m)-algebres, Linear Alg. Appl., to appear.
7. I. Correa, I. R. Hentzel, L. A. Peresi, Minimal identities of Bernstein algebras, preprint.

ON POWER ASSOCIATIVE COMPOSITION ALGEBRAS

JOSÉ MARÍA PÉREZ
Departamento de Matemáticas, Universidad de Zaragoza
50009 Zaragoza, Spain

Abstract. Okubo showed that a power associative composition algebra over a field of characteristic not 2 or 3 is Hurwitz. In this paper we extend Okubo's result to composition algebras satisfying $(x,x,x) = (x,x,x^2) = 0$ over a field of characteristic not 2 which contains at least four elements.

An algebra A with product xy over a field F of characteristic not 2 is said to be a composition algebra if there is defined a nondegenerate symmetric bilinear form $(\ ,\ )$ on it satisfying the composition law:

$$(xy, xy) = (x,x)(y,y) \tag{1}$$

A unital composition algebra is called a Hurwitz algebra ([4]). By the generalized Hurwitz theorem, a Hurwitz algebra is isomorphic to one of the following algebras: F, $F \oplus F$, a quadratic field extension of F, a generalized quaternion algebra or an octonion algebra ([7]).

Hurwitz algebras play an important role in different parts of Algebra, so it is interesting to give characterizations of them.

An algebra is termed a power associative algebra if any element generates an associative subalgebra. If under any scalar extension we obtain a power associative algebra then we will say A to be a strictly power associative algebra.

In [5] Okubo proved

Theorem A. *Any finite dimensional power associative composition algebra over a field of chacteristic not* 2 *or* 3 *is Hurwitz.*

An algebra is termed a nth-power associative algebra if, for any element x, all the possible monomials in x of degree n are equal. This monomial is denoted by x^n.

It is known ([1]) that, under certain restrictions in the characteristic of the base field or commutativity assumptions, a third and fourth power associative algebra is a power associative algebra. Our main objetive is to prove:

Theorem A'. *Any finite dimensional composition algebra satisfying $x^2x = xx^2$ and $(x^2x)x = x^2x^2$ over a field of characteristic not* 2 *which contains at least four elements is Hurwitz.*

A few words about notation. The subalgebra generated by an element x is denoted by $alg\langle x\rangle$. $F\langle x_1, \ldots, x_n\rangle$ is the subspace spanned by the set $\{x_1, \ldots, x_n\}$. The right and left multiplication operators on A are denoted by R_x and L_x respectively. A will be a composition algebra with associated bilinear form $(\ ,\ )$ and F the base field with $charF \neq 2$. $[x,y]$ is the standard commutator $xy - yx$ and (x,y,z) is the standard associator $(xy)z - x(yz)$.

S. González (ed.), Non-Associative Algebra and Its Applications, 330–333.

Some previous results that we will use are ([2]):

Lemma 1. *If A is finite dimensional then the following statements are equivalent:*

(a) R_x *is bijective.*
(b) $(x,x) \neq 0$.
(c) L_x *is bijective.*

Lemma 2. *If A is finite dimensional and $[x,y]=0$ then*

$$(y,y)x^2 + (x,x)y^2 = 2(x,y)yx \tag{2}$$

The next propositon allows us to simplify the hypothesis of Theorem A. In the following we will suppose that A verifies the hypothesis of Theorem A'.

Proposition 3. *For any x in A:*

(a) $(x,x)x^3 + (x,x)^2 x = 2(x,x^2)x^2$.
(b) $alg\langle x\rangle = F\langle x, x^2\rangle$.
(c) A *is strictly power associative.*

Proof. As, by the restrictions on F, the identities $x^2x = xx^2$ and $x^2x^2 = (x^2x)x$ hold under scalar extension, we will consider F to be algebraically closed. Hence F is infinite, and we will be able to introduce the Zarisky topology, where the set $S = \{x \in A \mid (x,x) \neq 0\}$ becomes dense.

(a) For any $x \in S$, (2) with $y = x^2$ reduces to

$$(x,x)x^2x^2 + (x,x)^2x^2 = 2(x,x^2)x^2x \tag{3}$$

Since $x^2x^2 = x^3x$, simplifying x by Lemma 1, we obtain $x^3(x,x) + (x,x)^2x = 2(x,x^2)x^2$ for all x in S. By density it holds for any x in A.

(b) By (3) and part (a), $alg\langle x\rangle = F\langle x,x^2\rangle$ $\forall x \in S$. Given a basis $\{e_i\}$ of A, an element x can be written as $x = \sum x_i e_i$, so the coordinates of x, x^2, x^3 and x^2x^2 in $\{e_i\}$ are polynomial functions in $\{x_i\}$. For any x in S, the matrix whose entries are these polynomial functions has all 3×3 minors zero, because $alg\langle x\rangle = F\langle x,x^2\rangle$. Part (b) follows by density.

Using part (b), to prove (c) it is enough to check, by (a) and the hypothesis, that $(x^i, x^j, x^k) = 0 \quad 1 \leq i,j,k \leq 2$, for any element in S. By density this gives part (c). □

This proposition allows us to change the original assumptions on A by A to be a finite dimensional strictly power associative composition algebra over a field F of characterictic not 2. First of all, we note that it is enough to prove theorem A' for an algebraically closed field. Let K be the algebraic closure of F and 1 the unity of $K \otimes_F A$. Choosing $x \in A$ such that $(x,x) \neq 0$, by Lemma 1, R_x is bijective. The element $e = R_x^{-1}(x)$ lies in A and $xe = x$, so that, in $K \otimes_F A$, $x(e-1) = 0$. Again by Lemma 1, $1 = e \in A$ and hence A has a unit element.

In the remainder of the paper F will be algebraically closed. We look for an appropiate element to be the unit element.

We mean an unit idempotent e to be an idempotent such that $(e, e) = 1$.

Lemma 4. *If $(x, x) \neq 0$ then $alg\langle x\rangle$ has unity e_x, which is a unit idempotent.*

Proof. Choose $e_x = 2(x, x)^{-2}(x, x^2)x - (x, x)^{-1}x^2$ and use Proposition 3. □

Let e be one of those unit idempotents. It is well known ([6]) that an idempotent e, in a power associative algebra over an algebraically closed field, gives a descomposition, Peirce descomposition, in subspaces relative to the multiplication operator by e in the following way:

$$A = A_1 \oplus A_{1/2} \oplus A_0$$

where $A_i = \{x \in A \mid ex + xe = 2ix\}$ $i = 0, 1/2, 1$.

Proposition 5. *We have*

(a) $A_1 = \{x \in A \mid x = ex = xe\}$.
(b) $A_0 = 0$.

Proof. Linearizing $[x, x^2] = 0$ gives

$$[y, x^2] + [y, xy + yx] = 0 \tag{3}$$

Replacing x by e and y by $x_i \in A_i$ we obtain $(2i - 1)[e, x_i] = 0$. This establishes (a). Moreover, any $x \in A_0$ satisfies $ex = xe$ and $ex + xe = 0$. Thus $A_0 = 0$. □

Finally

Proposition 6. *It holds*

(a) $(A_{1/2}, A_1) = 0$.
(b) $A_{1/2} = 0$.

Proof. (a) Setting $x \in A_{1/2}$ and $y \in A_1$, $(x, y) = (ex, y) + (xe, y) = (ex, ey) + (xe, ye) = 2(x, y)$. So $(x, y) = 0$ and $(A_{1/2}, A_1) = 0$.

(b) Linearizing $(x, x, x^2) = 0$ we obtain

$$(y, y, x^2) + (y, x, yx + xy) + (x, y, yx + xy) + (x, x, y^2) = 0$$

This relation with $y = e$ and $x \in A_{1/2}$ yields

$$x^2e = e(ex^2)$$

On the other hand, the same substitution in (3) gives

$$ex^2 = x^2e$$

using Lemma 1 we obtain $x^2 = ex^2$.

If there exists x in $A_{1/2}$ such that $(x, x) \neq 0$, by lemma 4, $alg\langle x\rangle$ has a unity e_x. Thus $x^2 = e_x x^2$ and, by the previous, $(e_x - e)x^2 = 0$. Since $(x^2, x^2) = (x, x)^2 \neq 0$, by Lemma 1, $e_x = e$ and so $x \in A_1$. Thus $x = 0$, but this contradicts $(x, x) \neq 0$. Hence $(A_{1/2}, A_{1/2}) = 0$. By part (a) and Proposition 5, $(A_{1/2}, A) = 0$. Since $(\ ,\ )$ is nondegenerate, $A_{1/2} = 0$. □

Theorem A' follows from Propositions 5 and 6.

Third power associative composition algebras have been determined in [3] over fields of characteristic $\neq 2, 3$. The results there allow to suppress the restriction on the number of elements in Theorem A'.

References

1. A.A. Albert, Power-associative rings, Trans. Amer. Math. Soc. **64** (1948), 552-593.
2. A. Elduque and H.C. Myung, Flexible composition algebras and Okubo algebras, Commun. Algebra **19** (1991), 1197-1227.
3. A. Elduque and J.M. Pérez, Third power associative composition algebras, to appear.
4. H.C. Myung, *Malcev-admissible algebras*, Birkhäuser, Boston-Basel-Stuttgart, 1986.
5. S. Okubo, Dimension and classification of general composition algebras, Hadronic J. **4** (1981), 216-273.
6. R.D. Schafer, *An introduction to nonassociative algebras*, Academic Press, New York, 1966.
7. K.A. Zhevlakov, A.M. Slinko, I.P. Shestakov and A.I. Shirsov, *Rings that are nearly associative*, Academic Press, New York, 1982.

ENUMERATION AND CLASSIFICATION OF ALBERT ALGEBRAS: REDUCED MODELS AND THE INVARIANTS MOD 2

HOLGER P. PETERSSON
Fachbereich Mathematik, FernUniversität
Lützowstraße 125
D-58084 Hagen
Germany

and

MICHEL L. RACINE
Department of Mathematics, University of Ottawa
585 King Eduard
K1N 6N5 Ottawa, Ontario
Canada

Abstract.

Key words: Albert division algebras. Reduced models. Tits process. Invariants mod 2.

1. Introduction

In this report, we are concerned with the problem of enumerating and classifying Albert algebras over an arbitrary base field k, for simplicity assumed to be of characteristic not 2 and 3. This assumption, though actually unnecessary, allows us to keep prerequisites from the theory of Jordan algebras at a minimum; in particular, plain old *linear* Jordan algebras (instead of quadratic ones) form a perfectly adequate framework for the results we are interested in. The reader who cares also about the characteristics 2 and 3 may consult the literature cited as we go along.

The principal result to be announced here effectively attaches to any Albert algebra $\mathcal{J}$ over k its unique reduced model - a reduced Albert algebra which is characterized by the condition that it becomes isomorphic to $\mathcal{J}$ whenever scalars are extended to an arbitrary reducing field of $\mathcal{J}$ and which provides a convenient way of defining the invariants mod 2 recently introduced by Serre [Se1] and Rost [Ro1] using a different method. Proofs will appear elsewhere.

2. Albert algebras

We write C_{split} for the *split octonions*, that is, for the unique octonion algebra over k containing zero divisors, its most natural realization being the one by Zorn vector matrices [J1, p. 142]. C_{split} allows a unique involution, written as $^-$, which is central in the sense that, for all $x \in C_{\text{split}}$, $x = \bar{x}$ iff $x \in k1$. Writing M_3 for the functor of 3-by-3 matrices from nonassociative algebras to itself, and $^t-$ for the

S. González (ed.), Non-Associative Algebra and Its Applications, 334–340.

conjugate transpose, the k-vector space

$$\mathcal{J}_{\text{split}} = \{x \in \mathrm{M}_3(C_{\text{split}}) : x = {}^t\overline{x}\},$$

considered under the classical symmetric matrix product

$$(x, y) \mapsto x \cdot y = \frac{1}{2}(xy + yx),$$

is well known to become a central simple exceptional Jordan algebra of dimension 27, called the *split Albert algebra over k*. By an *Albert algebra over k* we mean a nonassociative k-algebra $\mathcal{J}$ which is a *k-form of $\mathcal{J}_{\text{split}}$*, so becomes isomorphic to it after extending scalars to the algebraic closure. If this is so the ordinary trace of matrices via descent induces a nondegenerate quadratic form on $\mathcal{J}$, called its *trace form* and written as $T = T_{\mathcal{J}}$.

3. Enumeration of reduced Albert algebras

As an ad hoc definition, we declare an Albert algebra $\mathcal{J}$ with bilinear multiplication $(x, y) \mapsto x \cdot y$ to be *reduced* if it contains zero divisors, i.e., if there are nonzero elements $a, b \in \mathcal{J}$ satisfying $a \cdot b = 0$. That this terminology, as well as the corresponding one of an Albert division algebra to be defined below, agrees with the established usage, follows from [P2, Corollary 3]. *Enumeration* of reduced Albert algebras, that is, the task of writing them all down, is accomplished by a classical result due to Schafer [S] (see also Racine [R, Theorem 1]) which may be described as follows. Let C be an *octonion algebra over k*, i.e., a k-form of C_{split}. As the algebra C_{split}, C as well allows a unique central involution, written as $^-$, and given any diagonal matrix $g \in \mathrm{GL}_3(k)$, the k-vector space

$$\mathrm{H}_3(C, g) = \{x \in \mathrm{M}_3(C) : x = g^{-1}{}^t\overline{x}g\},$$

again considered under the symmetric matrix product, is easily seen to be a reduced Albert algebra over k. Conversely, we have:

Theorem 1. (Schafer [S]). *Let $\mathcal{J}$ be a reduced Albert algebra over k. Then there exist an octonion algebra C over k as well as a diagonal matrix $g \in \mathrm{GL}_3(k)$ such that $\mathcal{J} \cong \mathrm{H}_3(C, g)$.*

4. Classification of reduced Albert algebras

Classification of reduced Albert algebras, that is, the task of deciding under what circumstances two reduced Albert algebras (given in the form of Theorem 1, for example) are isomorphic, is acomplished by attaching invariants. In this particular instance, a single invariant actually suffices, as may be seen from the following result. (cf. **2.**) (see also Racine [R, Theorem 3]).

Theorem 2. (Springer [Sp], Serre [Se2]). *Two reduced Albert algebras are isomorphic if and only if they have isometric trace forms.*

For a proof the reader is referred to Petersson-Racine [PR3]. In particular, if a reduced Albert algebra $\mathcal{J}$ is given in the form of Theorem 1, it follows that the octonion algebra C up to isomorphism is uniquely determined by $\mathcal{J}$. Hence we are allowed to call it the *coordinate algebra of* $\mathcal{J}$.

5. Enumeration of Albert division algebras

Albert algebras which are not reduced are division algebras in the sense that all equations $a \cdot x = b$ for arbitrary $a \neq 0 \neq b$ have a unique solution. The most convenient way to enumerate Albert division algebras is by means of the *Tits process* (cf. Petersson-Racine [PR1, PR2]), a general construction depending on four parameters $B, *, u, \beta$.
Here $(B, *)$ is a *central simple associative k-algebra of degree* 3 *with involution of the second kind*, i.e, a k-form of

$$(\ \mathrm{M}_3(k) \oplus \mathrm{M}_3(k)^{\ \mathrm{op}},\ \varepsilon),$$

where ε stands for the exchange involution. It follows that the center, L, of B is either a quadratic field extension of k or isomorphic to $k \oplus k$, that B is central separable of degree 3 and rank 9 over L (so, in particular, $\dim_k B = 18$) and that the generic norm of B exists, being a cubic form $N_B : B \to L$. By contrast,

$$A = \mathrm{H}(B, *) = \{a \in B : a = a^*\}$$

forms a central simple Jordan algebra of degree 3 and dimension 9 *over* k whose generic norm satisfies

$$N_A = N_B \mid_A : A \to k.$$

Furthermore, assume we are given elements $u \in A^\times$, $\beta \in L^\times$ related to each other by the condition

$$N_A(u) = \beta\beta^*\ .$$

Then the vector space

$$\mathcal{J} = A \oplus B$$

carries a unique commutative k-algebra structure such that (i) A is a unital subalgebra of $\mathcal{J}$ acting on B via

$$(a, b) \mapsto a \cdot b = \frac{1}{2}(T(a)1 - a)b,$$

T being the generic trace of A and B, and (ii) the square in $\mathcal{J}$ of any $b \in B$ is given by the expresion

$$(T(bub^*)1 - bub^*) \oplus (\beta^* b^{*2} u^{-1} - \beta^* T(b)^* b^* u^{-1} + \frac{1}{2}\beta^*(T(b)^{*2} - T(b, b)^*)u^{-1}).$$

We call this construction the *Tits process* and write the resulting algebra as $\mathcal{J}(B, *, u, \beta)$. The last part of the following theorem in particular takes care of the enumeration problem for Albert division algebras.

Theorem 3. (Tits [J2], Petersson-Racine [PR1, PR2]).
a) *$\mathcal{J} = \mathcal{J}(B, *, u, \beta)$ is an Albert algebra.*
b) *$\mathcal{J}$ is a division algebra if and only if $\beta \notin N_B(B^\times)$.*
c) *Every Albert algebra has the form $\mathcal{J}(B, *, u, \beta)$ as above.*

6. Cohomological invariants

Thanks to the work of Serre [Se1-3] and Rost [Ro1, Ro2] it is possible to attach three cohomological invariants to any Albert algebra $\mathcal{J}$, to wit,

- two invariants, closely related to the trace form and called the *invariants mod* 2, belonging to $H^3(k, \mathbf{Z}/\mathbf{Z}2)$, $H^5(k, \mathbf{Z}/\mathbf{Z}2)$, respectively,
- a single invariant, called the *invariant mod* 3, belonging to $H^3(k, \mathbf{Z}/\mathbf{Z}3)$.

In this report, we ignore the invariant mod 3 and address ourselves to the problem of describing a convenient way to define the invariants mod 2. Our approach is based on the following concept.

7. The reduced model of an Albert algebra

Let $\mathcal{J}$ be an Albert algebra over k. By a *reduced model of* $\mathcal{J}$ we mean an Albert algebra $\mathcal{J}_{\text{red}}$ over k which is reduced and satisfies $\mathcal{J} \otimes_k k' \cong \mathcal{J}_{\text{red}} \otimes_k k'$ for any reducing field k'/k of $\mathcal{J}$ (e.g., any cubic subfield of $\mathcal{J}$). We can now state our principal result.

Main Theorem. *Every Albert algebra admits a reduced model which is unique up to isomorphism.*

As an application of this, we may declare the invariants mod 2 of $\mathcal{J}$ to be the invariants mod 2 of $\mathcal{J}_{\text{red}}$ as defined in Serre [Se1] or Petersson-Racine [PR4]. This definition is equivalent to the one of the Serre-Rost approach [Se3, Ro2] since $\mathcal{J}$ always allows reducing fields k'/k of *odd* degree and then the restriction map

$$H^i(k, \mathbf{Z}/\mathbf{Z}2) \longrightarrow H^i(k', \mathbf{Z}/\mathbf{Z}2) \qquad (i \geq 0)$$

is injective. Conversely, one could use the invariants mod 2 to prove existence and uniqueness of a reduced model. We prefer to adopt a different approach here which not only leads to a proof of the Main Theorem but at the same time yields an explicit description of $\mathcal{J}_{\text{red}}$ in terms of the parameters used to build up $\mathcal{J}$ by means of the Tits process.

8. The octonion algebra of a 9-dimensional Jordan algebra of degree 3

Let $\mathcal{J}$ be a central simple Jordan algebra of dimension 9 and degree 3 over k. Then $\mathcal{J} \cong H(B, *)$ for some central simple associative k-algebra $(B, *)$ of degree 3 with insolution of the second kind, unique up to isomorphism (cf. Jacobson [J2, V Theorem 11].
By Theorem 3 b), the Tits process Albert algebra $\mathcal{J}(B, *, 1, 1)$ is reduced and hence

has a certain coordinate algebra (**4.**), which is some octonion algebra over k depending only on A. Following Petersson-Racine [PR4], we call it the *octonion algebra of* A and write it as Oct A. This set-up is closely connected to constructions due to Petersson [P1], Okubo [O] and Faulkner [F], see also Elduque-Myung [EM]. Following Petersson-Racine [PR2], A arises from the Tits process: Given any 3-dimensional separable commutative associative subalgebra $E \subset \mathcal{J}$, there exist a 2-dimensional composition algebra M over k with canonical involution $^-$ as well as invertible elements $v \in E, \gamma \in M$ satisfying $N_E(u) = \gamma\bar{\gamma}$ such that

$$A \cong \mathcal{J}(E \otimes_k M, \mathbf{1}_E \otimes {}^-, v, \gamma).$$

We now have:

Theorem 4. *Let A be a central simple Jordan algebra of dimension* 9 *and degree* 3 *over k, given as*

$$A \cong \mathcal{J}(E \otimes_k M, \mathbf{1}_E \otimes {}^-, v, \gamma)$$

by means of the Tits process. Let

$$h = \mathrm{diag}(h_1, h_2, h_3) \in \mathrm{GL}_3(k)$$

represent the symmetric bilinear form

$$E \times E \to k,\ (w, w') \mapsto T_E(vww')$$

relative to some basis of E over k. Then

$$\mathrm{Oct}\ A \cong \mathrm{Cay}(M; -d_{E/k}h_1, -d_{E/k}h_2),$$

Cay *referring to the Cayley-Dickson doubling process and $d_{E/k}$ being the discriminant of E/k.*

The proof of this result is based on the description of the norm of Oct A given in Petersson-Racine [PR4, 4.2] and on Frobenius reciprocity for the Scharlau transfer in the theory of quadratic forms [Sc, 2.5.6].

9. Reduced models and the Tits process

Let $\mathcal{J}$ be an Albert algebra over k, given as

$$\mathcal{J} = \mathcal{J}(B, *, u, \beta)$$

by means of the Tits process (Theorem 3b)). Observing that $u \in A^\times$ (**5.**) and that the u-isotope $A^{(u)}$ of $A = \mathrm{H}(B, *)$ is a central simple Jordan algebra of degree 3 and dimension 9, we have:

Theorem 5. *Notations being as above, the reduced model of $\mathcal{J}$ may be described explicitly as*

$$\mathcal{J}_{\mathrm{red}} = \mathrm{H}_3(C_{\mathrm{red}}, g_{\mathrm{red}})$$

where $C_{\text{red}} = \text{Oct } A^{(u)}$. *Also,* C_{red} *contains the center* L *of* B *as a composition subalgebra, hence may be realized as*

$$C_{\text{red}} = \text{Cay}(L; g_1, g_2)$$

by means of the Cayley-Dickson doubling process, and then

$$g_{\text{red}} = \text{diag}(-g_1, -g_2, 1).$$

Observe that an explicit description of $C_{\text{red}} = \text{Oct } A^{(u)}$ above may be read off from Theorem 4.

10. Open questions

The most important open question in this context has been raised by Serre [Se1]: Do the invariants mod 2 and 3 classify Albert algebras?
The main result of this report may be expressed by saying that, given any Albert algebra $\mathcal{J}$, there always exists another Albert algebra having trivial invariant mod 3 and the same invariants mod 2 as $\mathcal{J}$. Here it is natural to ask the dual question: Does there exist an Albert algebra having trivial invariants mod 2 and the same invariant mod 3 as $\mathcal{J}$?

References

[EM] A. Elduque and H. C. Myung. *On flexible composition algebras.* Comm. Algebra **21** (7) (1993), 2481 - 2505.

[F] J. R. Faulkner. *Finding octonion algebras in associative algebras.* Proc. Amer. Math. Soc. **104** (4) (1988) 1027 - 1030.

[J1] N. Jacobson. "Lie algebras". Interscience Publ., New York - London - Sydney, 1966.

[J2] – "Structure and representations of Jordan algebras". Amer. Math. Coll. Publ. **39**, Providence, RI, 1968.

[O] S. Okubo. *Pseudo-quaternion and pseudo-octonion algebras.* Hadronic J. **1** (1978), 1250 - 1278.

[P1] H. P. Petersson. *Eine Identität fünften Grades, der gewisse Isotope von Kompositionsalgebren genügen.* Math. Z. **109** (1969), 217 - 238.

[P2] – *On linear and quadratic Jordan division algebras.* Math. Z. **177** (1981), 541 - 548.

[PR1] H. P. Petersson and M. L. Racine. *Jordan algebras of degree 3 and the Tits process.* J. Algebra **98** (1986), 211 - 243.

[PR2] – *Classification of algebras arising from the Tits process.* J. Algebra **98** (1986), 244 - 279.

[PR3] – *Albert algebras.* In W. Kaup, K. McCrimmon, H. P. Petersson (eds.): Proc. of the Conference on Jordan algebras at Oberwolfach 1992. To appear.

[PR4] – *On the invariants* mod 2 *of Albert algebras*. Submitted.

[R] M. L. Racine. *A note on quadratic Jordan algebras of degree* 3. Trans Amer. Math. Soc. **164** (1972), 93 - 103.

[Ro1] M. Rost. *A* (mod 3) *invariant for exceptional Jordan algebras*. C. R. Acad. Sci. Paris **315** Série I (1991), 823 - 827.

[Ro2] – *A descent property for Pfisterforms*. Preprint, 1991.

[S] R. D. Schafer. *The exceptional simple Jordan algebras*. Amer. J. Math. **70** (1948), 82 - 94.

[Sc] W. Scharlau. "Quadratic and hermitian forms". Springer Verl., Berlin-Heidelberg, 1985.

[Se1] J.-P. Serre. *Résumé des cours de l'année* 1990-91. Annuaire du Collège de France.

[Se2] – Lettre à M. L. Racine de 25 April 1991.

[Se3] – Lettre à M. L. Racine de 5 Juin 1992.

[Sp] T. A. Springer. *The classification of reduced exceptional simple Jordan algebras*. Nederl. Akad. Wetensch. Proc. Ser. A **63** = Indag. Math. **22** (1960), 414 - 422.

COMPOSITION ALGEBRAS OVER OPEN DENSE SUBSCHEMES OF CURVES OF GENUS ZERO

SUSANNE PUMPLÜN
Fachbereich Mathematik, FernUniversität - Gesamthochschule -
Lützowstraße 125
58084 Hagen, Germany

Abstract. With the help of some basic results from algebraic geometry and the Riemann-Roch Theorem the structure of composition algebras over certain classes of rings is investigated.

Key words: Composition algebras, curves of genus zero

While the structure of composition algebras over fields is relatively well known, there are many questions left open concerning composition algebras over rings.

Let R be a commutative associative ring. Then an R-algebra C is said to be a *composition algebra* if it admits a quadratic form $N : C \to R$ satisfying the following two conditions: (1) Its induced symmetric bilinear form is nondegenerate, so it induces an R-module isomorphism from C to its dual $\check{C} = \mathrm{Hom}_R(C, R)$ and (2) N permits composition, that is $N(uv) = N(u)N(v)$ for all $u, v \in C$. Here, the term "R-algebra" refers to unital nonassociative algebras over R, which are finitely generated projective of rank > 0 as R-modules. The form N ist uniquely determined by these two conditions and is called the *norm* of C.
Composition algebras over R are quadratic and alternative of rank $1, 2, 4$ or 8 and invariant under base change ([P1]). A composition algebra of rank 2 (resp. 4, resp. 8) is called torus (resp. quaternion algebra, resp. octonion algebra).

In the special case of the polynomial ring $k[t]$ over a certain field k the composition algebras are well known:

Theorem 1. (Petersson [P1], 6.8) *Let C be a composition algebra of rank r over $k[t]$ and assume char $k \neq 2$ or $r > 2$ and k perfect, then C is defined over k.*

Petersson's proof of this theorem adapts a classical result of Harder [K] on symmetric bilinear forms over $k[t]$, char $k \neq 2$, to the setting of composition algebras.
One can generalize this proof to get information on composition algebras over certain classes of rings.
Fix an arbitrary base field k, char $k \neq 2$. To understand the idea for this generalization it is important to know the structure of the projective line over k, which is denoted by $\mathbf{P}^1_k$.
Following [P2] the projective line consists of

S. González (ed.), Non-Associative Algebra and Its Applications, 341–343.

- the *base set* X of all homogeneous prime ideals $P \subset S := k[x_0, x_1]$, which do not contain both variables x_0 and x_1;

- the *Zariski topology* on X, for which the *principal open sets*

$$\mathcal{D}(f) := \{P \in X : f \notin P\},\ f \in S \text{ homogeneous, form a base;}$$

- the *structure sheaf* $\mathcal{O}_X$ of unital commutative associative k-algebras on the topological space X with the sections given by

$$\Gamma(U, \mathcal{O}_X) = \bigcap_{P \in U} S_{(P)_0} \subset k(t),\ t = \frac{x_0}{x_1},$$

for open $\phi \neq U \subset X$ and with the restriction morphisms from U to an open $\phi \neq V \subset U$ simply given by the corresponding inclusion maps.
Here, $S_{(P)_0}$ denotes the homogeneous elements of degree 0 in the graded k-algebra obtained from S by homogeneous localization relative to P.

Let X' be a (geometrically integral, complete smooth) curve over k of genus 0. Hence X' is either isomorphic to $\mathbf{P}^1_k$ (then X' is called *rational*) or it is a form of the projective line. (It becomes isomorphic to $\mathbf{P}^1$ after passing to the algebraic closure of k or an appropriate quadratic extension). Then X' is called *nonrational*. If $X' \cong \mathbf{P}^1_k$, the function field $K(X')$ of X' is $k(t)$. If X' is nonrational, $K(X')$ is the function field of an anisotropic conic $< 1, -a, -b >$ over k, $K(X') = k(t)(\sqrt{at^2 + b})$. The fact that $\mathbf{P}^1_k - \{\infty\} = \mathbf{A}^1_k \cong \operatorname{Spec} k[t]$ for the closed point $\infty \in \mathbf{P}^1_k$ of degree 1 was used by Petersson in his above mentioned proof to get information on composition algebras over $k[t]$. Now, when one takes a closed point $P_0 \in X'$ of degree 2 (i.e., the residue field of P_0 is a quadratic extension of k) out of X', one gets the affine scheme of a ring R

$$X' - \{P_0\} \cong \operatorname{Spec} R.$$

It is obvious from the construction that

$$k \subset R \subset K(X'),\ \operatorname{Quot} R \cong K(X')$$

and R is Dedekind.

In case $X' \cong \mathbf{P}^1_k$, any such P_0 is represented by the principal ideal generated by an irreducible monic polynomial $f \in k[t]$ of degree 2 and

$$R \cong \left\{ \frac{g(t)}{f^n(t)} \in k(t) \mid g(t) \in k[t],\ n \geq 0,\ \deg g \leq 2n \right\}$$

is not a principal ideal domain.

If X' is nonrational, let for instance P_0 be the place of the unique extension of the place ∞ of $k(t)$ to $K(X') = k(t)(\sqrt{at^2 + b})$. Following Pfister [Pf] this place is also denoted by ∞. Then

$$R \cong k[t][\sqrt{at^2 + b}]$$

and R is a principal ideal domain ([Pf], Prop. 1).
For these rings the following holds:

Theorem 2. *Let C be a composition algebra without zero divisors of rank $r > 2$ over R. Then either C is defined over k or contains a composition algebra of rank $\frac{r}{2}$, which is defined over k.*

For the proof of this theorem one can use the theory of composition algebras in the language of locally ringed spaces by Petersson [P1] and, in particular, the categorical equivalence between the category of composition algebras over R and composition algebras over Spec R, as well as some elementary lattice theory and the Riemann-Roch Theorem. The following example is an application of this theorem.

Example 3. *Take $k = \mathbf{R}$ the real numbers. Then there exists - up to isomorphism - only one quaternion algebra, namely the split quaternion algebra of 2-by-2 matrices $Mat_2(R)$ over the principal ideal domain $R = \mathbf{R}[t][\sqrt{-t^2-1}]$, and also only one octonion algebra, namely the split octonion algebra of Zorn's vector matrices $Zor(R)$.*

If one takes a point of degree 3 out of any curve X' of genus 0, there is a similar result as above for the octonion algebras over the affine scheme. However, when taking out points of higher degree this technique does not give any new results. Proofs will appear elsewhere.

References

[K] M. Knebusch. *Grothendieck- und Wittringe von nicht-ausgearteten symmetrischen Bilinearformen.* Sitzungsber. Heidelb. Akad. Wiss. Math.-Natur.Kl., Springer-Verlag, Berlin, Heidelberg and New York, 1970.

[P1] H. P. Petersson. *Composition Algebras over Algebraic Curves of Genus Zero.* Trans. Amer. Math. Soc. 337 (1) (1993), 473 - 491.

[P2] - *Composition algebras over the projective line.* Contemp. Math. 131 (2) (1992), 673 - 677.

[Pf] A. Pfister. *Quadratic lattices in function fields of genus* 0. Proc. London Math. Soc. 66 (2) (1993) pp, 257 - 278.

SIMPLE JORDAN SUPERALGEBRAS

M. L. RACINE *
*Department of Mathematics,
University of Ottawa,
Ottawa, Ontario K1N 6N5,
Canada.*

and

E. I. ZEL'MANOV
*Department of Mathematics,
University of Wisconsin,
Madison, Wisconsin 53706-1388,
USA.*

1. Introduction

Let K be a field of characteristic not 2, $\Gamma =< 1, g_i \mid i = 1, 2, \ldots >$ the *Grassman* (or *exterior*) *algebra* over K on a countable number of generators g_i, with

$$g_i^2 = 0, \qquad g_i g_j = -g_j g_i, \qquad i \neq j.$$

The elements

$$1, \quad g_{i_1} g_{i_2} \cdots g_{i_r}, \qquad i_1 < i_2 < \ldots < i_r$$

form a K-basis of Γ. Letting Γ_0 (respectively Γ_1) be the span of the products of even length (respectively of odd length), Γ is the direct sum of its even and odd parts:

$$\Gamma = \Gamma_0 + \Gamma_1.$$

If $\mathcal{V}$ is a variety of algebras, a $\mathbf{Z}_2$-graded K-algebra

$$A = A_0 + A_1$$

is a *$\mathcal{V}$-superalgebra* if its *Grassman envelope*

$$\Gamma(A) := A_0 \otimes \Gamma_0 + A_1 \otimes \Gamma_1$$

belongs to $\mathcal{V}$.

An *associative superalgebra* is nothing but a $\mathbf{Z}_2$-graded associative algebra. For example, $n + m \ \times \ n + m$ matrices $M_{n+m}(K)$ can be viewed as an associative

* The research of the first author is supported in part by a grant from NSERC.

S. González (ed.), Non-Associative Algebra and Its Applications, 344–349.

superalgebra by taking the diagonal components $M_n(K)$ and $M_m(K)$ as the even part and the off-diagonal components as the odd part; this is an example of a simple associative superalgebra.

$$\left(\begin{array}{cc|c} \cdot & \cdot & \cdot \\ \cdot & \cdot & \cdot \\ \hline \cdot & \cdot & \cdot \end{array}\right).$$

For another example take

$$\left\{\begin{pmatrix} a & b \\ b & a \end{pmatrix} \mid a, b \in M_n(K)\right\}.$$

This algebra is simple as a superalgebra but not as an algebra; it is commutative if and only if $n = 1$. A *Jordan superalgebra* $J = J_0 + J_1$ satisfies *supercommutativity*

$$a_\alpha \mathrm{R}_{b_\beta} = (-1)^{\alpha\beta} b_\beta \mathrm{R}_{a_\alpha},$$

where $a_\alpha \in J_\alpha, \alpha, \beta \in \mathbf{Z}_2$ and R denotes multiplication on the right. It also satisfies the *linearized Jordan identity* in operator form

$$\begin{aligned} \mathrm{R}_{a_\alpha}\mathrm{R}_{b_\beta}\mathrm{R}_{c_\gamma} &+ (-1)^{\alpha\beta+\alpha\gamma+\beta\gamma}\mathrm{R}_{c_\gamma}\mathrm{R}_{b_\beta}\mathrm{R}_{a_\alpha} + (-1)^{\beta\gamma}\mathrm{R}_{(a_\alpha c_\gamma)b_\beta} \\ &= \mathrm{R}_{a_\alpha b_\beta}\mathrm{R}_{c_\gamma} + (-1)^{\beta\gamma}\mathrm{R}_{a_\alpha c_\gamma}\mathrm{R}_{b_\beta} + (-1)^{\alpha\beta+\alpha\gamma}\mathrm{R}_{b_\beta c_\gamma}\mathrm{R}_{a_\alpha} \\ &= (-1)^{\alpha\beta+\alpha\gamma}\mathrm{R}_{c_\gamma}\mathrm{R}_{a_\alpha b_\beta} + (-1)^{\alpha\beta}\mathrm{R}_{b_\beta}\mathrm{R}_{a_\alpha c_\gamma} + \mathrm{R}_{a_\alpha}\mathrm{R}_{b_\beta c_\gamma}, \end{aligned}$$

where $a_\alpha \in J_\alpha$, $b_\beta \in J_\beta$ and $c_\gamma \in J_\gamma$. Intuitively if two odd elements are transposed the sign changes. If the characteristic is not 3 then these two equations define a Jordan superalgebra. If the characteristic is 3, we also need the Jordan identity for J_0. Superalgebras were introduced by physicists. Jordan superalgebras were first studied by Kac and Kaplansky (see [2] for the chronology).

An element of J is said to be *homogeneous* if it belongs to J_0 or J_1. By supercommutativity the product of two homogeneous elements is commutative unless they are both odd in which case it anticommutes. As a way of remembering this we will denote the product by a *dot* in the commutative cases and by *brackets* in the anticommutative case. Moreover since J_0 is a Jordan algebra we denote it A and J_1 is an A-bimodule which we denote M.

2. Examples:

1) If $B = B_0 + B_1$ is an associative superalgebra then the product defined by $a_\alpha \mathrm{R}_{b_\beta} := \frac{1}{2}(a_\alpha b_\beta + (-1)^{\alpha\beta} b_\beta a_\alpha)$ defines a Jordan superalgebra structure on B which we denote by B^+. In particular, if $B = M_{n+m}(K)$ is an associative superalgebra as above, we denote B^+ by $M_{n,m}(K)$. Following Kac, we denote the plus superalgebra of

$$\left\{\begin{pmatrix} a & b \\ b & a \end{pmatrix} \mid a, b \in M_n(K)\right\}$$

by $Q(n)$. In general if B is a simple associative superalgebra which is not commutative then B^+ is a simple Jordan superalgebra. Thus $M_{n,m}(K)$ is simple and so is Q_n when $n > 1$. A Jordan superalgebra is said to be *special* if it is isomorphic to a subsuperalgebra of some B^+, B an associative superalgebra.

2) For B an associative superalgebra, let * be a graded linear map such that $a^{**} = a$ and $(a_\alpha b_\beta)^* = (-1)^{\alpha\beta} b_\beta^* a_\alpha^*$. Such a * is called a *superinvolution* and $H(B,^*) = \{b \in B \mid b^* = b\}$, the symmetric elements of B, form a Jordan subsuperalgebra of B^+. For $B = M_{n+2m}(K)$ as in example 1), let * be the involution induced by

$$\begin{pmatrix} I_n & 0 \\ 0 & S_m \end{pmatrix} \qquad \text{where } S = \begin{pmatrix} 0 & 1 \\ -1 & 0 \end{pmatrix}.$$

The simple Jordan superalgebra $H(B,^*)$ is called the *orthosymplectic superalgebra* and is denoted $osp(n, 2m)$.

3) For $B = M_{n+n}(K) = \left\{\begin{pmatrix} a & b \\ c & d \end{pmatrix} \mid a, b, c, d \in M_n(K)\right\}$, let * be the superinvolution given by $\begin{pmatrix} a & b \\ c & d \end{pmatrix}^* = \begin{pmatrix} d^t & -b^t \\ c^t & a^t \end{pmatrix}$, where t is the transpose involution. Then

$$H(B,*) = \left\{\begin{pmatrix} a & b \\ c & d \end{pmatrix} \in M_{n+n}(K) \mid d = a^t, b^t = -b, c^t = c\right\}$$

is a subsuperalgebra of $M_{n,n}(K)$ which, folowing Kac, we denote $P(n)$. For $n > 1$, $P(n)$ is simple.

4) Let $V = V_0 \oplus V_1$ be a graded vectorspace. A *superform* (,) on V is is a bilinear form on $V = V_0 \perp V_1$ whose restriction to V_0 is symmetric and whose restriction to V_1 is skew-symmetric. Let $A = Ke + V_0$, $M = V_1$ and $J = A + M$. Then $e.x := x$ and $v.w := (v, w)e$ for $x \in J$, $v, w \in V$ define a Jordan superalgebra structure on J. J is simple if and only if the form (,) is nondegenerate.

5) K_3, the *Kaplansky superalgebra*: $A = Ke$, $M = Kx + Ky$ with $e^2 = e$, $e.x = \frac{1}{2}x$, $e.y = \frac{1}{2}y$ and $[x, y] = e$.

6) D_t, $A = Ke_1 + Ke_2$, $M = Kx + Ky$ with $e_i^2 = e_i$, $e_1.e_2 = 0$, $e_i.x = \frac{1}{2}x$, $e_i.y = \frac{1}{2}y$ and $[x, y] = e_1 + te_2$. The Jordan superalgebra D_t is simple if and only if $t \neq 0$.

7) K_{10}, the *Kac superalgebra*: $A = (Ke_1 + \sum_{1\le i\le 4} Kv_i) + Ke_2$ and $M = \sum_{i=1,2}(Kx_i + Ky_i)$; the following basis for the Kac algebra $J = A + M$ is obtained by scaling the usual basis [1]: $e_1, v_1, v_2, v_3, v_4, e_2, x_1, y_1, x_2, y_2$, where

$$\begin{aligned}
e_i^2 &= e_i, \quad e_1.v_i = v_i, \quad && v_1.v_2 = 2e_1 \quad v_3.v_4 = 2e_1 \\
e_i.x_j &= \tfrac{1}{2}x_j, \ y_1.v_1 = x_2, \ y_2.v_1 = -x_1, \ && x_1.v_2 = -y_2, \ x_2.v_2 = y_1, \\
e_i.y_j &= \tfrac{1}{2}y_j, \ x_2.v_3 = x_1, \ y_1.v_3 = y_2, \ && x_1.v_4 = x_2, \quad y_2.v_4 = y_1,
\end{aligned}$$

$$[x_i,\ y_i] = e_1 - 3e_2,\ [x_1,\ x_2] = v_1,\ [x_1,\ y_2] = v_3,\ [x_2,\ y_1] = v_4, \quad [y_1, y_2] = v_2,$$

and every other product is zero or is obtained by the symmetry or skew-symmetry of one of the above products. One checks that that this superalgebra is simple if char.$K \neq 3$ and that in characteristic 3, it possesses a simple subsuperalgebra of dimension 9 spanned by e_1, v_i, $1 \leq i \leq 4$, x_j, y_j, $1 \leq j \leq 2$. We denote this superalgebra by K_9 and refer to it as the *degenerate Kac superalgebra*.

8) Denote by $H_n(K)$ and $S_n(K)$ the symmetric and skew-symmetric $n \times n$ matrices. For K a field of characteristic 3, let $A = H_3(K)$ and $M = \overline{S_3(K)} \oplus \overline{\overline{S_3(K)}}$, two copies of $S_3(K)$. To extend the Jordan algebra structure on A and A-bimodule structure on M to a Jordan superalgebra structure on $J = A + M$ one defines

$$[\overline{S_3(K)},\ \overline{S_3(K)}] = [\overline{\overline{S_3(K)}},\ \overline{\overline{S_3(K)}}] = \{0\},$$

and for any $a, b \in S_3(K)$,

$$[\overline{a},\ \overline{\overline{b}}] = ab + ba \in H_3(K) = A.$$

9) Let $B = B_0 + B_1$, with $B_0 = M_2(K)$, $B_1 = Km_1 + Km_2$, where K is a field of characteristic 3. If we define a B_0-bimodule structure on B_1 by

$$\begin{array}{llll} e_{11}m_1 = m_1, & e_{22}m_2 = 0, & e_{12}m_1 = m_2, & e_{21}m_1 = 0; \\ m_1e_{11} = 0, & m_1e_{22} = m_1, & m_1e_{12} = -m_2, & m_1e_{21} = 0; \\ e_{11}m_2 = 0, & e_{22}m_2 = m_2, & e_{12}m_2 = 0, & e_{21}m_2 = m_1; \\ m_2e_{11} = m_2, & m_2e_{22} = 0, & m_2e_{12} = 0, & m_2e_{21} = -m_1, \end{array}$$

and a multiplication from $B_1 \times B_1$ to B_0 by

$$m_1^2 = -e_{12}, \quad m_2^2 = e_{12}, \quad m_1m_2 = e_{11}, \quad m_2m_1 = -e_{22},$$

then B is a superalternative algebra with superinvolution $(a+m)^* := \bar{a} - m$, where $\bar{\ }$ is the symplectic involution of B_0. Then $H_3(B)$, the symmetric matrices with respect to the *-transpose superinvolution, form a simple Jordan superalgebra which is i-exceptional since B is not superassociative. This example is due to Shestakov.

3. The first classification theorem

Jordan superalgebras do not always behave as one might naively expect. The second author [6] has shown that finite dimensional semisimple Jordan superalgebras need not be a direct sum of simple superalgebras. Kantor [3] has constructed finite dimensional simple Jordan superalgebras whose even part is not semisimple. Using the Kantor-Koecher-Tits construction, Kac [2] has derived a classification of finite dimensional simple Jordan superalgebras over algebraically closed fields of characteristic 0 from his classification of the corresponding Lie superalgebras. Assuming that the even part is semisimple, we obtain the following classification over an algebraically closed field of arbitrary characteristic.

First Classification Theorem. *Let $J = A+M$ be a finite dimensional simple Jordan superalgebra over an algebraically closed field K of characteristic not 2. If A is semisimple and $M \neq \{0\}$ then J is isomorphic to one of the following superalgebras:*

i) K_3, *the Kaplansky superalgebra,*

ii) K_9, *the degenerate Kac superalgebra, char.*$K = 3$,

iii) K_{10}, *the Kac superalgebra, char.*$K \neq 3$,

iv) D_t, $t \neq 0$,

v) *the superalgebra of a nondegenerate superform,*

vi) $M_{n,m}(K)$,

vii) $Q(n)$, $n > 1$,

viii) $osp(n, 2m)$,

ix) $P(n)$, $n > 1$,

x) $H_3(K) + \overline{S_3(K)} \oplus \overline{\overline{S_3(K)}}$, char.$K = 3$.

xi) $H_3(B)$, char.$K = 3$, B *as in example 9).*

4. Main ingredients of the proof

While the proof will appear elsewhere, we wish to indicate its flavour. To use structure theory and represention theory we must show that if

$$J = A + M,$$

A a semisimple Jordan algebra, M an A-bimodule then A cannot have too many simple summands nor M too many irreducible components.

Proposition. *Let $J = A + M$ be a finite dimensional simple unital Jordan superalgebra with A semisimple. Then A does not contain 3 central non-zero idempotents. Therefore either A is simple or $A = A' \oplus A''$, a direct sum of two simple summands.*

Proposition. *Let $J = A + M$ be a finite dimensional simple unital Jordan superalgebra with A semisimple and let $M = \sum_{i=1}^{r} M_i$ be the decomposition of M into a direct sum of irreducible A-bimodules. Then either $r \leq 2$ or J is a Jordan superalgebra of a superform. Therefore, with the exception of the superalgebras of superforms, M is either an irreducible A-bimodule or $M = M' \oplus M''$, the direct sum of two irreducible A-bimodules.*

The following proposition which extends to Jordan superalgebras a result which is well known for other Jordan structures [5] allows us to use the Peirce decomposition to patch up larger superalgebras from smaller ones.

Proposition. *Let $J = A + M$ be a simple finite-dimensional unital Jordan superalgebra over an arbitrary field K, $e \in A$, an idempotent. Then $J\mathrm{U}_e$ is a simple superalgebra.*

A detailed analysis of the simple superalgebras with A of capacity ≤ 3 is then carried through. The following proposition is typical of the type of results one obtains.

Proposition. *Let $J = A + M$ be a simple unital Jordan superalgebra with $A = J(V, Q)$, the Jordan algebra of a nondegenerate quadratic form on a vector space V of dimension $n \geq 2$. Then one of the following holds:*

i) J is the superalgebra of a superform (in which case $M.V = \{0\}$ and $[M,\ M] \subseteq K1$, i.e., $[\ ,\ ]$ is a skew-symmetric form on M),

ii) $n = 3$ and J is isomorphic to $P(2)$,

iii) $n = 3$ and J is isomorphic to $Q(2)$,

iv) $n = 2$ and $J \cong D_t$.

References

1. Hogben , Leslie, and Kac, VictorDG., *The correct multiplication table for the exceptional Jordan superalgebra F*, Comm. in Alg., **11** (1983), 1155–1156.
2. Kac, VictorDG., *Classification of simple Z-graded Lie superalgebras and simple Jordan superalgebras*, Comm. in Alg. **5** (1977), 1375–1400.
3. Kantor, I.DI., *Connection between Poisson brackets and Jordan and Lie superalgebras*, in Lie Theory, Differential Equations and Representation Theory, Publications CRM, Montréal, 1990, 213–225.
4. Kaplansky, I., *Graded Jordan algebras I*, (preprint).
5. McCrimmon, K., *Peirce ideals in Jordan algebras*, Pacific J. Math., **78** (1978), 397–414.
6. Zel'manov, E.DI., *Semisimple finite-dimensional Jordan superalgebras*, preprint.

ABSOLUTE VALUED ALGEBRAS OF DEGREE TWO

ANGEL RODRIGUEZ PALACIOS
Departamento de Análisis Matemático, Facultad de Ciencias
Universidad de Granada, 18071-GRANADA (SPAIN).

To the memory of my mother-in-law Inès de la Pas López.

1. Introduction

Following [2], we will say that a (nonassociative) algebra A is algebraic if, for every x in A, the subalgebra $A(x)$ of A generated by x is finite-dimensional. If in fact $\dim(A(x)) \leq m$ for all x in A and some natural number m only depending on A, then the algebraic algebra A is called of *bounded degree*, and the smallest such a number m is called the *(bounded) degree* of A. From the fact that $\mathbf{C}$ is the only finite-dimensional absolute valued complex algebra it follows easily that also $\mathbf{C}$ is the only algebraic absolute valued complex algebra. In the same way, $\mathbf{R}$ is the only absolute valued real algebra of degree one.

However, it seems to be an interesting open problem if every algebraic absolute valued real algebra is finite-dimensional. This was affirmatively answered by A. A. Albert [2] under the additional assumption of the existence of a unit element. But, since 1960, Albert's result has only become an auxiliary tool for the proof of the Urbanik-Wright theorem [14] asserting that actually $\mathbf{R}, \mathbf{C}, \mathbf{H}$, and $\mathbf{O}$ *are the only absolute valued real algebras with a unit element* . Interesting partial answers to the question we are considering have been provided by M. L. El-Mallah, by showing that *an algebraic absolute valued real algebra A is finite dimensional whenever either A has a nonzero idempotent which commutes with all elements of A* [5] *or A has an algebra involution $*$ satisfying $\| x^* \| = \| x \|$ and $xx^* = x^*x$ for all x in A* [7] (note that the first requirement implies the former [8]). In any case, since finite-dimensional absolute valued real algebras are of dimension *1*, *2*, *4*, or *8* [1], algebraic absolute valued real algebras are of bounded degree equal *1*, *2*, *4*, or *8*.

The aim of this paper is to encourage the work on the problem quoted above by proving that absolute valued real algebras of degree two are finite-dimensional. We will even determine such algebras.

2. Examples of absolute valued algebras of degree two

From now on, by an absolute valued algebra we mean a nonzero real algebra (say A) endowed with a norm $\| \, . \, \|$ satisfying $\| xy \| = \| x \| \| y \|$ for all x, y in A. The most

S. González (ed.), *Non-Associative Algebra and Its Applications,* 350–356.

classical examples of absolute valued algebras are **R** (the reals), **C** (the complex), **H** (the quaternions), and **O** (the octonions), for which we refer to [4]. If **A** stands for anyone of the algebras **C**, **H**, or **O**, we can construct three new absolute valued algebras (denoted by $\overset{*}{\mathbf{A}}$, $^*\mathbf{A}$, and $\mathbf{A}^*$) by simply replacing the original product of **A** by the ones $\odot_1, \odot_2$, and $\odot_3$ given respectively by $x \odot_1 y := x^* y^*$, $x \odot_2 y := x^* y$, and $x \odot_3 y := xy^*$, where $*$ denotes the standard involution of **A**. Among the absolute valued algebras presented until now, only $\mathbf{R}, \mathbf{C}$, $\overset{*}{\mathbf{C}}$ (the para-complex), $\mathbf{H}, \overset{*}{\mathbf{H}}$ (the para-quaternions), **O**, and $\overset{*}{\mathbf{O}}$ (the para-octonions) are flexible. If we add the algebra **P** of pseudo-octonions [11] to this list, then we obtain all flexible absolute valued algebras (see [10] and [8]). This is interesting for our work because, since one-generated subalgebras of flexible algebras are commutative and the only commutative algebras among those listed above are $\mathbf{R}, \mathbf{C}$, and $\overset{*}{\mathbf{C}}$, it follows that every flexible absolute valued algebra different from **R** is of degree two. The following proposition provides other examples of absolute valued algebras of degree two.

Proposition 1. *$^*\mathbf{C}, \mathbf{C}^*, {}^*\mathbf{H}, \mathbf{H}^*, {}^*\mathbf{O}$, and $\mathbf{O}^*$ are absolute valued algebras of degree two.*

Proof. If A denotes either $^*\mathbf{C}, {}^*\mathbf{H}$, or $^*\mathbf{O}$, then a straightforward calculation shows that, for x in A we have

$$x^2x = \| x \|^2 x \tag{1}$$

$$(x^2)^2 = \| x \|^2 x^2 \tag{2}$$

and

$$xx^2 - 2(x \mid \iota)x^2 + \| x \|^2 x = 0 \;, \tag{3}$$

where ι is the left unit of A, and $(. \mid .)$ denotes the inner product on A from which the norm of A derives. From (1), (2), and (3) it follows that the linear hull of $\{x, x^2\}$ is a subalgebra of A. Therefore $^*\mathbf{C}, {}^*\mathbf{H}$, and $^*\mathbf{O}$ are of degree two. The same is true for $\mathbf{C}^*, \mathbf{H}^*$, and $\mathbf{O}^*$ because these algebras are isomorphic to the opposite algebras of $^*\mathbf{C}, {}^*\mathbf{H}$, and $^*\mathbf{O}$, respectively. □

3. Absolute valued algebras satisfying $x^2x = xx^2 = \| x \|^2 x$

$\mathbf{R}, \overset{*}{\mathbf{C}}, \overset{*}{\mathbf{H}}, \overset{*}{\mathbf{O}}$, and **P** are absolute valued algebras satisfying the conditions in the title of this section. We will see that these are the only ones.

Lemma 1. *Let A be a normed real algebra, and assume that the equality $x^2x = \| x \|^2 x$ holds for all x in A. Then the norm of A derives from an inner product.*

Proof. For x, y in A the mapping $\lambda \to \| x + \lambda y \|$ from **R** to **R** is convex, hence it has left and right derivatives $\tau^-(x,y)$ and $\tau^+(x,y)$ at zero. Writing the equality $(x+\lambda y)^2(x+\lambda y) = \| x + \lambda y \|^2 (x + \lambda y)$ in the form

$$\| x \|^2 x + \lambda F(x,y) + \lambda^2 F(y,x) + \lambda^3 \| y \|^2 y = \| x + \lambda y \|^2 (x + \lambda y) , \qquad (4)$$

we can compute left and right derivatives at $\lambda = 0$ to obtain

$$F(x,y) = \| x \|^2 y + 2 \| x \| \tau^-(x,y)x \text{ and } F(x,y) = \| x \|^2 y + 2 \| x \| \tau^+(x,y)x .$$

It follows that, if $x \neq 0$, then $\tau^-(x,y) = \tau^+(x,y)$, and it is well-known (see for example [3; Theorem 1 in p. 22]) that this last equality implies that the mapping $y \to \tau^+(x,y)$ is a linear functional on A. Now let x, y be arbitrary elements in A, take $\lambda = 1$ in (4) and substitute $F(x,y)$ and $F(y,x)$ by their values above computed. Then we have

$$[\| x \|^2 + 2 \| x \| \tau^+(x,y) + \| y \|^2]x + [\| x \|^2 + 2 \| y \| \tau^+(y,x) + \| y \|^2]y =$$

$$\| x + y \|^2 (x + y) .$$

In this way, if x and y are linearly independent, then

$$\| x \|^2 + 2 \| x \| \tau^+(x,y) + \| y \|^2 = \| x + y \|^2$$

and

$$\| x \|^2 + 2 \| y \| \tau^+(y,x) + \| y \|^2 = \| x + y \|^2 ,$$

hence

$$\| x \| \tau^+(x,y) = \| y \| \tau^+(y,x) .$$

Using the linearity of τ^+ in the second variable whenever the first one is nonzero, it follows that the last equality remains true if x and y are linearly dependent and, by defining $(x \mid y)$ to be the common value of both sides of that equality, $(. \mid .)$ becomes a symmetric bilinear form on A satisfying $(x \mid x) = \| x \|^2$ for all x in A. □

Theorem 1. *Let A be an absolute valued algebra satisfying $x^2 x = x x^2 = \| x \|^2 x$ for all x in A. Then A is isometrically isomorphic to* $\mathbf{R}, \overset{*}{\mathbf{C}}, \overset{*}{\mathbf{H}}, \overset{*}{\mathbf{O}}$ *or* $\mathbf{P}$.

Proof. From the assumption $x^2 x = \| x \|^2 x$ and Lemma 1 it follows that the norm of A derives from an inner product. This, together with the assumption $x^2 x = x x^2$, implies that A is finite-dimensional [6; Theorem 2.13]. By [8; Theorem 4.1], A is isometrically isomorphic to either $\mathbf{R}, \mathbf{C}, \overset{*}{\mathbf{C}}, \mathbf{H}, \overset{*}{\mathbf{H}}, \mathbf{O}, \overset{*}{\mathbf{O}}$, or $\mathbf{P}$. Finally we observe that the equality $x x^2 = \| x \|^2 x$ does not hold in either $\mathbf{C}, \mathbf{H}$, or $\mathbf{O}$. □

4. Absolute valued algebras satisfying $x^2 x = \| x \|^2 x$ and $(x^2)^2 = \| x \|^2 x^2$.

We have seen in the proof of Proposition 1 that the equalities in the title of this section are satisfied in either $^*\mathbf{C}$, $^*\mathbf{H}$, or $^*\mathbf{O}$ and of course also $\mathbf{R}$. We will see that in no more absolute valued algebras these equalities are satisfied.

Theorem 2. *For an absolute valued algebra A the following assertions are equivalent:*

i) The equalities $x^2x = \| x \|^2 x$ and $(x^2)^2 = \| x \|^2 x^2$ hold for all x in A.

ii) A has a left unit ι and, for every x in A, the equality $x^2 = \| x \|^2 \iota$ holds.

iii) A is isometrically isomorphic to $\mathbf{R}$,$^*\mathbf{C}$,$^*\mathbf{H}$, *or* $^*\mathbf{O}$.

Proof. $(i) \Rightarrow (ii)$. By the assumptions on A we have that, for every nonzero element x in A, $\| x \|^{-2} x^2$ is a nonzero idempotent (say ι) satisfying $\iota x = x$, so that it is enough to prove that there cannot exist two different nonzero idempotents in A. Let assume that e and f are different nonzero idempotents in A. Taking into account that the norm of A derives from an inner product (Lemma 1), identifying the coefficients of λ, λ^2, and λ^3 in both sides of the equality

$$[(e+\lambda f)^2]^2 = \| e+\lambda f \|^2 (e+\lambda f)^2,$$

and writing $z := ef + fe$, we obtain

$$ez + ze = 2(e \mid f)e + z \tag{5}$$

$$z^2 + z = e + f + 2(e \mid f)z \tag{6}$$

$$fz + zf = 2(e \mid f)f + z\ , \tag{7}$$

and, subtracting (7) from (5), we have also

$$(e-f)z + z(e-f) = 2(e \mid f)(e-f)\ . \tag{8}$$

The assumptions on A imply that, for every x in A and every y in the linear hull of $\{x, x^2\}$, we have $x^2 y = \| x \|^2 y$. Then $(e+f)^2(e+f) = \| e+f \|^2 (e+f)$ and, since by (6) $e+f$ belongs to the linear hull of $\{z, z^2\}$, we also have $z^2(e+f) = \| z \|^2 (e+f)$. It follows $\| z \|^2 (e+f)^2 = \| e+f \|^2 z^2$, or equivalently

$$\| z \|^2 (e+f+z) = 2[1+(e \mid f)]z^2. \tag{9}$$

Putting in (9) the value of z^2 given by (6), we obtain

$$[\| z \|^2 + 2(1-(e \mid f)) - 4(e \mid f)^2]z + [\| z \|^2 - 2(1+(e \mid f))](e+f) = 0.$$

Now observe that $e + f \neq 0$, hence $z \neq 0$ by (6), so if $\| z \|^2 + 2(1-(e \mid f)) - 4(e \mid f)^2 = 0$, then $\| z \|^2 + 2(1-(e \mid f)) - 4(e \mid f)^2 = \| z \|^2 - 2(1+(e \mid f))(=0)$ and therefore $(e \mid f)^2 = 1$, which is impossible because $e \neq \pm f$. It follows that there exists a real number μ such that $z = \mu(e+f)$. Substituting in (8) this value of z and taking into account that $e \neq f$, we obtain $\mu = (e \mid f)$. Replacing in (9) z by $(e \mid f)(e+f)$ and taking norms, we obtain $\| e+f \| = 2$, hence the contradiction $e = f$.

$(ii) \Rightarrow (iii)$. By the structure theorem for absolute valued algebras with a left unit (see [13; Theorem 2] or more precisely [13; Remark 4.(1)]), we know that the norm of A derives from an inner product $(. \mid .)$ and also, if for x in A we write $x^* := 2(x \mid \iota)\iota - x$, then we have $x^*(xy) = \| x \|^2 y$ for all x, y in A. Taking in particular $y = x$, since $x^2 = \| x \|^2 \iota$, it follows $x^*\iota = x$. Since the mapping $x \to x^*$ from A to A is a linear isometry, it follows that, if we replace the product of A by the one $\diamond$ given by $x \diamond y := x^* y$, then we obtain an absolute valued algebra $\mathbf{A}$ for which ι becomes a (two-sided) unit. By the noncommutative Urbanik-Wright theorem [14; Theorem 1], we conclude that $\mathbf{A}$ coincides with either $\mathbf{R}, \mathbf{C}, \mathbf{H}$, or $\mathbf{O}$, and clearly $*$ becomes the standard involution on $\mathbf{A}$. Hence A equals either $\mathbf{R},{}^* \mathbf{C},{}^* \mathbf{H}$, or ${}^*\mathbf{O}$.

(iii) $\Rightarrow$ *(i)*. As we said in the proof of Proposition 1, this is of straightforward verification. □

5. The main result

The following result is well-known and easy to prove.

Lemma 2. *Every two-dimensional absolute valued algebra is isometrically isomorphic to either* $\mathbf{C}, \overset{*}{\mathbf{C}},{}^* \mathbf{C}$, *or* $\mathbf{C}^*$.

Theorem 3. *The absolute valued algebras of degree two are* $\mathbf{C}, \overset{*}{\mathbf{C}},{}^* \mathbf{C}, \mathbf{C}^*, \mathbf{H}, \overset{*}{\mathbf{H}},{}^* \mathbf{H}, \mathbf{H}^*, \mathbf{O}, \overset{*}{\mathbf{O}},{}^* \mathbf{O}, \mathbf{O}^*$, *and* $\mathbf{P}$.

Proof. From Section 2 we know that all absolute valued algebras listed in the statement are of degree two. To see the converse, let A be an absolute valued algebra of degree two. For x in A we will denote by $A(x)$ the subalgebra of A generated by x, and we will write

$$\mathcal{X} := \{x \in A \ : \ A(x) \cong \mathbf{C}\}, \quad \overset{*}{\mathcal{X}} := \{x \in A \ : \ A(x) \cong \overset{*}{\mathbf{C}}\},$$

$${}^*\mathcal{X} := \{x \in A \ : \ A(x) \cong {}^* \mathbf{C}\}, \quad \mathcal{X}^* := \{x \in A \ : \ A(x) \cong \mathbf{C}^*\},$$

$$\mathcal{Y} := \{x \in A \ : \ \dim(A(x)) \leq 1\}.$$

Then from Lemma 2 we deduce

$$A = \mathcal{X} \cup \overset{*}{\mathcal{X}} \cup {}^*\mathcal{X} \cup \mathcal{X}^* \cup \mathcal{Y}. \tag{10}$$

We will also consider the closed subsets of A given by

$$\mathcal{Z} := \{x \in A \ : \ x^2 x = \| x \|^2 x\}, \quad \mathcal{T} := \{x \in A \ : \ x x^2 = \| x \|^2 x\},$$

$$\mathcal{U} := \{x \in A \ : \ (x^2)^2 = \| x \|^2 x^2\}.$$

First assume that $\mathcal{X} \neq \emptyset$. Clearly $\mathcal{X} \subseteq A \backslash (\mathcal{Z} \cup \mathcal{T})$ and, conversely, if x is in $A \backslash (\mathcal{Z} \cup \mathcal{T})$, then, since $\overset{*}{\mathcal{X}} \cup {}^*\mathcal{X} \cup \mathcal{X}^* \cup \mathcal{Y} \subseteq \mathcal{Z} \cup \mathcal{T}$, by (10) x lies in $\mathcal{X}$. Now $\mathcal{X} = A \backslash (\mathcal{Z} \cup \mathcal{T})$ is a nonempty open subset of A. For x in A and n in $\mathbf{N}$, define inductively $x^{(n}$ by $x^{(1} := x$ and $x^{(n+1} := x x^{(n}$. If we fix y in $\mathcal{X}$, then, for every x in A, we have that $y + \lambda x$ lies in $\mathcal{X}$ for infinitely many values of λ in $\mathbf{R}$, hence $(y + \lambda x)^{(n+m} = (y + \lambda x)^{(n} (y + \lambda x)^{(m}$ for infinitely many values of λ and all m, n in $\mathbf{N}$. It follows $x^{(n+m} = x^{(n} x^{(m}$, so A is a power-associative algebra. Since $\mathbf{R}, \mathbf{C}, \mathbf{H}$, and $\mathbf{O}$ are the only power-associative absolute valued algebras [9; Theorem 2.2] and $\mathbf{R}$ is not of degree two, we obtain that A must be either $\mathbf{C}, \mathbf{H}$, or $\mathbf{O}$.

Now assume $\mathcal{X} = \emptyset$ and $\overset{*}{\mathcal{X}} \neq \emptyset$. If x is in $\overset{*}{\mathcal{X}} \cap \mathcal{U}$, then routine calculations in $A(x) = \overset{*}{\mathbf{C}}$ show that x is a real multiple of an idempotent in $A(x)$, which is impossible because $\dim(A(x)) = 2$. Therefore $\overset{*}{\mathcal{X}} \subseteq A \backslash \mathcal{U}$. Conversely, if x is in $A\backslash\mathcal{U}$, then, since ${}^*\mathcal{X} \cup \mathcal{X}^* \cup \mathcal{Y} \subseteq \mathcal{U}$, it follows from (10) that x lies in $\overset{*}{\mathcal{X}}$. Now $\overset{*}{\mathcal{X}} = A\backslash\mathcal{U}$ is a nonempty open subset of A. As above, for fixed y in $\overset{*}{\mathcal{X}}$ and every x in A, $y + \lambda x$ lies in $\overset{*}{\mathcal{X}}$ for infinitely many values of λ in $\mathbf{R}$, so $(y+\lambda x)^2(y+\lambda x) = (y+\lambda x)(y+\lambda x)^2$ for infinitely many values of λ, and so $x^2x = xx^2$. Because ${}^*\mathbf{C}$ and $\mathbf{C}^*$ do not satisfy the last equality, we obtain ${}^*\mathcal{X} = \mathcal{X}^* = \emptyset$, so, by (10), $A = \overset{*}{\mathcal{X}} \cup \mathcal{Y}$, and so $x^2x = xx^2 = \| x \|^2 x$ for all x in A. By Theorem 1, A equals $\overset{*}{\mathbf{C}}, \overset{*}{\mathbf{H}}, \overset{*}{\mathbf{O}}$, or $\mathbf{P}$.

Now assume $\mathcal{X} = \overset{*}{\mathcal{X}} = \emptyset$ and ${}^*\mathcal{X} \neq \emptyset$. If x is in ${}^*\mathcal{X} \cap \mathcal{T}$, then routine calculations in $A(x) = {}^*\mathbf{C}$ show that x is a real multiple of the left unit of $A(x)$ which is impossible. Therefore ${}^*\mathcal{X} \subseteq A\backslash\mathcal{T}$. The converse inclusion is also true in view of (10) and the clear fact $\mathcal{X}^* \cup \mathcal{Y} \subseteq \mathcal{T}$. Now ${}^*\mathcal{X} = A\backslash\mathcal{T}$ is a nonempty open subset of A. Because the equality $x(x^2x) = (x^2x)x$ is true in ${}^*\mathbf{C}$ (hence in ${}^*\mathcal{X}$), as above this equality remains true for every x in A. Since $\mathbf{C}^*$ does not satisfy this equality, we have $\mathcal{X}^* = \emptyset$, so, by (10), $A = {}^*\mathcal{X} \cup \mathcal{Y}$, and therefore the equalities $x^2x = \| x \|^2 x$ and $(x^2)^2 = \| x \|^2 x^2$ hold in A. By Theorem 2, A equals either ${}^*\mathbf{C}, {}^*\mathbf{H}$, or ${}^*\mathbf{O}$.

In view of (10), the only remaining case we have to consider is the one in which $A = \mathcal{X}^* \cup \mathcal{Y}$. Then the opposite algebra A^0 of A satisfies the equalities $x^2x = \| x \|^2 x$ and $(x^2)^2 = \| x \|^2 x^2$. By Theorem 2, A^0 equals either ${}^*\mathbf{C}, {}^*\mathbf{H}$, or ${}^*\mathbf{O}$. Therefore A equals either $\mathbf{C}^*, \mathbf{H}^*$, or $\mathbf{O}^*$. □

6. Power-commutative absolute valued algebras

Following [12] we will say that an algebra A is power-commutative if the subalgebra of A generated by an arbitrary element of A is commutative. Flexible algebras and power-associative algebras are examples of power-commutative algebras. The following corollary is an easy consequence of our main result and the commutative Urbanik-Wright theorem [14].

Corollary 1. *For an absolute valued algebra A the following assertions are equivalent:*

i) A is power-commutative.

ii) A equals either $\mathbf{R}, \mathbf{C}, \overset{*}{\mathbf{C}}, \mathbf{H}, \overset{*}{\mathbf{H}}, \mathbf{O}, \overset{*}{\mathbf{O}}$, *or* $\mathbf{P}$.

iii) A is flexible.

Proof. $(i) \Rightarrow (ii)$. Since one-generated subalgebras of A are commutative, by [14; Theorem 3] A is of degree one or two. By Theorem 3, A equals $\mathbf{R}, \mathbf{C}, \overset{*}{\mathbf{C}}$, ${}^*\mathbf{C}, \mathbf{C}^*, \mathbf{H}, \overset{*}{\mathbf{H}}, {}^*\mathbf{H}, \mathbf{H}^*, \mathbf{O}, \overset{*}{\mathbf{O}}, {}^*\mathbf{O}, \mathbf{O}^*$, or $\mathbf{P}$. But ${}^*\mathbf{C}, \mathbf{C}^*, {}^*\mathbf{H}, \mathbf{H}^*, {}^*\mathbf{O}$, and $\mathbf{O}^*$ are not power-commutative.

$(ii) \Rightarrow (iii) \Rightarrow (i)$. These implications are clear. □

It must be noted that, as we have pointed out in Section 2, the equivalence between assertions (*ii*) and (*iii*) above is known (see [10; Theorem 3.8] and [8; Theorem 4.1]). The implication (*i*) $\Rightarrow$ (*ii*) in Corollary 1 contains the following result, previously proved by M. L. El-Mallah and A. Micali in [9], and already quoted and used in the proof of Theorem 3.

Corollary 2. $\mathbf{R}, \mathbf{C}, \mathbf{H}$, *and* $\mathbf{O}$ *are the only power-associative absolute valued algebras.*

Acknowledgements

The author would like to express their gratitude to M. Cabrera and J. Martínez for several useful suggestions concerning the results in this paper.

References

1. A. A. Albert, Absolute valued real algebras, Ann. Math. 48 (1947), 495-501.
2. A. A. Albert, Absolute valued algebraic algebras, Bull. Amer. Math. Soc. 55 (1949), 763-768. A note of correction. Ibid. 55 (1949), 1191.
3. J. Diestel, *Geometry of Banach spaces - Selected topics,* Lecture Notes in Math. 485, Springer-Verlag, Berlin, 1975.
4. H. D. Ebbinghaus, H. Hermes, F. Hirzebruch, M. Koecher, K. Mainzer, J. Neukirch, A. Prestel and R. Remmert, *Numbers,* Graduate texts in Mathematics 123, Springer-Verlag, New York, 1990.
5. M. L. El-Mallah, Quelques rsultats sur les algbres absolument values, Arch. Math. 38 (1982), 432-437.
6. M. L. El-Mallah, Sur les algbres absolument values qui vrifient l'identit $(x, x, x) = 0$, J. Algebra 80 (1983), 314-322.
7. M. L. El-Mallah, Absolute valued algebras with an involution, Arch. Math. 51 (1988), 39-49.
8. M. L. El-Mallah, Absolute valued algebras containing a central idempotent, J. Algebra 128 (1990), 180-187.
9. M. L. El-Mallah and A. Micali, Sur les algbres normes sans diviseurs topologiques de zro, Boletín de la Sociedad Matemática Mexicana 25 (1980), 23-28.
10. M. L. El-Mallah and A. Micali, Sur les dimensions des algbres absolument values, J. Algebra 68 (1981), 237-246.
11. S. Okubo, Pseudo-quaternion and pseudo-octonion algebras. Hadronic J. 1 (1978), 1250-1278.
12. R. Raffin, Anneaux a puissances commutatives et anneaux flexibles. C. R. Acad. Sci. París 230 (1950), 804-806.
13. A. Rodríguez, One-sided division absolute valued algebras. Publ. Mat. 36 (1992), 925-954.
14. K. Urbanik and F. B. Wright, Absolute valued algebras. Proc. Amer. Math. Soc. 11 (1960), 861-866.

ON A GENERALIZATION OF THE JORDAN INVERSE

MICHAEL ROTH
Mathematisches Institut, Technische Universität München, 80290 München, Germany.

Abstract. In the work presented here we consider rational maps on finite dimensional vector spaces that generalize Jordan inversions in characteristic zero.

Key words: Jordan inversion, Hua's identity, structure algebra

1. Introduction

There are several classes of nonassociative algebras that are generalizations of Jordan algebras, like power-associative algebras, Lie triple algebras or the algebras investigated by M. Osborn in [6]. A common feature of these algebras is that they satisfy an identity that generalizes the *Jordan identity*. In this survey we want to present a class of algebras whose properties come from a generalization of the *Jordan inverse*. (These algebras, the so-called R-algebras, were introduced by Sebastian Walcher and Rolf Niemczyk [5].)

The presented results were obtained in joint work by S. Walcher and the author. A more complete account of the results, including proofs, will be given elsewhere.

Let us start with an axiomatic characterization of Jordan inversions due to T.A. Springer. Throughout this note let V denote a finite dimensional vector space over a field K of characteristic zero.

Definition: (T.A. Springer [8]) *A j-structure is a triple (V, Φ, e), consisting of a finite-dimensional vector space V, a rational map Φ of V into itself and a non-zero element $e \in V$, satisfying the following conditions:*

$(J1)$ *Φ is rational, homogeneous of degree -1 and an involution,* $\Phi(e) = e$, $D\Phi(e) = -Id$.

$(J2)$ *Hua's identity at the unit:* $\Phi(e - \Phi(e - \Phi(e - x))) = x$

$(J3)$ $\mathfrak{g}_\Phi \cdot e = V$

Here $\mathfrak{g}_\Phi$ is the *structure algebra* of Φ, which is defined as

$$\mathfrak{g}_\Phi := \{B \in gl(V) : D\Phi(x)Bx = -B^\# \Phi(x) \text{ for a } B^\# \in gl(V), \forall x \in U\},$$

with $U \subset V$ open and $D\Phi(x)$ the usual derivative of Φ. Furthermore $\mathfrak{g}_\Phi \cdot e$ denotes the orbit of $\mathfrak{g}_\Phi$ through e.

Remark: We have replaced Springer's original third axiom (which involves the structure group) by its Lie algebra version, which is equivalent in characteristic zero.

S. González (ed.), Non-Associative Algebra and Its Applications, 357–360.

Indeed there is a 1-1-correspondence between j-structures and Jordan inversions:

Theorem 1 (T.A. Springer) a) *Let (V, Φ, e) be a j-structure, then the multiplication $y \cdot z := \frac{1}{2}D^2\Phi(e)(y,z)$ defines a Jordan algebra A_Φ with unit e, in which $\Phi(x)$ is the inverse.*

b) *Conversely, if $A = (V, \cdot)$ is a Jordan algebra with a unit e and $\Phi_A(x) := x^{-1}$ is the multiplicative inverse in A, then (V, Φ_A, e) is a j-structure.*

(In b) Springer obtained the same algebra using a different formula.)

2. Generalization

Analyzing the effect of the three axioms shows how we can obtain a reasonable generalization: The first axiom assures that $y \cdot z := \frac{1}{2}D^2\Phi(e)(y,z)$ defines a commutative algebra with unit e at all. The second axiom guarantees that the repeated change from an algebra A to the j-structure Φ_A to the algebra A_{Φ_A} always leads back to the algebra we started with. It's the third axiom (J3) which forces the algebra A_{Φ_A} to be Jordan.

Inversions satisfying only the first two axioms (so-called R-inversions) are still related to a class of commutative algebras, called R-algebras. This class of commutative nonassociative algebras cannot be described by an identity.

Its characteristic property is revealed by looking at the *system of differential equations* $\dot{x} = x^2$ where x^2 is the square in the nonassociative algebra. (This approach works at first for $K = \mathbb{R}$ or $\mathbb{C}$, then by standard tricks also for arbitrary base fields of characteristic 0, cf. Lefschetz [3].) L. Markus [4] was the first to relate homogeneous quadratic differential equations to algebras.

Let $G(z,t)$ denote the solution of the differential equation for the initial value z at "time" $t = 0$. Now a commutative algebra is called an R-algebra, if the solution $G(x,t)$ is rational in x and t. Examples of R-algebras are the power associative algebras and the nilpotent algebras with a unit adjoint.

The main link between inversions and differential equations is the following

Proposition: *Let Φ be a (rational) inversion with a unit e satisfying the axioms (J1) and (J2) and A_Φ defined as above. Then*

$$G(x,t) = \frac{1}{t}(\Phi(e - tx) - e) .$$

This equation is the key to the following theorem:

Theorem 2 (Niemczyk/Walcher [5])

a) *Let Φ be an R-inversion, (i.e. a (rational) inversion satisfying (J1) and (J2)) with unit e. Then A_Φ, defined as above, is an R-algebra with unit e.*

b) *Conversely, let $A = (V, \cdot)$ be a R-algebra with a unit e and define*

$$\Phi_A(x) := G(e - x, 1) + e.$$

Then Φ_A is an R-inversion with unit e.

Note the definition of Φ_A in b). This term is, in general, different from the multiplicative inverse.

3. More generalization

Over the base field $K = \mathbb{R}$ or $\mathbb{C}$ there is a correspondence of the same kind ever for *arbitrary* commutative unital algebras, if we replace rationality in $(J1)$ by analyticity near e to get an axiom denoted as $(J1^-)$.

Theorem 3 *Let $K = \mathbb{R}$ or $K = \mathbb{C}$.*

a) *If Φ is an inversion with unit e satisfying $(J1^-)$ and $(J2)$, then A_Φ, defined as above, is an commutative algebra with unit e.*

b) *Conversely, let $A = (V, \cdot)$ be a commutative algebra with unit e and define*

$$\Phi_A(x) := G(e - x, 1) + e.$$

Then Φ_A is an inversion with unit e, satisfying $(J1^-)$ and $(J2)$.

Now having worked out the skeleton of Springer's axioms we should direct our attention to the omitted axiom (J3):

Let $A := (V, \cdot)$ be an algebra with a unit e. The square $p(x) := x^2$ in A can be considered as an element of the Lie-algebra $Pol(V)$ of all polynomial vector fields of V. (The Lie bracket is given by $[q, r](x) := Dr(x) \cdot q(x) - Dq(x) \cdot r(x)$.)

The centralizer $C(p) = \{q \in Pol(V) | [p, q] = 0\}$ of p is of interest, since its elements produce *solution preserving maps* for the differential equation $\dot{x} = p(x)$, in detail: A polynomial vector field generates a one-parameter group of solution preserving maps if and only if it lies in the centralizer of p. Koecher ([1], [2]) proved :

Theorem 4 (M. Koecher) *If $A = (V, \cdot)$ has a unit e and $p(x) = x^2$, then*

a) $C(p) = \mathrm{Der}(p) \oplus C_1$.

b) $q \in C_1 \Longrightarrow q = q_a$ *with* $q_a(x) := 2x(xa) - x^2 a$ *and* $a = q(e)$, *hence* $C_1 = \{q_a | a \in J(A)\}$ *with* $J(A) := \{q(e) | q \in C_1\}$.

d) *$J(A)$ is a Jordan subalgebra of A.*

d) $J(A) = A \iff A$ *is Jordan.*

4. The inversion and Koecher's subalgebra

Now the main question about the omitted axiom (J3) is: Given an inversion Φ satisfying the axioms $(J1^-)$ and (J2), what can be said about the orbit $\mathfrak{g}_\Phi \cdot e$ of the structure algebra through the unit? The answer is: $\mathfrak{g}_\Phi \cdot e$ is identical with Koecher's Jordan subalgebra $J(A_\Phi)$ in the commutative algebra A_Φ corresponding to Φ:

$$\mathfrak{g}_\Phi \cdot e \quad = \quad J(A_\Phi)$$

More precisely, the following theorem holds (cf. [7]):

Theorem 5 *Let Φ be an inversion satisfying the axioms $(J1^-)$ and $(J2)$ and let $A_\Phi = (V, \cdot)$ be the corresponding commutative algebra. For every $a \in V$ we have:*

$$L(a) \in \mathfrak{g}_\Phi \iff a \in \mathfrak{g}_\Phi \cdot e \iff a \in J(A_\Phi)$$

Remark: As one application this result shows how (J3) enforces the algebra to be Jordan: (J3) describes the "maximal situation" $J(A_\Phi) = A$, from which the Jordan property follows in a natural way by theorem 4.d).

References

1. M. Koecher: Die Riccatische Differentialgleichung und nicht-assoziative Algebren. *Abh. Math. Sem. Univ. Hamburg* **46** (1977), 129–141.
2. M. Koecher: On commutative non associative algebras. *J. Algebra* **62** (1980), 479–493.
3. S. Lefschetz: *Algebraic geometry.* Princeton Univ. Press, Princeton 1953.
4. L. Markus: Quadratic differential equations and nonassociative algebras. *Ann. Math. Studies* **45** (1960), Princeton Univ. Press, 185–213.
5. R. Niemczyk, S. Walcher: Birational maps and a generalization of power-associative algebras. *Comm. Algebra* **19** (1991), 2169-2194.
6. J. M. Osborn: A generalization of power-associativity. *Pacific J. Math.* **14** (1964), 1367–1379.
7. M. Roth: *Zentralisator und Normalisator quadratischer Polynome.* Dissertation, TU München, 1992.
8. T. A. Springer: *Jordan algebras and algebraic groups.* Springer-Verlag, Berlin, Heidelberg, New York, 1973.

ON GEODESIC LOOPS OF TRANS-SYMMETRIC SPACES

LIUDMILA SABININA
Instituto de Matemáticas, UNAM.
Area de la Investigación Científica, Circuito Exterior, C. U.
04510 México D.F., México

Abstract. One can consider the class of trans-symmetric spaces that we study below as a subclass of the class of spaces with reductive affine connections. Our interest in trans-symmetric spaces is based on the fact that they generalize the well known construction of symmetric spaces of E.Cartan [1] and that of s-spaces [2, 3, 4]. The aim of the present work is to propose a description of trans-symmetric spaces in terms of geodesic loops of their canonical reductive connections.

Let us consider a C^w-manifold M equipped with a family $\{\sigma_x\}_{x\epsilon M}$ of local C^w-diffeomorphisms and with an additional C^w-smooth mapping $e : M \longrightarrow M$ such that the mapping $(x, y) \longrightarrow \sigma_x y$ is also C^w-smooth and is defined in a neighborhood of the set $\{(x, y) \mid y = e(x)\}$ in $M \times M$. Then $(M, \{\sigma_x\}_{x\epsilon M}, e)$ is called a trans-symmetric space (ts-space [5]) if the family of trans-symmetries $\{\sigma_x\}_{x\epsilon M}$ satisfies the following conditions:

$$\sigma_x \cdot \sigma_y = \sigma_{\sigma_x y} \cdot \sigma_{e(x)} \tag{1}$$

$$(\sigma_x \circ e)_{\star,x} - Id_{\star,x} \text{ - is non-singular} \tag{2}$$

or, equivalently, for an arbitrary point x from M the mapping $\rho_{ex} : y \longrightarrow \sigma_y ex$ defined locally in a neighborhood of the point x is a local C^w-diffeomorphism at the point x. In [5] we have proved that their exists on the manifold M with a trans-symetric structure $\{\sigma_x\}_{x\epsilon M}$ a connection ∇ such that

$$\text{all } \sigma_x \text{ are local affine morphisms of the connection } \nabla \tag{3}$$

and

$$\nabla S = 0 \tag{4}$$

where $S_x = (\sigma_x)_{\star,e(x)} : T_{e(x)}M \longrightarrow T_x M$ and $(\nabla_X S)Y = S(\nabla_{e_\star X} Y) - \nabla_X(SY)$.

The connection ∇ satisfying the conditions (3)-(4) is unique and is reductive, i.e. $\nabla T = 0, \nabla R = 0$, where T and R are respectively the torsion and the curvature of the connection ∇. This connection is called the canonical connection of the trans-symmetric space $(M, \{\sigma_x\}_{x\epsilon M}, e)$. One can describe it in the following way [5]. Let us introduce transvections $P_x^b = \sigma_{(\rho_{e(b)})^{-1}x} \circ (\sigma_x)^{-1}$, then for arbitrary C^w-smooth vector fields X, Y on M

$$(\nabla_X Y)_b = \frac{d}{dt}([(P_{x(t)}^b)_{\star,b}]^{-1} Y_{x(t)})|_{t=0}$$

where $x(t)$ is a smooth curve in M passing through the point b ($b\epsilon M$) at $t = 0$ with $x'(0) = X_b$. Let us consider the set of geodesic loops $< M, (\cdot_a), a >$, $a\epsilon M$, of the

S. González (ed.), Non-Associative Algebra and Its Applications, 361–366.

affinely connected space (M,∇) with local analytic composition laws $(\cdot_a)$ defined by the formulas [6, 7]

$$x\cdot_a y = (Exp_x \circ \tau_{ax} \circ (Exp_a)^{-1})y \tag{5}$$

where $Exp_z : T_zM \longrightarrow M$ is the exp-mapping and $\tau_{ax} : T_aM \longrightarrow T_xM$ is the ∇-parallel transport of vectors along the geodesic curve joining points a and x (which is unique for all the points x sufficiently close to a).

Let us denote by ψ_a the C^ω-mapping $\sigma_a \circ e : M \longrightarrow M$ defined in a neighborhood of the point a with $\psi_a a = a$.

The mappings σ_a, $\forall a \epsilon M$, and e are morphisms of the affine connection ∇, that is ([5] and [6, 7])

$$\begin{aligned} \sigma_a(x\cdot_b y) &= \sigma_a x \cdot_{\sigma_a b} \sigma_a y, \\ e(x\cdot_b y) &= ex \cdot_{eb} ey. \end{aligned} \tag{6}$$

In particular, for $\psi_a = \sigma_a \circ e$ we get

$$\psi_a(x\cdot_b y) = \psi_a x \cdot_{\psi_a b} \psi_a y \tag{7}$$

or in the operator form

$$\psi_a \circ L_x^b = L_{\psi_a x}^{\psi_a b} \circ \psi_a,$$

where $L_x^b : M \longrightarrow M, y \longrightarrow x\cdot_b y$ is a local C^ω- diffeomorphism defined locally in a neighborhood of the point b [6, 7].

Let a be a fixed point from M. Then the geodesic composition law $(\cdot_a)$, its local C^ω-smooth endomorphism ψ_a and the diffeomorphism σ_a determine the initial trans-symmetric structure $\{\sigma_x\}_{x\epsilon M}$ in a neighborhood of the point a uniquely up to isomorphism. Indeed, by virtue of (4)

$$\nabla_X(SY) = S(\nabla_{e_\star X}Y),$$

where X and Y are C^ω-smooth vector fields on M. Thus for the geodesic curve $x(t)$, joining $a = x(0)$ with a sufficiently close point $b = x(1)$

$$\nabla_{x'(t)}(SY) = (\sigma_{x(t)})_{\star,ex(t)} \nabla_{(ex)'(t)} Y$$

and for $Y_{ex(t)}$ parallel along $ex(t)$,

$$(SY)_{x(t)} = (\sigma_{x(t)})_{\star,ex(t)} Y_{ex(t)}$$

is parallel along $x(t)$. So we have

$$\begin{aligned} Y_{ex(t)} &= \tau_{ea,ex(t)} Y_{ea}, \\ (\sigma_{x(t)})_{\star,ex(t)} Y_{ex(t)} &= \tau_{a,x(t)}(\sigma_a)_{\star,ea} Y_{ea} \end{aligned}$$

and

$$(\sigma_{x(t)})_{\star,ex(t)} \circ \tau_{ea,ex(t)} = \tau_{a,x(t)} \circ (\sigma_a)_{\star,ea}$$

or, equivalently

$$(\sigma_b)_{\star,eb} \circ \tau_{ea,eb} = \tau_{a,b} \circ (\sigma_a)_{\star,ea},$$

then

$$\sigma_b \circ L^{ea}_{eb} = \sigma_b \circ Exp_{eb} \circ \tau_{ea,eb} \circ (Exp_{ea})^{-1} = Exp_{\sigma_b eb} \circ (\sigma_b)_{\star,eb} \circ \tau_{ea,eb} \circ (Exp_{ea})^{-1} =$$
$$= Exp_b \circ \tau_{a,b} \circ (\sigma_a)_{\star,ea} \circ (Exp_{ea})^{-1} = Exp_b \circ \tau_{a,b} \circ (Exp_a)^{-1} \circ \sigma_a = L^a_b \circ \sigma_a$$

and

$$\sigma_b \circ L^{ea}_{eb} = (\sigma_b \circ \sigma_a^{-1}) \circ \sigma_a \circ L^{ea}_{eb} = (\sigma_b \circ \sigma_a^{-1}) \circ L^a_{\psi_a b} \circ \sigma_a.$$

Finally, we get

$$\sigma_b = L^a_b \circ (L^a_{\psi_a b})^{-1} \circ \sigma_a, \tag{8}$$

i.e. the initial ts-structure $\{\sigma_x\}_{x \in M}$ in a neighborhood of the point a is uniquely determined by the geodesic composition low $(\cdot_a)$, its local C^w-smooth endomorphism ψ_a and an additional C^w-smooth diffeomorphism σ_a such that its inverse σ_a^{-1} is defined in a neighborhood of the point a. The following question arises naturally: describe the class of loops that are isomorphic to geodesic loops of canonical connections of trans-symmetric spaces. Theorem 1 and Theorem 2 below give an answer to this question.

Let us continue to consider the geodesic loop $< M, \cdot_a, a >$. Firstly, we note that being a geodesic loop with respect to the reductive affine connection, the loop $< M, \cdot_a, a >$ satisfies the following identities (9)-(10):

$$x^k \cdot_a (x^l \cdot_a y) = x^{k+l} \cdot_a y, \quad \forall k, l \in Z \tag{9}$$

(left monoalternative property),

$$l_{x,y}(z \cdot_a w) = l_{x,y}(z) \cdot_a l_{x,y}(w), \text{ where } l_{x,y} = (L^a_{x \cdot_a y})^{-1} \circ L^a_x \circ L^a_y \tag{10}$$

(left special property).

Further, by virtue of (7)

$$\psi_a(x \cdot_a y) = (\psi_a x) \cdot_a (\psi_a y), \tag{11}$$

i.e. ψ_a is a local analytic endomorphism of $< M, \cdot_a, a >$ and

$$(\psi_a)_{\star,a} - Id_{\star,a}\text{- is non-singular }. \tag{12}$$

Now let us substitute into the identety

$$\sigma_x \circ \sigma_y^{-1} = L^y_x \circ (L^y_{\psi_y x})^{-1}$$

the expressions

$$\sigma_z = L^a_z \circ (L^a_{\psi_a z})^{-1} \circ \sigma_a, \quad L^z_w = L^a_z \circ L^a_{(L^a_z)^{-1} w} \circ (L^a_z)^{-1}, \quad \psi_y = \sigma_y \circ e.$$

Then we get

$$l_{\psi_a y, (L^a_{\psi_a y})^{-1} \psi_a x} = l_{y, (L^a_y)^{-1} x}$$

or, equivalently

$$l_{\psi_a y, \psi_a z} = l_{y,z}. \tag{13}$$

The local diffeomorphism σ_a by virtue of (1) and (8) is related to ψ_a

$$\psi_a \circ \sigma_a = \sigma_a \circ L^a_{e(a)} \circ (L^a_{\psi_a e(a)})^{-1} \circ \psi_a \tag{14}$$

and satisfies the following identity

$$\sigma_a \circ L^x_y = L^{\sigma_a x}_{\sigma_a y} \circ \sigma_a, \text{ where } L^x_y = L^a_x \circ L^a_{(L^a_x)^{-1} y} \circ (L^a_x)^{-1}.$$

Setting $x = a,\ y = x$ we get

$$\sigma_a \circ L^a_x = L^a_{\sigma_a a} \circ L^a_{(L^a_{\sigma_a a})^{-1} \sigma_a x} \circ (L^a_{\sigma_a a})^{-1}, \tag{15}$$

which is equivalent to the identity above.

There is an additional identity that is derived by putting into (1) written in the form

$$(\sigma_x \circ \sigma_a^{-1}) \circ \sigma_a \circ (\sigma_y \circ \sigma_{ea}^{-1}) = (\sigma_{\sigma_x y} \circ \sigma_a^{-1}) \circ \sigma_a \circ (\sigma_{ex} \circ \sigma_{ea}^{-1})$$

the identity (8) written in the form

$$\sigma_x \circ \sigma_a^{-1} = L^a_x \circ (L^a_{\psi_a x})^{-1}.$$

We get

$L^a_x \circ (L^a_{\psi_a x})^{-1} \circ \sigma_a \circ L^{ea}_y \circ (L^{ea}_{\psi_{ea} y})^{-1} = L^a_{\sigma_x y} \circ (L^a_{\psi_a \sigma_x y})^{-1} \circ \sigma_a \circ L^{ea}_{ex} \circ (L^{ea}_{\psi_{ea} ex})^{-1}$,

or, equivalently

$L^a_x \circ (L^a_{\psi_a x})^{-1} \circ L^a_{\sigma_a y} \circ (L^a_{\sigma_a \psi_{ea} y})^{-1} = L^a_{\sigma_x y} \circ (L^a_{\psi_a \sigma_x y})^{-1} \circ L^a_{\psi_a x} \circ (L^a_{\sigma_a \psi_{ea} ex})^{-1}$.

Setting $\sigma_a y = z$, we get

$L^a_x \circ (L^a_{\psi_a x})^{-1} \circ L^a_z \circ (L^a_{\psi_a z})^{-1} = L^a_{(\sigma_x \sigma_a^{-1}) z} \circ (L^a_{(\psi_a \sigma_x \sigma_a^{-1}) z})^{-1} \circ L^a_{\psi_a x} \circ (L^a_{\psi_a(\psi_a x)})^{-1}$

(we note that $e \circ \sigma_a = \sigma_{ea} \circ e$ by virtue of (1), see [5]).

Further, by virtue of (8) we can rewrite the last identity in the following form

$L^a_x \circ (L^a_{\psi_a x})^{-1} \circ L^a_z \circ (L^a_{\psi_a z})^{-1} = L^a_{L^a_x (L^a_{\psi_a x})^{-1} z} \circ (L^a_{\psi_a L^a_x (L^a_{\psi_a x})^{-1} z})^{-1} \circ L^a_{\psi_a x} \circ (L^a_{\psi_a^2 x})^{-1}$,

or, equivalently

$$L^a_x \circ (L^a_{\psi_a x})^{-1} \circ L^a_z \circ (L^a_{\psi_a z})^{-1} = L^a_{L^a_x (L^a_{\psi_a x})^{-1} z} \circ (L^a_{L^a_{\psi_a x} (L^a_{\psi_a^2 x})^{-1} \psi_a z})^{-1} \circ L^a_{\psi_a x} \circ (L^a_{\psi_a^2 x})^{-1} \tag{16}$$

Finally, we get the following theorem

Theorem 0.1. *A geodesic loop* $< M, \cdot_a, a >$ *of the canonical affine connection of the trans-symmetric space* $(M, \{\sigma_x\}_{x \epsilon M}$ *satisfies the properties (9)-(16).*

Conversely, if we are given an arbitrary local analitic loop $< M, \cdot, a >$ *on a manifold* M *satisfying (9)-(10) equipped with a local analytic endomorphism* $\psi (= \psi_a) : M \longrightarrow M$ *(i.e.* (11) *is true) satisfying* (12), (13), (16) *and with an additional analytic diffeomorphism* $\sigma (= \sigma_a)$ *such that its inverse* σ^{-1} *is defined in a neighborhood of the point* a *and (14)-(15) hold, then we can construct in a neighborhood of the point* a *in* M *a trans-symmetric structure* $\{\sigma_x\}_{x \epsilon M}$ *by putting*

$$\sigma_x = L_x \circ (L_{\psi x})^{-1} \circ \sigma, \tag{17}$$

$$e = \sigma^{-1} \circ \psi$$

and the local analytic loop $< M, \cdot, a >$ *will be isomorphic to the geodesic loop* $< M, \cdot_a, a >$ *centered at the point* a *of the canonical affine connection corresponding to the trans-symmetric structure* (17).

We start the proof of this proposition by introducing in the space M the reductive affine connection ∇ uniquely determined [6, 7] by the loop $< M, \cdot, e >$ with properties (9)-(10). The corresponding geodesic multiplications $(\cdot_x)$ are of the following form

$$L^x_y = L_x \circ L_{(L_x)^{-1}y} \circ (L_x)^{-1} \tag{18}$$

where $L_x = L^a_x : M \longrightarrow M,\ y \longrightarrow x \cdot y$.

The mappings ψ and σ are local morphisms of the connection ∇ by virtue of (11), (15) and (18), i.e.

$$\sigma_a \circ L^x_y = L^{\sigma_a x}_{\sigma_a y} \circ \sigma_a \tag{19}$$

and all the mappings σ_x and e are also (see (17)).

By virtue of (17) and (13) the following identity holds

$$\sigma_x \circ (\sigma_y)^{-1} = L^y_x \circ (L^y_{(\sigma_y \circ e)x})^{-1}, \tag{20}$$

therefore starting from (16) with (14), (19), (20) we get (1). The tangent mapping $(\sigma_a \circ e)_{\star,a} - Id_{\star,a} = (\psi_a)_{\star,a} - Id_{\star,a}$ – is non-singular by virtue of (12). Thus $(\sigma_x \circ e)_{\star,x} - Id_{\star,x}$ is non-singular for x sufficiently close to a. The property (2) holds and $(M, \{\sigma_x\}_{x \epsilon M})$ is a trans-symmetric space.

Further, since σ_a is a local morphism of the connection ∇ we have

$$\sigma_x = L^a_x \circ (L^a_{\psi x})^{-1} \circ \sigma_a = L^a_x \circ \sigma_a \circ (L^{ea}_{ex})^{-1}$$

or, equivalently

$$\sigma_x \circ (L^{ea}_{ex}) = L^a_x \circ \sigma_a.$$

Then, for arbitrary analytic vector fields X and Y on M we have in an arbitrary point b in M

$$(\sigma_b)_{\star,eb} \nabla_{e_\star,_b X_b} Y = \{\tfrac{d}{dt}[(\sigma_b)_{\star,eb}(L^{eb}_{ex(t)})^{-1}_{\star,eb} Y_{ex(t)}]\}_{t=0} =$$

$$= \{\tfrac{d}{dt}[(\sigma_b)_{\star,eb}(L^{ex(t)}_{eb})_{\star,ex(t)} Y_{ex(t)}]\}_{t=0} = \{\tfrac{d}{dt}[(\sigma_b \circ L^{ex(t)}_{eb})_{\star,ex(t)} Y_{ex(t)}]\}_{t=0} =$$

$$= \{\tfrac{d}{dt}[(L^{x(t)}_b \circ \sigma_{x(t)})_{\star,ex(t)} Y_{ex(t)}]\}_{t=0} = \{\tfrac{d}{dt}[(L^{x(t)}_b)_{\star,x(t)}(\sigma_{x(t)})_{\star,ex(t)} Y_{ex(t)}]\}_{t=0} =$$

$$= \{\tfrac{d}{dt}[(L^{x(t)}_b)_{\star,x(t)}(SY)_{x(t)}]\}_{t=0} = \{\tfrac{d}{dt}[(L^b_{x(t)})^{-1}_{\star,b}(SY)_{x(t)}]\}_{t=0} = \nabla_{X_b}(SY)$$

where $x(t)$ is an integral curve of the vector field X such that x(0)=b. Thus

$$S(\nabla_{e_\star X} Y) = \nabla_X(SY), \text{ where } S_x = (\sigma_x)_{\star,ex}$$

and the connection ∇ coincides with the canonical connection of the trans-symmetric space $(M, \{\sigma_x\}_{x \epsilon M})$ by virtue of Proposition 5 from [5]. Therefore, the initial loop $< M, \cdot, a >$ is isomorphic to the geodesic loop of the trans-symmetric space $(M, \{\sigma_x\}_{x \epsilon M})$ centered at the point a. Hence we have the following theorem

Theorem 0.2. *A local analytic loop* $< M, \cdot, a >$ *satisfying (9)-(10) equipped with a local analytic endomorphism* $\psi (= \psi_a) : M \longrightarrow M$, $\psi a = a$, *satisfying (11)-(13), (16) and with an analytic diffeomorphism* $\sigma (= \sigma_a) : M \longrightarrow M$ *such that its inverse* σ^{-1} *is defined in neighborhood of the point* a *and (14)-(15) hold, is isomorphic to a geodesic loop of a trans-symmetric space. The corresponding trans-symmetric structure is determined up to isomorphism by formulas (17).*

References

1. S.Helgason, *Differential Geometry, Lie Groups, and Symmetric Spaces*, Academic Press, Inc. 1978.
2. A.J.Ledger, Espaces de Riemann Symetriques Generalises, C.R.Acad.Sci.Paris, **264** (1967), 947–948.
3. A.S.Fedenko, *Spaces with Symmetries* (Russian), Minsk, State Univ., 1977.
4. O.Kowalski, *Generalized symmetric spaces*, Lecture Notes in Mathematics no. 805, Berlin-Heidelberg-New York, 1980.
5. L.V.Sabinin, L.L.Sabinina, On Geometry of Trans-symmetric Spaces, Webs and quasigroups, Tver Univ., 1991, 117–122.
6. L.V.Sabinin, Odules as a new approach to a Geometry with Connection, Soviet Math. Dokl. **18** (1977), no. 2, 515–518.
7. P.O.Mikheev, L.V.Sabinin, Quasigroups and Differential Geometry. In: Quasigroups and loops: Theory and application. Ed.: O.Chein, H.O.Pflugfelder, J.D.H.Smith. Berlin: Heldermann Verlag, 1990, 357–430.

QUADRATIC SYSTEMS, BLOW-UP, AND ALGEBRAS

ARTHUR A. SAGLE
University of Hawaii
Hilo, HI 96720 USA

and

MICHAEL K. KINYON
Indiana University South Bend
South Bend, IN 46634 USA

Abstract. Systems of quadratic ordinary differential equations such as the predator-prey, Riccati, and Euler systems may be expressed as $\dot{X} = \beta(X, X) \equiv X^2$ where β is the multiplication in an algebra A. A common problem is to determine the blow-up of solutions in finite time. It is shown that solutions which blow up may sometimes be expressed in terms of idempotents E and equilibria N which occur in the blow-up. Examples using a Jordan algebra are given. Blow-up often occurs in reaction-diffusion-convection systems of partial differential equations. Using Fourier methods, this leads to the study of blow-up of quadratic systems of ordinary differential equations occurring in an infinite dimensional algebra.

1. Basics

Systems of (autonomous) quadratic ordinary differential equations in real or complex commutative algebras A can be expressed in the form $\dot{X} = C + TX + X^2$ in A where $C \in A$ is a constant vector, $T : A \to A$ is a linear transformation, and $X^2 = \beta(X, X)$ is squaring in the multiplication β in A [4] [5] [7]. For example, the predator-prey model is given by

$$\dot{x_1} = ax_1 - bx_1x_2, \qquad \dot{x_2} = -dx_2 + cx_1x_2$$

where x_1 represents the prey and x_2 represents the predator. The quadratic interaction terms define a multiplication on $\mathbf{R}^2$ via $\beta(X, X) = \begin{bmatrix} -bx_1x_2 \\ cx_1x_2 \end{bmatrix}$ for $X = \begin{bmatrix} x_1 \\ x_2 \end{bmatrix}$ and bilinearization. This and the definition $T = diag(a, -d)$ allow us to write the system in the desired form.

Blow-up [1] [2] [3] of solutions to quadratic systems can occur in a finite (escape) time. For example, the solution to $\dot{x} = x^2$ in $\mathbf{R}$ with initial condition $x(0) = x$ is $x(t) = (1 - tx)^{-1}x$ and blow-up occurs at $t_0 = 1/x$. More generally, $\dot{X} = X^2$ occurring in a power-associative algebra A has the solution $X(t) = (I - tL(X))^{-1}X$ [4] [7] where $L(X) : A \to A : Y \mapsto \beta(X, Y)$. The blow-up of this solution depends on the invertibility of the transformation $I - tL(X)$. In a later example, we shall consider this solution in a Jordan algebra, and its blow-up is related to the structure of A.

Typical structural results include the following [4] [5] [7].

S. González (ed.), Non-Associative Algebra and Its Applications, 367–371.

Theorem: *(1) If $A = A_1 \oplus \cdots \oplus A_n$ is semisimple where each $A_k^2 = A_k$ is simple, and $X = X_1 + \cdots + X_k$ in A, then $\dot{X} = X^2$ in A decouples into $\dot{X}_k = X_k^2$ in each A_k.*
(2) If $A = S + R$ where S is semisimple and R is the radical of A, and $X = X_S + X_R$ in A, then $\dot{X} = X^2$ can be solved by first solving $\dot{X}_S = X_S^2$ in S (autonomous), and then solving $\dot{X}_R = 2X_S X_R + X_R^2$ in R (nonautonomous).

Proof: (1) $\dot{X}_1 + \cdots + \dot{X}_n = \dot{X} = X^2 = (X_1 + \cdots + X_n)^2 = X_1^2 + \cdots + X_n^2$ using $X_j X_k \in A_j A_k = \{0\}$; thus $\dot{X}_k = X_k^2$ in A_k. For (2), see the references.

The structure of A (as determined by its simple components (if any) and radical) sometimes give idempotents $E = E^2 \neq 0$, or nilpotents $N \neq 0$. These special elements determine special ray solutions and their qualitative behavior. If E is an idempotent of A and $TE = 0$, then the unique solution to $\dot{X} = TX + X^2$, $X(0) = E$ is given by $x(t)E$ where $x(t)$ is the unique solution to $\dot{x} = x^2$, $x(0) = 1$ in $\mathbf{R}$. This solution blows up at time $t_0 = 1$. An *equilibrium* of $\dot{X} = X^2$ is a constant solution $X(t) \equiv N$. Thus $0 = \dot{X} = N^2$; that is, an element is an equilibrium if and only if it is a nilpotent of index 2. The stability of equilibria are sometimes determined by idempotents as follows [4].

Theorem: *Let $F_t(X)$ denote the solution to $\dot{X} = TX + X^2$ in A such that $F_0(X) = X$. Let $E \in A$ be an idempotent satisfying $TE = 0$. Then*
(1) If $x \neq 0$, then $F_t(xE)$ blows up in finite time.
(2) The origin is an unstable equilibrium.

Proof: (1) Let $x(t)$ be the unique solution to $\dot{x} = x^2$, $x(0) = x$ in $\mathbf{R}$. Then $(x(t)E)' = x(t)^2 E = T(x(t)E) + (x(t)E)^2$ and $x(0)E = xE$. Thus $F_t(xE) = x(t)E$. Since $x(t)$ blows up at time $t_0 = 1/x$, so does $F_t(xE)$. For (2), observe that any neighborhood of the origin contains a point xE for some $x \neq 0$; thus an unbounded solution starts in that neighborhood.

2. More on Blow-up

We now consider solutions of the form $X(t) = (1/f(t))Y(t)$ where f is a real-valued differentiable function with $f(t_0) = 0$ and $Y(t_0) \neq 0$; that is, t_0 is a regular singularity of the solution.

Theorem: *Let $X(t) = (1/f(t))Y(t)$ be a solution to $\dot{X} = TX + X^2$ as above.*
(1) If $f'(t_0) \neq 0$, then $E \equiv (-1/f'(t_0))Y(t_0)$ is an idempotent.
(2) If $f'(t_0) = 0$, then $N \equiv Y(t_0) \neq 0$ satisfies $N^2 = 0$

Remark: In particular, the blow-up of a solution can imply the existence of equilibria.

Proof: Plug $X(t) = (1/f(t))Y(t)$ into $\dot{X} = TX + X^2$ to obtain

$$-f'(t)Y(t) + f(t)Y'(t) = f(t)TY(t) + Y(t)^2.$$

Thus at t_0 we have $-f'(t_0)Y(t_0) = Y(t_0)^2$. Now if $f'(t_0) \neq 0$, then

$$E^2 = (1/f'(t_0)^2)Y(t_0)^2 = (1/f'(t_0)^2) - f'(t_0)Y(t_0) = E.$$

If $f'(t_0) = 0$, then clearly $N^2 = 0$.

Examples (1) Let A be 2-dimensional and let $X(t) = (1/(t-t_0)^2)Y(t)$ where $\{Y(t_0), Y'(t_0)\}$ is a basis for A. Set $N = Y(t_0)$ and $E = -Y(t_0)$, and a straightforward calculation gives
(a) $N^2 = 0$, $NE = EN = N$, $E^2 = E$ in A. Also A is associative and has the 1-dimensional radical $\mathbf{R}N$.
(b) The blowing-up solution to $\dot{X} = X^2$ is $X(t) = (1/(t-t_0)^2)(N-(t-t_0)E)$, and for all $a \in \mathbf{R}$, aN is an unstable equilbrium.

(2) Let A be the Jordan algebra of 2×2 real symmetric matrices with multiplication $\beta(X,Y) = (1/2)(XJY + YJX)$ where $J = diag(1,-1)$. More general choices for J are possible, but this will suffice to illustrate the basic calculations. Now we seek an element $X \in A$ such that the solution $F_t(X) = (1/f(t))Y(t)$ blows up at time t_0 with $f(t_0) = f'(t_0) = 0$. From the solution form $F_t(X) = (I - tL(X))^{-1}X$, we obtain $f(t)X = (I - tL(X))Y(t)$. Thus at time t_0, we have for $N = Y(t_0)$ that $0 = f(t_0)X = (I - t_0L(X))N = N - t_0L(X)N$, or $L(X)N = (1/t_0)N$. Consequently, we shall find X so that N is an eigenvector for $L(X)$, and we shall take t_0 to be the reciprocal of the corresponding eigenvalue. Now the equation $NJN = 0$ has two solutions in A: $N_1 = a\begin{pmatrix} 1 & 1 \\ 1 & 1 \end{pmatrix}$ and $N_2 = a\begin{pmatrix} 1 & -1 \\ -1 & 1 \end{pmatrix}$. It is straightforward to determine X in either case; $X = \begin{pmatrix} x_1 & x_2 \\ x_2 & x_3 \end{pmatrix}$, where $x_2 = (-1)^j(1/2)(x_1 + x_3)$ if $N = N_j$. In both cases, $t_0 = 2/(x_1 - x_3) = 3/trace(L(X))$. Assume $trace(X) \neq 0$, otherwise $X = xJ$ since J is the identity element of A, and the solution $F_t(xJ)$ in this case is given in section 1. For fixed N and the corresponding X, the solution $F_t(X)$ is computed using Cramer's rule $(I - tL(X))^{-1} = adj(I - tL(X))/det(I - tL(X))$. Tedious calculations show that $det(I - tL(X)) = (1 - t/t_0)^3$ and $adj(I - tL(X))X/(1 - t/t_0) \equiv G(t,X)$ is a matrix polynomial in t and X. Set $Y(t) = t_0^2G(t,X)$ and $f(t) = (t-t_0)^2$. Then one can verify that $(1/f(t))Y(t) = (I - tL(X))^{-1}X$.

Remarks (1) Computer results suggest that $\{X : trace(X) = 0\}$ is contained in the domain of attraction of the origin.
(2) Jordan algebras appear in the matrix Riccati equation of the classical problem of a linear control system with quadratic cost [4].

3. Reaction-Diffusion-Convection (RDC) Models and Algebras

The RDC partial differential equations consdiered arise in population dynamics, moving chemical fluids, combustion theory, and continuum mechanics. For example, the temperature $u = u(x,t)$ of a fluid moving in a tube with an autocatalytic reaction (given by quadratic interaction) satisfies and equation fo the form $u_t = au_{xx} + bu^2 + cu_x\alpha$. Here the diffusion au_{xx} involves the second derivative relative to the space variable, the reaction bu^2 involves the quadratic interaction, and the convection $cu_x\alpha$ involves the velocity α of the fluid.

Higher dimensional models give analogous equations. For example the temperature of two reacting chemical fluids moving in a subset W of $\mathbf{R}^2$ satisfy $Z_t = A\nabla^2 Z + B(Z,Z) + C\nabla Z a$, for $Z = \begin{pmatrix} u(X,t) \\ v(X,t) \end{pmatrix}$ where $X = (x,y) \in \mathbf{R}^2$ and the

2×2 matrix A mixes the separate diffusions given by $\nabla^2 Z = \begin{pmatrix} au_{xx} + bu_{yy} \\ cv_{xx} + dv_{yy} \end{pmatrix}$; the quadratic reaction is given by

$$B(Z,Z) = \begin{pmatrix} b_{111}u^2 + b_{112}uv + b_{122}v^2 \\ b_{211}u^2 + b_{212}uv + b_{222}v^2 \end{pmatrix};$$

the convection is given by the 2×2 matrix C operating on

$$\nabla Z \cdot a = \begin{pmatrix} u_x & u_y \\ v_x & v_y \end{pmatrix} \begin{pmatrix} \alpha \\ \beta \end{pmatrix}$$

where α (resp. β) is the velocity of the total fluid in the x (resp. y) direction, and ∇Z is the Jacobian relative to the space variables.

We now show that RDC systems give quadratic systems of ordinary differential equations. Thus using basics of fluid dynamics (or linear feedback), we illustrate this for the 1-dimensional system $u_t = au_{xx} + bu^2 + du_x u$ (here $\alpha = ku$). For $(x,t) \in \mathbf{R} \times \mathbf{R}$, expand the solution into its (complex) Fourier series $u(x,t) = \sum_{n=-\infty}^{\infty} U_n(t)\exp(in\pi x)$. Next consider the convolution algebra of sequences $C = \{V = \{V_n\}_{n=-\infty}^{\infty}\}$ with the product $\beta(V,V) = \{\beta_n(V,V)\}$ where $\beta_n(V,V) = \sum_{p+q=n} V_p V_q$, the convolution product. Compute the derivatives and product $u_t = \sum \dot{U}_n(t)\exp(in\pi x)$,

$$u^2 = (\sum U_n(t)\exp(in\pi x))^2 = \sum(\sum_{p+q=n} U_p(t)U_q(t))\exp(in\pi x),$$

etc., and substitute into $u_t = au_{xx} + bu^2 + du_x u = \sum(\dot{U}_n - (-a\pi^2 n^2 U_n + b\beta_n(U,U) + i\pi d\sum_{p+q=n} pU_pU_q)\exp(in\pi x) = (\dot{U} - (a\pi^2 D^2 U + b\beta(U,U) + i\pi d\beta(DU,U)))\exp(i\pi x)$ where $U = \{U_n\}$, $D = diag(n)$, and $\exp(i\pi x) = \{\exp(in\pi x)\}$. Thus properties such as blow-up of the RDC equation are reflected in the quadratic ordinary differntial equation $\dot{U} = -a\pi^2 D^2 U + b\beta(U,U) + i\pi d\beta(DU,U) \equiv TU + U^2$ in A where $X^2 \equiv b\beta(X,X) + i\pi d\beta(DX,X)$ in a reaction-convection algebra A.

The initial condition for the RDC equation at $t = 0$ is given by $u(x,0) \equiv u_0(x) = \sum c_n \exp(in\pi x)$. This gives the initial condition for the ordinary differential equation $U(0) = C = \{c_n\}$. The boundary conditions give relations on the sequence $\{U_n(t)\}$. Thus "no flux" through the boundary occurs when $U_{-n}(t) = U_n(t)$, $n > 0$; consequently $U(t) = \{\ldots, U_1(t), U_0(t), U_1(t), \ldots\}$ is symmetric about $U_0(t)$.

Theorem: *Let $u_t = au_{xx} + bu^2 + du_x u$ have a smooth initial condition and no flux boundary conditions as above. If $b \neq 0$, then there exist solutions starting close to $u(x,t) \equiv 0$ which blow up in finite time; thus the equilibrium solution $u \equiv 0$ is unstable.*

Proof: Let A be the reaction-convection algebra of sequences so that X^2 is defined as above; let $T = -a\pi^2 D^2$ as above. Let $E = (1/b)\{\ldots, 0, 1, 0, \ldots\}$ with 1 in the U_0 position and 0 elsewhere. Then $DE = 0$ so that $TE = 0$ and $E^2 = b\beta(E,E) + i\pi d\beta(DE,E) = E$; now use the results of section 1.

Remarks: (1) Similar results hold for the higher dimensional systems provided there exists $b \neq 0$ with $\sum_{ij} b_{kij} = b$ for $k = 1, \ldots, n$.

(2) The equation $u_t + u_x u = au_x x + bu^2$ is a variation of the Navier-Stokes equation where $u(x,t)$ represents the velocity of the fluid, and $u_t + u_x u \equiv Du/Dt$ is the "material derivative" used in the Lagrangian viewpoint of fluid dynamics. For a steady flow, $Du/Dt = 0$ and therefore $au_{xx} + bu^2 = 0$ which is the equation whose solutions are Jacobian elliptic functions [6]. This extends to higher dimensional systems using algebras.

References

1. C. Budd, J. Dold, A. Stuart, Blow-up in a system of partial differential equations with conserved first integral, preprint
2. P. Grindrod, "Patterns and Waves: the theory and applications of reaction-diffusion equations", Oxford Univ. Press, 1991.
3. M. Herroro and J. Valazquez, Some results on blow-up for semilinear parabolic problems, IMA Preprint Series #1000, Univ. of Minnesota, July 1992
4. M. K. Kinyon and A. A. Sagle, Quadratic dynamical systems and algebras, to appear in J. Diff. Eq.
5. L. Markus, Quadratic differential equations and nonassociative algebras, in "Contributions to the Theory of Nonlinear Oscillations", Vol. V, L. Cesari, J.P. LaSalle, and S. Lefschetz (eds.), Princeton Univ. Press, Princeton, 1960, pp. 185-213
6. M. Tabour, Chaos and Integrability in Nonlinear Dynamics, Wiley, 1989
7. S. Walcher, "Algebras and Differential Equations", Hadronic Press, Palm Harbor, 1991

QUANTIZATION OF POISSON SUPERALGEBRAS AND SPECIALITY OF JORDAN POISSON SUPERALGEBRAS

IVAN SHESTAKOV*
Institute of Mathematics,
630090, Novosibirsk, RUSSIA

Departamento de Matemáticas,
Universidad de Oviedo
33007, Oviedo, SPAIN

Abstract. The problem of speciality and i-speciality is considered for Jordan superalgebras related with Poisscn superalgebras. The quantizations of Poisson superalgebras play an important role in the consideration. The i-speciality of Jordan Poisson superalgebras is obtained by means of quantization of Poisson superalgebras. The criterium for speciality of Jordan Poisson superalgebras is given also.

In 1989 I.Kantor [5] introduced a functor from the category of Poisson superalgebras to the category of Jordan superalgebras. The class of Jordan superalgebras that can be obtained by means of this functor has proved very interesting for Jordan theory: it has produced new simple superalgebras (even in the finite dimensional case) and new examples of prime degenerate Jordan algebras (see [8], [11], [15]). We will consider the problem of speciality for these Jordan Poisson superalgebras. It is proved in [7], [6] that the most interesting examples of Jordan Poisson superalgebras are not special. The main result of the given article states that any Jordan Poisson superalgebra is i-special, i.e., is a homomorphic image of a special one. We also give the speciality criterium for Jordan Poisson superalgebras.

All the algebras and superalgebras under consideration will be defined, if otherwise is not stated, over an arbitrary unital associative and commutative ring of scalars Φ; when dealing with Jordan superalgebras, we also will assume that $\frac{1}{2} \in \Phi$.

1. Poisson superalgebras

Let us remind that a *superalgebra* means a Z_2-graded algebra $A = A_0 + A_1$. We will consider only homogeneous elements of a superalgebra A, i.e., either *even* (belonging to A_0) or *odd* (belonging to A_1); if $a \in A_i$, then $p(a) = i$ will denote a parity index of a.

A superalgebra $P = P_0 + P_1$ is called a *Poisson superalgebra* if
1) P is associative and *supercommutative*, i.e. satisfies the identity

$$ab = (-1)^{p(a)p(b)}ba;$$

2) P has another multiplication (a *Poisson bracket*) $\{a, b\}$, under which P (with the given Z_2-grading) becomes a Lie superalgebra;

* Supported by the FICYT

S. González (ed.), Non-Associative Algebra and Its Applications, 372–378.

3) P satisfies the *Leibniz identity*

$$\{ab,c\} = a\{b,c\} + (-1)^{p(b)p(c)}\{a,c\}b.$$

Let us consider several examples of Poisson superalgebras.

1.1. Classical Poisson brackets. Let $P = P_{even}$ be the algebra $\Phi[p_1,...,p_n;q_1,...,q_n]$ of polynomials on (even) variables $p_1,...,p_n;q_1,...,q_n$, with the bracket

$$\{f,g\} = \sum_i \left(\frac{\partial f}{\partial p_i}\frac{\partial g}{\partial q_i} - \frac{\partial f}{\partial q_i}\frac{\partial g}{\partial p_i}\right).$$

This bracket is used for describing the dynamics of observables in classical mechanics. Note that P_{even} is a trivial superalgebra, i.e., its odd part is zero.

1.2. Grassmann Poisson brackets. Now, let $P = P_{odd}$ be the (supercommutative) superalgebra $\Phi[z_1,z_2,...]$ of polynomials on odd (or *Grassmann*) variables $z_1,z_2,\ldots$. In view of supercommutativity these variables satisfy the relations $z_iz_j = -z_jz_i, z_i^2 = 0$; hence, the superalgebra P_{odd} is isomorphic to the *Grassmann algebra* $G = G_0 + G_1$ with a unique grading defined by the condition that the elements z_i are odd: a product $z_{i_1}z_{i_2}\ldots z_{i_n}$ is even or odd according as n is even or odd. A bracket on P_{odd} is defined by the rule

$$\{f,g\} = (-1)^{p(f)}\sum_j \frac{\partial f}{\partial z_j}\frac{\partial g}{\partial z_j},$$

where $\frac{\partial}{\partial z_j}(z_1\ldots z_{j-1}z_jz_{j+1}\ldots z_n) = (-1)^{(j-1)}z_1\ldots z_{j-1}z_{j+1}\ldots z_n$. For example, $\{z_i,z_j\} = -\delta_{ij}$, $\{z_1z_2,z_1z_2z_3\} = 0$, $\{z_1z_2,z_2z_3\} = -z_1z_3$.

1.3. Tensor product of Poisson superalgebras. Let P_1, P_2 be Poisson superalgebras. Consider their tensor product $P = P_1 \otimes P_2$ and define in it multiplication and Poisson bracket by

$$\begin{aligned} a_1 \otimes b_1 \cdot a_2 \otimes b_2 &= (-1)^{p(a_2)p(b_1)}a_1a_2 \otimes b_1b_2, \\ \{a_1 \otimes b_1, a_2 \otimes b_2\} &= (-1)^{p(a_2)p(b_1)}(a_1a_2 \otimes \{b_1,b_2\} + \{a_1,a_2\} \otimes b_1b_2). \end{aligned}$$

Then P becomes a Poisson superalgebra. For example, we can obtain in this way the Poisson superalgebra $P_{mixed} = P_{even} \otimes P_{odd}$ of mixed Poisson brackets on polynomials on even variables $x_1,\ldots,x_n,y_1,\ldots,y_n$ and odd variables $z_1,z_2,\ldots$. Note that relative to ordinary multiplication P_{mixed} is just a free supercommutative superalgebra on even variables $x_1,\ldots,x_n,y_1,\ldots,y_n$ and odd variables $z_1,z_2,\ldots$.

1.4. Poisson-Lie superalgebras. Let L be a Lie superalgebra which is a free module over Φ, and $S(L)$ be the supersymmetric superalgebra of the Φ-supermodule $L = L_0 + L_1$, i.e., $S(L)$ is the quotient algebra $T(L)/I$, where $T(L)$ is the tensor superalgebra of L, and I is the ideal of $T(L)$ generated by the set $\{a \otimes b - (-1)^{p(a)p(b)}b \otimes a | a,b \in L_0 \cup L_1\}$. The superalgebra $S(L)$ is isomorphic to the free supercommutative superalgebra over the set $E_0 \cup E_1$, where E_i is a base of the Φ-module $L_i, i = 0, 1$. Now we define a bracket in $S(L)$. For elements e_i, e_j from the base of L we set

$$\{e_i, e_j\} = [e_i, e_j] \in L,$$

where the product on the right side is the one in L, and then extend the bracket to $S(L)$ by means of the Leibniz identity. (See [3] for the details.)

There is another way of introducing this bracket in $S(L)$. Let $U(L)$ be the universal associative enveloping superalgebra of L. It is well known that $U(L)$ has the canonical filtration

$$\Phi = U_0(L) \subset U_1(L) = L \subset \ldots \subset U_i(L) \subset \ldots,$$

where the Φ-module $U_i(L)$ is generated by all products of i elements from L. Now, let $gr(U(L)) = \oplus_i U_i/U_{i+1}$ be a graded superalgebra, associated with the filtration. By the Poincare-Birkhoff-Witt theorem (see [14]) we have an isomorphism $gr(U(L)) \cong S(L)$. Let $\tilde{a} = a + U_{i-1} \in gr_i(U(L)), \tilde{b} = b + U_{j-1} \in gr_j(U(L))$, then we set

$$\{\tilde{a}, \tilde{b}\} = (ab - (-1)^{p(a)p(b)}ba) + U_{i+j-2} \in gr_{i+j-1}(U(L)).$$

The supercommutators $[a, b]_s = ab - (-1)^{p(a)p(b)}ba$ satisfy the (super)identities of Jacobi and Leibniz, therefore, the bracket $\{a, b\}$ satisfies these identities too. Thus, $S(L)$ is a Poisson superalgebra.

2. Jordan Poisson superalgebras

An algebra J is called a *Jordan algebra* if it satisfies the identities

$$\begin{aligned} xy &= yx, \\ (x^2y)x &= x^2(yx). \end{aligned}$$

A superalgebra $J = J_0 + J_1$ is called a *Jordan superalgebra* if its *Grassmann envelope* $G_0 \otimes J_0 + G_1 \otimes J_1$ for the Grassmann algebra $G = G_0 + G_1$ is an ordinary Jordan algebra.

An important example of a Jordan superalgebra is the superalgebra $A^{(+)}$ for an associative superalgebra $A = A_0 + A_1$, which coinsides with A as a Φ-module and has a new supersymmetric multiplication

$$a \circ b = \frac{1}{2}(ab + (-1)^{p(a)p(b)}ba).$$

A Jordan superalgebra J is called *special* if it can be imbedded in a superalgebra $A^{(+)}$ for a suitable associative superalgebra A. It is well known that there exist Jordan superalgebras which are not special; they are called *exceptional.* Let $Jord$ and $SJord$ denote the classes of all Jordan superalgebras and of all special Jordan superalgebras respectively. Consider "the closure" $\overline{SJord}$ of the class $SJord$ in a sense of identities; i.e., $\overline{SJord}$ denotes the variety of superalgebras, generated by the class $SJord$. A superalgebra J belongs to the class $\overline{SJord}$ if and only if it satisfies all the superidentities which are valid for all special superalgebras; in this case J is called *i-special*. The two strict inclusions are true

$$SJord \subset \overline{SJord} \subset Jord.$$

Given any class of Jordan superalgebras, one of the first and natural questions about its structure is whether it consists of special (or of i-special) superalgebras; and if the answer is negative, the next step would be to look for some criteria of speciality for superalgebras from this class.

We will consider these problems for a class of Jordan superalgebras obtained from Poisson superalgebras by means of a doubling process going back to I.Kantor [5].

Let P be a Poisson superalgebra, and $J(P) = P \oplus \overline{P}$, where $\overline{P}$ is an isomorphic copy of the Φ-module P with the opposite parity of elements (i.e., $p(\bar{a}) = 1 - p(a)$). Define multiplication $\bullet$ on $J(P)$ by

$$a \bullet b = ab,\ \bar{a} \bullet b = (-1)^{p(b)}\overline{ab},\ a \bullet \bar{b} = \overline{ab},\ \bar{a} \bullet \bar{b} = (-1)^{p(b)}\{a, b\},$$

where $a, b \in P$ and ab is the product in P.

The superalgebra $J(P)$, with the grading $J(P)_0 = P_0 \oplus \overline{P_1},\ J(P)_1 = P_1 \oplus \overline{P_0}$, is a Jordan superalgebra. We will call it the *Jordan Poisson superalgebra*.

In [7],[6] K.McCrimmon and D.King have proved that the superalgebras $J(P_{even})$ and $J(P_{mixed})$ are not special, except in the trivial case of $J(P_{odd}[z_1])$. It was conjectered also in [7] that the superalgebra $J(P_{even})$ is not i-special.

The answer to the conjecture has proved negative – in 1992 the author proved the i-speciality of the $J(P_{even})$. Next the author [12] and, at the same time, V.G.Skosyrskii [15] have proved the i-speciality of the superalgebra $J(P_{mixed})$. Finally, in [13] the author have proved that any Jordan Poisson superalgebra is i-special. In the next section we will give some details of the proof of this result.

3. Quantization and i-speciality

Let us formulate the theorem:

I-speciality Theorem. *For any Poisson superalgebra P the corresponding Jordan Poisson superalgebra $J(P)$ is i-special.*

The proof used essentially the notion of quantization deformation of Poisson superalgebras, which was considered, for the case of algebras, in a lot of papers (see, for example, [1],[2], [4],[10]). Roughly speaking, a quantization is a change of a commutative multiplication in an algebra A to a noncommutative one, which depends on a parameter t in such a way that by setting $t = 0$ we obtain the original multiplication in A. In Poisson (super)algebra case the new multiplication should satisfy also some condition for the brackets.

Let P be a Poisson superalgebra. Consider the Φ-module $P[t]$ of polynomials on t with the coefficients from P. Assume that there exists an associative multiplication $*$ on $P[t]$, which agrees with Z_2-grading and satisfies the conditions

$$\begin{aligned} a * b &= ab(mod\ t), \\ a * b - (-1)^{p(a)p(b)} b * a &= t\{a, b\}(mod\ t^2), \\ a * t &= t * a = at, \end{aligned}$$

for any $a, b \in P$. Then we will say that P admits an (algebraic) *quantization deformation*. (If to consider formal power series instead of polynomials, the same definition will give a notion of *formal* quantization deformation.)

The importance of the notion of quantization for our problem shows the following

Theorem 1. *If a Poisson superalgebra P admits an algebraic quantization deformation, then the Jordan Poisson superalgebra $J(P)$ is i-special.*

We'll try to illustrate the idea of the proof with some physical interpretations.

Consider the case of classical Poisson brackets. As we have mentioned already, the Poisson algebra P_{even} can be considered as an algebra of observables in classical mechanics: it is the algebra of commutative polynomials on coordinates p_i and impulses q_i with the Poisson brackets $\{p_i, p_j\} = \{q_i, q_j\} = 0, \{p_i, q_j\} = \delta_i^j$.

In the quantum mechanics formalism the observables are the bounded hermitian operators on a Hilbert space, and they form a Jordan algebra H relative to the vector space operations and symmetric multiplication $a \circ b = \frac{1}{2}(ab + ba)$. Unlike the algebra P_{even}, this algebra has only one multiplication; but we may introduce another one by setting

$$\{b_1, b_2\} = \frac{i}{\hbar}(b_1 b_2 - b_2 b_1),$$

where i is a complex number and $\hbar$ is the Plank constant. (Note that this new multiplication plays the same role for the quantum mechanics formalism as the Poisson bracket does for classical mechanics.) Now both algebras P_{even} and H have two multiplications: the Jordan one (associative and commutative in the P_{even} case) and the Lie one (the bracket $\{,\}$).

The idea of considering the algebras P_{even} and H as algebras with two multiplications is going back to the Russian fysicist G.Zaitsev, who called these algebras in [16] the *Lie-Jordan algebras of observables.* From contemporary point of view this means that we have to consider instead of algebras P_{even} and H the *Jordan superalgebras* $J(P_{even})$ and its quantum analogue $J(H) = H + \overline{H}$, with the brackets $\{,\}$ as a product of odd elements and all other products induced by multiplication in H.

Now, it is well known that the classical mechanics can be considered as a "limit case" of the quantum one when $\hbar \to 0$ (or the light velocity $c \to \infty$). Thus, the superalgebra $J(P_{even})$ of classical mechanics should also be some kind of a "limit" of its quantum counterpart $J(H)$. But the superalgebra $J(H)$ is evidently special; so, it is naturally to expect that $J(P_{even})$ should be at least i-special. These informal arguments can be formally realized by means of quantizations.

With theorem 1 at hand, the next thing we have to do is to quantify Poisson superalgebras. For the algebra P_{even} in the characteristic 0 case we can use the well known *Moyal quantization* [9]

$$a * b = \mu \exp(t(D \otimes D))(a \otimes b) = \mu \sum_{k=0}^{\infty} \frac{t^k}{k!}(D \otimes D)^k (a \otimes b),$$

where $a, b \in P_{even}, \mu(a \otimes b) = ab$, $D \otimes D = \sum_i \left(\frac{\partial}{\partial p_i} \otimes \frac{\partial}{\partial q_i} - \frac{\partial}{\partial q_i} \otimes \frac{\partial}{\partial p_i}\right)$. (In fact, the *Moyal sine* $a * b - b * a$ is more known, which gives the quantization of the

Poisson brackets.) With some modifications this quantization can be generalized for the superalgebras P_{odd} and P_{mixed} (see [12]). For general case we have no such nice formulas. Moreover, we even don't know whether any Poisson superalgebra admits quantization? Fortunately, for our purpose it suffices to quantify Poisson-Lie superalgebras. For this case we have

Theorem 2. *For any Lie superalgebra L which is a free Φ-module the Poisson superalgebra $S(L)$ admits an algebraic quantization deformation.*

A scheme of the proof is the following. By the Poincare-Birkhoff-Witt theorem $S(L)[t]$ is isomorphic, as a Φ-module, to $U(L)[t]$, where $U(L)$ is the universal associative enveloping superalgebra for L. Consider the new Lie superalgebra $L^{<t>}$ over $\Phi[t]$ with the same base as L and multiplication

$$[l_1, l_2]^{<t>} = t[l_1, l_2].$$

Then $U(L^{<t>}) \cong U(L)[t] \cong S(L)[t]$ as Φ-modules. For $a, b \in S(L)$ denote by $a * b$ their product in $U(L^{<t>})$; then $*$ is the needed quantization.

Now, Poisson superalgebras form a variety, and there exist free objects in this variety; a Poisson superalgebra $P[X]$ is called a *free Poisson superalgebra* on a graded set of generators $X = X_0 \cup X_1$ if for any Poisson superalgebra P any parity preserving mapping of X into P can be extended in a unique way to a Poisson superalgebra homomorphism of $P[X]$ to P.

Proposition. *For any graded set of generators X the free Poisson superalgebra $P[X]$ is isomorphic to the Poisson-Lie superalgebra $S(L[X])$, where $L[X]$ is a free Lie superalgebra on X.*

The proof of the i-speciality theorem becomes now quite obvious. Any Poisson superalgebra P can be represented as a homomorphic image of a suitable free Poisson superalgebra $P[X]$. It is easy to see that a homomorphism of Poisson superalgebras always can be extended to a homomorphism of related Jordan Poisson superalgebras. Thus, the Jordan superalgebra $J(P)$ is a homomorphic image of the superalgebra $J(P[X])$. But $J(P[X])$ is i-special, according to the Proposition and theorems 2 and 1; therefore, $J(P)$ is i-special as well.

4. Speciality criteria

Consider now the next question: when Jordan Poisson superalgebras are special? Quite unexpectedly, the answer proves very simple.

Speciality criterium. *Let P be a unital Poisson superalgebra. Then, the Jordan superalgebra $J(P)$ is special if and only if P satisfies the condition $\{\{P, P\}, P\} = 0$.*

By applying this theorem to Poisson-Lie superalgebras, we obtained the following

Speciality criterium for Poisson-Lie superalgebras. *Let L be a Lie superalgebra over a field of characteristic not 2; then, the Jordan superalgebra $J(S(L))$ is special if and only if L satisfies one of the following conditions:*
1) $L^2 = 0$;
2) $L_0^2 = L_1^2 = 0$, the action of L_0 on L_1 is defined by

$$ax = -xa = \varphi(a, x)e_1,$$

where $\varphi : L_0 \times L_1 \to F$ *is a bilinear form,* $e_1 \in L_1$, $\varphi(L_0, e_1) = 0$;
3) $L_0^2 = L_0 L_1 = 0$, *the multiplication in* L_1 *is defined by*

$$xy = \varphi(x)\varphi(y)e_0,$$

where $\varphi \in L_1^*$, $e_0 \in L_0$;
4) $L_0^2 = 0$, *the remaining products are defined by*

$$xy = \varphi(x)\varphi(y)e_0, \quad ax = -xa = \psi(a)\varphi(x)e_1,$$

where $x, y \in L_1$, $a \in L_0$, $\varphi \in L_1^*$, $\psi \in L_0^*$, $e_0 \in \ker\psi$, $e_1 \in \ker\varphi$.

References

1. J.C.Baez, R-Commutative Geometry and Quantization of Poisson Algebras, Adv. in Math., **95** (1992), 61–91.
2. F.Bayen, M.Flato, C.Fronsdal, A.Lichnerowicz and D.Sternheimer, Deformation Theory and Quantization. I. Deformations of Symplectic Structures, Annals of Physics, **111** (1978), 61–110.
3. F.A.Berezin, *An introduction to algebra and analysis with anticommuting variables*, Izdat. Moskov. Gos. Univ., Moscow, 1983; English transl., Introduction to superanalysis, Reidel, Dordrecht, 1987.
4. *Deformation Theory of Algebras and Structures and Applications*, Edited by M.Hazewinkel and M.Gerstenhaber, NATO ASI Series C: Mathematical and Physical Sciences, **247**, Kluwer, Dordrecht, 1986.
5. I.L.Kantor, Jordan and Lie superalgebras defined by Poisson algebra, Algebra and analysis (Tomsk, 1989), 55–80, Amer. Math. Soc. Transl. Ser.2, **151**, Amer. Math. Soc., Providence, RI, 1992.
6. D.King and K.McCrimmon, The Kantor Construction of Jordan Superalgebras, Communs in Algebra, **20**, №1 (1992), 109–126.
7. K.McCrimmon, Speciality and non-speciality of two Jordan superalgebras, J.Algebra, **149**, №2 (1992), 326–351.
8. Y.A.Medvedev and E.I.Zel'manov, Some Counter-examples in the Theory of Jordan Algebras, *Nonassociative Algebraic Models*, 1–16, Nova Science, New York, 1992.
9. J.E.Moyal, Quantum mechanics as a statistical theory, Proc. Cambridge Philos. Soc., **45** (1949), 99–124.
10. H.Omori, Y,Maeda and A.Yoshioka, Weyl Manifolds and Deformation Quantization, Adv. in Math., **85** (1991), 224–255.
11. I.P.Shestakov, Superalgebras and Counter-examples, Siberian Math. Z., **32**, №6 (1991), 188–196.
12. I.P.Shestakov, Quantization of Poisson algebras and weak speciality of related Jordan superalgebras, Doklady RAN, to appear.
13. I.P.Shestakov, Quantization of Poisson superalgebras and speciality of Jordan Poisson superalgebras, Algebra and Logic, **32**, №5 (1993).
14. M.Sheunert, *The Theory of Lie Superalgebras, An Introduction*, Lecture Notes in Mathematics, **716**, Springer, Berlin – Heidelberg, 1979.
15. V.G.Skosyrskii, Prime Jordan algebras and Kantor Construction, Algebra and Logic, to appear.
16. G.A.Zaitsev, *Algebraic problems of mathematical and theoretical physics*, Nauka, Moscow, 1974.

DUPLICATED ALGEBRAS OF ALGEBRAS

YOSHIAKI TANIGUCHI
Department of Mathematics,
Nishinippon Institute of Technology,
Kanda-machi, Fukuoka-ken 800-03, Japan

Abstract. A duplicated Malcev algebra is constructed from a Malcev algebra M and a nonassociative algebra N by using a representation of M into N and an alternative bilinear mapping of N into M satisfying some conditions. It is shown that the representations of alternative algebra A, Malcev algebra M, Lie triple system T induce the representations of duplicated alternative algebra $A \oplus A$, Malcev algebra $M \oplus M$, Lie triple system $T \oplus T$ respectively.

For a Malcev algebra M, a nonassociative algebra N, a representation of M into N and an alternative bilinear mapping satisfying some conditions, it is shown that the direct sum $M \oplus N$ as a vector space becomes a Malcev algebra with respect to a certain product (Theorem 1.1). By this means, we show that an alternative algebra A, a Malcev algebra M and a Lie triple system T induce the duplicated alternative algebra $A \oplus A$, Malcev algebra $M \oplus M$ and Lie triple system $T \oplus T$ respectively. It is known that an alternative algebra induces a Malcev algebra [2] and a Malcev algebra induces a Lie triple system [1]. We consider the relations among the representations of these algebras in §2, for which we refer [6]. Throughout the paper, it is assumed that any vector space is a finite dimensional vector space over a field of characteristic not two.

1. Duplicated algebras

We consider a quadruple (M, N, ρ, f), where M, N are finite dimensional nonassociative algebras with bilinear products xy and pq respectively over a field F of characteristic not two, ρ is a linear mapping of M into the algebra $\mathrm{End}(N)$ of linear endomorphisms of a vector space N, and f is a bilinear mapping of N into M. Let $M \oplus N$ be a direct sum as a vector space. An element $x \oplus p$ of $M \oplus N$ is also denoted as $\begin{pmatrix} x \\ p \end{pmatrix}$ in column vector form. For fixed elements α, β in F, define a bilinear product of $x \oplus p$, $y \oplus q$ in $M \oplus N$ as

$$(*) \qquad \begin{pmatrix} x \\ p \end{pmatrix} \begin{pmatrix} y \\ q \end{pmatrix} := \begin{pmatrix} \alpha xy + \beta f(p, q) \\ \alpha\{\rho(x)q - \rho(y)p\} \end{pmatrix}.$$

If M is anticommutative and f is alternative, then the algebra $M \oplus N$ is anticommutative and the converse holds provided $\alpha\beta \neq 0$.

S. González (ed.), Non-Associative Algebra and Its Applications, 379–383.

A *Malcev algebra* M is an anticommutative algebra satisfying the condition

$$(**) \qquad (xz)(yw) = (xy \cdot z)w + (yz \cdot w)x + (zw \cdot x)y + (wx \cdot y)z,$$

$x, y, z, w \in M$ [3].

A linear mapping ρ of a Malcev algebra M into the algebra $\mathrm{End}(V)$ of linear endomorphisms of a vector space V is called a *representation* of M if

$$\rho(xy \cdot z) = \rho(x)\rho(yz) + \rho(xz)\rho(y) + [\rho(z), \rho(y)\rho(x)],$$

$x, y, z \in M$ [8].

Theorem 1.1. *Let M be an anticommutative algebra and f be an alternative mapping of N into M. Assume the quadruple (M, N, ρ, f) satisfies the following conditions:*

(1) *M satisfies the Malcev condition $(**)$,*

(2) *ρ is a representation of the Malcev algebra M into the vector space N,*

(3) $(xy)f(p,q) = xf(\rho(y)p, q) + yf(\rho(x)q, p) - f(\rho(y)\rho(x)p, q) - f(\rho(x)\rho(y)q, p)$,

(4) $f(\rho(x)p, \rho(y)q) = f(\rho(xy)p, q) + f(\rho(y)p, q)x + (f(p,q)x)y + f(\rho(y)\rho(x)q, p)$,

(5) $\rho(f(p,q))\rho(x)r = \rho(x)\rho(f(p,r))q - \rho(f(r,q)x)p + \rho(f(\rho(x)q, p))r - \rho(f(\rho(x)p, r))q$,

(6) $f(p,r)f(q,s) = f(\rho(f(p,q))r, s) + f(\rho(f(q,r))s, p) + f(\rho(f(r,s))p, q)$
$\qquad + f(\rho(f(s,p))q, r)$,

$x, y \in M$, $p, q, r, s \in N$, then the algebra $M \oplus N$ is a Malcev algebra. Conversely, if $M \oplus N$ is a Malcev algebra with respect to the product $(*)$, then M is a Malcev algebra and ρ is a representation of M if $\alpha \neq 0$, (3) to (6) hold if $\alpha\beta \neq 0$.

Proof. In order to show that $M \oplus N$ satisfies the Malcev condition $(**)$ for all $X, Y, Z, W \in M \oplus N$, it suffices to verify this in case: (i) $X = x \oplus 0, Y = y \oplus 0, Z = z \oplus 0, W = w \oplus 0$; (ii) $X = x \oplus 0, Y = y \oplus 0, Z = z \oplus 0, W = 0 \oplus s$; (iii) $X = x \oplus 0, Y = 0 \oplus q, Z = z \oplus 0, W = 0 \oplus s$; (iv) $X = x \oplus 0, Y = y \oplus 0, Z = 0 \oplus r, W = 0 \oplus s$; (v) $X = x \oplus 0, Y = 0 \oplus q, Z = 0 \oplus r, W = 0 \oplus s$; (vi) $X = 0 \oplus p, Y = 0 \oplus q, Z = 0 \oplus r, W = 0 \oplus s$, since $M \oplus N$ is anticommutative. But this follows from the conditions (1) to (6), and the converse is also verified simultaneously.

Corollary 1.2. *Let N be an ideal of a Malcev algebra M, ρ be the left multiplication in M and f be the multiplication of M, then $M \oplus N$, especially $M \oplus M$ becomes a Malcev algebra with respect to the product*

$$(***) \qquad \begin{pmatrix} x_1 \\ x_2 \end{pmatrix} \begin{pmatrix} y_1 \\ y_2 \end{pmatrix} := \begin{pmatrix} \alpha\, l(x_1) & \beta\, l(x_2) \\ \alpha\, l(x_2) & \alpha\, l(x_1) \end{pmatrix} \begin{pmatrix} y_1 \\ y_2 \end{pmatrix},$$

$x_1, y_1 \in M, x_2, y_2 \in N$, where $l(x)$ is a left multiplication in M.

Let $\alpha = 1$, $f = 0$. Regarding a vector space V over F as an algebra with trivial product, we have

Corollary 1.3. *If ρ is a representation of a Malcev algebra M into a vector space V, then $M \oplus V$ is a Malcev algebra with respect to the product*

$$\begin{pmatrix} x \\ v \end{pmatrix} \begin{pmatrix} y \\ w \end{pmatrix} := \begin{pmatrix} xy \\ \rho(x)w - \rho(y)v \end{pmatrix},$$

$x, y \in M$, $v, w \in V$.

An *alternative algebra* A is a nonassociative algebra satisfying the conditions $x(xy) = x^2y$ and $(xy)y = xy^2$, $x, y \in A$. Corollary 1.2 is also valid for an alternative algebra.

Proposition 1.4. *If A is an alternative algebra, then $A \oplus A$ is an alternative algebra with respect to the product $(***)$.*

A *Lie triple system* T is a ternary algebra with a trilinear product $[xyz]$ satisfying the following conditions:

$$[xxy] = 0,$$

$$[xyz] + [yzx] + [zxy] = 0,$$

$$[L(x,y), L(z,w)] = L(L(x,y)z, w) + L(z, L(x,y)w),$$

$x, y, z, w \in T$, where $L(x,y)$ is a left multiplication $z \mapsto [xyz]$ in T. As a generalization of [5, 10] we have the following

Theorem 1.5. *Let T be a Lie triple system with a product $[xyz] := L(x,y)z$, then the direct sum $T \oplus T$ as a vector space becomes a Lie triple system with respect to a trilinear product*

$$\left[\begin{pmatrix} x_1 \\ x_2 \end{pmatrix}\begin{pmatrix} y_1 \\ y_2 \end{pmatrix}\begin{pmatrix} z_1 \\ z_2 \end{pmatrix}\right] := \alpha \begin{pmatrix} \alpha\, L(x_1,y_1) + \beta\, L(x_2,y_2) & \beta\, L(x_1,y_2) + \beta\, L(x_2,y_1) \\ \alpha\, L(x_1,y_2) + \alpha\, L(x_2,y_1) & \alpha\, L(x_1,y_1) + \beta\, L(x_2,y_2) \end{pmatrix} \begin{pmatrix} z_1 \\ z_2 \end{pmatrix},$$

$x_i, y_i, z_i \in T$ $(i = 1, 2)$.

It is known that any alternative algebra A with a product xy is a Malcev algebra with respect to a product $xy - yx$ [2], and any Malcev algebra M with a product xy is a Lie triple system with respect to a trilinear product $[xyz] := x(yz) - y(xz) + 2(xy)z$ [1].

Proposition 1.6. *Let M_A be a Malcev algebra induced from an alternative algebra A, and T_M be a Lie triple system induced from a Malcev algebra M. Then the duplicated Malcev algebra $M_A \oplus M_A$ coincides with $M_{A \oplus A}$, and the duplicated Lie triple system $T_M \oplus T_M$ coincides with the Lie triple system $T_{M \oplus M}$.*

2. Representations of duplicated algebras

We recall the definitions of representations and a weak representation of algebras concerned.

A pair (λ,σ) of linear mappings of an alternative algebra A into the algebra $\mathrm{End}(V)$ of linear endomorphisms of a vector space V is called a *representation* of A into V if

$$[\lambda(x),\sigma(y)] = [\sigma(x),\lambda(y)] = \lambda(y)\lambda(x) - \lambda(yx) = \sigma(y)\sigma(x) - \sigma(xy),$$

$x,y \in A$ [4].

A linear mapping ρ of a Malcev algebra M into $\mathrm{End}(V)$ is called a *weak representation* of M into V if

$$[\Lambda(x,y),\rho(z)] = \rho(< xyz >),$$

$x,y,z \in M$, where $\Lambda(x,y) := [\rho(x),\rho(y)] - \rho(xy)$ and $< xyz > := x(yz) - y(xz) + (xy)z$.

A pair (λ,θ) of bilinear mappings of a Lie triple system T into $\mathrm{End}(V)$ is called a *representation* of T into V if

$$\lambda(x,y) = \theta(y,x) - \theta(x,y),$$

$$[\lambda(x,y),\theta(z,w)] = \theta([xyz],w) + \theta(z,[xyw]),$$

$$\theta(z,w)\theta(x,y) - \theta(y,w)\theta(x,z) - \theta(x,[yzw]) + \lambda(y,z)\theta(x,w) = 0,$$

$x,y,z,w \in T$ [7].

The following theorems show that representations of an alternative algebra, a Malcev algebra and a Lie triple system induce the representations of the duplicated alternative algebra, Malcev algebra and Lie triple system respectively.

Theorem 2.1. *Let (λ,σ) be a representation of an alternative algebra A into a vector space V. Define the linear mapping λ^*, σ^* of the duplicated alternative algebra $A\oplus A$ into $\mathrm{End}(V\oplus V)$ by*

$$\lambda^*\left(\begin{pmatrix} x_1 \\ x_2 \end{pmatrix}\right) := \begin{pmatrix} \alpha\,\lambda(x_1) & \beta\,\lambda(x_2) \\ \alpha\,\lambda(x_2) & \alpha\,\lambda(x_1) \end{pmatrix}, \quad \sigma^*\left(\begin{pmatrix} x_1 \\ x_2 \end{pmatrix}\right) := \begin{pmatrix} \alpha\,\sigma(x_1) & \beta\,\sigma(x_2) \\ \alpha\,\sigma(x_2) & \alpha\,\sigma(x_1) \end{pmatrix},$$

$x_1,x_2 \in M$. Then a pair (λ^*,σ^*) is a representation of the alternative algebra $A\oplus A$ into the vector space $V\oplus V$.

Theorem 2.2. *Let ρ be a weak representation (resp. representation) of a Malcev algebra M into a vector space V. Define the linear mapping ρ^* of the duplicated Malcev algebra $M\oplus M$ into $\mathrm{End}(V\oplus V)$ by*

$$\rho^*\left(\begin{pmatrix} x_1 \\ x_2 \end{pmatrix}\right) := \begin{pmatrix} \alpha\,\rho(x_1) & \beta\,\rho(x_2) \\ \alpha\,\rho(x_2) & \alpha\,\rho(x_1) \end{pmatrix},$$

$x_1,x_2 \in M$. Then ρ^* is a weak representation (resp. representation) of the Malcev algebra $M\oplus M$ into the vector space $V\oplus V$.

Theorem 2.3. *Let (λ, θ) be a representation of a Lie triple system T into a vector space V. Define the bilinear mappings λ^*, θ^* of the duplicated Lie triple system $T \oplus T$ into* $\mathrm{End}(V \oplus V)$ *by*

$$\lambda^*\left(\begin{pmatrix} x_1 \\ x_2 \end{pmatrix}, \begin{pmatrix} y_1 \\ y_2 \end{pmatrix}\right) := \alpha \begin{pmatrix} \alpha\,\lambda(x_1, y_1) + \beta\,\lambda(x_2, y_2) & \beta\,\lambda(x_1, y_2) + \beta\,\lambda(x_2, y_1) \\ \alpha\,\lambda(x_1, y_2) + \alpha\,\lambda(x_2, y_1) & \alpha\,\lambda(x_1, y_1) + \beta\,\lambda(x_2, y_2) \end{pmatrix},$$

$$\theta^*\left(\begin{pmatrix} x_1 \\ x_2 \end{pmatrix}, \begin{pmatrix} y_1 \\ y_2 \end{pmatrix}\right) := \alpha \begin{pmatrix} \alpha\,\theta(x_1, y_1) + \beta\,\theta(x_2, y_2) & \beta\,\theta(x_1, y_2) + \beta\,\theta(x_2, y_1) \\ \alpha\,\theta(x_1, y_2) + \alpha\,\theta(x_2, y_1) & \alpha\,\theta(x_1, y_1) + \beta\,\theta(x_2, y_2) \end{pmatrix},$$

$x_1,\ x_2,\ y_1,\ y_2 \in T$. *Then a pair (λ^*, θ^*) is a representation of the Lie triple system $T \oplus T$ into $V \oplus V$.*

If (λ, σ) is a representation of an alternative algebra A, then $\rho : x \mapsto \lambda(x) - \sigma(x)$ is a weak representation of a Malcev algebra M_A associated with A [8].

If ρ is a representation of a Malcev algebra M into V, define the bilinear mappings $\lambda,\ \theta$ of a Lie triple system T_M associated with M into $\mathrm{End}(V)$ as

$$\lambda(x, y) := [\rho(x), \rho(y)] + 2\rho(xy),$$

$$\theta(x, y) := \rho(x)\rho(y) + 2\rho(y)\rho(x) - \rho(xy),$$

$x,\ y \in M$. Then a pair (λ, θ) is a representation of T_M [9]. We have the following result concerned with Proposition 1.6.

Proposition 2.4. *Let (λ, σ) be a representation of an alternative algebra A, and let $\tilde{\rho}^* : X \mapsto \lambda^*(X) - \sigma^*(X)$, $X \in M_{A\oplus A}$ be a weak representation of the Malcev algebra $M_{A\oplus A}$ associated with $A \oplus A$. Then $\tilde{\rho}^*$ coincides with the weak representation ρ^* of the duplicated Malcev algebra $M_A \oplus M_A$ defined in Theorem 2.2.*

Let ρ be a representation of a Malcev algebra M and let $(\tilde{\lambda}^*, \tilde{\theta}^*)$ be a representation of Lie triple system $T_{M\oplus M}$ associated with $M \oplus M$. Then $(\tilde{\lambda}^*, \tilde{\theta}^*)$ coincides with the representation (λ^*, θ^*) of the Lie triple system $T_M \oplus T_M$ defined in Theorem 2.3.

References

1. O.Loos, Über eine Beziehung zwischen Malcev-Algebren und Lie-Tripel-systemen, Pacific J. Math., **18** (1966), 553-562.
2. A.I.Malcev, Analytic loops, Mat. Sb. (N.S.), **36(78)** (1955), 569-576 (Russian).
3. A.A.Sagle, Malcev algebras, Trans. Amer. Math. Soc., **101** (1961), 426-458.
4. R.D.Schafer, Representations of alternative algebras, Trans. Amer. Math. Soc., **72** (1952), 1-17.
5. Y.Taniguchi, On a kind of pairs of Lie triple systems, Math. Japon., **24** (1980), 605-608.
6. Y.Taniguchi and K.Yamaguti, Induced representations of duplicated Malcev algebras, Mem. Nishinippon Inst. Tech., Ser. Sci. Tech., **23** (1993), 1-8.
7. K.Yamaguti, On the cohomology space of Lie triple system, Kumamoto J. Sci., Ser. A, **5** (1960), 44-52.
8. K.Yamaguti, On the theory of Malcev algebras, Kumamoto J. Sci., Ser. A, **6** (1963), 9-45.
9. K.Yamaguti, A remark on Lie triple systems associated with Malcev algebras, Mem. Fac. Gen. Ed., Kumamoto Univ., Ser. Nat. Sci., No.**10** (1975), 1-4.
10. K.Yamaguti and Y.Taniguchi, A construction of general Lie triple systems from general Lie triple systems, Bull. Fac. Sch. Educ., Hiroshima Univ., Part II, **14** (1992), 73-80.

INTRODUCTION AUX ALGÈBRES DE BERNSTEIN PÈRIODIQUES

(cas Moufang, idempotents, caractéristique 2)

R. VARRO
Département de Mathématiques et Informatique Appliquées, Université de Montpellier III, BP 5043, 34032 Montpellier cedex 1, France

Abstract. Nous présentons ici une généralisation de la notion d'algèbre de Bernstein d'ordre n en introduisant la notion de périodicité. Après la définition et un exemple d'algèbre de Bernstein d'ordre n et de période p, nous étudions les propriétés des idempotents et nous montrons que dans le cas d'une algèbre associative ou qu'en caractéristique 2, les seules algèbres de Bernstein périodiques sont les quasi-constantes.

1. Définition et exemple

Soient K est un corps commutatif, (A,ω) une K-algèbre commutative pondérée, et n, p deux entiers tels que $n \geq 0$ et $p \geq 1$.

1.1. Définition

L'algèbre A est de Bernstein d'ordre n et de période p, notée $B(n,p)$-algèbre, si pour tout x dans A on a : $x^{[n+p+1]} = \omega(x)^{2^n(2^p-1)}x^{[n+1]}$, le couple (p,n) vérifiant cette relation, étant minimal pour l'ordre lexicographique. ◇

1.2. Proposition

La pondération d'une $B(n,p)$-algèbre est unique.

Démonstration.

Soient (A,ω) une $B(n,p)$-algèbre et $\omega' : A \to K$ une pondération. Pour tout $x \in \ker(\omega)$, de $x^{[n+p+1]} = 0$ on a $\omega'(x) = 0$ d'où $\ker(\omega) = \ker(\omega')$. ◇

1.3. Remarque

En particulier les $B(n,1)$-algèbres sont les algèbres de Bernstein à l'ordre n introduites par Abraham ([1]) et développées par Mallol ([6]). A notre connaissance, des exemples à des ordres et dimensions quelconques de ce type d'algèbre n'ont pas encore été construits. Nous en donnons ci-dessous.

1.4. Exemple

On suppose que $car(K) \neq 2$. Soit p, m et n des entiers non nuls, $n \geq 2$, σ une permutation d'ordre p opérant sur l'ensemble $\{1,\ldots,m\}$ et A une algèbre dont la multiplication par rapport à la base $\{e_1,\ldots,e_m,\varepsilon_1,\ldots,\varepsilon_n\}$ est : $e_ie_j = \frac{1}{2}(e_{\sigma(i)} + e_{\sigma(j)})$,

S. González (ed.), Non-Associative Algebra and Its Applications, 384–388.

$\varepsilon_r\varepsilon_s = \frac{1}{2}(\varepsilon_{r+1} + \varepsilon_{s+1})$, $e_i\varepsilon_r = \varepsilon_{r+1}$ et $\varepsilon_r\varepsilon_n = \varepsilon_{r+1}$ avec $1 \leq i, j \leq m$ et $1 \leq r, s \leq n-1$, les autres produits étant nuls. L'algèbre A, pondérée par $\omega(e_i) = 1$ et $\omega(\varepsilon_r) = 0$, est $B(n,p)$. En effet, soit $x = \sum_{1\leq i\leq m} \alpha_i e_i + \sum_{1\leq r\leq n} \beta_r\varepsilon_r$; pour tout $1 \leq q \leq n-1$ on a $x^{[q+1]} = \omega(x)^{2^q-1}\sum_{1\leq i\leq m}\alpha_i e_{\sigma^q(i)} + \lambda_q\sum_{1\leq r\leq n-q}\beta_r\varepsilon_{r+q}$ où les $\lambda_q \in K$ vérifient la relation : $\lambda_{q+1} = \lambda_q^2(2\beta_{n-q} + \sum_{1\leq r\leq n-q-1}\beta_r) + 2\omega(x)\lambda_q$ et $\lambda_0 = 1$. Ensuite on a $x^{[n+1]} = \omega(x)^{2^n-1}\sum_{1\leq i\leq m}\alpha_i e_{\sigma^n(i)}$, et pour $k \geq 1$: $\left(\sum_{1\leq i\leq m}\alpha_i e_{\sigma^k(i)}\right)^2 = \omega(x)\sum_{1\leq i\leq m}\alpha_i e_{\sigma^{k+1}(i)}$ d'où récursivement $(x^{[n+1]})^{[p+1]} = (\omega(x)^{2^n-1}\sum_{1\leq i\leq m}\alpha_i e_{\sigma^n(j)})^{[p+1]} = \omega(x)^{2^{n+p}-1}\sum_{1\leq i\leq m}\alpha_i e_{\sigma^{n+p}(i)} = \omega(x)^{2^n(2^p-1)}x^{[p+1]}$. Si on pose $e = \frac{1}{m}\sum_{1\leq i\leq m} e_i$, on a $e^2 = e$ et la sous-algèbre de A engendré par $\{e, \varepsilon_1, \ldots, \varepsilon_n\}$ est $B(n,1)$. De même, le sous-espace engendré par $\{e_1, \ldots, e_m\}$ est une $B(0,p)$-sous-algèbre de A. ◇

La définition d'une $B(n,p)$-algèbre ne nécessite pas de considération sur l'associativité. Dans ce qui suit on traite ce cas particulier en étudiant :

2. Les algèbres $B(n,p)$ de Moufang

Une algèbre A est de Moufang commutative si elle vérifie les relations de Moufang ([7]) : $(xzx)y = x(z(xy))$ et $(xz)(xy) = x(zy)x$: où $xzx = x(zx)$ et $x, y, z \in A$.

Il est clair qu'une algèbre de Moufang commutative est de Jordan. On montre sans peine que dans une algèbre de Moufang commutative, pour tout $x \in A$ l'application $L_x : y \to xy$ vérifie : $(L_x)^k = L_{x^k}$, où $k \geq 3$.

Dans ce paragraphe K un corps commutatif tel que $|K| \geq 2^{n+p}$.

2.1. Théorème

Si une algèbre de Moufang est $B(n,p)$ alors elle est quasi constante d'ordre n.

Démonstration.

Si l'algèbre A est de Moufang et $B(n,p)$, A étant de Jordan il existe $e \in A$, $e^2 = e$, et on a $A = Ke \oplus U_0 \oplus U_{1/2}$ où $U_\lambda = \{x \in \ker(\omega)/ex = \lambda x\}$. On va montrer que $U_0 = \ker(\omega)$. En effet, pour tout $x \in \ker(\omega)$, de la polarisation de $(\alpha e + x)^{[n+p+1]} = \alpha^{2^n(2^p-1)}(\alpha e + x)^{[n+1]}$ on a $2^p L_e^{n+p}(x) = L_e^n(x)$ d'où $2^p ex = ex$ et $ex = 0$ car $p \geq 1$. Ensuite il exite $x \in U_0$ tel que $x^{[n+1]} = 0$ et $x^{[n]} \neq 0$. En effet, pour tout $(\alpha, x) \in K^* \times U_0$ on a : $(\alpha e + x)^{[n+p+1]} = (\alpha e + x)^{2^{n+p}} = \alpha^{2^{n+p}}e^{2^{n+p}} + x^{2^{n+p}} = \alpha^{2^{n+p}}e + e^{[n+p+1]} = \alpha^{2^{n+p}}e$ et $\alpha^{2^n(2^p-1)}(\alpha e + x)^{[n+1]} = \alpha^{2^{n+p}}e + \alpha^{2^n(2^p-1)}x^{[n+1]}$, d'où $x^{[n+1]} = 0$. Et de même si on suppose que pour tout $x \in U_0$ on a $x^{[n]} = 0$ on a $(\alpha e + x)^{[n+p]} = \alpha^{2^{n-1}(2^p-1)}(\alpha e + x)^{[n]}$ ce qui contredit la minimalité de (p,n). Il en découle que $(\alpha e + x)^{[n+1]} = \alpha^{2^n}e$ avec n minimal, donc A est quasi-constante d'ordre n. ◇

L'existence d'idempotents n'est pas un privilègre du cas Moufang. Dans ce qui suit nous étudions ceci sans imposer aucune relation d'associativité et désormais dans toute la suite A désignera une $B(n,p)$-algèbre.

3. Idempotents d'une $B(,n,p)$-algèbre

3.1. DÉFINITION

Un élément $e \in A$, $e \neq 0$ est un idempotent s'il existe un entier $s \geq 1$ tel que $e^{[s+1]} = e$. ◇

On note $Ip(A)$ l'ensemble des idempotents de A et pour tout $x \in A$ on pose $O(x) = \{x^{[k]}/k \geq 1\}$. On montre sans difficulté que pour tout $x \in A$, l'ensemble $O(x)$ est fini si et seulement si il existe $k \geq 1$ tel que $x^{[k]} \in Ip(A)$ ou $x^{[k]} = 0$.

3.2. PROPOSITION

Dans une $B(n,p)$-algèbre :

(a) il existe au moins un idempotent ;

(b) tout idempotent est de poids non nul ;

(c) $e \in Ip(A) \Longleftrightarrow O(e) = O(e^{[k+1]})$ pour tout $k \geq 1$.

Démonstration

(a) Il existe $e \in A$, $\omega(e) = 1$ d'où $e^{[n+p+1]} = e^{[n+1]}$ et $e^{[n+1]} \in Ip(A)$. (b) Supposons qu'il existe $e \in Ip(A)$ tel que $\omega(e) = 0$. On a $e^{[s+1]} = e$. Or, de la définition de $B(n,p)$-algèbre on a $e^{[n+p+1]} = 0$. Si $s > n+p$, de $(e^{[n+p+1]})^{[s-n-p+1]} = 0$ on obtient $e = 0$, ce qui est contradictoire. Si $s \leq n+p$, de $(e^{[n+p+1]})^{[s-r+1]} = 0$, où $n+p = ms+r$, $0 \leq r < s$, on trouve à nouveau $e = 0$. (c) Si $e \in Ip(A)$, soit $s \geq 1$ le plus petit entier tel que $e^{[s+1]} = e$, on a $O(e) = \{e, \ldots, e^{[s]}\} = O(e^{[k+1]})$. La réciproque est immédiate. ◇

3.3. DÉFINITION

Si $e \in Ip(A)$, l'ordre de e est défini par : $o(e) = |O(e)|$. ◇

3.4. PROPOSITION

Si $|K| \geq 2^{n+p}$ alors une $B(n,p)$-algèbre admet un idempotent d'ordre p.

Démonstration.

Si $x^{[n+p]} = 0$ pour tout $x \in \ker(\omega)$, de la minimalité de (p,n), il existe $e \in A$ tel que $e^{[n+p+1]} = \omega(e)^{2^n(2^p-1)} e^{[n+1]}$ et $e^{[n+p]} \neq \omega(e)^{2^n(2^{p-1}-1)} e^{[n+1]}$. On a $\omega(e) \neq 0$ et $(\omega(e)^{-1}e)^{[n+1]}$ est un idempotent d'ordre p. Donc, s'il n'y a pas d'idempotent d'ordre p il existe $x \in \ker(\omega)$ tel que $x^{[n+p]} \neq 0$. Soit $z \in A$, $\omega(z) = 1$; pour tout $\alpha \in K$, $z_\alpha = (z+\alpha x)^{[n+1]}$ est un idempotent d'ordre $o(z_\alpha) < p$. En élevant $z_\alpha^{[o(z_\alpha)+1]} = z_\alpha$ à la puissance pleine $p - o(z_\alpha)$ on a $z_\alpha^{[p]} = z_\alpha^{[p-o(z_\alpha)]}$ et en identifiant les coefficients de $\alpha^{2^{n+p-1}}$, on obtient $x^{[n+p]} = 0$. Contradiction. ◇

3.5. PROPOSITION

Soit $e \in Ip(A)$.

(a) Pour tout entier $k \geq 1$, $e^{[k]} \in Ip(A)$ et $o(e^{[k]}) = o(e)$.

(b) $c = \omega(e)^{-1}e \in Ip(A)$ et $o(c)$ divise $o(e)$ et p.

Démonstration

(a) Résulte des résultats (b) et (c) de 3.2

(b) Soient $r = o(e)$ et $s = o(c)$. De $e^{[r+1]} = e$ on a $\omega(e)^{2^r} = \omega(e)$, d'où $c^{[r+1]} = c$ et donc $s \leq r$. Si $r = qs+t$ avec $0 \leq t < s$, on a $c^{[t+1]} = c$ d'où $t = 0$. Pour montrer que $o(c)$ divise p, on montre que $s \leq p$, pour cela il suffit que $e^{[p+1]} = e$. Soient u et v

deux entiers vérifiant $us > n+p$ et $v = us-(n+p)$. Posons $\varepsilon = c^{[n+1]}$ on a $\varepsilon^{[p+1]} = \varepsilon$ et $c = c^{[us+1]} = c^{[n+p+v+1]} = \varepsilon^{[p+v+1]} = \varepsilon^{[v+1]}$ d'où $c^{[p+1]} = \varepsilon^{[p+v+1]} = \varepsilon^{[v+1]} = c$. Ensuite, si $p = ms+t$, $0 \leq t < s$, de $c^{[p+1]} = c$ on déduit $c^{[t+1]} = c$, donc $t = 0$. ◇

On peut déterminer l'ensemble des idempotents de poids $\neq 1$ à l'aide du sous-ensemble des idempotents de poids 1. Pour tout $u \geq 2$, on pose $M_u = 2^u - 1$ et on dit qu'un élément λ de K est une M_u-racine primitive de l'unité si $\lambda^{M_u} = 1$ et $\lambda^{M_k} \neq 1$, pour tout $1 \leq k \leq u-1$.

3.6. PROPOSITION

Soit s un entier ≥ 2. Les énoncés suivants sont équivalents :

(a) il existe un idempotent e d'ordre s et $\omega(e) \neq 1$.

(b) il existe un idempotent c, $\omega(c) = 1$, et λ une M_u-racine primitive de l'unité, où ppcm $(u, o(c)) = s$.

Démonstration

(a) $\Longrightarrow$ (b). Posons $\lambda = \omega(e)$, alors $c = \lambda^{-1}e \in Ip(A)$ et $o(c)$ divise s. On a $\lambda \neq 1$ et $\lambda^{M_s} = 1$. Soit $u \geq 2$ le plus petit entier vérifiant $\lambda^{M_u} = 1$, on a $u \leq s$ et si $s = uq + r$, $0 \leq r < u$, alors $\lambda = \lambda^{2^s} = (\lambda^{2^{uq}})^{2^r} = \lambda^{2^r}$, d'où $r = 0$, donc l'entier u divise s. Alors, si $t = ppcm(u, o(c))$, t divise s et $e^{[t+1]} = \lambda^{2^t} c^{[t+1]} = \lambda c = e$, d'où $t = s$.

(b) $\Longrightarrow$ (a) s'établit sans peine. ◇

Nous allons établir qu'en caractéristique 2, les seules algèbres de Bernstein périodiques sont de période 1 (qui sont aussi les quasi-constantes).

4. $B(n,p)$-algèbres en caractéristique 2

Dans ce paragraphe K est un corps commutatif infini de caractéristique 2. Pour $s \geq 1$, on pose $Ip_s(A) = \{e \in Ip(A)/o(e) = s\}$.

4.1. PROPOSITION

Soit A une $B(n,p)$-algèbre.

(a) Il existe $y \in \ker(\omega)$ tel que $y^{[n+1]} = 0$ et $y^{[n]} \neq 0$.

(b) Pour tout e, $e' \in Ip(A)$ on a : $\omega(e)e' = \omega(e')e$.

(c) Les ensembles $Ip_p(A)$ et $\{\alpha \in K/\alpha^{2^p-1} + 1 = 0\}$ sont équipotents.

Démonstration.

Soit $e \in Ip(A)$, $\omega(e) = 1$, on a $e^{[p+1]} = e$ (cf. 3.4 et 3.5b). On a $y^{[n+1]} = 0$ pour tout $y \in \ker(\omega)$: en effet, partant de $(e+y)^{[n+p+1]} = (e+y)^{[n+1]}$, comme $y^{[n+p+1]} = 0$ on a $e^{[n+1]} = e^{[n+1]} + y^{[n+1]}$ d'où $y^{[n+1]} = 0$. Et si $y^{[n]} = 0$ pour tout $y \in \ker(\omega)$, pour tout $\alpha \in K$ on a $(\alpha e + x)^{[n]} = \alpha^{2^{n-1}} e^{[n]}$ et donc $(\alpha e + x)^{[n+p]} = \alpha^{2^{n+p-1}} e^{[n]} = \alpha^{2^{n-1}(2^p-1)}(\alpha e + x)^{[n]}$, ce qui contredit la définition d'une $B(n,p)$-algèbre.

(b) Soient e, $e' \in Ip(A)$, $o(e) = r$ et $o(e') = s$. Posons $q = ppcm(r,s)$ et $e' = \alpha e + x$ avec $\alpha = \omega(e)^{-1}\omega(e')$, $x \in \ker(\omega)$. On a $e' = e'^{[q+1]} = \alpha^{2^q} e + x^{[q+1]} = \alpha e + x^{[q+1]}$ et $x^{[q+1]} = x$. Ainsi x est un idempotent de poids nul, donc $x = 0$ (cf. 3.2 b) et $e' = \alpha e$.

(c) Notons $K_p = \{\alpha \in K/\alpha^{2^p-1}+1 = 0\}$. Soit $e \in Ip_p(A)$, la restriction à $Ip_p(A)$ de l'application $x \rightarrow \omega(e)^{-1}\omega(x)$, a pour image K_p et d'après (b) elle est injective.

◇

Soient $\mathbf{F}_2 = \mathbf{Z}/2\mathbf{Z}$ et $\mathbf{F}_4 = {}^{\mathbf{F}_2[X]}/(X^2 + X + 1)$.

4.2. LEMME

(a) Pour qu'il existe dans K une racine cubique non triviale de l'unité, il faut et il suffit que K soit une extension de $\mathbf{F}_4$.
(b) Quel que soit l'entier $n \geq 1$, les racines dans K du polynôme $X^n + 1$ sont des racines dans K du polynôme $X^3 + 1$.

Démonstration.

(a) La condition suffisante est triviale. Elle est nécessaire car si on a $\alpha \in K$, $\alpha \neq 1$, $\alpha^3 = 1$, alors $\mathbf{F}_2(\alpha)$ est un sous-corps de K isomorphe à $\mathbf{F}_4$.

(b) Le résultat est immédiat si n est pair. Soit $\alpha \in K$, $\alpha^{2m+1} = 1$, en calculant $\alpha(\alpha+1)^{2m+2} = \alpha(\alpha+1)^{2m}(\alpha+1)^2$ on aboutit à $\alpha^2 + 1 = \alpha$, donc $\alpha^3 + 1 = 0$. ◇

4.3. THÉORÈME

Si A une $(B(n,p)$-algèbre alors A est quasi-constante d'ordre n.

Démonstration.

On va montrer que $Ip_1(A) \neq \emptyset$, pour cela on considère deux cas.

(i) Si K n'est pas une extension de $\mathbf{F}_4$: tout idempotent de A est de poids 1 et donc A contient un unique idempotent e (cf. 4.1.b) et on a $o(e) = 1$.

(ii) Si K est une extension de $\mathbf{F}_4$: K possède deux racines cubiques non triviales de l'unité, $\zeta \in K$ et $\zeta^2 = \zeta + 1$. Soit $e \in Ip(A)$. Si $e \neq e^2$, alors $e \in Ip_2(A)$. En effet, on a $\omega(e) \neq 1$ car sinon $\omega(e^2) = \omega(e)$ d'où $e = e^2$ (cf 4.1.b). Si $\omega(e) = \zeta$, alors $e^2 = \zeta e$ et $e^{[3]} = \zeta^2 e^2 = \zeta^3 e = e$, i.e., e, $e^2 \in Ip_2(A)$ et $\zeta^{-1} e \in Ip_1(A)$. On a le même résutlat si on suppose que $\omega(e) = \zeta + 1$.

Dans les deux cas on a $Ip_1(A) = \{\varepsilon\}$, alors d'après 4.1 (a) : $x^{[n+1]} = \omega(x)^{2^n}\varepsilon$, pour tout $x \in A$ avec n minimal, donc A est quasi-constante d'ordre n. ◇

5. Bibliographie

1. V. M. Abraham, The genetic algebra of polyploids, Proc. London Math. Soc. (3) **40**, 385-429 (1980).
2. S. Bernstein, Démonstration mathématique de la loi d'hérédité de Mendel, C.R. Ac. Sc. Paris **177**, 528-531 (1923).
3. S. Bernstein, Principe de stationnarité et généralisation de la loi de Mendel, C. R. Ac. Sc. Paris, **177**, 581-584 (1923).
4. P. Holgate, Genetic algebras satisfying Bernstein's stationary principle, J. London Math. Soc. **9**, 613-623 (1975).
5. Yu. I. Lyubich, Basic concepts and theorems of evolution genetics of free populations, Russian Math. Surveys **26**, 51-123 (1971).
6. C. Mallol, A propos des algèbres de Bernstein, Thèse d'Etat, Université de Montpellier II, France, 1989.
7. R.D. Schafer, An introduction to non associative algebras, Academic Press, New York,(1966).
8. R. Varro, Algèbres de Bernstein périodiques, Thèse, Université de Montpellier II, France, 1992.
9. S. Walcher, Algebras which satisfy a train equation for the first three plenary powers, Arch. Math. **56**, 547-551 (1991).
10. A. Worz-Busekros, Algebras in genetics, Lect. Notes in Biomathematics **36**, Springer-Verlag, Berlin-Heildelberg-New York (1980).

RANDOM JORDAN DERIVATIONS

MARIA VICTORIA VELASCO and ARMANDO R. VILLENA
Departamento de Análisis Matemático, Facultad de Ciencias, Universidad de Granada, 18071-GRANADA (SPAIN).

1. Introduction

A classical nonassociative operators topic is the continuity of Jordan derivations on Banach algebras which have some aditional property. We recall that a *Jordan derivation* on a Banach algebra A is a linear mapping $D : A \to A$ such that $D(a^2) = D(a)a + aD(a)$, $\forall a \in A$, or equivalently satisfying that $D(a \cdot b) = D(a) \cdot b + a \cdot D(b)$, $\forall a, b \in A$, (where, as it is ussual, $a \cdot b = \frac{1}{2}(ab + ba)$).

In 1970, Sinclair ([6]) conjectured that *every Jordan derivation on a semisimple Banach algebra is continuous*, and besides he proved that *every continuous Jordan on a Banach algebra A is a derivation i. e. $D(ab) = aD(b) + D(a)b$, $\forall a, b \in A$.* At this time it was also known a celebrated theorem of Johnson and Sinclair [5] asserting that *every derivation on a semisimple Banach algebra is continuous.* Therefore, for a semisimple Banach algebra A, the problem of find out if a Jordan derivation on A is continuous is equivalent to investigate if D is a derivation.

In 1975 Cusak [4] proved that *any Jordan derivation on a semiprime Banach algebra is a derivation.* This result, obtained years later by Brešar [3], allows to prove, as the last author observe, that *Jordan derivations on semisimple Banach algebra are continuous* .

Now, our purpose is to randomize the above result about automatic continuity for Jordan-derivations on semisimple Banach algebras, defining the *stochastic Jordan derivation* concept as a simple generalization of the Jordan derivation one, and proving that *stochastic Jordan derivations on semisimple Banach algebras are continuous*, understanding this continuity in the more logical sense, the stochastic one. Besides we weaken the stochastic Jordan derivation property considering more general operators, the *probable Jordan derivations*, and we inquire about the continuity that they have, getting another automatic continuity theorem. As a particular case of these results we show the Sinclair conjecture.

2. Some topics about random operators

We consider a fixed probability space $(\Omega, \Sigma, \mathbb{P})$ and given a Banach space, Y, we denote by $\mathcal{L}_0(\mathbb{P}, Y)$ the linear space of all *Y-valued Bochner random variables* on $(\Omega, \Sigma, \mathbb{P})$. We endow $\mathcal{L}_0(\mathbb{P}, Y)$ with the convergence in probability topology and identifying, as it is custom, the elements that coincide almost surely (*abbr. a.s.*) we obtain the space $L_0(\mathbb{P}, A)$ which is an metrizable complete topological vectorial

S. González (ed.), Non-Associative Algebra and Its Applications, 389–394.

space with respect to this topology.

Let X, Y be Banach spaces. As it is known, a *random operator* fron X to Y is a mapping from X to $\mathcal{L}_0(\mathbb{P}, Y)$. A random operator T from X to Y is a *linear* one if $\mathbb{P}[T(\alpha x+\beta y)=\alpha T(x)+\beta T(y)]=1$, $\forall x, y \in X$, α, β constants.

Random operators arise extensively in Random Equations Theory ([1], [7]). Of course, every classic operator $T : X \to Y$ can be regarded as a random operator from X to Y since $\mathcal{L}_0(\mathbb{P}, Y)$ contains trivially Y as a subspace.

A random operator T from X to Y is *stochastically continuous* at a point x_0 in X if for every sequence $\{x_n\}_{n\in\mathbb{N}}$ in X converging to x_0 it is satisfied that $\{T(x_n)\}_{n\in\mathbb{N}}$ converges to $T(x_0)$ in probability, *i.e.* $\lim_{n\to\infty} \mathcal{P}[\|Tx_n - Tx_0\| > \varepsilon] = 0$, $\forall \varepsilon > 0$. It is said that T is *stochastically continuous* when the above property is satisfied for every element belonging to X.

Let Ω_0 be a measurable set, with $\mathbb{P}[\Omega_0] > 0$, and T a linear random operator from X to Y. Considering Ω_0 as a new probability space with the inherited structure from Ω, (the induced probability on Ω_0 is the conditional probability $\mathbb{P}_{\Omega_0}$), we can define the operator $T_{\Omega_0} : X \to \mathcal{L}_0(\mathbb{P}_{\Omega_0}, Y)$ by $T_{\Omega_0}(x) = T(x)/\Omega_0$, and the linear random operator T_{Ω_0} is said to be a *conditional operator* of D.

Definition 2.1. We say that a linear random operator, D, is *probably continuous* when some of its conditional operators are stochastically continuous (*i.e.* the operator behave as a stochastically continuous one on some measurable subset of Ω with positive measure). For every probably continuous linear random operator, D, we define the real number

$$\alpha(D) = \sup\{\mathbb{P}[\Omega_0] : D_{\Omega_0} \textit{ is stochastically continuous}\}.$$

This supremum is indeed a maximum (see [9]) so it can be judged as *the probability of D being stochastically continuous.*

3. Random Jordan derivations

Definition 3.1. A *stochastic Jordan derivation* on a Banach algebra A is a linear random operator D, from A to A, such that $\mathbb{P}[D(a^2) = D(a)a + aD(a)] = 1$, $\forall a \in A$ or equivalently $\mathbb{P}[D(a\cdot b) = D(a)\cdot b + a\cdot D(b)]$, $\forall a, b \in A$.

If D is a linear random operator on A satisfying only that *there exists δ in* $]0,1[$ *such that*

$$\mathbb{P}[D(a^2) = D(a)a + aD(a)] \geq \delta, \ \forall a \in A, \tag{1}$$

then we say that D is a *probable Jordan derivation.*

Particularly, when the probability considered is the degenerate one then stochastic Jordan derivations, probable Jordan derivations and Jordan derivations (in the classical sense) are the same concept.

Our goal is to inquire about the continuity property for probable Jordan derivations. But at the first place we observe that we do not know in advance if the property (1) is equivalent to this natural other

$$\mathbb{P}[D(a\cdot b) = D(a)\cdot b + a\cdot D(b)] \geq \delta, \ \forall a, b \in A. \tag{2}$$

The next theorem, that will allows us to prove the equivalence between the above conditions, (1) and (2), is just the result that will discover the most deepest relation between probable Jordan derivations and stochastic Jordan derivations. We will see how this concepts are statements of the same property with the only difference that for probable Jordan derivations the property has a "local" declaration.

A random operator Q from a Banach space X to another Banach space Y will be called *quadratic* if there exists a bilinear random operator T from $X \times X$ to Y such that

$$Q(x) = T(x, x), \ \forall x \in X.$$

Theorem 3.2. *Let X, Y be Banach spaces, and T a simetric bilinear random operator from $X \times X$ to Y whose associated cuadratic random operator is Q. Then the following properties hold:*

1. *Let assume that $0 < \delta < 1$ is such that $\mathbb{P}[Q(x) = 0] \geq \delta$, $\forall x \in X$. Then $\mathbb{P}[T(x, y) = 0] \geq \delta$, $\forall x, y \in X$. In fact there exists a measurable set Ω_0, with $\mathbb{P}[\Omega_0] \geq \delta$, such that $T_{\Omega_0} \equiv 0$.*
2. *The set*

$$\{\mathbb{P}[\Omega_0] : \ \Omega_0 \textit{ is measurable and } T_{\Omega_0} \equiv 0\}$$

has a maximum, the set

$$\{\mathbb{P}[Q(x) = 0] : \ x \in X\}$$

has a minimum, and these values coincide with the quantity

$$\min\{\mathbb{P}[T(x, y)] = 0 : \ x, y \in X\}.$$

Proof. For every x in X, y in Y, and λ in $\mathbb{R}^+$ we have that

$$\delta \leq \mathbb{P}[Q(x + \lambda y) = 0] = \mathbb{P}[Q(x) + \lambda T(x, y) + \lambda^2 Q(y) = 0] =$$

$$\mathbb{P}[Q(x) + \lambda T(x, y) + \lambda^2 Q(y) = 0, Q(x) = 0] + \mathbb{P}[Q(x) + \lambda T(x, y) + \lambda^2 Q(y) = 0, Q(x) \neq 0].$$

On the one hand we observe that

$$\mathbb{P}[Q(x) + \lambda T(x, y) + \lambda^2 Q(y) = 0, Q(x) = 0] \leq$$

$$\mathbb{P}[\lambda T(x, y) + \lambda^2 Q(y) = 0] = \mathbb{P}[T(x, y) + \lambda Q(y) = 0] =$$

$$\mathbb{P}[T(x, y) = \lambda Q(y)] \leq \mathbb{P}[\|T(x, y)\| = \lambda \|Q(y)\|].$$

On the other hand we have that

$$\mathbb{P}[Q(x) + \lambda T(x, y) + \lambda^2 Q(y) = 0, Q(x) \neq 0] = \mathbb{P}[Q(x) = -\lambda T(x, y) - \lambda^2 Q(y), Q(x) \neq 0] \leq$$

$$\mathbb{P}[\|Q(x)\| \leq \lambda \|T(x, y)\| + \lambda^2 \|Q(y)\|, Q(x) \neq 0].$$

In consecuence

$$\delta \leq \mathbb{P}[\|T(x, y)\| = \lambda \|Q(y)\|] + \mathbb{P}[\|Q(x)\| \leq \lambda \|T(x, y)\| + \lambda^2 \|Q(y)\|, Q(x) \neq 0],$$

so, making $\lambda \to 0$, it follows that

$$\delta \leq \mathbb{P}[T(x,y) = 0],\ \forall x \in X,\ y \in Y.$$

Now the multilinear uniform randomization principle [8], Theorem 4.1, proves the result. □

Is very easy to check that for a Jordan derivation D, the random operator

$$Q(a) = D(a^2) - D(a)a - aD(a)$$

is a quadratic one whose associated bilinear random operator is

$$\frac{1}{2}T(a,b) = D(a \cdot b) - D(a) \cdot b - a \cdot D(b).$$

A straightforward application of the last result to this bilinear operator shows the equivalence between the two reasonable ways of defining the probable Jordan derivability condition.

Corollary 3.3. *Let D be a random operator on a Banach algebra A and $0 < \delta < 1$. Then the following conditions are equivalent:*
1. $\mathbb{P}[D(a^2) = aD(a) + D(a)a] \geq \delta,\ \forall a \in A,$
2. $\mathbb{P}[D(a \cdot b) = a \cdot D(b) + D(a) \cdot b] \geq \delta,\ \forall a, b \in A,$

Now, associated to every probable Jordan derivation, D, we define the quantity

$$\delta_J(D) := \inf\{\mathbb{P}[D(a^2) = aD(a)+D(a)a],\ a \in A\} = \inf\{\mathbb{P}[D(a\cdot b) = a\cdot D(b)+D(a)\cdot b] :\ a, b \in A\}$$

which is considered as *the probability of D being a Jordan derivation.* Obviously: *Every probable Jordan derivation, D is a stochastic Jordan derivation if, and only if, $\delta_J(D) = 1$.*

For getting examples of probable Jordan derivations we can consider a derivation D, on a Banach algebra A, with some conditional operator D_{Ω_0} being a stochastic derivation, (*i.e.* $D(a^2) = D(a)a + aD(a)$, *(a. s.)* on Ω_0 $\forall a \in A$.) However, reciprocally, if D is a probable derivation on A, then we only know that every element a in A has associated a measurable set Ω_a, with $\mathbb{P}[\Omega_a] \geq \delta(D)$, such that $D(a^2) = D(a)a + aD(a)$ $(a.s.)$ on Ω_a, but taking the measurable set Ω_b associated to another element b in A, we can not know, in advance, any relation between Ω_a and Ω_b. Surpresively, for every a, b in A, the sets Ω_a and Ω_b can be identified with a commun set Ω_0, as we obtaing from Theorem 3.2(ii).

Corollary 3.4. *A linear random operator D on a Banach algebra A is a probable Jordan derivation if, and only if, there exists a measurable set Ω_0, with $\mathbb{P}[\Omega_0] \geq \delta_J(D)$, such that the conditional operator D_{Ω_0} is a stochastic Jordan derivation.*

Therefore the continuity of probable Jordan derivations is reduced to the continuity of stochastic Jordan derivations.

4. Continuity of random derivations

In the study of automatic continuity of stochastic Jordan derivations we are follow the classic pattern showing, at the first place, that *every stochastic Jordan derivation defined on a semisimple Banach algebra is a stochastic derivation.* For that we check that the semiprimitivity property of the Banach algebra A is transferred to the algebra $L_0(\mathbb{P}, A)$. We recall that the Banach algebra A is semiprime if $\{0\}$ is the only two sided ideal ideal J of A with $J^2 = \{0\}$, or equivalently if, and only if, $aAa = 0 \Rightarrow a = 0,\ \forall a \in A$.

Lema 4.1. *Let A be a semiprime Banach algebra. If $\mathbf{y}$ in $\mathcal{L}_0(\Omega, A)$ is such that $\mathbf{y}\mathcal{L}_0(\Omega, A)\mathbf{y} = 0$ then $\mathbf{y} = 0$ (a.s.). In particular the algebra $L_0(\Omega, A)$ is semiprime.*

Proof. Let $\mathbf{y}$ be an element of $\mathcal{L}_0(\Omega, A)$ such that $\mathbf{y}\mathcal{L}_0(\Omega, A)\mathbf{y} = 0$ $(a.s.)$. Since A is trivially contained in $\mathcal{L}_0(\Omega, A)$, it is satisfied that $\mathbf{y}a\mathbf{y} = 0$ $(a.s.)$, $\forall a \in A$. We define, for every element a in A, the set $C_a = \{b \in A : bab = 0\}$. Because A is semiprime, we have that $\bigcap_{a\in A} C_a = 0$, so, by [9, Lemma 3.1],

$$\mathbb{P}[\mathbf{y} = 0] = \mathbb{P}[\mathbf{y} \in \bigcap_{a\in A} C_a] = \inf\{\mathbb{P}[\mathbf{y} \in \bigcap_{a\in F} C_a] :\ F \subseteq I, F\, finite\} = 1.$$

□

Definition 4.2. We say that a linear random operator on A is a stochastic derivation if $\mathbb{P}[D(ab) = aD(b) + D(a)b] = 1,\ \forall a, b \in A$.

A linear random operator on A is said to be *probable derivation* if it has a conditional operator being a stochastic derivation (*i.e.* the operator behaves as a stochastic derivation on some measurable set of positive measure).

Now we observe, bearing the last lemma in mind, that the whole of the way crossed by Brešar in [3], for showing that *every stochastic Jordan derivation defined on a semiprime Banach algebra is a derivation*, can be followed step by step "stochastically" (*i.e.* repeating every argument exposed here for a stochastic Jordan derivation). Therefore we conclude the next result.

Corollary 4.3. *A linear random operator on a semiprime Banach algebra is a stochastic Jordan derivation if, and only if, is a stochastic derivation.*

From Corollary 3.4 it follows this other.

Corollary 4.4. *Let D be linear random operator on a semiprime Banach algebra. Then, D is a probable Jordan derivation if and only if, D is probable derivation.*

In [8], the celebrated Johnson-Sinclair result about automatic continuity of derivations on semisimple Banach algebras [5], was randomized, establishing the following theorem.

Theorem 4.5. *Every stochastic derivation on a semisimple Banach algebra is stochastically continuous.*

Since, as is well known, semisimple algebras are semiprime ones [2, Theorem 30.5], from Corollary 4.3 and Theorem 4.5, we deduce the expected result.

Corollary 4.6. *Every stochastic Jordan derivation on a semisimple Banach algebra is stochastically continuous.*

Now, applying Corollary 4.3 (see also Corollary 3.4), we have the next result.

Corollary 4.7. *Every probable Jordan derivation, D, on a semisimple Banach algebra is probably continuous. Moreover the probability of D being continuous ($\alpha(D)$) is at least the probability of D being a Jordan derivation ($\delta_J(D)$).*

Therefore the conjecture of Sinclair is established in a more general sense.

References

1. A.T. Bharucha-Reid, *Random Integral Equations*, Academic Press, New York, 1972.
2. F. F. Bonsall, J. Duncan, *Complete Normed Algebras*, Springer-Verlag, New York, 1973.
3. M. Brešar, Jordan derivations on semisimple rings, Proc. Amer. Math. Soc., vol 104, n. 4, 1988, 1003-1006.
4. J. M. Cusak, Jordan derivations on rings, Proc. Amer. Math. Soc., vol 53, n. 2, 1975, 321-324
5. B. E. Johnson, A. M. Sinclair, Continuity of derivations and a problem of Kaplansky, Amer. J. Math. 90 (1968), 1067-1973.
6. A. M. Sinclair, Jordan homomorphisms and derivations on semisimple Banach algebras, Proc. Amer. Math. Soc., vol. 24, 1970, 209-214.
7. A. V. Skorohod, *Random Linear Operator*, D. Reidel Publishing Company, Holland, 1984.
8. M. V. Velasco, A R. Villena, Continuity of random derivations, to appear in Proc. Amer. Math. Soc..
9. M. V. Velasco, A R. Villena, A random closed graph theorem, submitted to publication.

RANDOM DERIVATIONS ON H^*-ALGEBRAS

ARMANDO R. VILLENA
Departamento de An lisis Matem tico, Facultad de Ciencias, Universidad de Granada, 18071-GRANADA (SPAIN).

Abstract. We obtain the continuity of stochastically derivative linear random operators on (nonassociative) H^*-algebras with zero annihilator. Moreover, we investigate the stochastic size of the separating subspace for linear random operators, on H^*-algebras, which have some probability of being derivative.

The continuity of derivations is a basic automatic continuity problem which has drawn the attention of many authors. The fundamental work which started investigation into the continuity of derivations on Banach algebras is due to Johnson and Sinclair [4] who established a number of fundamental principles in what is now known as the Theory of Automatic Continuity.

Even in the nonassociative context it has taken a great interest about these problems. For instance, continuity of derivations is proved by the following authors for the following kinds of algebras:

de la Harpe [3], *for classical Banach Lie algebras of operators on a Hilbert space,*

Youngson [11], *for JB^*-algebras,*

Zalar [12], *for Mal'cev H^*-algebras,*

by the author [8], *for general nonassociative H^*-algebras,*

and also by the author [7], *for semisimple complete normed alternative algebras.*

In this talk we shall be mainly concerned with random derivations on nonassociative H^*-algebras and the continuity problem for these operators.

¿From now on we let A be a fixed *H^*-algebra*, that is a complex algebra with a conjugate-linear algebra involution $*$ whose underlying vector space is a Hilbert space in which equalities

$$(ab \mid c) = (a \mid cb^*) = (b \mid a^*c)$$

hold for all a, b and c in A.

As it is usual a derivation on A will be a lincar map D from A to itself satisfying

$$D(ab) = D(a)b + aD(b).$$

MSC (1991): 46H40.

S. González (ed.), Non-Associative Algebra and Its Applications, 395–399.

The study of Banach algebra valued random variables and random operators on Banach algebras are of great importance in the random equations theory. For this reason these have been considered extensively mainly for C^*-algebras and H^*-algebras.

Fron now on we let $(\Omega, \Sigma, \mathbb{P})$ a fixed probability space and we consider the algebra $L_0(A)$ of all *A-valued Bochner random variables* on Ω, with the usual almost surely identification, and endowed with the usual probability convergence topology which is a complete metrizable linear topology. Now we recall that a random operator on A is an operator from A into $L_0(A)$ and a such operator $\mathcal{D}$ is said to be *stochastically derivative* if it is linear and equality

$$\mathcal{D}(ab) = \mathcal{D}(a)b + a\mathcal{D}(b)$$

holds for all a and b in A. The *continuity* of $\mathcal{D}$, also called *stochastic continuity*, means that for every a in A

$$\lim_{n\to\infty} \mathbb{P}[\| Tx_n - Tx \| \geq \varepsilon] = 0,$$

for every sequence $\{a_n\}$ in A converging to a and every positive ε.

Now we can measure discontinuity of $\mathcal{D}$ by considering its so-called *separating subspace*

$$\mathcal{S}(T) = \{\mathbf{a} \in \mathcal{L}_0(A) : \exists\{a_n\} \to 0 \textit{ in } A \textit{ with } \{Da_n\} \to \mathbf{a} \textit{ in probability}\}.$$

Closed Graph Theorem shows that $\mathcal{D}$ is stochastically continuous if and only if $\mathcal{S}(\mathcal{D}) = 0$.

It is not surprising that in order to prove our continuity theorems we first require a number of results from H^*-algebra theory being *central closeability* of topologically simple H^*-algebras the crucial fact that we use in our proofs.

As it is well known a normed algebra is said to be *topologically simple* if it has nonzero product and not nonzero proper closed ideals. A prime algebra is said to be *centrally closed* if scalar multiples of the identity operator are the only partially defined centralizers.

The following result by Cabrera and Rodriguez is a essential one:

Theorem 1 [1]. *Every topologically simple H^*-algebra is a centrally closed prime algebra.*

Our goal is to show how techniques of central closeability provide suitable sequences for the treatment of our automatic continuity problem.

The investigation into this techniques was started in an earlier work [9] showing the following result.

Proposition 1 [9]. *In a centrally closed prime algebra A one of the following assertions holds:*

1. There exists sequences $\{a_n\}$ in B and $\{T_n\}$ in the usual multiplication algebra of B (from now on denoted by $M(B)$) such that:

$$T_n \cdots T_1 a_n \neq 0 \text{ and } T_{n+1} T_n \cdots T_1 a_n = 0, \forall n \in \mathbb{N}.$$

2. There exists some operator T in $M(B)$ such that:

$$\dim(T(B)) = 1.$$

Thanks to this result and structrure theorems for H^*-algebras we obtained this other.

Theorem 2 [9]. *The separating subspace $\mathcal{S}(D)$ for a derivation D on an H^*-algebra A is contained in the annihilator, $Ann(A)$, of A.*

We recall that the annihilator, $Ann(A)$, of A is defined as the set of those elements a in A such that

$$ab = ba = 0$$

for all b in A.

In contrast, in the case where the derivation is a random one a major obstacle is the assertion *2* from Proposition *1*. Thus we need an improvement of this assertion. For that we show a very useful result based on the central closeability and the topological simplicity.

Theorem 3 [10]. *Let B be a centrally closed topologically simple algebra with infinite dimension and G a nonempty open subset of B which does not contain the zero. Then, one of the following assertions holds:*

1. There exist sequences $\{a_n\}$ in B and $\{T_n\}$ in $M(B)$ such that:

$$T_n \cdots T_1 a_n \in G \text{ and } T_{n+1} T_n \cdots T_1 a_n = 0, \ \forall n \in \mathbb{N}.$$

2. There exists a sequence $\{S_n\}$ in $M(B)$ such that:

$$dim S_n(B) = 1,$$

$$S_n^2(B) \cap G \neq \emptyset,$$

$$S_n S_m = 0 \text{ if } m < n.$$

Finally we can obtain the following.

Theorem 4 [10]. *Let $\mathcal{D}$ be a stochastically derivative random operator on an H^*-algebra A. Then every element in the separating subspace for $\mathcal{D}$ is almost surely valued in the annihilator of A.*

Sketch of the proof. We divide the proof of the above result in three steps.

In the first one we use Theorems 1 and 3 and familiar constructions in automatic continuity theory to obtain that:

for every minimal closed ideal I of A and every b in I, the linear random functional $\mathcal{D}_b$ on A defined by

$$a \mapsto (\mathcal{D}a \mid b)$$

is stochastically continuous.

It must be pointed out that in the proof of the above assertion a crucial role is played by the stochastic graduation for the discontinuity of linear random operators that Velasco and myself obtained in two previous papers [6, 7].

In a second step we obtain from the stochastic continuity of $\mathcal{D}$ when A has zero annihilator as a consequence of the following structure theorem due to Cuenca and Rodr!guez:

Theorem 5 [2]. *Every H^*-algebra with zero annihilator is the closure of the orthogonal sum of its minimal closed ideals which are topologically simple H^*-algebras themselves.*

Finally, a few of measure theory and the following basic structure theorem allows us to finish the proof.

Theorem 6. *Every H^*-algebra is the orthogonal sum of its annihilator and an $H*$-algebra with zero annihilator.* □

Now is reasonable to weaken the condition of being stochastically derivative by considering that the linear random operator $\mathcal{D}$ has only some *probability of being derivative* in the following sense:

$$\inf\{\mathbb{P}[\mathcal{D}(ab) = \mathcal{D}(a)b + a\mathcal{D}(b) : a, b \in A\} = \delta(\mathcal{D}) > 0.$$

Operators satisfying the above condition will be called *probably derivative.*

For a such operator we apply a uniform randomization principle [5] to obtain a measurable subset Ω' with $\mathbb{P}[\Omega'] = \delta(\mathcal{D})$ such that the conditional operator $\mathcal{D}_{\Omega'}$ defined by $\mathcal{D}_{\Omega'}(a) = \mathcal{D}(a)_{|\Omega'}$ is stochasically derivative. Therefore we obtain the following result.

Theorem 7 [10]. *For a linear random operator $\mathcal{D}$ on an H^*-algebra A, with probability δ of being derivative, every element in the separating subspace for $\mathcal{D}$ is valued in the annihilator of A with a probability at least δ.*

References

1. M. Cabrera and A. Rodr!guez, Extended centroid and central closure of semiprime normed algebras: a first approach. Commun. Algebra 18 (1990), 2293-2326.
2. J. A. Cuenca and A. Rodr!guez, Structure theory for noncommutative Jordan H^*-algebras, J. Algebra 106 (1987), 1-14.
3. P. de la Harpe, *Classical Banach-Lie algebras and Banach Lie groups of operators in Hilbert space*, Lect. Notes in Math. 285, Springer-Verlag, Berlin 1972.
4. B.E. Johnson and A.M. Sinclair, Continuity of derivatios and a problem of Kaplansky, Amer. J. Math. 90 (1968), 1067-1073.
5. M. V. Velasco and A. R. Villena, Continuity of random derivations, Proc. Amer. Math. Soc. (to appear).

6. M. V. Velasco and A. R. Villena, A random closed graph theorem, submitted for publication.
7. M. V. Velasco and A. R. Villena, A random Banach-Steinhaus theorem, submitted for publication.
8. A.R. Villena, Continuity of derivation on a complete normed alternative algebra, J. Inst. Math. & Comp. Sci. 3 (1990), 99-106.
9. A.R. Villena, Continuity of derivations on H^*-algebras, Proc. Amer. Math. Soc. (to appear).
10. A.R. Villena, Stochastic continuity of random derivations on H^*-algebras, Proc. Amer. Math. Soc. (to appear).
11. M.A. Youngson, Hermitian operators on Banach Jordan algebras, Proc. Edimburgh Math. Soc. (2) 22 (1979), 93-104.
12. B. Zalar, Continuity of derivation on Malcev H^*-algebras, Math. Proc. Cambridge Philos. Soc. 110 (1991), 455-459.

ALGEBRAS OF RANK THREE

SEBASTIAN WALCHER
Mathematisches Institut
TU München,
80290 München
Germany.

Abstract. We give a few results on the class of algebras indicated in the title. Among these algebras are Bernstein-Jordan algebras and the pseudo-composition algebras recently investigated by Meyberg and Osborn. Some applications to ordinary differential equations are discussed.

Key words: quadratic algebra, pseudo-composition algebra, Peirce decomposition

1. Introduction

Let A be a commutative algebra of finite dimension over an infinite field K of characteristic not 2 or 3. Then A is called an *algebra of rank* 3 if

$$x^3 = \gamma_1(x)x^2 + \gamma_2(x)x$$

for all $x \in A$, with γ_1 a linear form and γ_2 a quadratic form on A.

These algebras were first discussed in Röhrl/Walcher [5], where the following was proved.

Proposition *The following are equivalent for an algebra A:*
(i) A is of rank 3.
(ii) The elements a, a^2 and a^3 are linearly dependent for all $a \in A$.
(iii) Every cyclic subalgebra of A has dimension 2 or less.

This last property was the initial motivation for the investigation of these algebras in the light of a more general question: How do conditions on subalgebras influence the structure of an algebra? In the following some results from joint work in progress with H. Röhrl will be presented. Proofs will be given elsewhere.

There is one exceptional class of rank 3 algebras satisfying $x^2 = \lambda(x)x$ for all x, with λ a linear form. (Equivalently, a and a^2 are linearly dependent for all $a \in A$.) This class will be excluded from consideration. For every other rank 3 algebra the forms γ_1 and γ_2 are uniquely determined.

2. Examples

To see that there are interesting algebras of rank 3 let us look at a few examples.

- Two-dimensional algebras.

S. González (ed.), Non-Associative Algebra and Its Applications, 400–404.

- More generally, quadratic algebras are of rank 3. Here an algebra is called *quadratic* if it satisfies an identity

$$x^2 = \rho_1(x)x + \rho_2(x)c$$

with ρ_1 a linear form, ρ_2 a quadratic form on A and $c \in A$. In particular, Jordan algebras associated to bilinear forms are in this class; it is not hard to see that these are precisely the unital algebras of rank 3.
- Train algebras of rank 3, satisfying an identity

$$x^3 = (1+\delta)\omega(x)x^2 - \delta\omega(x)^2 x$$

with ω a homomorphism from A onto K and $\delta \in K$. In particular, $\delta = 0$ yields Bernstein-Jordan algebras.
- Algebras satisfying the identity $x^3 = 0$. These are more troublesome than one might expect. It is easy to see that they are Jordan, and obviously they are nil and therefore nilpotent. But obtaining more detailed information seems to be very hard. A recent paper of Hentzel et al. [3] provides some nilpotency results which were found with the assistance of computer algebra.

3. Structure (in a sense)

In a certain sense, the structure of rank 3 algebras is quite well-understood. Let us first introduce the notion of modification of an algebra $(A, \cdot)$ by a linear form λ (briefly: λ-modification). This is defined as the commutative algebra $(A, \circ)$ with $x \circ x = x^2 + \lambda(x)x$. Obviously any λ-modification preserves subalgebras and therefore a λ-modification of a rank 3 algebra is again of rank 3. The following result from [5] shows why these modifications are useful.

Lemma *Every algebra of rank 3 is a λ-modification of an algebra satisfying $x^3 = \gamma_2(x)x$.*

The identity $x^3 = \gamma_2(x)x$ is known to imply the identity $\gamma_2(x^2) = \gamma_2(x)^2$ (cf. [5]). For this reason algebras satisfying this identity were christened *pseudo-composition algebras* (briefly: *PC algebras*) by Meyberg/Osborn [4]. Meyberg and Osborn provided a classification of simple PC algebras over an algebraically closed field: These are either quadratic or coordinatized by an "ordinary" composition algebra; in the latter case there is a close connection to simple Jordan algebras of degree 3.

Actually, there is a more general result behind this connection, as is shown by the following result proved in [2].

Theorem *Let A be a PC algebra. On $\hat{A} := K \cdot 1 + A$ define a commutative product by*

$$(\xi \cdot 1 + x) * (\xi \cdot 1 + x) := (\xi^2 + 2\gamma_2(x)) \cdot 1 + (2\xi x + x^2).$$

Then $(\hat{A}, *)$ is a unital power-associative algebra with generic minimum polynomial of degree 3.

This result ist stated differently (under more restrictive hypotheses) in [2], but an analysis of the proof reveals that one really gets the theorem stated above.

On the other hand, one can construct a PC algebra from any unital power-associative algebra with generic minimum polynomial of degree 3: On the subspace of trace zero elements define the "new" product as the trace zero part of the "old" product. This construction is quite familiar; it has been used decades ago by Springer [6] and Tits, for example. More recently, Faulkner [1] used a variant of this in order to show that there is an octonion algebra hidden in a 3×3 matrix algebra (over an appropriate base field).

It is not hard to see that simple PC algebras give rise to semisimple power-associative algebras and conversely. Thus one has a quite good understanding of the structure of PC algebras (and the result of Meyberg/Osborn [4]).

The problem is to carry this understanding over to arbitrary rank 3 algebras, since λ-modifications (while preserving subalgebras) may destroy or create ideals. Therefore a study of the general case is still necessary.

4. Peirce decomposition and ideals

The primary tool for the investigation of rank 3 algebras is the Peirce decomposition with respect to an idempotent.

Lemma *Let A be of rank 3 and c an idempotent of A. Then the eigenvalues of the left multiplication $L(c)$ are among $1, \frac{1}{2}$ and $\alpha := \gamma_1(c) - 1$. If $\alpha \notin \{1, \frac{1}{2}\}$, then $L(c)$ is semisimple.*

In the latter case we call c a *regular idempotent*. Of course, there are rank 3 algebras which do not have idempotents even when the base field is extended, and these have to be taken care of separately. Fortunately, this can be done in a quite satisfactory manner.

Proposition *Let K be algebraically closed and A of rank 3 over K. If A contains no idempotent then either A satisfies $x^3 = \gamma_1(x)x^2$ with $\gamma_1(A^2) = 0$, or A is a rank 3 train algebra for $\delta = \frac{1}{2}$.*

Proposition *As a consequence we find that no such algebra (of dimension > 1) is simple. More precisely, algebras of the first type are their own radical (in the sense of Albert) while the Albert radical for an algebra of the second type is equal to the kernel of ω.*

In Meyberg/Osborn [4] it is illustrated how useful the Peirce decomposition is for the structure theory of PC algebras, where the multiplication rules for the Peirce spaces are rather nice. In general, things become more complicated. But the following result on ideals of rank 3 algebras shows (among other things) that there is still valuable information to be gained from the Peirce decomposition.

Proposition *Let A be a rank 3 algebra which contains no ideal of codimension 1.*

(a) Then $\gamma_1(I) = \gamma_2(A, I) = 0$ for every proper ideal I of A, and $x^3 = 0$ for all $x \in I$.

(b) A contains a unique maximal ideal J which is the Albert radical of A.

(c) If $J \neq 0$ and if A contains a regular idempotent then there is a $\mathbf{Z}_2$-grading $J = J_0 + J_1$ of J, and furthermore $J_0^2 = 0$.

The grading is of the same type as observed in the radical of Bernstein-Jordan algebras, with the familiar consequences.

Finally we note that a semisimple algebra of rank 3 is either simple or isomorphic to $K \oplus K$.

5. The Riccati equation $\dot{x} = x^2$ in rank 3 algebras

Here we assume that K is the field of real or complex numbers. Discussing quadratic differential equations of this kind (characterized by the geometric property that every solution curve remains in a subspace of dimension ≤ 2) was another motivation for the discussion of rank 3 algebras. It turns out that here we have a class of differential equations which is not (at least not immediately) accessible to the methods created by Lie but still amenable to algebraic methods.

Consider the special case of PC algebras first. Since these have very nice properties, one almost expects nice features of the associated differential equation. This is indeed the case, and is a direct consequence of the identities $x^3 = \gamma_2(x)x$ and $\gamma_2(x^2) = \gamma_2(x)^2$.

Proposition *Let A be a PC algebra. Then the map*

$$A \to K^2, x \mapsto \begin{pmatrix} \gamma_2(x) \\ \gamma_2(x, x^2) \end{pmatrix}$$

maps solutions of $\dot{x} = x^2$ in A to solutions of $\dot{x}_1 = 2x_2, \dot{x}_2 = 3x_1^2$ in K^2.

The second equation yields $\dot{x}_1^2 = 4x_1^3 - c$, with c a constant depending on the initial values. For $c \neq 0$ this is the differential equation of a Weierstrass $\wp$-function for a very particular lattice (with $\frac{\pi}{3}$ rotational symmetry).

In the general case one cannot expect such precise information, but still the "local" condition on solutions of initial value problems yields a lot of global information. The following result is a special case of a theorem contained in [7].

Proposition *Let λ and μ be linear forms and $\phi(x) := \lambda(x)\mu(x^2) - \lambda(x^2)\mu(x)$ for all $x \in A$. Then the identity $D\phi(x)x^2 = 2\gamma_1(x)\phi(x)$ holds for all x, and therefore the set of zeros of ϕ is an invariant set for $\dot{x} = x^2$.*

Thus we obtain a rather detailed "geometric" picture about the behavior of the solutions, and furthermore a systematic method of constructing first integrals of the given differential equation.

If A is a PC algebra then ϕ is itself a first integral (as follows from $\gamma_1 = 0$), and this implies that $\dot{x} = x^2$ can be solved by employing the $\wp$-function from above and elementary functions.

References

1. J. Faulkner: Finding octonion algebras in associative algebras. *Proc. Amer. Math. Soc.* **104** (4), 1027-1030 (1988).
2. H. Gradl, K. Meyberg, S. Walcher: R-algebras with an associative trace form. *Nova J. of Algebra and Geometry* (to appear).
3. I.R. Hentzel, D.P. Jacobs, L.A. Peresi, S.R. Sverchkov: Solvability of the ideal of all weight zero elements in Bernstein algebras. *Comm. Algebra* (to appear).
4. K. Meyberg, J.M. Osborn: Pseudo-composition algebras. *Math. Z.* (to appear).
5. H. Röhrl, S. Walcher: Algebras of complexity one. *Algebras Groups Geom.* **5**, 61-107 (1988).
6. T.A. Springer: On a class of Jordan algebras. *Indag. Math.* **21**, 254-264 (1959).
7. S. Walcher: Algebras and differential equations. Hadronic Press, Palm Harbor (1991).

INNER PRODUCT CHARACTERIZATIONS OF CLASSICAL CAYLEY-DICKSON ALGEBRAS

BORUT ZALAR*
Institute of Mathematics, Jadranska 19
61000 Ljubljana, Slovenija

Abstract. Classical Cayley-Dickson algebras are inner product algebras with certain additional properties. We investigate which combinations of these properties characterize classical Cayley-Dickson algebras among inner product algebras.

1. Introduction

Let $\mathbb{R}$, $\mathbb{C}$, $\mathbb{H}$ and $\mathbb{D}$ denote the algebras of real numbers, complex numbers, quaternions and octonions respectively. These four algebras will be called *classical Cayley-Dickson algebras.* A nonassociative real algebra $\mathcal{A}$ equipped with an inner product will be called an *inner product algebra.* We use notation $\langle x, y\rangle$ for the inner product and $|x| = \sqrt{\langle x, x\rangle}$ for the norm derived from it. Every classical Cayley-Dickson algebra is an inner product algebra while the converse is far from being true. Therefore we are interested in the following problem:

> Which additional properties characterize classical Cayley-Dickson algebras among inner product algebras?

One such characterization is well-known and appears in [3].

Characterization 0. *Every absolute-valued algebra with identity element is isomorphic to one of the classical Cayley-Dickson algebras.*

In the present paper we prove three new results concerning this problem.

Characterization 1. *Every absolute-valued inner product algebra with power-associative symmetrization is isomorphic to one of the classical Cayley-Dickson algebras.*

Characterization 2. *Every alternative inner product algebra satisfying the norm identity $|x^2| = |x|^2$ is isomorphic to one of the classical Cayley-Dickson algebras.*

Characterization 3. *Every inner product algebra with identity element, in which the subset of the real line $\{\frac{|x|^2|y|^2-|xy|^2}{\langle x,y\rangle^2}; \langle x, y\rangle \neq 0\}$ is bounded, is isomorphic to one of the classical Cayley-Dickson algebras.*

Note that $\mathcal{A}$ is an absolute-valued algebra if $|xy| = |x|\,|y|$ holds for all $x, y \in \mathcal{A}$ while the symmetrization $\mathcal{A}^+$ is the same vector space as $\mathcal{A}$ with a new product $x \cdot y = \frac{1}{2}(xy + yx)$ for $x, y \in \mathcal{A}$.

---

* Supported in part by the Slovenian Ministry of Science.

S. González (ed.), Non-Associative Algebra and Its Applications, 405–409.

2. Proofs

Lemma. *Let $\mathcal{A}$ be a commutative associative inner product algebra satisfying a norm identity $|x^2| = |x|^2$. Then $\mathcal{A}$ is isometrically isomorphic to $\mathbb{R}$ or $\mathbb{C}$.*
Proof. First we prove the inequality

$$|\langle x, y\rangle| \leq |xy| \leq |x||y| \tag{1}$$

for $x, y \in \mathcal{A}$. From

$$|x+y|^2 = |x^2 + y^2 + 2xy| \leq |x^2| + |y^2| + 2|xy| =$$

$$= |x|^2 + |y|^2 + 2|xy|$$

we obtain $\langle x, y\rangle \leq |xy|$. If we replace x with $-x$, we get $|\langle x, y\rangle| \leq |xy|$. Notice that (1) is valid without the assumption of associativity.

If $x = 0$ or $y = 0$, then $|xy| \leq |x||y|$ is obvious. Otherwise we have for $a = \frac{x}{|x|}$ and $b = \frac{y}{|y|}$

$$4|ab| = |(a+b)^2 - (a-b)^2| \leq$$

$$\leq |a+b|^2 + |a-b|^2 = 2|a|^2 + 2|b|^2 = 4$$

and so

$$\frac{1}{|x||y|}|xy| \leq 1.$$

Therefore $\mathcal{A}$ is a normed algebra and by [1] we only have to prove that $\mathcal{A}$ does not contain any topological zero divisors. Suppose that $|a| = 1$, $|x_n| = 1$ and $ax_n \to 0$. We shall see that this leads to a contradiction. Using (1) we have first $|\langle a, x_n\rangle| \leq |ax_n| \to 0$. Since $\mathcal{A}$ is associative

$$|\langle a^2, x_n^2\rangle| \leq |a^2 x_n^2| = |(ax_n)^2|^2 \to 0.$$

Thus

$$|a + x_n|^4 = (1 + 2\langle a, x_n\rangle + 1)^2 \to 4.$$

On the other hand

$$|a + x_n|^4 = |a^2 + 2ax_n + x_n^2|^2 = 1 + 4|ax_n|^2 + 1+$$

$$+2\langle a^2, x_n^2\rangle + 4\langle a^2, ax_n\rangle + 4\langle x_n^2, ax_n\rangle \to 2$$

establishes a desired contradiction. The isometry is obvious.

Proof of Characterization 1. Since $x \cdot x = \frac{1}{2}(x^2 + x^2) = x^2$, we have $|x \cdot x| = |x^2| = |x|^2$ and so $\mathcal{A}^+$ is a commutative power-associative algebra satisfying the norm identity $|x^2| = |x|^2$. If we take a nonzero $b \in \mathcal{A}^+$, then the subalgebra generated by b is associative, commutative and satisfies the requirements of the Lemma. This means that $\mathcal{A}^+$ contains a nonzero idempotent which we denote by e. Next we prove that e is the identity element of $\mathcal{A}^+$.

Consider the Pierce decomposition $\mathcal{A}^+ = \mathcal{A}_1^+ \oplus \mathcal{A}_{\frac{1}{2}}^+ \oplus \mathcal{A}_0^+$ where $\mathcal{A}_i^+ = \{x; e \cdot x = ix\}$. We prove our result by establishing $\mathcal{A}_0^+ = \mathcal{A}_{\frac{1}{2}}^+ = (0)$.

Let $x \in \mathcal{A}_0^+$ or equivalently $e \cdot x = 0$. By (1) we have $\langle e, x\rangle = 0$. Since $e \cdot x^2 = 0$ also holds (it is well-known that $\mathcal{A}_0^+$ is a subalgebra of $\mathcal{A}^+$ as can be seen in [2]), we have $\langle e, x^2\rangle = 0$. Since $|x + \lambda e|^4 = |(x + \lambda e)^2|^2$ holds for all real λ, we have

$$|x|^4 + \lambda^4 + 2|x|^2\lambda^2 = |x^2|^2 + \lambda^4$$

which implies $x = 0$ and so $\mathcal{A}_0^+ = (0)$. Now take $x \in \mathcal{A}_{\frac{1}{2}}^+$. If we use the same equality $|x + \lambda e|^4 = |(x + \lambda e)^2|^2$, we obtain

$$(|x|^2 + 2\lambda\langle e, x\rangle + \lambda^2)^2 = |x^2 + \lambda x + \lambda^2 e|^2$$

which implies $\langle e, x\rangle = 0$ and $\langle x^2, x\rangle = 0$. The subalgebra generated by x is associative and commutative. If $x \neq 0$, then, by the Lemma, this subalgebra has a nonzero idempotent f and $x^2 - 2\langle f, x\rangle x + |x|^2 f = 0$ holds since $\langle f, x\rangle$ is a "real part" of x because of isometry. Therefore

$$-2\langle x, f\rangle\langle x, x\rangle + |x|^2\langle f, x\rangle = 0$$

and so $\langle f, x\rangle = 0$. This further implies $x^2 = -|x|^2 f$ and so $f \in \mathcal{A}_{\frac{1}{2}}^+ \cdot \mathcal{A}_{\frac{1}{2}}^+ \subset \mathcal{A}_1^+ + \mathcal{A}_0^+ = \mathcal{A}_1^+$. Thus $e \cdot f = f$. If we expand both sides of the equality $|e + \lambda f|^4 = |(e + \lambda f)^2|^2$, we obtain $\langle e, f\rangle = 1$ and so $e = f$ follows. This is a contradiction since $e \cdot x = \frac{1}{2}x$ and $f \cdot x = x$.

We proved that e is the identity element of $\mathcal{A}^+$ and therefore $ex + xe = 2x$ holds for all $x \in \mathcal{A}$. Now $|xe + ex|^2 = 4|x|^2$ implies $\langle xe, ex\rangle = |x|^2$ and since $|ex| = |xe| = |x|$, we have $ex = xe$. This finally implies that e is also the identity element of $\mathcal{A}$ and Characterization 0 completes the proof.

Proof of Characterization 2. Given a nonzero $x \in \mathcal{A}$, the subalgebra generated by x is associative and commutative. According to the Lemma, $\mathcal{A}$ contains a nonzero idempotent e. We prove that e is the identity element of A. Obviously $e(ex - x) = (xe - x)e = 0$, so it suffices to prove that $\mathcal{A}$ does not contain any zero divisors.

Suppose that $xy = 0$, $|x| = |y| = 1$. Then $|yx|^2 = |(yx)^2| = |yxyx| = 0$ and so $yx = 0$. Next $|x + y|^2 = |x^2 + y^2| = |x - y|^2$ implies $|x + y|^2 = 2$. Therefore

$$4 = |x^2 + y^2|^2 = 2 + 2\langle x^2, y^2\rangle$$

and $x^2 = y^2$ follows. But then $1 = |x^4| = |x^2y^2| = 0$ is a contradiction.

Now we know that $\mathcal{A}$ has identity and we will simply denote it by 1. If x and 1 are orthogonal, then by the Lemma, the subalgebra generated by $\{1, x\}$ is isomorphic to $\mathbb{C}$ and so $x^2 = -|x|^2$ holds. Our next goal is to prove that $\mathcal{A}$ is in fact an absolute-valued algebra so that $|xy| = |x||y|$ holds for all $x, y \in \mathcal{A}$.

If $x = 0$, there is nothing to prove so from now on we take $x \neq 0$. Suppose first that $1, x, y$ are pairwise orthogonal. Since $x + y \in \{1\}^\perp$, we have

$$(x + y)^2 = -|x + y|^2 = -|x|^2 - |y|^2 = x^2 + y^2.$$

This tells us that $xy = -yx$ and so the alternativity of $\mathcal{A}$ implies

$$|xy|^2 = |(xy)^2| = |xy \cdot xy| =$$
$$= |xy \cdot yx| = |xy^2x| = |x|^2|y|^2.$$

In our next step we take x, y which are both orthogonal to 1. Then we have

$$(x+y)^2 = -|x+y|^2 = x^2 + y^2 - 2\langle x, y\rangle$$

and so $xy + yx = -2\langle x, y\rangle$. We may decompose

$$xy = \langle 1, xy\rangle + z, \qquad yx = \langle 1, yx\rangle - z$$

with $\langle 1, z\rangle = 0$ and therefore $\langle 1, xy\rangle + \langle 1, yx\rangle = -2\langle x, y\rangle$. Since $\mathcal{A}$ is alternative, $(xy)x = x(yx)$ holds. If we expand this equality, we obtain

$$\langle 1, xy\rangle = \langle 1, yx\rangle = -\langle x, y\rangle, \tag{2}$$

$$\langle x, z\rangle = \langle x, xy\rangle = \langle x, yx\rangle = 0. \tag{3}$$

Note that $xz + zx = -2\langle x, z\rangle$ and $x, 1$ are linearly independent. Now we define $y_1 = -\frac{\langle x,y\rangle}{|x|^2}x + y$. Then x and y_1 are orthogonal and by the above paragraph we have $|xy_1|^2 = |x|^2|y_1|^2$. ¿From this and (2) we obtain $|xy| = |x|\,|y|$.

Finally let x, y be arbitrary. We decompose $x = \alpha + a$ and $y = \beta + b$, with a, b orthogonal to 1. According to (2) and (3), we have

$$|xy|^2 = |\alpha\beta + \beta a + \alpha b + ab|^2 =$$
$$= \alpha^2\beta^2 + \beta^2|a|^2 + \alpha^2|b|^2 + |a|^2|b|^2 +$$
$$+2\alpha\beta(\langle 1, ab\rangle + \langle a, b\rangle) = (\alpha^2 + |a|^2)(\beta^2 + |b|^2) = |x|^2|y|^2.$$

Proof of Characterization 3. Let M be a positive bound of the mentioned subset, i.e. $M \geq 0$ and

$$\{\frac{|x|^2|y|^2 - |xy|^2}{\langle x, y\rangle^2}; \langle x, y\rangle \neq 0\} \subset [-M, M].$$

If $\mathcal{A}$ is not isomorphic to $\mathbb{R}$, then it is at least 2-dimensional. Therefore there exists $x \in \{1\}^\perp$ with the property $|x| = |1| = \alpha$. Take nonzero $\lambda \in \mathbb{R}$. Since

$$|\frac{|1+\lambda x|^2|x|^2 - |(1+\lambda x)x|^2}{\langle 1+\lambda x, x\rangle^2}| \leq M,$$

we obtain

$$|\frac{\lambda^2(\alpha^4 - |x^2|^2) - 2\lambda\langle x, x^2\rangle + \alpha^4 - \alpha^2}{\alpha^4\lambda^2}| \leq M$$

and thus $\lambda = 0$ is not a pole of the rational function on the left side. This implies that $\alpha^4 - \alpha^2 = 0$ and since $\alpha \notin \{0, -1\}$ it follows that $|1| = 1$.

Next we assume that $\{1, x, y\}$ is an orthonormal subset of $\mathcal{A}$ and prove that $\{1, x, y, xy\}$ is also an orthonormal subset of $\mathcal{A}$. Take $\lambda \in \mathbb{R}$ different from ± 1. Since $\langle 1, x\rangle = \langle 1, y\rangle = \langle x, y\rangle = 0$, we have

$$\left|\frac{|1+\lambda x|^2|1-\lambda y|^2 - |(1+\lambda x)(1-\lambda y)|^2}{\langle 1+\lambda x, 1-\lambda y\rangle}\right| \leq M,$$

$$\left|\frac{\lambda^4(1-|xy|^2) + 2\lambda^3(\langle y, xy\rangle - \langle x, xy\rangle) - 2\lambda^2\langle 1, xy\rangle}{(1-\lambda^2)^2}\right| \leq M$$

and thus $|xy| = 1$, $\langle 1, xy\rangle = 0$ and $\langle y, xy\rangle = \langle x, xy\rangle$. We can replace x by $-x$ and then the last equality gives us $xy \in \{x, y\}^\perp$.

Our next goal is to prove that $\mathcal{A}$ is an absolute-valued algebra. Take first some $x, y \in \{1\}^\perp$. Suppose that $x \neq 0$ for a moment. Then we can form an element $z = -\frac{\langle x,y\rangle}{|x|^2}x + y$ which is orthogonal to x. From the above paragraph it follows $\langle zx, 1\rangle = 0$, $\langle zx, x\rangle = 0$ and $|xz| = |x||z|$. This in turn implies $\langle 1, xy\rangle = -\langle x, y\rangle$, $\langle xy, x\rangle = 0$ and $|xy| = |x||y|$. The last three equalities also hold for $x = 0$ and so they are valid without any restrictions for $x, y \in \{1\}^\perp$.

Now take $a, b \in \mathcal{A}$ and decompose $a = \alpha + x$, $b = \beta + y$ where $\alpha, \beta \in \mathbb{R}$ and $x, y \in \{1\}^\perp$. Then

$$|ab|^2 = |(\alpha + x)(\beta + y)|^2 = |\alpha\beta + \alpha y + \beta x + xy|^2 =$$

$$= \alpha^2\beta^2 + \alpha^2|y|^2 + \beta^2|x|^2 + |xy|^2 + 2\alpha\beta\langle 1, xy\rangle +$$

$$+ 2\alpha\beta\langle x, y\rangle = \alpha^2\beta^2 + \alpha^2|y|^2 + \beta^2|x|^2 + |x|^2|y|^2 =$$

$$= (\alpha^2 + |x|^2)(\beta^2 + |y|^2) = |\alpha + x|^2|\beta + y|^2 = |a|^2|b|^2$$

and Characterization 0 now completes the proof.

Acknowledgments. I would like to express my gratitude to Prof. Rodriguez for his helpful comments and suggestions. I would also like to thanks Prof. Vukman from our Institute for the financial support from his personal grant within the Slovenian Ministry of Science. Without his help my journey to Oviedo would not be possible.

References

1. I. Kaplansky, Normed algebras, Duke. Math. J. **16** (1949), 399–418.
2. R.D. Schafer, *An Introduction to Nonassociative Algebras*, Academic Press, 1966.
3. K. Urbanik and F.B. Wright, Absolute valued algebras, Proc. Amer. Math. Soc. **11** (1960), 861 –866.

SUR LES $\mathbb{R}$-ALGÈBRES DE JORDAN NON COMMUTATIVES, DE DIVISION, DE DIMENSION 8, POSSÉDANT UN AUTOMORPHISME OU UNE DÉRIVATION NON TRIVIAUX

ABDELLATIF ROCHDI*
Departamento de Álgebra, Geometría y Topología
Universidad de Málaga, Facultad de Ciencias. Campus de Teatinos.
29071 Málaga (Spain).

Abstract. In this paper we study the noncommutative Jordan real division algebras of dimension 8, whose Lie derivation algebra is nontrivial, and characterize those whose automorphism group is nontrivial. We give also an affirmative answer of a question in [1].

1. Introduction

Une algèbre A sur un corps commutatif K est dite de division, si pour tout $x \in A; x \neq 0$, L_x et R_x (2.1) sont inversibles dans l'algèbre $End_K A$. Il est bien connu qu'une algèbre réelle associative de division, de dimension finie, est isomorphe a $\mathbb{R}$, $\mathbb{C}$ ou $\mathbb{H}$ (l'algèbre réelle des quaternions de Hamilton) [8]. Bruck et Kleinfeld ont montré qu'une algèbre réelle alternative de division, de dimension finie, est ou bien associative, ou bien isomorphe a $\mathbb{O}$ (l'algèbre réelle des octonions de Cayley-Dickson) [6]. Un des résultats fondamentaux pour les algèbres réelles non nécéssairement associatives, de division, de dimension finie, est celui qui a été démontré par Kervaire [11] et Milnor-Bott [4], qui affirme que si l'espace vectoriel réel $\mathbb{R}^n$ possede un produit bilinéaire sans diviseurs de zéro, alors n = 1, 2, 4 ou 8. Cependant, la détérmination des algèbres réelles non associatives, de division, de dimension finie, est encore un problème ouvert même en dimension 4. Osborn [12] a établi en 1962, la base de la théorie des algèbres quadratiques et a détérminé toutes les algèbres réelles quadratiques de division, de dimension 4, et une classe particulière d'algèbres réelles quadratiques (non alternatives) de division, de dimension 8. Ceci est a l'origine de nos travaux ultérieurs. Kaidi montre dans [9] qu'une algèbre réelle de Jordan non commutative, de division, de dimension finie et vérifiant l'identité : $(x, x, [x, y]) = 0$, est quadratique et alternative. En particulier, une algèbre réelle de Jordan, de division, de dimension finie, est isomorphe a $\mathbb{R}$ ou $\mathbb{C}$. Benkart, Britten et Osborn [3] réduisent en 1982 la détérmination des algèbres réelles flexibles, de division, de dimension finie, a celles de Jordan n. c. Une algèbre réelle de Jordan n. c. de division, de dimension finie, telle que deux éléments qui n'appartiennent pas a la même sous-algèbre de dimension 2, engendrent une sous-algèbre de dimension 4, est isomorphe a : $\mathbb{R}$, $\mathbb{C}$, $\mathbb{H}^{(\lambda)}$ ou $\mathbb{O}^{(\lambda)}$, $\lambda \neq \frac{1}{2}$ [13]. Nous avons donné dans [10] un procédé qui

* Je remercie le Professeur A. Kaidi pour ses remarques et suggestions, et également le Professeur J. A. Cuenca pour son soutient et ses encouragements.

S. González (ed.), Non-Associative Algebra and Its Applications, 410–416.

généralise celui de Cayley-Dickson, qui a permi l'obtention d'une nouvelle famille d'algèbres réelles de Jordan n. c. de division, de dimension 8, qui contient strictement les algèbres qui peuvent s'obtenir a partir de $\mathbb{R}$, par mutations et par extension cayleyenne, mais ce dernier s'avère insuffisant pour la détérmination de toutes les algèbres réelles de Jordan n. c. de division, de dimension 8 [14]. Récemment, nous avons donné un nouveau procédé qui permet leur détérmination [7]. Dans ce papier, nous traitons le cas où il existe une dérivation ou un automorphisme non triviaux.

2. Notations et prérequis

2.1 Définitions. Soient K un corps commutatif de caractéristique nulle, $\alpha \in K$ et A une K-algèbre. On appelle mutation α de A et on note $A^{(\alpha)}$, l'algèbre ayant pour espace vectoriel sous-jacent A et pour produit : $x^\alpha y = \alpha xy + (1-\alpha)yx$; $x, y \in A$. On note $Aut(A)$ le groupe des automorphismes de A (celui de l'algèbre réelle $\mathbb{O}$ de Cayley-Dickson est noté G_2), et $Der(A)$ l'algèbre de Lie des dérivations de A. Il est bien connu que si A est une $\mathbb{R}$-algèbre de dimension finie, alors l'algèbre de Lie du groupe de Lie $Aut(A)$ et l'algèbre $Der(A)$ coincident. Pour toute K-algèbre et pour tout $\lambda \in K$, on a: $Der(A) \subset Der(A^{(\lambda)})$. Une K-algèbre A est dite de division, si pour tout $x \in A, x \neq 0$, les opérateurs de multiplications $L_x : y \mapsto xy$ et $R_x : y \mapsto yx$ sont inversibles dans l'algèbre $End_K A$. Si x, y, z sont des éléments de A, on désigne par $[x,y]$ et (x,y,z) respectivement, les éléments: $xy - yx$ et $(xy)z - x(yz)$. A est dite flexible si elle vérifie l'identité: $(x,y,x) = 0$ pour tout $x, y \in A$. Elle est dite de Jordan non commutative, si de plus elle vérifie l'identité: $(x^2, y, x) = 0$ pour tout $x, y \in A$. Si A possede un élément unité 1, elle est dite quadratique si $1, x, x^2$ sont linéairement dépendants pour tout $x \in A$. Il est clair qu'une K-algèbre quadratique et flexible est de Jordan n. c. Si A est quadratique, elle est dite cayleyenne si elle est munie d'une involution $\bar{\ }$ multiplicative telle que: $x + \overline{x}, x\overline{x} \in K1$ pour tout $x \in A$. Il est bien connu qu'une K-algèbre quadratique A, s'obtient a partir d une K-algèbre anti-commutative $(V, \wedge)$, et d'une forme bilinéaire $(,)$ sur V, par: $A = K \oplus V$ muni de sa structure naturelle de K espace vectoriel, et du produit : $(\alpha + x)(\beta + y) = (\alpha\beta + (x,y)) + (\alpha y + \beta x + x \wedge y)$. On note de la même facon que pour V, la forme bilinéaire: $A \times A \to K \quad (\alpha + x, \beta + y) \mapsto \alpha\beta + (x,y)$ qu'on appelle la forme bilinéaire associée a A. $(V, \wedge)$ est appelée l'algèbre anti-commutative associée a A, ses éléments sont appelés vecteurs, ceux de K scalaires. On note alors A par $(V, (,), \wedge)$. Dans [9] on montre qu une $\mathbb{R}$-algèbre de Jordan n. c. de division, de dimension finie est quadratique.

2.2 Procédé de Cayley-Dickson Généralisé. Soient $(B, \bar{\ })$ une K -algèbre cayleyenne et $\gamma, \alpha, \delta, \in K$. On appelle extension cayleyenne généralisée de $(B, \bar{\ })$ d'indice (γ, α, δ) et on note $E_{\gamma,\alpha,\delta}(B)$, l'algèbre cayleyenne [10] , ayant pour espace vectoriel sous-jacent $B \times B$ et pour produit : $(x,y)(x,y) = (x^\alpha x + \gamma\overline{y}.y, y\overline{x} + yx + \frac{\delta}{2}[y,y])$.

2.3 Remarques. 1) Si ∂ est une dérivation de B qui commute avec l'involution $\bar{\ }$, alors l application : $D_\partial : A = E_{\gamma,\alpha,\delta}(B) \to A \quad (x,y) \mapsto (\partial x, \partial y)$ est une dérivation de A. Si de plus, les dérivations de B commutent avec $\bar{\ }$, alors l'application: $Der(B) \to Der(A) \quad \partial \to D_\partial$ est un monomorphisme d'algèbres de Lie.

2) Si $(B, \div)$ est une algèbre des quaternions, de division sur $K, \alpha \neq \frac{1}{2}$ et $\gamma\delta \neq 0$, alors l'application : $Der(B) \rightarrow Der(A) \quad \partial \rightarrow D_\partial$ est un isomorphisme.

2.4 Proposition. *L'algèbre réelle* $(E_{-1,\alpha,\delta}(H^{(\lambda)}))^{(\mu)}$ *est de Jordan n. c. Elle est de division, si et seulement si :* $\lambda, \mu \neq \frac{1}{2}$, $\alpha > \frac{1}{2}$ *et* $(2\alpha - 1)\delta^2 < 4$.

On pourra consulter [14] pour les demonstrations concernant (2.3) et (2.4).

On note par C la classe de toutes les $\mathbb{R}$-algèbres de Jordan n. c. de division, de dimension 8, G la sous-classe de C des algèbres qui s'obtiennent a partir de $\mathbb{R}$ par mutations et par extension cayleyenne généralisée (2.4), D la sous-classe de C des algèbres qui possedent une dérivation non triviale, et enfin A la sous-classe de C des algèbres qui possedent un automorphisme non trivial. Nous avons les inclusions suivantes : $G \subset D \subset A \subset C$. Considerons l'algèbre réelle A qui possede une base $1, u; y_1, z_1; y_2, z_2; y_3, z_3$ pour laquelle la multiplication de A est donnée par la table I suivante:

TABLE I

	1	u	y_1	z_1	y_2	z_2	y_3	z_3
1	1	u	y_1	z_1	y_2	z_2	y_3	z_3
u	u	-1	az_1	$-ay_1$	bz_2	$-by_2$	cz_3	$-cy_3$
y_1	y_1	$-az_1$	-1	au	αy_3	μz_3	$-\alpha y_2$	$-\mu z_2$
z_1	z_1	ay_1	$-au$	-1	μz_3	μy_3	$-\mu z_2$	$-\mu y_2$
y_2	y_2	$-bz_2$	$-\alpha y_3$	$-\mu z_3$	-1	bu	αy_1	μz_1
z_2	z_2	$-by_2$	$-\mu z_3$	$-\mu y_3$	$-bu$	-1	μz_1	μy_1
y_3	y_3	$-cz_3$	αy_2	μz_2	$-\alpha y_1$	$-\mu z_1$	-1	cu
z_3	z_3	cy_3	μz_2	μy_2	$-\mu z_1$	$-\mu y_1$	$-cu$	-1

où a, b, c sont des paramètres > 0 et $\alpha, \mu \in \mathbb{R}$. Alors A est de Jordan n. c., de plus A est de division, si et seulement si : $\alpha\mu < 0$.

Si $\alpha = -\mu = 1$ et $c > a = b > 1$, alors $A \in D$ et $A \notin G$. Ainsi, ce premier procédé est insuffisant pour la détérmination de toutes les $\mathbb{R}$-algèbres de Jordan n. c. de division, de dimension 8. Nous verrons par la suite, moyennant d'autres exemples, que les deux dernières inclusions sont également strictes. Ce qui montre l'immensité de la classe C.

2.5 Procédé d'isotopie vectorielle. Soit $(W, \times)$ une $\mathbb{R}$-algèbre anti-commutative, de dimension finie, munie d'une forme bilinéaire symétrique $(/)$ définie négative et soit φ un automorphisme de l'espace vectoriel réel W. On pose: $x \wedge y = {}^t\varphi$ $(\varphi(x) \times \varphi(y)); x, y \in W$ (${}^t\varphi$ étant l'automorphisme transposé de φ). L'algèbre réelle $A = (W, (/), \times)$ est flexible, de division, si et seulement si $(W, (/), \wedge)$ est flexible, de division. On dira que $(W, (/), \wedge)$ est obtenue a partir de A et φ par isotopie vectorielle, et on la note $A(\varphi)$. Nous avons obtenu le résultat suivant :

2.6 Théorème. *Les $\mathbb{R}$-algèbres de Jordan n. c. de division, de dimension 8, s'obtiennent a partir de l'algèbre réelle $\mathbb{O}$ de Cayley-Dickson par isotopie vectorielle*

(i. e. $A = (W, (/), \times) \in C$ si et seulement si, il existe un automorphisme φ, de l'espace vectoriel W tel que $A = \mathbb{O}(\varphi)$) et sont à isomorphisme près $\mathbb{O}(s)$ ou s est un automorphisme symétrique de l'espace euclidien $(W, -(/))$, défini positif. De plus : $\mathbb{O}(s') \cong \mathbb{O}(s)$ (s' et s étant deux automorphismes symétriques de $(W, -(/))$, définis positifs) si et seulement si, il existe $\varphi \in G_2$ tel que : $\tilde{s}' = \varphi \tilde{s}^t \varphi$ ($\tilde{s}$ étant le prolongement naturel de s à $\mathbb{O}$).

On pourra consulter [7] pour les demonstrations concernant (2.5) et (2.6).

3. Dérivations

3.1 Théorème. [14] *Soit A une algèbre réelle de Jordan n. c. de division, de dimension 8, dont l'algèbre de Lie des dérivations n'est pas triviale. Alors, il existe une base $1, u; y_1, z_1; y_2, z_2; y_3, z_3$ de A, trois parametres $a, b, c > 0$ et quatre autres $\alpha, \lambda, \eta, \rho$ pour lesquels la multiplication de A est donnée par la table II suivante :*

TABLE II

	1	u	y_1	z_1	y_2	z_2	y_3	z_3
1	1	u	y_1	z_1	y_2	z_2	y_3	z_3
u		-1	az_1	$-ay_1$	bz_2	$-by_2$	cz_3	$-cy_3$
y_1			-1	au	αy_3	$-\alpha z_3$	$-\alpha y_2$	αz_2
z_1				-1	$\lambda y_3 + \eta z_3$	$\eta y_3 - \lambda z_3$	$-\lambda y_2 - \eta z_2 + \rho z_3$	$-\eta y_2 + \lambda z_2 - \rho y_3$
y_2					-1	bu	$\alpha y_1 + \lambda z_1$	ηz_1
z_2						-1	ηz_1	$-\alpha y_1 - \lambda z_1$
y_3							-1	$cu + \rho z_1$
z_3								-1

De plus, une algèbre réelle dont la multiplication est donnée par la table II est de Jordan n. c. et possede une dérivation non triviale. Elle est de division, si et seulement si : $\alpha\eta < 0$ et $b\rho^2 < 4c\eta^2$.

Pour $D \neq A$ on considere l'algèbre réelle A dont la multiplication est donnée par la table 1 avec $\mu = -1$, $\alpha > 0$, $b < 1 < c$ et $\alpha \neq 1, b, c$. Alors $A \in A$ et $A \notin D$.

Benkart et Osborn détérminent dans [2] toutes les possibilités pour l'algèbre de Lie des dérivations, d'une algèbre réelle de division, établissant le théoreme de classification suivant :

Théorème. *Soit A une R-algèbre de division, de dimension finie. Alors :*
1) Si $dimA = 1$ ou 2, alors $Der(A) = 0$.
2) Si $dimA = 4$, alors $Der(A)$ est isomorphe a $su(2)$ où $dim(Der(A)) \leq 1$.
3) Si $dimA = 8$, alors $Der(A)$ est isomorphe a l'une des algèbres de Lie suivantes:
i) G_2 compacte
ii) $su(3)$
iii) $su(2) \oplus su(2)$

iv) $su(2) \oplus N$, où N est une algèbre abélienne de dimension ≤ 1
v) N, une algèbre abélienne de dimension ≤ 2
De plus, toutes ces dernieres possibilités se réalisent.

Benkart et Osborn donnent ensuite dans [1] une classification complete pour les algèbres réelles de division, de dimension 4, dont l'algèbre de Lie des dérivations est $su(2)$, et pour les algèbres réelles de division, de dimension 8, dont l'algèbre de Lie des dérivations est G_2 compacte, $su(3)$ et $su(2) \oplus su(2)$. Ils donnent également des exemples d'algèbres réelles de division pour chacun des autres cas de l'algèbre de Lie des dérivations.

3.2 Remarque. Si A est une algèbre réelle de Jordan n. c. de division, de dimension 8. Alors $Der(A)$ ne peut être $su(3)$.

3.3 Théorème. *Soit A une $\mathbb{R}$-algèbre de Jordan n. c. de division, de dimension 8. Alors $Der(A) = G_2$ compacte, si et seulement si $A \cong \mathbb{O}^{(\lambda)}$;$\lambda \neq \frac{1}{2}$.*

3.4 Théorème. *Soit A une $\mathbb{R}$-algèbre de Jordan n. c. de division, de dimension 8. Alors $Der(A) = su(2) \oplus su(2)$ si et seulement si $A \cong (E_{-1,\alpha,0}(H))^{(\lambda)}; 1 \neq \alpha > \frac{1}{2}$ et $\lambda \neq \frac{1}{2}$.*

Pour les demonstrations de (3.2), (3.3) et (3.4), voir [14].

3.5 Remarque. Dans [1], Benkart et Osborn posent le probleme de l'existence d'une $\mathbb{R}$-algèbre de division, de dimension 8, dont l'algèbre de Lie des dérivations est $su(2)$, et dont la décomposition en $su(2)$-modules irréductibles est de la forme: 1+1+3+3. Nous donnons ici une réponse affirmative avec une algèbre de Jordan n. c. En effet, nous avons vu dans (2.3 ii)) que si $\alpha \neq \frac{1}{2}$ et $\delta \neq 0$, alors $Der(E_{-1,\alpha,\delta}(\mathbb{H})) = su(2)$. De plus $< 1 >, < f = (0,1) >, < i, j, k >$ et $< if, jf, kf >$ sont des $su(2)$-modules irréductibles de $E_{-1,\alpha,\delta}(\mathbb{H})$, qui est leur somme directe. Les conditions supplémentaires : $\alpha > \frac{1}{2}$ et $(2\alpha - 1)\delta^2 < 4$. assurent que l'algèbre $E_{-1,\alpha,\delta}(\mathbb{H})$ est de division (2.4).

4. Automorphismes

4.1 Définition. Soit A une K-algèbre. On appelle reflexion de A tout automorphisme involutif non identique.

Nous avons caractérisé les algèbres réelles de Jordan non commutatives de division, de dimension 8, qui possedent un automorphisme non trivial:

4.2 Proposition. *Soient $\mathbb{O} = (W, (/), \wedge)$ l algèbre réelle de Cayley-Dickson et s un automorphisme symétrique de l'espace euclidien $(W, -(/))$, défini positif. Alors les trois propriétés suivantes sont équivalentes :*
i) $Aut(\mathbb{O}(s))$ n'est pas trivial.
ii) $Aut(\mathbb{O}(s))$ contient une reflexion.
iii) s laisse stable une sous-algèbre de $\mathbb{O}$ de dimension 4.

4.3 Théorème. *Soit A une algèbre réelle de Jordan n. c. de division, de dimension 8, dont le groupe des automorphismes n'est pas trivial. Alors, il existe une base $1, u; y_1, z_1; y_2, z_2; y_3, z_3$ de A, trois parametres $a, b, c > 0$ et neuf autres $\alpha, \beta, \gamma, \mu, \lambda, \eta, \sigma, \delta$ et ρ pour lesquels la multiplication de A est donnée par la table 3 suivante :*

TABLE III

	1	u	y_1	z_1	y_2	z_2	y_3	z_3
1	1	u	y_1	z_1	y_2	z_2	y_3	z_3
u		-1	az_1	$-ay_1$	bz_2	$-by_2$	cz_3	$-cy_3$
y_1			-1	au	$\alpha y_3 + \beta z_3$	$\gamma y_3 + \mu z_3$	$-\alpha y_2 - \gamma z_2$	$-\beta y_2 - \mu z_2$
z_1				-1	$\lambda y_3 + \eta z_3$	$\sigma y_3 + \delta z_3$	$-\lambda y_2 - \sigma z_2 + \rho z_3$	$-\eta y_2 - \delta z_2 - \rho y_3$
y_2					-1	bu	$\alpha y_1 + \lambda z_1$	$\beta y_1 + \eta z_1$
z_2						-1	$\gamma y_1 + \sigma z_1$	$\mu y_1 + \delta z_1$
y_3							-1	$cu + \rho z_1$
z_3								-1

De plus, une algèbre réelle dont la multiplication est donnée par la table 3 de Jordan n. c. et possede un automorphisme non trivial. Elle est de de division, si et seulement si : $\beta\gamma - \alpha\mu, \beta\lambda - \alpha\eta, \gamma\lambda - \alpha\sigma > 0$ et $c(\alpha\delta - \beta\sigma - \lambda\mu + \gamma\eta)^2 + b(\beta\gamma - \alpha\mu)\rho^2 < 4c(\beta\lambda - \alpha\eta)(\mu\sigma - \gamma\delta)$.

Voir [14] pour une demonstration des deux résultats ci-dessus.

Finalement, pour $A \neq C$, on fixe des parametres $\lambda_1, .., \lambda_7$ positifs et distincts deux a deux. Soient alors $\mathbb{O} = (W, (/), \wedge)$, l'algèbre réelle de Cayley-Dickson et $e_1, ..., e_7$ la base canonique de l'espace euclidien $(W, -(/))$, on pose : $x_1 = e_1 + e_2, x_2 = e_1 - e_2, x_3 = e_3 + e_4, x_4 = e_3 - e_4, x_5 = e_5, x_6 = e_6 + e_7$ et $x_7 = e_6 - e_7$. On obtient une nouvelle base, $x_1, ..., x_7$ orthogonale de $(W, -(/))$, telle que $x_i \wedge x_j$ n'est colinéaire a aucun x_k, pour tout couple (i, j) d'indices distincts de 1,...,7. On définit alors l'automorphisme (symétrique) s par : $s(x_i) = \lambda_i x_i$ pour tout $i = 1, .., 7$. s ne laisse invariante aucune sous-algèbre de $\mathbb{O}$ de dimension 4 car sinon s laisserait invariante une sous-algèbre de $(W, \wedge)$ de dimension 3, qui serait engendrée linéairement par trois vecteurs propres de s, et l'un d'eux, colinéaire au produit vectoriel des deux autres. Ceci contredit le choix de la base $x_1, ..., x_7$. En vertu de (4.2) $Aut(\mathbb{O}(s))$ est trivial.

References

1. G.M. Benkart and J.M. Osborn, An Investigation of Real Division Algebras using Derivations, Pacific Journal of Mathematics **96**, no. 2 (1981), 265–300.
2. G.M. Benkart and J.M. Osborn, The Derivation Algebra of a Real Division Algebras, American Journal of Mathematics **103**, no. 6, (1981), 1135–1150.
3. G.M. Benkart, D.J. Britten and J.M. Osborn, Real Flexible Division Algebras, Can. J. Math. **XXXIV**, no. 3, (1982), 550–588.

4. R. Bott and J. Milnor, On the parallelizability of the spheres, Bull. Amer. Math. Soc. **64** (1958), 87–89.
5. N. Bourbaki, Eléments de Mathématiques -algèbre -, Hermann, Chapitre I-III, 1970.
6. R.H. Bruck and E. Kleinfeld, The structure of alternative division rings, Bull. Proc. A. M. S. **2** (1951), 878–890.
7. J.A. Cuenca Mira, R. de los Santos Villodres, A. Kaidi et A. Rochdi, Classification of the Real Quadratic Flexible Division Algebras of Dimension 8, Universidad de Málaga, Spain. Preprint.
8. N. Jacobson, *Basic Algebra I*, Freeman and company, New York 1985.
9. A. Kaidi, Bases para una teoria de las álgebras no asociativas normadas, Universidad de Granada, Spain (1977).
10. A. Kaidi et A. Rochdi, Sur les algèbres Réelles de Jordan non commutatives de Division Linéaire, de Dimension 8, Nonassociative Algebraic Models. Nova Sciences Publishers, Inc. New York 1992, 183–193.
11. M.A. Kervaire, Non-parallelizability of the n-sphere for $n \geq 7$, Proc. Nat. Acad. Sci. **44** (1958), 280–283.
12. J.M. Osborn, Quadratic Division Algebras, Trans. Amer. Math. Soc. **105** (1962), 202–221.
13. A. Rochdi, Sur les algèbres non associatives normées de division, Thèse de 3e cycle, Faculté des sciences de Rabat. Maroc 1987 .
14. A. Rochdi, Etude des algèbres Réelles de Jordan non commutatives de Division, de Dimension 8, dont l'algèbre de Lie des dérivations n'est pas triviale. Universidad de Málaga, Spain. Preprint.

Zeitfracht Medien GmbH
Ferdinand-Jühlke-Straße 7
99095 Erfurt, Deutschland
produktsicherheit@kolibri360.de